PLANT BIOSTIMULANTS AND GROWTH REGULATORS IN SUSTAINABLE AGRICULTURE

PLANT BIOSTIMULANTS AND GROWTH REGULATORS IN SUSTAINABLE AGRICULTURE

Deepak Kumar, PhD
Sanjay K. Singh, PhD
Nutan Kaushik, PhD
Mahaveer P. Sharma, PhD
Editors

AAP | APPLE ACADEMIC PRESS

First edition published 2027

Apple Academic Press Inc.
1265 Goldenrod Circle, NE,
Palm Bay, FL 32905 USA

760 Laurentian Drive, Unit 19,
Burlington, ON L7N 0A4, Canada

CRC Press
2385 NW Executive Center Drive,
Suite 320, Boca Raton FL 33431

4 Park Square, Milton Park,
Abingdon, Oxon, OX14 4RN, UK

Library and Archives Canada Cataloguing in Publication

CIP data on file with Canada Library and Archives

Library of Congress Cataloging-in-Publication Data

CIP data on file with US Library of Congress

ISBN: 978-1-77964-144-1 (hbk)
ISBN: 978-1-77964-145-8 (ebk)
DOI: 10.1201/9781779641458

About the Editors

Deepak Kumar, PhD
Executive Director, Nextnode Bioscience Pvt. Ltd., Opposite GEB Office, Kadi, Gujarat, India; Director (R&D), Shri Ram Solvent Extractions Pvt. Ltd., Jaspur, Uttarakhand, India

Dr. Deepak Kumar is presently working as Executive Director at Nextnode Bioscience Pvt. Ltd., Kadi, and Director (R&D) of the Department of Scientific and Industrial Research (DSIR)-recognized laboratory of Shri Ram Solvent Extractions Pvt. Ltd. He is the author or co-author of 68 peer-reviewed publications (journal articles, book chapters, and books). He has been involved in the quality production of biofertilizers and biopesticides for the last 12 years. Dr. Kumar is a Certified Farm Advisor (Organic Farming), certified by the National Institute of Agricultural Extension Management [MANAGE], Government of India [GOI], and a Food Safety Mitra (certified by the Food Safety and Standards Authority of India [FSSAI], GOI). He also served as a technical expert member (biofertilizer) in the establishment of the Agribusiness Incubation Center (ABI) at the ICAR–National Soybean Research Institute (ICAR–NSRI), Indore, Madhya Pradesh.. Additionally, Dr. Kumar was selected as an AYE mentor by IFCI Venture Capital Funds Limited (a Government of India undertaking) for the SC/ST mentorship program under the Ministry of Social Justice and Empowerment, Government of India (from February 2022 to date). Moreover, he is also serving as a guest faculty for the course "PG Diploma in Biotechnology Law and Policy (PGDBLP)" at Gujarat National Law University. He obtained his MSc in Biotechnology from C.C.S. University, Meerut; his MBA in Human Resource Management from Punjab Technical University, Jalandhar; and his PhD in Agricultural Biotechnology from Acharya Narendra Deva University of Agriculture & Technology, Ayodhya, Uttar Pradesh, India. Dr. Kumar has received many prestigious awards from various scientific and market research organizations, including the Indo-Global Award for Innovation in Biotech

(2021) from the Indian Economic Development & Research Association (IEDRA), New Delhi; the Distinguished Industrial Microbiologist Award (2020) from the Microbiologists Society of India (MSI); the India Leadership Award (2019) in the Best Agribiotech Entrepreneur category from Blindwink, Bangalore; the Science Entrepreneur Award (2018) from Biologix Research and Innovation Center (BRICPL), India; and the Young Scientist Award (2018) from the Doctor's Agricultural and Horticultural Development Society, Lucknow, India. He has completed and guided six indigenous and externally funded (DSIR and DST) research projects. Dr. Kumar recently developed and licensed a neem endophytic mycoflora-based mycofertilizer, AsperPhos, which has shown potential in mineral solubilization. Moreover, he serves as a technical consultant for more than a dozen companies, providing customized technical support. He is an active member of several associations and professional bodies, including the Fertilizer Association of India (FAI), the Society for Technology Management (STEM), the International Competence Center for Organic Agriculture (ICCOA), and the Association of Microbiologists of India (AMI).

Sanjay K. Singh, PhD
Scientist 'G' and In-Charge, Biodiversity and Paleobiology Group, Agharkar Research Institute, Pune, Maharashtra, India

Dr. Sanjay K. Singh obtained his postgraduate degree in Botany from Kanpur University and doctorate in Botany (Mycology) from Gorakhpur University, Gorakhpur, Uttar Pradesh, India. He availed himself of several postdoctoral fellowships, including one from the Council of Scientific and Industrial Research (CSIR), New Delhi (1996) and a Research Associateship under a Planning Commission-sponsored project at Goa University (2000–2001). He received a Young Scientist Award from the Department of Science and Technology (DST), New Delhi (1997–1999), He received training in fungal biology, biodiversity, conventional and genomic identification, systematics, germplasm conservation, and secondary metabolite profiling, with applications in agriculture, medicine, and industry. With vast experience in fungal biology and plant pathology, Dr. Singh has made significant contributions, most notably the establishment of a state-of-the-art National Facility for Culture Collection of Fungi in 2008, supported by DST, Government of India. Under this flagship program, he founded the exclusive repository known as the National Fungal Culture Collection of India (NFCCI), affiliated with the World Federation for Culture Collections (WFCC). This facility serves more than 700 academic research institutions and approximately 100 industries across 28 states in India. Under the DBT Associate Program for young scientists (under age 40) in the niche area of agri-biotechnology, Dr. Singh received specialized training from world-renowned researcher Prof. Gary A. Strobel at Montana State University (MSU), USA, focusing on endophytes and their bioactive metabolites. He also attended the International Advanced Training on Genetic Resources and Intellectual Property Rights (GRIP), sponsored by the Swedish International Development Authority (SIDA), at the Swedish University of Agricultural Sciences (SLU), Sweden, and in Johannesburg, South Africa. In 2018, he supervised Dr. Frank Ackiah from Ghana, who was selected for the C. V. Raman International Fellowship for African Researchers, sponsored by DST, Government of India. Dr. Singh has served as a member of the Research Advisory Committees of the CSIR Institute of Microbial Technology (IMTECH), Chandigarh, and the ICAR–National Bureau of Agriculturally Important Microorganisms (NBAIM), Mau, under the Indian Council of Agricultural Research (ICAR), New Delhi. He is an elected Fellow of the Maharashtra Academy of Sciences (FMASc), the Indian Phytopathological Society (FIPS), and an honorary fellow of the Society for Biotechnology (India). He was elected President, West Zone, of the Indian Phytopathological Society (2008–2009), and served as a Council Member of both the Indian Phytopathological Society (IPS) and the Mycological Society of India (MSI). He is the recipient of the A. K. Sarbhoy Memorial Award from the Indian Phytopathological Society (ICAR), New Delhi, and the Dr. S. K. Shome Memorial Award from the Mycological Society of India. He is a life member of several prestigious scientific societies, including the Indian Science Congress Association (ISCA), IPS, MSI, and the Association of Fungal Biologists (AFB). Dr. Singh has led 14 research projects, published more than 115 research papers, authored three books, contributed 20 book chapters, and delivered over 30 invited lectures at symposia, seminars, and workshops. He has organized three conferences and more than 15 national-level capacity-building workshops. In addition, he serves as a reviewer and editorial board member for several prestigious journals. His international visits include countries such as Nepal, Fiji, Brazil, South Africa, Sweden, and the United States.

Nutan Kaushik, PhD
Director General, Food and Agriculture Foundation, Amity University, Noida, Uttar Pradesh, India

Dr. Nutan Kaushik is a professional with over 25 years of experience in research and development (R&D) and more than 16 years in postgraduate (PG) teaching in the fields of biotechnology, plant protection, natural products, analytical chemistry, and agricultural policy. As the head of an R&D research group, she has been actively involved in project development and management, laboratory oversight, supervision and teaching of postgraduate and PhD students and researchers, researcher training, publication development, and grant writing. Dr. Kaushik has led cross-functional and research teams through interactive and motivational leadership. She is recognized for her ability to coach, guide, and mentor team members to optimize resource utilization and enhance productivity. She possesses strong skills in data analysis, comparing diverse research datasets, and ensuring the implementation of safe research practices. Her specialization in agro-industrial research includes the development of biopesticides and methods for monitoring environmental pollution. She actively encourages researchers to develop skills and experience in various aspects of research management, including budgeting, project planning, teaching, publishing, conference participation, outreach activities, and committee membership. Dr. Kaushik is the author or co-author of 65 peer-reviewed journal articles, one book, 20 patents, and 15 book chapters. She has served as a reviewer for numerous reputed journals, including *PLOS ONE, Industrial Crops and Products, Biomass and Bioenergy, PLOS Neglected Tropical Diseases, 3 Biotech, Journal of the American Oil Chemists' Society, Canadian Journal of Plant Pathology, Plant Cell Reports, BioControl,* and many others. She has managed 44 international and national research projects and supervised six PhD and

46 postgraduate students. Dr. Kaushik has presented her research across the globe, including in the United States, France, Germany, Australia, Belgium, Bangladesh, China, Dubai, Israel, Norway, Lithuania, the Netherlands, Poland, the Republic of Korea, Spain, Sweden, Thailand, Turkey, the United Kingdom, Vietnam, and other countries.

Mahaveer P. Sharma, PhD
Principal Scientist (Agricultural Microbiology), ICAR–National Soybean Research Institute, Indore, Madhya Pradesh, India

Dr. Mahaveer P. Sharma is currently working in the Agricultural Research Service (ARS) as Principal Scientist (Agricultural Microbiology) at the ICAR–Indian Institute of Soybean Research (ICAR–IISR), Indore, under the Department of Agricultural Research and Education (DARE), Ministry of Agriculture & Farmers Welfare, Government of India. Dr. Sharma holds a graduate degree in agriculture from Rajasthan Agricultural University, Bikaner, and a doctorate in microbiology. He was awarded a gold medal during his master's program. Dr. Sharma began his career in mycorrhizal research at the University of Delhi and The Energy and Resources Institute (TERI), New Delhi, before joining ARS under ICAR. He specializes in soil microbiological research, particularly the use of plant growth-promoting microbes, including arbuscular mycorrhizal fungi and soybean rhizobia, to enhance plant growth, soil carbon sequestration, drought tolerance, and overall crop productivity. He has received several honors, including best paper presentation and travel grant awards at various international scientific meetings and conferences. He has also been the recipient of multiple external research grants from the Department of Biotechnology (DBT), Department of Science and Technology (DST), and the Indian Council of Agricultural Research (ICAR). In recognition of his contributions, he was awarded the DBT–CREST Award 2013 by the DBT, Government of India, to conduct research at the United States Department of Agriculture–Agricultural Research Service (USDA–ARS), USA, on signature fatty acid biomarkers. Dr. Sharma is a fellow of the Mycological Society of India (2020) and a recipient of the National Senior Scientist Award 2021 from the Microbiologists Society of India. He also serves as an associate editor for several high-impact journals. At ICAR–IISR, Indore, he has established an agri-business incubation facility to support and nurture agri-startups focused on soybean and biofertilizer production. He has authored approximately 130 research articles in refereed journals, magazines, and review volumes of international repute. Additionally, numerous microbial accessions submitted to the National Center for Biotechnology Information (NCBI) database and cultures deposited in international microbial repository authorities are credited to his research.

Contents

Contributors

Shachi Agrawal
Department of Botany, Dr. Manmohan Singh Bengaluru City University, Bangaluru, Karnataka, India

Ayesha Ansari
Department of Biosciences, Graphic Era Deemed to be University, Dehradun, Uttarakhand, India

Mrinalini Banerjee
National Law University Odisha, Cuttack, Odisha, India

Vaidehi Bhagwat
Nanobioscience Group, Agharkar Research Institute, Pune, Maharashtra, India

Subir Bose
Department of Biotechnology and Microbiology, MIET College, Meerut, Uttar Pradesh, India

Sonbeer Chack
Department of Agriculture, Himalayan University, Itanagar, Arunachal Pradesh, India

Hemraj Chhipa
College of Horticulture and Forestry, Agriculture University, Kota, Jhalawar, Rajasthan, India

Deepali Choudhari
Nanobioscience Group, Agharkar Research Institute, Pune, Maharashtra, India

Prince Choyal
ICAR–National Soybean Research Institute, Indore, Madhya Pradesh, India

Naceur Djébali
Laboratory of Bioactive Substances, Center of Biotechnology of Borj Cedria, BP 901 Hammam-Lif 2050, Tunisia

Saurabh Dubey
RS Formulation Development and Consultant, Bhiwadi, Rajasthan, India

Pallavi Gaur
Department of Microbiology, SGRR University, Dehradun, Uttarakhand, India

Pankaj Gautam
Department of Life Sciences, Graphic Era University, Dehradun, Uttarakhand, India

Vandana Ghormade
Nanobioscience Group, Agharkar Research Institute, Pune, Maharashtra, India

Azucena Gonzalez-Coloma
Institute of Agricultural Sciences, CSIC, Madrid, Spain

Swapnaja Gulawani
Bioprospecting Group, Agharkar Research Institute, Pune, Maharashtra, India

Harsh Gupta
Faculty of Agriculture, Mahatma Gandhi Chitrakoot Vishwavidyalaya, Chitrakoot, Satna, Madhya Pradesh, India

Payal Gupta
Department of Biotechnology, Graphic Era University, Dehradun, Uttarakhand, India

Manindra S. Hanspal
Presidency School of Law, Presidency University, Bengaluru, Karnataka, India

Raja Husain
Department of Agriculture, Himalayan University, Itanagar, Arunachal Pradesh, India

Nusrat Iqbal
Formulation Division, Institute of Pesticide Formulation Technology, Udyog Vihar, Gurugram, Haryana, India

Anjana Jajoo
School of Life Sciences, Devi Ahilya University, Indore, Madhya Pradesh, India

Pallavi Joshi
School of Agriculture, Graphic Era Hill University, Dehradun, Uttarakhand, India

Nutan Kaushik
Amity Food and Agriculture Foundation, Amity University, Noida, Uttar Pradesh, India

N. A. Khan
Department of Plant Molecular Biology and Genetic Engineering, Acharya Narendra Deva University of Agriculture and Technology (ANDUAT), Ayodhya, Uttar Pradesh, India

Somanjana Khatua
Department of Botany, Faculty of Science, University of Allahabad, Prayagraj, Uttar Pradesh, India

Deepak Kumar
R&D Division, Nextnode Bioscience Pvt. Ltd., Gujarat, India; Shri Ram Solvent Extraction Pvt. Ltd., Jaspur, Uttarakhand, India

Narendra Kumar
Faculty of Agriculture, Oriental University, Indore, Madhya Pradesh, India

Navin Kumar
Department of Biotechnology, Graphic Era University, Dehradun, Uttarakhand, India

Asha Kumari
Amity Food and Agriculture Foundation, Amity University, Noida, Uttar Pradesh, India

Giriraj Kumawat
ICAR–National Soybean Research Institute, Indore, Madhya Pradesh, India

Sanchali Kundu
Department of Botany, Krishnagar Government College, Krishnagar, West Bengal, India

Sanghati Kundu
Department of Botany, Santipur College, Nadia, West Bengal, India

G. Maheswari
Nanobioscience Group, Agharkar Research Institute, Pune, Maharashtra, India

Takwa Marzouk
Laboratory of Bioactive Substances, Center of Biotechnology of Borj Cedria, BP 901 Hammam-Lif 2050, Tunisia

Koustav Mondal
Department of Botany, Krishnagar Government College, Krishnagar, West Bengal, India

Dexter Achu Mosoh
Center for Biodiversity Exploration and Conservation (CBEC), Jabalpur, Madhya Pradesh, India; Indian Institute of Technology Gandhinagar, Gujarat, India; School of Science, SAGE University, Bhopal, Madhya Pradesh, India

Sonika Pandey
Division of Plant Protection, ICAR–Indian Institute of Pulses Research (IIPR), Kanpur, Uttar Pradesh, India

Ekta Pandre
Faculty of Agriculture, Mahatma Gandhi Chitrakoot Vishwavidyalaya, Chitrakoot, Satna, Madhya Pradesh, India

Tulika Panigrahi
Faculty of Agriculture Science and Technology, AKS University, Satna, Madhya Pradesh, India

Veerendra K. Patel
Faculty of Agriculture, Mahatma Gandhi Chitrakoot Vishwavidyalaya, Chitrakoot, Satna, Madhya Pradesh, India

Umesh Patle
Faculty of Agriculture, Mahatma Gandhi Chitrakoot Vishwavidyalaya, Chitrakoot, Satna, Madhya Pradesh, India

Prabha Rai Kalal
School of Life Sciences, Devi Ahilya University, Indore, Madhya Pradesh, India

Vishnu D. Rajput
Academy of Biology and Biotechnology, Southern Federal University, Rostov-on-Don, Russia

Lipi Rina
Department of Agriculture, Himalayan University, Itanagar, Arunachal Pradesh, India

Mohd. Samama
Department of Biotechnology, Graphic Era University, Dehradun, Uttarakhand, India

Gyanesh Kumar Satpute
ICAR–Indian Institute of Soybean Research, Indore, Madhya Pradesh, India

Md. Shamim
Department of Molecular Biology and Genetic Engineering, Dr. Kalam Agricultural College, Bihar Agricultural University, Sabour, Bihar, India

S. Shanthakumar
Gujarat National Law University (GNLU), Gandhinagar, Gujarat, India

Ayushi Sharma
Amity Food and Agriculture Foundation, Amity University, Noida, Uttar Pradesh, India

Mahaveer P. Sharma
ICAR–National Soybean Research Institute, Indore, Madhya Pradesh, India

Sanjay K. Singh
Biodiversity and Paleobiology Group, MACS' Agharkar Research Institute, Pune, Maharashtra, India

Shivangni Singh
Nanobioscience Group, Agharkar Research Institute, Pune, Maharashtra, India

Tanvi Singh
Department of Pathology and Laboratory Medicine, Perelman School of Medicine, University of Pennsylvania, Philadelphia, Pennsylvania, USA

Vivek Kumar Singh
Faculty of Agriculture, Mahatma Gandhi Chitrakoot Vishwavidyalaya, Chitrakoot, Satna, Madhya Pradesh, India

Perumalla Srikanth
Amity Institute of Horticulture Studies and Research (AIHSR), Amity University, Noida, Uttar Pradesh, India

Deepti Srivastava
Department of Agriculture, Integral Institute of Agricultural Science and Technology, Integral University, Lucknow, Uttar Pradesh, India

Pratibha Srivastava
Bioprospecting Group, Agharkar Research Institute, Pune, Maharashtra, India

S. P. Tiwari
Faculty of Agriculture, Mangalayatan University, Jabalpur, Madhya Pradesh, India

Rakesh Kumar Verma
ICAR–National Soybean Research Institute, Indore, Madhya Pradesh, India

Nitin Vikram
Zila Parishad Krishi Mahavidyalaya, Banda, Uttar Pradesh, India

Priyank Vyas
Department of Biotechnology, Graphic Era University, Dehradun, Uttarakhand, India

Tara Chand Yadav
Department of Computer Science and Biosciences, Marwadi University, Rajkot, Gujarat, India

Shaniya Zehra
School of Life Sciences, Devi Ahilya University, Indore, Madhya Pradesh, India

Abbreviations

AA	amino acid
AAPFCO	American Association of Plant Food Control Officials
ABA	abscisic acid
AChE	acetylcholineesterase
ACO	1-aminocyclopropane-1-carboxylic acid oxidase
ACS	1-aminocyclopropane-1-carboxylate synthase
ACTs	artemisinin-based combination therapies
ACX	acyl-CoA oxidase
ALS	acetolactate synthase
AM	arbuscular mycorrhizal
AMF	arbuscular mycorrhizal fungi
AMS	Agricultural Marketing Service
ARF	auxin response factor
BA	benzyladenine
BCA	biological control agent
BCR	bubble column reactor
BL	brassinolide
BMR	barley malt residue
BPH	brown planthopper
BR	brassinosteroid
BSs	biostimulants
CA	carbonic anhydrase
Ca^{2+}	calcium
CAGR	compound annual growth rate
CAT	catalase
CEC	cation exchange capacity
CHSA	chitin synthetase A
CIB	Central Insecticides Board
CIB&RC	Central Insecticides Board and Registration Committee
CK	cytokinin
CLS	*Curvularia* leaf spot
CMCS	carboxymethyl chitosan
CNP	chitosan nanoparticle
CNT	carbon nanotube
CO_2	carbon dioxide
CuCNP	Cu-chitosan nanoparticle

DD	degree of deacetylation
DGWG	dee-geo-woo-gen
DH	degree of hydrolysis
DHA	dehydrogenase activity
DHAR	dehydroascorbate reductase
EBIC	European Biostimulant Industry Council
EC	European Commission
EF	endophytic fungi
EM	ectomycorrhizal
ENM	engineered nanomaterial
EPA	Environmental Protection Agency
EPS	exopolysaccharides
ERF	ethylene response factor
EU	European Union
FA	fulvic acids
FAD	fatty acid desaturase
FFA	Federal Fertilizer Act
FIFRA	Federal Insecticide, Fungicide, and Rodenticide Act
GA1	gibberellins A1
GA_3	gibberellic acids
GAI	gibberellic acid-insensitive
GAs	gibberellins
GDC	glycine decarboxylase
GFM	gene family member
GGDP	geranylgeranyl diphosphate
GMP	good manufacturing practices
GPOX	guaiacol peroxidase
GRAS	generally recognized as safe
GSFC	Gujarat State Fertilizers & Chemicals Ltd.
H_2O_2	hydrogen peroxide
HA	humic acids
HM	heavy metal
HPLC	high-performance liquid chromatography
HTC	hydrothermal carbonization
IAA	indole-3-acetic acid
IAM	indole-3-acetamide
IAOx	indole-3-acetaldoxime
IAR3	IAA-Ala-Resistance-3
iP	isopentenyladenine
IPA	indole-3-pyruvic acid
IPrA	indole-3-propionic acid
iPRDP	iP riboside 5'-diphosphate

iPRTP	iP riboside 5'-triphosphate
IPT	intermittent preventive treatment
ISR	induced systemic resistance
ITRC	Indian Toxicology Research Center
JA	jasmonic acid
JAs	jasmonates
JHAMT	juvenile hormone acid methyltransferase
K^+	potassium
KAT	3-ketoacyl-CoA thiolase
LMV	low molecular weight
LOG	Lonely-Guy phosphoribohydrolase
LOX	13-lipoxygenase
LSF	liquid surface fermentation
MAE	microwave-assisted extraction
MATE	multidrug and toxic compound extrusion
MDHAR	monodehydroascorbate reductase
MeJA/MJ	methyl jasmonate
MFP	multifunctional protein
Mg^{2+}	magnesium
MIC	minimal inhibitory concentration
MRL	maximum residue level
MS	mass spectrometry
MTT	methyl thiazolyl tetrazolium
MW	molecular weight
NAA	naphthaleneacetic acid
NBF	nano-biofertilizer
NGS	next-generation sequencing
NH_3	ammonia
NH_4^+	ammonium
NO	nitric oxide
NPK	nitrogen, phosphorus, and potassium
NPs	nanoparticles
NR	nitrate reductase
NUE	nutrient use efficiency
PAL	phenylalanine ammonia-lyase
PAMP	pathogen-associated molecular pattern
PBs	plant biostimulants
PDI	percent disease severity
PGPB	plant growth-promoting bacteria
PGPF	plant growth-promoting fungi
PGPM	plant growth-promoting microbes
PGPR	plant growth-promoting rhizobacteria

PGRs	plant growth regulators
PHs	plant hormones
PHs	protein hydrolysates
POD	peroxidase
PPPs	plant protection products
PR	pathogenesis-related
PRRs	pattern recognition receptors
PSB	phosphate-solubilizing bacteria
PSF	phosphate-solubilizing fungi
RA	rosmarinic acid
RC	Registration Committee
RIG-I	retinoic acid-inducible gene I
ROS	reactive oxygen species
RRB	raw rice bran
RWC	relative water content
SA	salicylic acid
SAM	S-adenosyl-methionine
SAR	systemic acquired resistance
SCP	single-cell protein
SEM	structural equation model
SEs/SWEs	seaweed extracts
SiO_2	silicon dioxide
SMC	seasonal malaria chemoprevention
SMCs	small-molecule compounds
SmF	submerged fermentation
SOC	soil organic carbon
SOD	superoxide dismutase
SSF	solid-state fermentation
SSSF	semi-solid-state fermentation
STR	stirred tank reactor
STR	strictosidine synthase
SWB	seaweed-based biostimulants
TAM	tryptamine
TDZ	thidiazuron
TiO_2	titanium dioxide
TLC	thin-layer chromatography
TLR	toll-like receptor
TRIA	triacontanol
TTC	triphenyltetrazolium chloride
tZ	trans-zeatin

USDA	U.S. Department of Agriculture
VAM	vesicular-arbuscular mycorrhizal
VLP	virus-like particle
VOC	volatile organic compound
VSP	vegetative storage protein
WB	wheat bran
ZnO	zinc oxide
ZR	zeatin riboside

Preface

In sustainable agriculture, the integration of plant biostimulants and growth regulators offers a holistic approach to crop management that prioritizes environmental health, soil fertility, and long-term agricultural sustainability, while minimizing reliance on synthetic chemicals and inputs. The global biostimulants market was valued at USD 4.30 billion in 2024 and is projected to reach USD 7.60 billion by 2029, exhibiting a compound annual growth rate (CAGR) of 12.0% during the forecast period. Plant growth, nutrient uptake, and stress tolerance are all enhanced by the compounds found in biostimulants, which increase crop yields and improve crop quality. In an effort to maximize crop productivity and profitability, farmers are increasingly adopting biostimulants. A large number of biostimulants are compatible with organic agricultural practices because they are derived from natural sources. These products provide organic farmers with effective tools to enhance crop resilience and yield while meeting the standards of organic certification. In 2022, Europe led the global biostimulants market with a 40.6% share, followed by North America and Asia-Pacific. This high share is attributed to government support for organic agriculture—which aims to convert 25.0% of agricultural land to organic by 2030—and the high level of awareness among local farmers. Meanwhile, the global plant growth regulator market size was estimated to be around USD 4.06 billion in 2022 and is expected to grow at a CAGR of 9.94%, reaching USD 9.53 billion by 2031 (from USD 4.46 billion in 2023). In agriculture, plant growth regulators are frequently employed to control plant growth, enhance crop quality, and boost agricultural productivity. They have emerged as essential tools in modern agricultural practices for maximizing yield while minimizing negative environmental impacts.

Growth regulators and biostimulants are now the two primary agri-input components employed in the sustainable development of various crops worldwide. Different regulatory frameworks govern the marketing and commercialization of growth regulators and biostimulants. In the United States, the biostimulants industry aims to market its products under the term *biostimulants,* using language such as "supporting nutrient uptake," "natural nutritional processes," and "abiotic stress." To achieve this, a few fundamental guidelines for biostimulants must be established in the U.S. These guidelines aim to provide uniformity in the data required to demonstrate product safety and efficacy, particularly for items companies wish to label as plant biostimulants. In India, biostimulants have recently been included in Schedule VI of the Fertilizer Control Order. According to the proposed guidelines, biostimulants must first be registered with the government and demonstrate efficacy before being introduced to the market. Proper labelling is mandatory and must include the manufacturer's name, list of ingredients, and expiration date. In contrast, growth regulators fall under the purview of the Central Insecticides Board and Registration Committee (CIB&RC) for registration and are subsequently governed by state-level manufacturing regulations.

This edited book, *Plant Biostimulants and Growth Regulators in Sustainable Agriculture,* is a comprehensive compilation highlighting the applications of plant biostimulants and growth regulators in resilient agriculture, production systems, and their regulatory frameworks for commercialization. The book is divided into five sections, in which specialists in the field critically explore various facets of plant biostimulants and growth regulators. The first part (Chapters 1–2) focuses on an overview and regulatory frameworks of plant biostimulants and growth regulators. The second part (Chapters 3–7) discusses different types of biostimulants (viz., humic acids, biochar, chitosan, and fungal-based biostimulants), their categories, and production systems. Plant biostimulants for bioactive compound production are examined in the third part (Chapter 8) of the book. The rapidly expanding global market for plant growth regulators in the horticulture sector, particularly for abiotic stress management, is critically discussed in the fourth part (Chapters 9–13). The fifth part (Chapters 14–15) compiles the roles of nanomaterials in agriculture, including nanofertilizers, nanobiofertilizers, and nano-biostimulants. Moreover, this book will be of great interest to researchers, industry professionals, educators, climate change scientists, capacity builders, and policymakers.

We sincerely hope that readers will find this book both motivating and inspiring. Our mission is to disseminate the most up-to-date knowledge on plant growth regulators and biostimulants for sustainable agricultural production. It is our earnest hope that scientists and industry professionals will find this book valuable, and we warmly welcome any novel suggestions or feedback that readers may offer. We are confident that libraries around the world will include this book in their collections. We would like to extend our sincere thanks to all the chapter contributors for their invaluable support in bringing this book together. The editors also acknowledge the many books, experts, magazines, journals, and other sources that were consulted during the compilation of the chapters, and we are grateful for their guidance and insights. We deeply appreciate Apple Academic Press for accepting our proposal and publishing this book. To improve the content and scope of future editions, the editors' welcome constructive suggestions and thoughtful criticism from readers. Finally, we express our heartfelt gratitude to the Almighty "Prabhu," who provided the inspiration, wisdom, encouragement, and direction needed to complete this book endeavor.

—***Editors***

PART I

Plant Biostimulants and Growth Regulators: Regulatory Framework

CHAPTER 1

The Global Presence and Regulatory Framework of Plant Biostimulants and Growth Regulators: An Overview

DEEPAK KUMAR,[1,2,*] MAHAVEER P. SHARMA,[3] S. SHANTHAKUMAR,[4] MRINALINI BANERJEE,[5] TANVI SINGH,[6] and MANINDRA S. HANSPAL[7]

[1]*R&D Division, Nextnode Bioscience Pvt. Ltd., Kadi, Gujarat, India*

[2]*Shri Ram Solvent Extraction Pvt. Ltd., Jaspur, Uttarakhand, India*

[3]*ICAR–National Soybean Research Institute, Indore, Madhya Pradesh, India*

[4]*Gujarat National Law University (GNLU), Gandhinagar, Gujarat, India*

[5]*National Law University Odisha, Cuttack, Odisha, India*

[6]*Department of Pathology and Laboratory Medicine, Perelman School of Medicine, University of Pennsylvania, Pennsylvania, USA*

[7]*Presidency School of Law, Presidency University, Bengaluru, Karnataka, India*

*Corresponding author

ABSTRACT

The agricultural industry is witnessing a growing reliance on plant biostimulants (PBs) and growth regulator-based products to enhance crop productivity and resilience while reducing reliance on chemical fertilizers and pesticides. These products, derived from biological or chemical sources, play a vital role in modern agricultural practices. However, their

Plant Biostimulants and Growth Regulators in Sustainable Agriculture.
Deepak Kumar, Sanjay K. Singh, Nutan Kaushik, and Mahaveer P. Sharma (Eds.)

DOI: 10.1201/9781779641458-2

global presence and regulatory framework present intricate challenges and opportunities. This chapter explores the current landscape of PBs and growth regulator-based products on a global scale, focusing on their utilization, regulatory frameworks, and market dynamics. It examines the diverse range of products available, including microbial inoculants, seaweed extracts, humic substances, and synthetic growth regulators, and their respective roles in improving plant growth, uptake of nutrients, and resilience to stress. Furthermore, the presented topic delves into the regulatory frameworks governing these products across different regions and countries. It highlights variations in regulatory approaches, ranging from product categorization to registration requirements for commercialization. The complex interplay between regulatory bodies, industry stakeholders, and scientific communities shapes the market access and commercialization pathways for these products.

1.1 INTRODUCTION

The use of plant biostimulants (PBs) and growth regulator-based products in agriculture has increased dramatically in the past few years. These innovative formulations play a pivotal role in enhancing crop productivity, improving plant health, and mitigating the adverse effects of environmental stressors (Bartucca et al., 2022; Magnabosco et al., 2023). As a result, they have garnered significant attention from farmers, agronomists, and policymakers worldwide. The global presence of PBs and growth regulator-based products is characterized by an intricate web of manufacturers, distributors, and end-users spanning across diverse geographical regions. From the expansive farmlands of North America to the lush plantations of Southeast Asia, these products have transcended borders, offering sustainable solutions to agricultural challenges (Allied Market Research, 2023a, b; MarketsandMarkets, 2023a, b). However, amidst this proliferation, the regulatory frameworks governing the production, distribution, and application of PBs and growth regulators vary greatly among jurisdictions. While some regions have established robust guidelines and standards to ensure product efficacy and safety, others grapple with fragmented or nascent regulatory regimes, leading to ambiguity and inconsistency in product oversight (Caradonia et al., 2019; Du Jardin, 2015; Kumar et al., 2023b). This dynamic regulatory landscape underscores the importance of understanding the compliance requirements and market access pathways for stakeholders involved in the production and trade of these agricultural inputs. Navigating through this complex terrain demands a comprehensive comprehension of regulatory nuances, product registration procedures, labeling requirements, and adherence to quality control measures (Kumar, 2023a; Kumar et al., 2023b).

Against this backdrop, this chapter aims to provide an insightful exploration into the global presence and regulatory framework governing PBs and growth regulator-based products. By examining the types of PBs and growth regulators, key regulatory bodies, legislative frameworks, and emerging trends, it seeks to elucidate the challenges and opportunities inherent in this rapidly evolving sector. Furthermore, it endeavors to offer practical insights and recommendations to facilitate informed decision-making and promote regulatory harmonization across international borders.

1.2 PLANT BIOSTIMULANTS AND GROWTH REGULATORS: DEFINITION

PBs and growth regulators are substances or microorganisms applied to plants to enhance growth, development, and overall health. They differ from fertilizers, which primarily provide essential nutrients, in that they exert their effects through various mechanisms that promote physiological processes in plants (Du Jardin, 2015; Rademacher, 2015; Sabagh et al., 2021; Sible et al., 2021). BS can be derived from organic or synthetic sources and may contain a variety of compounds such as humic acids, seaweed extracts, amino acids, vitamins, and beneficial microbes (Gupta et al., 2021). The term "biostimulant" is not well understood, and many definitions have been put out in the past. BS are defined in a broad and imprecise manner. The majority of these definitions aimed to encourage the approval of BS by future legislation by making an effort to distinguish between BS and fertilizers as well as between pesticides and biocontrol agents (Caradonia et al., 2019; Du Jardin, 2015). The EU regulation on fertilizers, specifically Regulation (EU) 2019/1009, which came into effect on July 16, 2019, outlines the definition and regulatory framework for BS. According to this regulation, BS are distinct from fertilizers and plant protection products (PPPs). They are intended to improve the nutrient use efficiency (NUE) of plants, enhance tolerance to abiotic stress, and improve crop quality traits such as yield, color, and flavor (Sanjuán et al., 2023; SEIPASA, 2022). "A fertilizing product whose purpose is to stimulate plant nutrition in soil and plant rhizosphere with the sole goal of improving NUE, quality traits, tolerance to biotic and abiotic stress, or enhancing the absorption of bound nutrients is defined as a plant biostimulant" (https://www.legislation.gov.uk/eur/2019/1009/contents).

In India, the Fertilizer (Control) Amendment Order, 2021, sets down the regulatory framework for bio-stimulant registration in India. India is now among the few nations in the world that have separate regulations governing BS (Press Trust of India, 2023). The commercialization of BS falls under the Fertilizer (Control) Order, 1985, Schedule VI. As per the Indian Fertilizer Order (FCO) Amendment Order, 2021, "Biostimulant" means a material, microorganism, or mixture of both whose main application to seeds, plants, or the rhizosphere is to promote plant physiological processes and improve crop quality, growth, yield, and stress tolerance, irrespective of the amount of nutrients present. It excludes pesticides and plant growth regulators (PGR), which are governed by the Insecticide Act, 1968 (https://www.fertcon.net/wp-content/uploads/2021/06/6-AUXI.pdf).

The Agriculture Improvement Act of 2018, or the 2018 Farm Bill, was passed by the United States Congress in March 2019, which included provisions related to BS (USDA, 2019). The bill directed the USDA to study the regulation of BS and report back to Congress on potential regulatory pathways. This study was intended to inform future regulatory decisions regarding BS (USDA, 2019). As per the bill, a plant biostimulant is defined as "any material or microbe that, when introduced to plants, seeds, or the rhizosphere, encourages natural processes to improve or enhance the absorption of nutrients, tolerance to abiotic stress, nutrient efficiency, or quality of crops and yield" (https://www.congress.gov/115/bills/hr2/BILLS-115hr2enr.pdf). PBs are now controlled at the state level as fertilizers, soil inoculants, or soil amendments. This legal definition of the term serves as a first step towards developing a regulatory framework, suitable evaluation, approval, and standard national labeling of these products. The definition offered in the Farm Bill of 2018 is in line with the definition that the European Union is now recommending (http://www.biostimulants.eu/).

On the other hand, PGRs are compounds that regulate plant growth processes such as cell division, elongation, differentiation, and flowering. They can be synthetic chemicals or naturally occurring substances that mimic or interfere with the action of plant hormones (PHs), influencing plant growth and development in specific ways (Rademacher, 2015; Sabagh et al., 2021). Moreover, PGRs are also essential under adverse conditions by acting as thermoprotectants, scavengers of reactive oxygen species (ROS), enhancers of photosynthesis, promoters of the buildup of stress proteins, and performing several other regulatory roles associated with metabolism (Ma et al., 2017; Sabagh et al., 2021; Sharma et al., 2020). The definition of PGRs may vary depending on the country and its specific laws and regulations. The European Union (EU) does not have a single, universally recognized definition for PGRs as it pertains to regulatory frameworks (Traon et al., 2014). However, PGRs generally refer to a diverse group of natural or synthetic substances that control the development and growth of plants at low concentrations (1–10 ng/mL) (Jan et al., 2020). In India, PGRs are regulated under the "Insecticides Act, 1968" and the "Insecticides Rules, 1971." As per the laws, PGRs are substances that are employed to alter the way plants grow. Some of the changes they can make include boosting branching, decreasing shoot growth, increasing return bloom, removing extra fruit, and changing the maturity of the fruit (https://ppqs.gov.in/major-uses-pesticides-plant-growth-regulators-01062023). However, specific definitions and regulations pertaining to PGRs may vary depending on amendments and updates to these laws (Tewari and Arora, 2016).

1.3 TYPES OF PLANT BIOSTIMULANTS AND GROWTH REGULATORS

PBs and growth regulators are substances applied to plants to enhance growth, development, and overall performance. They can be natural or synthetic compounds that influence physiological processes in plants (Sabagh et al., 2021; Sible et al., 2021). It is also important to consider that the effectiveness of BS and growth regulators can differ based on variables such as plant species, environmental conditions, application methods, and the specific formulation of the product (Drobek et al.,

2019; Li et al., 2022; Sabagh et al., 2021). Additionally, proper dosage and timing of application are essential to attaining the intended outcomes without negative consequences. There are the following types of PBs and growth regulators, which are categorized as per their chemical nature and mode of action.

1.3.1 MAIN CATEGORIES OF PLANT BIOSTIMULANTS

BS can be categorized based on their origin, composition, or mode of action (Du Jardin, 2015; Gupta et al., 2021). As per the Indian Fertilizer Order (FCO) Amendment Order, 2021, the following categories apply to the biostimulants listed in Schedule VI of the FCO, 1985:

i. Humic and fulvic acids and their derivatives
ii. Amino acids and protein hydrolysates
iii. Seaweed extracts and other botanical extract
iv. Chitosan and other biopolymers (anti-transpirants)
v. Cell free microbial products
vi. Inorganic compounds like silicon
vii. Antioxidants
viii. Biochemicals
ix. Vitamins

Biostimulant-based products were viewed as "snake oils" for a long time, and doubts about their ability to positively impact plant development still exist. Many scientific studies have demonstrated, however, that many crop systems—particularly cereals and row crops—respond to these materials with increased productivity and enhanced resilience to biotic and abiotic challenges, including disease (Calvo et al., 2014; Sible et al., 2021). Induction of systemic resistance, enhanced root architecture and lateral root growth, enhanced water and nutrient uptake, and enhanced water and nutrient usage efficiency are additional beneficial impacts (Drobek et al., 2019; Li et al., 2022).

1.3.1.1 HUMIC SUBSTANCES

The market for humic-based BS is expected to attain $1,184 million USD by 2027, growing at a CAGR of 11.5% from its anticipated USD

686 million in 2022. The need for humic-based BS has increased dramatically on a global scale, and businesses and research have successfully integrated a variety of cutting-edge goods and ingredients to fulfil the needs of the populace. The growing market is a consequence of consumer and farmer demand for organic products that offer substitutes for synthetic inputs (Marketsandmarkets, 2023c). Humic materials are intricate organic mixtures that are found in soil, peat, coal, streams, lakes, and oceans (Jarukas et al., 2021). The breakdown of organic matter yields a combination of acids with phenolate and carboxyl groups, including fulvic and humic acids (Bulgari et al., 2015) (Table 1.1). Humic materials are essential for carbon sequestration, nitrogen cycling, and soil fertility (Ampong et al., 2022). These substances, which collectively make up to 60% of the organic matter in soils, are frequently categorized as fulvic acids (FA) (alkali and acid soluble), humic acids (HA) (alkali soluble but acid insoluble), and humin (alkali and acid insoluble) (Lamar, 2020).

1. **Humic Acids:** Among humic substances, these are the biggest molecules. With high molecular weights, they possess a dark color. Low pH values render humic acids insoluble in water; higher pH values cause them to become soluble (Lamar, 2020; Perminova et al., 2003). Their functions include bolstering soil water retention, improving soil structure, and increasing plant availability of nutrients (Ampong et al., 2022).

2. **Fulvic Acids:** Compared to humic acids, fulvic acids are smaller molecules and typically have a lighter color. They dissolve in both alkaline and acidic conditions (Jarukas et al., 2021). Fulvic acids are essential for chelating minerals, increasing their availability to plants, and promoting the uptake of nutrients by plant roots (Kumar et al., 2020). They also support the microbiological activity of the soil (Zhang et al., 2021).

3. **Humins:** The insoluble portion of humic materials is known as humins. These are heavy-molecular-weight, dark-colored substances that persist in the soil for a very long period. The final result of the process of humification, which turns organic matter into humus, is humins (Lamar, 2020).

TABLE 1.1 Chemical Composition of Fulvic and Humic Acids

Characteristic	Fulvic Acids	Humic Acids
Molecular weight (MW)	1000–10,000 Daltons	10,000–100,000 Daltons
	% of Organic Components	
Carbon	40–50	50–60
Oxygen	44–50	30–35
Hydrogen	4–6	4–6
Nitrogen	1–3	2–6
Sulfur	0–2	0–2

Source: Adapted from Sible et al. (2021).

Many studies suggest that these compounds may enhance plants' ability to withstand abiotic stress. Seaweed extract and humic acid pretreatment of *Agrostis palustris* Huds. and *Festuca arundinacea* enhanced root, shoot, and antioxidant capacity in dry soil conditions as well as leaf moisture (Van Oosten et al., 2017; Zhang and Ervin, 2008; Zhang et al., 2002). The quantity of potassium in shoot tissue was significantly higher in plants treated with fulvic and humic acids. Through encouraging vegetative growth, boosting flower counts, and shortening the time to flower, humic and fulvic acids enhanced the quality of the petunia crop (Martins et al., 2024). Bell pepper (*Capsicum annuum* L. cv. Demre) plants treated with humic acid and phosphorous had lower Na contents and higher concentrations of N, K, P, Fe, Ca, S, Mg, Mn, and Cu ions in their roots and shoots. These plants also showed a general protective effect against mild salt stress (Çimrin et al., 2010).

1.3.1.2 SEAWEED EXTRACTS

The seaweed extracts market growth is anticipated to reach a CAGR of 6.1% from 2023 to 2032, from an estimated $2 billion in 2022 to $3.5 billion by 2032 (Allied Market Research, 2023a). The abundance of bioactive compounds found in seaweed extracts, including vitamins, amino acids, polysaccharides, minerals (including micronutrients), and growth hormones (auxins, cytokinins, and gibberellins), has drawn attention to them as effective PBs in agriculture (Gupta et al., 2021; Van Oosten et al., 2017). The majority of industrial seaweed extract-based biostimulant products are made from red macroalgae (such as *Lithothamnium calcareum*), brown macroalgae (like *Ascophyllum nodosum, Durvillaea potatorum*, *Ecklonia maxima, Fucus vesiculosus, Sargassum* spp., *Laminaria digitata*, and *Turbinarias* spp.), and other types of seaweed (Battacharyya et al., 2015; Khan et al., 2009) (Table 1.2). The variables (nutrient availability, depth, light, salinity, and presence of organic pollutants) of the environment in which seaweeds grow and their geographic position determine their biochemical composition (Hurtado et al., 2021; Kumar and Sharma, 2021). Moreover, the composition of commercial formulations of seaweed extracts (SWEs) is primarily determined by the extraction process and is frequently proprietary (Van Oosten et al., 2017). Several extraction techniques are used to remove the active compounds found in algae from the cells and transfer them to a liquid medium in order to create a commercial product (Valverde et al., 2022). The final product composition can vary greatly depending on the technique employed (Povero et al., 2016). The two most commonly used extraction techniques are cold extraction and conventional extraction, which uses chemical agents. When strong acids or alkalis (KOH) are used in chemical extraction, the resultant product is dark because of the high oxidation of the active ingredients, which denatures the active chemicals and drastically reduces their characteristics (Calvo et al., 2014). On the other side, in the case of the cold extraction process, seaweeds are treated with the micronized method under low pH (i.e., 4.5) and high pressure at their physiological level, which results in active ingredients without any loss (Crampon et al., 2011; Valverde et al., 2022).

TABLE 1.2 An Overview of the Main Macroalgae Species Used in Seaweed Extract Biostimulants that are Sold Commercially

Macroalgae Species	Type	Percentage of Algal Species as Active Ingredients (A.I.) in Surveyed Commercial Products
Lithothamnium calcareum	Red	1.4
Fucus spp.	Brown	2.9
Macrocystis pyrifera	Brown	2.9
Ecklonia spp.	Brown	4.3
Durvillaea spp.	Brown	7.2
Laminaria spp.	Brown	8.6
Ascophyllum nodosum	Brown	63.7
Unspecified	–	15.9

Source: Data compiled from Sible et al. (2021); Sharma et al. (2014); Khan et al. (2009).

Seaweed extracts improve plant performance; they can be used as BS in agricultural farming. Consequently, in order to assess their ability to promote root growth and morphology, nutrition, and sugar accumulation, five extracts from *Ascophyllum nodosum* (B–F) and one from *Laminaria* (A) were tested on maize (*Zea mays* L.) plants. The outcomes showed that while all seaweed extracts functioned as BS in maize, their chemical characteristics seemed to be very important in determining the physiological response that each seaweed extract preferentially induced (Ertani et al., 2018). Additionally, seaweed extracts have primarily been employed for the same purposes in horticulture crops and trees. Reduced ferrous ion chelating ability had some negative consequences on nutritional value, although *Ascophyllum nodosum* seaweed extract boosted leaf relative water content (RWC), dry weight, and fresh weight in spinach (*Spinacia oleracea* L.) plants under drought stress (Xu and Leskovar, 2015). Similar to potassium fertilization, seaweed extracts (SWE) increased cotyledon growth when administered to lettuce (*Lactuca sativa* L.) seedlings (Möller and Smith, 1998). In light of their capacity to promote development and withstand stress, seaweed extracts (SEs) have found widespread application as BS in crop management. Chen et al. (2021) studied how sugarcane responded to SE treatment at various growth stages in terms of sugar content and yield. The findings showed that when sugarcane is stressed by drought, seaweed extracts boost cane production and encourage sucrose accumulation. Due to the phytohormones included in the product, stress reduction, and/or increased plant metabolism, the application of seaweed extracts has been widely reported to improve nutrient uptake, plant growth, and grain yield (González et al., 2013; Sharma et al., 2014; Sible et al., 2021).

1.3.1.3 PROTEIN HYDROLYSATES AND AMINO ACIDS

Protein hydrolysate (PHs) market size was estimated at USD 25.5 billion in 2023 and is anticipated to increase at a 7.1% CAGR to reach USD 38.84 billion by 2030 (Verified Market Research, 2024). PHs are a class of PBs comprising a combination of amino acids and peptides that are produced through chemical and/or enzymatic hydrolysis. In addition, they may include other substances such as phenols, carbohydrates, minerals, phytohormones, and other chemical ingredients that may enhance their biostimulant effect (Colla, 2020; Nasri, 2017). Legumes, alfalfa hay, vegetable byproducts, chicken feathers, fish byproducts, blood meal, and casein are examples of animal proteins that are hydrolyzed chemically or enzymatically to produce PHs and amino acids (AAs) (Colla et al., 2015, 2016). Different PHs have different chemical properties based on where the raw materials come from and how they are produced (Baroccio et al., 2017; Malécange et al., 2023).

The composition of peptides and basic AAs in each biostimulant is determined by the degree of hydrolysis (DH) and production procedure. Comparatively to PHs produced by chemical hydrolysis, PHs produced by enzymatic hydrolysis have a lower percentage of free amino acids (AAs) but a higher spectrum of AAs and more peptides overall (Islam et al., 2021; Malécange et al., 2023). For example, acid hydrolysis is the normal chemical process that destroys tryptophan (Colla et al., 2017). The energy-efficient enzymatic proteolysis approach is still more beneficial because proteases' selectivity and activity determine the final hydrolysate's composition in terms of amino acids and peptides (Moreno-Hernández et al., 2020). Thus, the primary bioactive components of enzymatically generated plant-derived phytohemagglucosides (PHs) are signaling peptides, while the primary bioactive components of chemically generated animal-derived PHs are free amino acids (Rouphael et al., 2021). As essential molecules with a variety of physiological functions, amino acids are crucial for seed germination (Sierras et al., 2016; Sun et al., 2024). In conditions of salinity stress, they can also act as osmolytes, which can support the regulation of transport, enzyme activation, stomatal opening control, redox homeostasis maintenance, heavy metal detoxification, and gene expression (Shahrajabian and Sun, 2022; Shahrajabian et al., 2021a; b; Sun et al., 2024). The use of inorganic fertilizers can be reduced by providing plants with ecologically acceptable amino acid BS (Cheng et al., 2016; Shahrajabian et al., 2022).

A multitude of studies have demonstrated that isolated or combined amino acids (AAs) and PHs showed positive effects on several crops, including lettuce, tomato, and beetroot (Alfosea-Simón et al., 2020; Dewang and Devi, 2022; El-Sherbeny and da Teixeira Silva, 2013; Naroozlo et al., 2019). PHs have the ability to stimulate biomass and root growth in a variety of crops, including maize, lilies, tomatoes, and potted snapdragon (Cristiano et al., 2018; Ertani et al., 2019; Malécange et al., 2023; Polo and Mata, 2018; Sestili et al., 2018). A common theory of how

PHs work suggests that they have an auxin-like effect (Colla et al., 2014). One of the main hormones influencing the growth and development of roots is auxin (Ubeda-Tomás et al., 2012). The confluence of multiple physiological changes pertaining to phytohormones, ROS scavenging, particular primary and secondary metabolic pathways, cytoskeletal reorganization, and transport may be the cause of this beneficial effect, based on transcriptomic (mRNAseq) and proteomic analyses (Ebinezer et al., 2020; Trevisan et al., 2017). Certain horticultural plants and fruit trees have higher yields and leaf area after applying PH treatments (Carillo et al., 2020; Kisvarga et al., 2022; Tanou et al., 2017). An external source of PH derived from animals (feather hydrolysate) has been shown to improve the development and growth of two crops (Chilli and brinjal plants), resulting in early flowering and an increased yield (Gurav et al., 2020). When a combination of 20% amino acids (AAs) and 12% algal extracts (called "Primo") was applied externally to grapevines, the outcome was more bunches per cane as well as larger and heavier berries than the control (Khan et al., 2014).

1.3.1.4 CHITOSAN

The market for chitosan was estimated to be worth USD 10.88 billion in 2022, and between 2023 and 2030, it is projected to increase at a compound annual growth rate (CAGR) of 20.1%. Over the course of the projected period, industrial-grade chitosan held a substantial market share due to its widespread application in water treatment, agrochemicals, and other industries and its inexpensive cost (Grand View Research, 2023). The biopolymer chitin, which is found in insect exoskeletons, fungal cell walls, and crustacean shells naturally, is deacetylated to produce chitosan. More than 80% of the acetyl groups of the N-acetyl-d-glucosamine residues are eliminated to generate chitosan from chitin, a co-polymer of N-acetyl-d-glucosamine and d-glucosamine (Pichyangkura and Chadchawan, 2015). Products based on chitosan have a number of unique qualities that make them useful in a variety of industries, including agriculture, where they are employed as BS (Román-Doval et al., 2023). Chitosan's characteristics are directly influenced by its extraction process and source. Molecular weight, particle size, crystalline structure, degree of deacetylation, and surface area are the main characteristics that dictate the effectiveness and range of uses for chitosan (Román-Doval et al., 2023). The number of monomeric units in the biopolymer determines the molecular weight (MW) of chitosan. Molecular weight (MW) affects viscosity and solubility; hence, it's important to control, evaluate, and modify these properties. Chitosan's MW normally ranges from 20 to 1200 kDa (Kulikov et al., 2006). Depending on their molecular weight ranges, chitosans can be divided into three fundamental types: high (>700 kDa), medium (150–700 kDa), and low (less than 150 kDa) molecular weight chitosan. The raw material sources and processing method have an impact on the molecular weight of chitosan (Minh et al., 2020; Román-Doval et al., 2023) (Table 1.3).

TABLE 1.3 Chitosan Properties on the Basis of Molecular Weight

	Chitosan		
Properties	**High Molecular Weight (>700 kDa)**	**Medium Molecular Weight (150–700 kDa)**	**Low Molecular Weight (>150 kDa)**
Physicochemical	Low nitrogen content	Medium nitrogen content	High nitrogen content
	Limited solubility	Limited solubility	Great solubility
	Low tensile strength	High tensile strength	High tensile strength
	Low permeability	Medium permeability	High permeability
	Pseudoplastic behavior	Pseudoplastic behavior	Possesses complexing and chelating abilities.
	Low-reactive groups for chemical activation and crosslinking	Good reactive groups for chemical activation and crosslinking	Great reactive groups for chemical activation and crosslinking
Biological	Biocompatible	Biocompatible	Biocompatible
	Antimicrobial (bacteria, fungi, viruses)	Antimicrobial (bacteria, fungi, viruses)	Antimicrobial (bacteria, fungi, viruses)
	Lower antioxidant	Higher antioxidant	Antioxidant
	Nontoxic	Nontoxic	Nontoxic
	Biodegradable	Biodegradable	Biodegradable
	–	–	Antitumor, blood anticoagulants, hypolipidemic and hypocholesteromic activities

Source: Adapted from Román-Doval et al. (2023).

Many genes associated with pathogenicity, such as glucanase and chitinase, are among the defensive genes that plants express in response to chitosan (Pichyangkura and Chadchawan, 2015). It also activates a large number of enzymes, including peroxidase, catalase, and superoxide dismutase (SOD), which are involved in the system that scavenges ROS. Together with nitric oxide and hydrogen peroxide signals, chromatin interactions may also directly regulate gene expression through the signal transduction pathway from chitosan, leading to its activities (Hidangmayum et al., 2019; Mukarram et al., 2023). Chitosan has been explored as a biostimulant to induce pathogen resistance, stimulate abiotic stress, and promote plant growth. However, these stimuli are very complex in nature and depend on different factors like chitosan structures, molecular weights, concentrations, plant species, and developmental stage (Chakraborty et al., 2020; Mukarram et al., 2023; Pichyangkura and Chadchawan, 2015). According to Amin et al. (2007), effects on physiological systems in plants, such as food absorption, cell elongation, cell division, enzyme activation, and protein synthesis, can be directly connected to chitosan's plant growth-enhancing actions. In a variety of crops, including cabbage, potatoes, beans, radish, soybeans, gerbera, and other crops, chitosan promotes plant growth (Chakraborty et al., 2020; Hidangmayum et al., 2019; Sharif et al., 2018). It also increases yield due to the encouragement of plant growth. The growth rates of roots, shoots, flowers, and the quantity of flowers are all significantly influenced by chitosan. Because chitosan molecules are highly hydrophilic, they decrease the amount of water in plant cells and speed up the actions of a number of biological macromolecules (Chakraborty et al., 2020; Minh et al., 2020).

1.3.1.5 SILICON

The second most common mineral element found in soil is silicon (Si), and between 50 and 70% of the soil mass is made up of silicon dioxide (Epstein, 1994, 1999; Ma and Yamaji, 2006). Silicon (Si) is mostly found in the soil solution as monomeric silicic acid (H_4SiO_4), with values ranging from 0.01 to 2.0 mM. Because H_4SiO_4 does not dissolve at pH values below 9, plants can absorb silicon (Si) in this non-ionic form either actively or passively, based on their internal needs and the concentration of silicon (Si) in the environment (Savvas and Ntatsi, 2015).

Concentrations of accumulated silicon (Si) in plant tissues can range from 0.1% to 10% in dry weight, primarily based on the type of plant (Epstein, 1999; Ma and Yamaji, 2008). On absorption, silicon (Si) builds up in a variety of tissues, mostly as hydrated amorphous silica polymers (Azad et al., 2021). Although silicon (Si) is now used to treat some commercial crops in an effort to increase their resilience to abiotic stressors, illnesses, and viruses, there is still a possibility to expand on the usage of this silicon (Si) as a biostimulant in horticulture (Asif et al., 2023; Wang et al., 2021; Zargar et al., 2019). The ecological roles of silicon (Si) are diverse and include intricate roles in plant processes as well as moderating relationships with other organisms and the environment (Cooke et al., 2016; Coskun et al., 2019). A wide variety of essential genes that reduce stress and control plant growth and development are activated as a result of silicon (Si) uptake. There are a number of specialized and non-specific transporters that carry silicon (Si) into the aerial sections of agricultural plants. The absorption of silicon (Si) in roots and aerial organs is attributed to genes like *Lsi2*, *Lsi1*, and *Lsi6* in a number of crop plants, including maize, rice, cucumber, and barley (Ouellette et al., 2017; Ratcliffe et al., 2017; Wang et al., 2015). Of these, the anion transporter *Li2* is located in the membranes of the root endodermis, while the aquaporin-related transporters *Li1* and *Li6* are widely dispersed in the tissues of the shoot and root (Mitani et al., 2011). Crop plants exposed to silicon (Si) have increased expression of many genes that are stress-tolerant. According to the research of Song et al. (2014), silicon (Si), for example, causes transcriptional control of genes involved in photosynthesis in rice plants under zinc stress. According to Lenka et al. (2011), rice is able to withstand a variety of stressful situations because of the expression of LEA genes, which encode proteins that confer stress tolerance. Silicon (Si) promotes changes in the C/N ratio in the source-sink relationship in rice, a silicon-accumulator, by encouraging the remobilization of amino acids to meet the increased nitrogen (N) requirement during grain development in unstressed conditions (Detmann et al., 2012, 2013).

Silicon (Si) activates a number of defense mechanisms against abiotic stresses, including those involving phytohormone production and suitable solutes (Majeed Zargar et al., 2012; Mir et al., 2020). When exogenous silicon (Si) is applied to okra (*Abelmoschus esculentus*) genotypes that are susceptible to salinity stress, significant proline accumulation is seen (Abbas et al., 2015). Silicon (Si) increased tolerance to salt stress in two

Capsicum annuum varieties by inducing an excess of total free amino acids, particularly glycine betaine and proline (Pereira et al., 2013). According to Neumann and Zur Nieden (2001), the administration of silicon has a positive impact on the yield, growth, and quality of distinct fruit crops. Moreover, silicon (Si) promotes improved fruit quality, increased yield, and the capacity to save energy and water (Mir et al., 2022). In the context of Cd toxicity, root morphology recovered photosynthetic efficiency, and the shoot of silicon (Si)-treated rice plants showed considerable improvement (Bari et al., 2020). Furthermore, silicon (Si) has been shown to mediate a broad range of physiological advantages, including improved growth, mechanical strength, uptake of nutrients, and resistance to biotic stresses such as fungal infections (Ahammed et al., 2021; Bakhat et al., 2018; Fauteux et al., 2005; Mir et al., 2022; Souri et al., 2021).

1.3.1.6 MICROBIAL BIOSTIMULANTS

The market for microbial BS was valued at approximately USD 3.13 billion in 2022 and is anticipated to expand between 2023 and 2031 at a CAGR of 10.30%, reaching USD 7.57 billion. The increasing demand for organic food items is a result of people's growing awareness of nutritional sustainability, which is driving market expansion (Growth Market Reports, 2023). BS seem to be a viable way to help plants grow better, even in stressful environments (Drobek et al., 2023; Franzoni et al., 2022). Microbial BS, among other types of BS, are made up of microorganisms like plant growth-promoting rhizobacteria (PGPR) and/or other microbes that promote the uptake of nutrients, produce secondary metabolites (i.e., organic acids, siderophores), nitrogen fixation, stress tolerance, and improve crop quality and yield when administered to plants (Hijri, 2023; Kaushal et al., 2023). Within this broad class of BS, microbial BS include bacteria that promote plant growth (*Azospirillum* spp., *Arthrobacter* spp., *Azotobacter* spp., *Acinetobacter* spp., *Rhizobium* spp., *Pseudomonas* spp., *Bacillus* spp., *Rhodococcus* spp., *Enterobacter* spp., *Serratia* spp., *Streptomyces* spp., *Ochrobactrum* spp., and *Stenotrophomonas* spp.) and have emerged as highly effective and reasonably priced agricultural inputs for enhancing plant yield (Ayed et al., 2022; Baltazar et al., 2021; Fadiji et al., 2022; Kaushal et al., 2023; Miceli et al., 2021; Orozco-Mosqueda et al., 2020).

Nitrogen fixation, solubilization of minerals (P, K, and Zn), production of phytohormones such as indole-3-acetic acid (IAA), cytokinin (CK), abscisic acid (ABA), ethylene (ET), and jasmonic acid (JA), enzymes (cellulose, chitinases, glucanase, protease, etc.), secondary metabolites (siderophores, lipopeptides, cyclic lipopeptides, rhamnolipids, N-acyl homoserine lactone), volatile organic compounds (VOCs), volatile inorganic compounds, terpenoids, carbohydrates (alcohols, acids, aldehydes, benzenoids, lactones, etc.), and nitrogen (amines, metalloid, amides, and imines) (Caulier et al., 2019; Chaudhary et al., 2022; Hamid et al., 2021; Kaushal et al., 2023; Lephatsi et al., 2021). The biomass of roots, their surface area, shoot length, leaf area, newly formed lateral roots, soil structure (water holding capacity, nutrients, water filtration, and porosity), enhanced nutrient and mineral acquisition (N, K, P, Zn, Fe, Mn, etc.), and a plant's ability to photosynthesize are all factors that affect the morphology of roots (Aremu et al., 2022; Backer et al., 2018; Hamid et al., 2021; Lin and Jones, 2022; Vocciante et al., 2022). Additionally, they improve resistance to both biotic and abiotic stress by turning on genes that produce amino acids, organic acids, phenolics, and enzymes, as well as the antioxidant defense system; however, the precise workings of the biomolecular and cellular mechanisms are still unknown (Castiglione et al., 2021; Lephatsi et al., 2021; Othibeng et al., 2022).

1.3.2 MAIN CATEGORIES OF PLANT GROWTH REGULATORS

Other than nutrients, organic substances known as PGRs have an impact on a plant's physiological functions when they are supplied in small quantities (Jan et al., 2020; Sabagh et al., 2021). More targeted reactions include changing the way carbon is partitioned, increasing the ratio of roots to shoots, improving photosynthesis, changing nutrient intake, improving water status, and changing crop canopy. The interplay of input, environment, and genotype is reflected in these responses (Meshram et al., 2022). Two general categories can be used to group PGRs: (i) plant growth promoters and (ii) plant growth inhibitors.

1. **Plant Growth Promoters:** Plant growth promoters are substances or compounds that enhance the growth, development, and overall health of plants. They can be natural or synthetic and are used in agriculture to improve crop yield and quality. There are

some common types of plant growth promoters, including: PHs (gibberellins, auxins, cytokinins, and ethylene); plant growth-promoting substances (brassinosteroids, triacontanol, polyamines, and jasmonates); synthetic growth regulators; natural extracts; and other organic amendments (Gianfagna et al., 1995; Rashid et al., 2022; Shaukat et al., 2021; Zhang et al., 2021).

2. **Plant Growth Inhibitors:** Plant growth inhibitors, also known as PGRs or PHs, are chemicals that influence the growth and development of plants. They can either promote or inhibit various physiological processes such as cell elongation, cell division, fruit ripening, and flowering. There are some common types of plant growth inhibitors, including ABA, salicylic acid, and benzoic acid (Brookbank et al., 2021; Kaya et al., 2023; Koo et al., 2020; Windisch et al., 2021).

1.3.2.1 GIBBERELLINS

The size of the gibberellins market was estimated at USD 703.6 million in 2021, and it is anticipated to increase at a CAGR of 8.8% from USD 765.52 million in 2022 to USD 1503.09 million by 2030. Since they are used in horticulture and agriculture to encourage plant growth and development, gibberellins are becoming more and more popular (Skyquest, 2023). Gibberellins provide a number of benefits, including promoting plant growth, boosting seed germination, and enhancing fruit quality and size (Castro-Camba et al., 2023). Gibberellins are a class of PHs that control a number of plant developmental processes, such as fruit ripening, blooming, leaf expansion, stem elongation, and seed germination (Castro-Camba et al., 2022 ; Ghosh and Halder, 2018). Gibberellins were initially identified in Japan in the 1920s when scientists noticed strange growth in rice plants that was afflicted by the fungus *Gibberella fujikuroi*. Three distinct gibberellins—named gibberellin A1, gibberellin A2, and gibberellin A_3—were isolated and characterized from the gibberellin A sample by Tokyo University scientists in the 1950s. Based on this first nomenclature of gibberellins A_1 (GA1), GA_2, and GA_3, the numbering scheme for gibberellins utilized over the past 50 years (Binenbaum et al., 2018; Gupta and Chakrabarty, 2013). Gibberellins (GAs) are tetracyclic, diterpenoid endogenous PGRs (Gao and Chu, 2020).

The proper location of bioactive GA in plants or tissues that bioactive GAs target to start their activity has not yet been determined, despite significant attempts to study GA production and movements. GA was found in actively growing tissues, such as shoot apices, young leaves, and flowers, according to the dwarf plant bioassay and related quantitative analysis (Binenbaum et al., 2018; Hedden and Sponsel, 2015; Hu et al., 2008; Zhang et al., 2019).

In agriculture, GA biosynthesis and signaling play a crucial role in plant growth and development (Gao and Chu, 2020). Gibberellins (GAs) are essential for agronomic characteristics and crop stress tolerance (Gao and Chu, 2020; Waqas et al., 2019). GAs function as phytohormones that stimulate growth and are crucial for various aspects of plant development, such as plant stature, seed germination, nutrient availability response, and stress tolerance (Ahmad et al., 2022; El-Sabagh et al., 2022). The emergence of semidwarf cultivars, which served as a pillar of the so-called "Green Revolution" that resulted in a sharp rise in rice and wheat yields, was the most important advance in agricultural methods throughout the 20th century (Evenson and Gollin, 2003; Ordonio et al., 2014). The GA biosynthesis gene semidwarf1 (sd1) in rice and the GA signaling gene Rht in wheat control the semidwarf habit. The semidwarf cultivar IR8, which achieved record yields throughout Asia and served as the foundation for the development of other high-yielding semidwarf plant types, was created in the early 1960s by crossing the Chinese variety Dee-geo-woo-gen (DGWG), which had the sd1 gene. SD1 has been widely included in contemporary rice cultivars since the 1960s (Asano et al., 2007; Gao and Chu, 2020; Tu and Wang, 2019). In the 1960s, the insertion of mutant dwarfing alleles at the Rht-B1 and Rht-D1 loci in wheat improved both the harvest index and lodging resistance, which resulted in significant increases in grain yields globally. Since then, developing current wheat cultivars has made extensive use of Rht-1 dwarfism genes. The "Wheat Green Revolution" genes Rht-B1 and Rht-D1 produce mutant DELLA proteins with a decreased affinity for the GA receptor GID1 (Peng et al., 1999). GA destabilizes DELLA proteins, which function as repressors of GA signaling. Semidominant GA-insensitive dwarfism is caused by a mutation in the DELLA protein. Similarly, Sln1 in barley was found to be associated with mutations in the DELLA proteins (Chandler et al., 2002; Gao and Chu, 2020; Peng et al., 1999).

1.3.2.2 AUXINS

The size of the global auxins market is expected to increase at a compound annual growth rate (CAGR) of 4.5% from 2022 to 2027, when it is expected to reach US$1.1 billion. The market is growing due to the rising need for auxins on a global scale in the plant tissue culture and PGR industries (Industryarc, 2023). Auxins are a class of PHs that play vital roles in controlling different facets of the growth and development of plants. They are primarily responsible for cell elongation, cell division, apical dominance, phototropism, gravitropism, and root initiation (Balzan et al., 2014; Woodward and Bartel, 2005; Zhang et al., 2022). The most common naturally occurring auxin is indole-3-acetic acid (IAA), although synthetic auxins such as 2,4-dichlorophenoxyacetic acid (2,4-D) and naphthaleneacetic acid (NAA) are also widely used in agriculture as growth regulators (Sugawara et al., 2014; Tan et al., 2024). Several studies have indicated that auxins and cyokinins (CKs) cooperate in a number of cellular and physiological processes (Žižková et al., 2017). These physiological events include the formation of leaves, advancement of the cell cycle, cell expansion, apical dominance, and embryonic development during the maturation of seeds (Jurado et al., 2010; Tromas et al., 2009).

According to *in vitro* assays and genetic studies, there are four Trp-dependent and one Trp-independent pathways in the auxin biosynthesis pathways: indole-3-pyruvic acid (IPA) pathway, tryptamine (TAM) pathway, indole-3-acetamide (IAM) pathway, and indole-3-acetaldoxime (IAOx) pathway. Among these, it is claimed that the IPA and TAM pathways aid in the development of plants (El-Sabagh et al., 2022; Taiz et al., 2015; Woodward and Bartel, 2005). Different abiotic stresses influence auxins' metabolism, homeostasis, and distribution in the cellular environment. Such changes in auxin distribution under stress conditions have been attributed to two molecular mechanisms: (i) inhibition of polar auxin transport caused by phenolic substances accumulated during exposure to stress and (ii) changes in PIN gene expression, which is required for polar auxin transport (El-Sabagh et al., 2022; Kovtun et al., 2000; Potters et al., 2009). According to Jain and Khurana (2009), peroxidases catalyze the breakdown of auxins (IAA), which controls stress-induced auxin metabolism. Therefore, auxins can be viewed as stress hormones that alter the expression of some stress-responsive genes through direct or indirect mediation (El-Sabagh et al., 2022).

1.3.2.3 CYTOKININS

Over the span of the projected period, the cytokinin market is expected to grow at a 7.9% CAGR, from US$1.7 billion in 2022 to US$3.64 billion by 2032. Advances and technological innovation in plant tissue culture and growth regulators are driving up demand for cytokinins in the global market (Future Market Insights, 2024). Cytokinins are a class of PGRs (phytohormones) that play crucial roles in various aspects of plant growth and development. They are involved in cell division, shoot and root growth, leaf senescence, nutrient mobilization, and responses to environmental stresses (Akhtar et al., 2020; Wu et al., 2021). There are two categories of cytokinins: (i) phenyl-urea-type and (ii) adenine-type (Ahmad et al., 2018).

1. **Phenyl-Urea-Type Cytokinins**: These cytokinins have a phenyl-urea moiety in their chemical structure. Examples include kinetin, benzyladenine (BA), and thidiazuron (TDZ). Phenyl-urea-type cytokinins are synthetic derivatives that are commonly used in plant tissue culture due to their high activity and stability (Ahmad et al., 2018; Fathy et al., 2022).
2. **Adenine-Type Cytokinins**: These cytokinins have an adenine base in their chemical structure. Examples include isopentenyladenine (iP), zeatin, and zeatin riboside (ZR). Adenine-type cytokinins are naturally occurring and play essential roles in various plant growth and development processes (Ahmad et al., 2018; Fathy et al., 2022).

These categories represent the major structural classes of cytokinins, but there are also other derivatives and analogues that exhibit cytokinin-like activity in plants. Cytokinins (CKs) are frequently regarded as auxin antagonists, synergists, and ABA antagonists in a variety of plant activities, including growth, development, and resistance to diverse abiotic stresses (Danilova et al., 2016; Pospíšilová, 2003). When a plant experiences salt stress, in addition to other hormonal pathways (like ABA), CKs are triggered. CKs have the ability to significantly boost salt stress tolerance through interactions with other PHs (i.e., ABA, auxins) (El-Sabagh et al., 2022; Iqbal et al., 2006). In agriculture, cytokinins are often used as PGRs to manipulate various aspects of plant growth and development. They are applied in the form of synthetic chemicals or through the use of cytokinin-rich fertilizers to enhance crop yields, improve stress tolerance, and regulate plant morphology (Akhtar et al., 2020; Wu et al., 2021). However, the application of cytokinins must be carefully controlled to avoid detrimental effects on plant growth or unintended environmental consequences (Mandal et al., 2022).

1.3.2.4 ABSCISIC ACID

The market for abscisic acid (ABA) in 2023 was estimated to be worth USD 100 million and is anticipated to grow at a CAGR of 10.41% to surpass USD 200 million by the end of 2030. Early in the 1960s, researchers discovered that ABA was a phytohormone that disrupted seed dormancy (Cornforth et al., 1965). Subsequent research has revealed ABA is a phytohormone that is necessary for several plant physiological processes, stress adaptations, and controlling plant growth (El-Sabagh et al., 2022; Iqbal, 2015). ABA is synthesized in roots and transferred to leaves via xylem in water-deficient environments, raising the concentration of ABA in the leaves. Guard cells use this rise in ABA as a signal to initiate a signaling cascade that controls the motion of guard cells (Awan et al., 2017; Bharath et al., 2021). The production and distribution of ABA result in stomatal closure and an overall reduction in transpiration rate during the abiotic stress response, which limits cell development (Arkhipova et al., 2020; Saradadevi et al., 2017; Tuteja, 2007). According to Dar et al. (2017), phosphatases and kinases in ABA signaling help mediate quick reactions to an array of abiotic stresses and up-regulate hormone-responsive transcription factors. Moreover, in plants subjected to abiotic stresses, ABA interacts with other PHs to modulate physiological responses in addition to mediating them via its own signaling (He et al., 2018). While ABA doesn't directly promote growth in the traditional sense, its regulatory roles are essential for ensuring plant survival and adaptation to various environmental conditions. As such, it is a crucial component of plant growth regulation, particularly under stress conditions (Gupta et al., 2022; Vishwakarma et al., 2017).

1.3.2.5 ETHYLENE

The size of the worldwide ethylene market in 2019 was worth USD 166,520 million and is expected to rise at a 5.6% CAGR to attain USD 245,005 million in 2027 (Fortune Business Insights, 2020). In view of consumer convenience, there is an increasing trend in the need for growth regulators, which is opening up new business opportunities for both small and major fertilizer companies. Ethylene is a gaseous phytohormone that plays a pivotal role in various aspects of plant growth and development (Dubois et al., 2018). Originally identified as a ripening agent in fruits, its regulatory functions extend across numerous physiological processes, making it a key player in plant biology (Awan et al., 2017; El-Sabagh et al., 2022). Ethylene influences plant growth and development through complex signaling pathways involving both ethylene-responsive genes and interactions with other hormones such as auxins, cytokinins, and ABA. It regulates diverse processes, including seed germination, seedling growth, root hair development, leaf and flower senescence, fruit ripening, and responses to biotic and abiotic stresses (Binder, 2020; Fatma et al., 2022; Iqbal et al., 2017; Padmanabhan et al., 2018). Understanding the mechanisms underlying ethylene signaling pathways has facilitated the development of strategies to modulate ethylene levels for agricultural purposes (Binder, 2020; Chang, 2016). Additionally, ethylene (ET) controls a number of the biochemical reactions that plants undergo in response to abiotic stressors such as heat, drought, waterlogging, cold, heavy metals, salinity, floods, or submerged situations (Awan et al., 2017; Azhar et al., 2023; El-Sabagh et al., 2022). For example, *Arabidopsis* and *Médicago truncatula* showed a strong correlation between freezing and cold stress and high ethylene (ET) levels. This suggests that maintaining ethylene (ET) homeostasis is essential for being tolerant of suboptimal temperature stress, such as freezing and chilling (Shi et al., 2012; Zhao et al., 2014). Similar to salt-tolerant *Arabidopsis* plants, high ethylene (ET) levels aid in the tolerance to salt stress. By preventing ROS generation and preserving Na^{+}/K^{+} equilibrium, the ethylene overproducer (ETIO1) responds favorably to salt stress (Yang et al., 2017). Ethylene biosynthesis inhibitors and receptor antagonists are commonly used to delay fruit ripening, prolong shelf life, and reduce post-harvest losses (Botton et al., 2019; Iqbal et al., 2017; Schaller and Binder, 2017). Conversely, exogenous application of ethylene or its precursors can promote flowering, enhance fruit ripening, and induce stress responses to improve crop yield and quality (Cocetta and Natalini, 2022; Khan et al., 2024; Manjunatha et al., 2012; Yin et al., 2017).

1.3.2.6 SALICYCLIC ACID

The market for salicylic acid was estimated to be worth $431.7 million in 2020 and is expected to increase at a CAGR of 7.5% from 2020 to 2030, when it is expected to reach $886.9 million. Salicylic acid has the chemical formula $C_3H_2O_2$. It is a lipophilic monohydroxybenzoic acid, a kind of phenolic acid, and a beta-hydroxy acid. It is a colorless, odorless, crystalline organic acid in terms of physical characteristics. Salicylic acid serves as an intermediary in the agriculture, personal care, and pharmaceutical sectors. The need for agricultural and cosmetic products is rising, which is fueling the salicylic acid market (Allied Market Research, 2022). Salicylic acid is one of the most important phenolic endogenous growth regulators and is present in practically all plant species that belong to a wide range of groupings (Dempsey and Klessig, 2017; Koo et al., 2020; War et al., 2011). The freshly harvested biomass of barley, rice, and soybean crops has been found to have a salicylic acid (SA) concentration of 1 $\mu g\ g^{-1}$. Its function in controlling a wide range of biological and physiological functions in plants has been determined. It has been extensively researched in plants and aids in triggering the development of systemic acquired resistance (SAR) to a variety of diseases (Liu et al., 2022; Misra and Saxena, 2009; Urmi et al., 2023). SAR is a phenomenon in which a plant's immune system is primed for a more robust response to subsequent pathogen attacks (Durrant and Dong, 2004). Salicylic acid is a key signaling molecule in SAR, helping plants develop resistance to pathogens (Klessig et al., 2018). Salicylic acid influences various aspects of plant growth and development, including seed germination, root development, flowering, and fruit ripening. It can modulate hormone signaling pathways, particularly auxin and ethylene, to regulate growth processes (Bagautdinova et al., 2022). Salicylic acid plays a crucial role in plant defense against various pathogens such as bacteria, fungi, and viruses. It activates the plant's defense mechanisms, leading to the production of pathogenesis-related proteins and other defense compounds (Benjamin et al., 2022; Lu, 2009). Moreover, salicylic acid helps plants cope with various environmental stresses, including drought, salinity, extreme temperatures, and heavy metal toxicity. It enhances the antioxidant defense system, reduces oxidative damage, and improves stress tolerance (Chen et al., 2021; Liu et al., 2022). Salicylic acid can delay senescence (ageing) in plants by inhibiting the breakdown of chlorophyll and other macromolecules. This prolongs the period of active growth and productivity in crops, ultimately leading to higher yields (Guo et al., 2021; Huang et al., 2022). Overall, the exogenous application of salicylic acid can be used to manipulate plant growth, enhance stress tolerance, and improve crop productivity in agriculture. However, the effects of salicylic acid can vary depending on factors such as concentration, timing of application, and plant species, so careful optimization is necessary for practical applications (Abdelaal et al., 2020; Ali et al., 2023; Khan et al., 2015).

1.3.2.7 JASMONATES

The size of the global market for methyl dihydro jasmonate (a synthetic form of the PH jasmonic acid) was estimated at USD 158.2 million in 2022, and throughout the projected period, the market is expected to grow at a CAGR of 5.6% to reach USD 258.24 million by 2031 (Business Research Insights, 2023). The collective term for jasmonic acid and its derivative compounds is jasmonates. Originally, it was isolated from *Jasminum grandiflorum* essential oil (Demole et al., 1962). There are several physiological processes in which jasmonates (JAs)—jasmonic acid (JA) and methyl jasmonates (MeJAs)—are known to be involved (Ahmad et al., 2016). Jasmonates control a number of physiological functions, including development, growth, and defense against external biotic and abiotic stresses (Ahmad et al., 2016; Parthier, 1991; Sood et al., 2023). Cyclopentanone molecules called jasmonates (JAs) are produced from α-linolenic acid. According to Ho et al. (2020), it is the primary precursor to a number of compounds in this category, including methyl jasmonate (MJ). *Arabidopsis thaliana* and *Solanum lycopersicum* have been used as model plants to investigate the production of jasmonates (JAs) in a variety of dicotyledonous and monocotyledonous plants (Ruan et al., 2019). Plants typically exhibit jasmonic acid (JA) concentrations between 0.01 and 3.0 ng/g FW (fresh weight); however, *Artemisia tridentata* is an exception, exhibiting methyl jasmonates (MeJA) levels as high as 95 $\mu g\ g^{-1}$ fresh weight (Preston et al., 2004). The chloroplasts of illuminated plants and their blossoms are the usual places to find high levels of jasmonic acid (JA) (Yan et al., 2013).

It is believed that jasmonate (JAs) and methyl jasmonates (MeJA) actively contribute to senescence (Reyes-Díaz et al., 2016). Several other jasmonic acid (JA)-induced physiological, developmental, and plant growth processes have also been documented. These processes include sex determination, fruit ripening, root elongation, and senescence; biotic and abiotic stress tolerance; fertility; storage organ formation; and interactions with other hormones (Ahmad et al., 2016; Hewedy et al., 2023; Kim et al., 2021; Nafie et al., 2011; Sohn et al., 2022). On exogenous application of jasmonates (JAs), plants react to biotic and abiotic environmental stresses in a variety of ways, ranging from the production of signaling molecules to the final death of individual cells (Ahmad et al., 2016; Avanci et al., 2010). Plants treated with jasmonate have been observed to exhibit increased pest resistance (Ahmad et al., 2016; Sohn et al., 2022). Thaler et al. (2001) reported that tomato plants treated with JA exhibited a decrease in flea beetles, *Frankliniella occidentalis* (thrips), and aphids. This was due to a rise in the levels of proteinase inhibitors and polyphenol oxidase. Similarly, methyl jasmonate application to two cultivars of *Gossypium hirsutum* caused them to express genes linked to disease (Zambounis et al., 2012).

1.3.2.8 TRIACONTANOL

Over the projected period of 2023–2030, the triacontanol market size is anticipated to increase exponentially in terms of revenue and market share at a remarkable rate of growth (CAGR). Due to its widespread application in the worldwide agriculture sector, triacontanol is becoming more and more in demand, which is responsible for the market's expansion (Verified Market Reports, 2023). Triacontanol, a long-chain fatty alcohol derived from natural plant waxes, was initially discovered in alfalfa hay (Ries et al., 1977). Natural epicuticular waxes from a variety of plant species, including *Croton californicus*, *Medicago sativa, Oryza sativa, Copernica cerifera,* and *Vaccinium ashei*, contain triacontanol (TRIA) (Islam and Mohammad, 2020; Naeem et al., 2012; Uchiyama and Ogasawara, 1981). It has garnered significant attention in agricultural research due to its potential as a growth regulator (Islam and Mohammad, 2020). Applying TRIA exogenously to most crops at relatively low concentrations promotes plant growth (Naeem et al., 2012; Verma et al., 2022). Triacontanol has been reported to positively influence various physiological processes in plants, including photosynthesis, nutrient uptake, enzyme activity, and hormone regulation (Sarwar et al., 2021, 2022; Weremczuk-Jeżyna et al., 2022). Exogenous administration of triacontanol (TRIA) improves the biomass of plants, gas exchange parameters, photosynthetic pigments, uptake of mineral nutrients, nitrate reductase (NR), leaf carbonic anhydrase (CA), activity, modulation of antioxidant enzyme activities, accumulation of osmolytes, quality attributes, and yield (Naeem et al., 2009, 2019; Waqas et al., 2016; Zaid et al., 2020). Through its interaction with plant membranes and signaling pathways, triacontanol modulates gene expression, leading to improved growth parameters such as increased biomass, enhanced root and shoot growth, and accelerated flowering and fruiting (Alharbi et al., 2021; Bhandari et al., 2021). The mode of action of triacontanol involves its incorporation into plant cell membranes, where it alters membrane fluidity and permeability, thereby enhancing nutrient absorption and photosynthetic efficiency (Islam and Mohammad, 2020; Verma et al., 2022; Sarwar et al., 2022). Additionally, triacontanol acts as a signaling molecule, triggering the expression of genes involved in growth promotion and stress tolerance pathways (Islam and Mohammad, 2020; Weremczuk-Jeżyna et al., 2022). Numerous studies have demonstrated the potential of triacontanol as a growth regulator in various crop species, including cereals, legumes, vegetables, and fruit trees (Chen et al., 2022; Pang et al., 2020; Verma et al., 2022). Application methods such as foliar spray, seed treatment, and soil drenching have been explored to optimize the efficacy of triacontanol in different agricultural systems (Ali and Perveen, 2020; Asadi-Karam and Keramat, 2017; El-Beltagi et al., 2022). Furthermore, triacontanol has been found to alleviate abiotic and biotic stresses, including drought, salinity, and pathogen attacks, thus enhancing plant resilience and productivity under adverse environmental conditions (Faiza et al., 2024; Islam and Mohammad, 2020; Zaid et al., 2020).

1.3.2.9 BRASSINOSTEROIDS

The term "Brassins" refers to the group of polyhydroxylated steroidal PHs known as brassinosteroids (BRs), which were first identified from the pollen grains of *Brassica napus* (Mitchell et al., 1970). Almost every plant part and the entire kingdom of plants contain them. The two richest sources of BRs are pollen and immature seed (Tang et al., 2016). The most important breakthrough was the identification of Brassinosteroid

Insensitive 1 (BRI1), a receptor kinase that, upon extracellular BRs detection, initiates an intracellular signaling cascade (Li and Chory, 1997). Since the initial identification of Brassinolide (BL), a vast array of chemically distinct BRs have been found in land plants and green algae alike, indicating that BR development occurred early in the history of plants (Kim, 2020). Brassinolide (BL), Castasterone (CS), Typhasterol (TY), 6-deoxocastasterone (6-deoxoCS), Teasterone (TE), and 28-norcastasterone (28-norCS) are among the BRs that have been investigated so far and are frequently found in a variety of plant species and habitats (Hussain et al., 2020; Manghwar et al., 2022). The market for 28-homobrassinolide was estimated to be worth USD 100 million in 2023 and is projected to grow at a CAGR of 13.98% to reach USD 250 million by the end of 2030 (Verified Market Report, 2024). Based on the various side chain alkyl-substitution patterns, BRs are categorized as C_{27}, C_{28}, or C_{29}. In general, two hydroxyl groups at ring A and a 6-ketone or 7-oxa-6-ketone system at ring B with a trans-fused A/B ring system are required for active BRs (Taiz and Zeiger, 2010; Tang et al., 2016). BRs have significant potential for application in agriculture as PGRs (Manghwar et al., 2022; Sharma et al., 2017). BRs can be used to enhance crop yield, improve stress tolerance, and optimize plant architecture. Research into the biosynthesis, signaling pathways, and physiological functions of BRs continues to advance our understanding of their roles in plant growth and development (Khan et al., 2023; Nolan et al., 2020; Wei and Li, 2020). BRs support plants in combating a variety of environmental stressors, including salinity, drought, extreme temperatures, and heavy metal toxicity. They enhance stress tolerance by modulating physiological and biochemical processes, including antioxidant defense mechanisms (Chaudhuri et al., 2022; Fariduddin et al., 2014; Manghwar et al., 2022). Moreover, BRs can enhance photosynthetic efficiency by regulating chlorophyll content, stomatal conductance, and photosynthetic enzyme activity, thereby improving overall plant productivity (Lima et al., 2017; Siddiqui et al., 2018).

1.4 PLANT BIOSTIMULANTS AND GROWTH REGULATORS: GLOBAL INSIGHT

The market for PBs and growth regulators has witnessed significant growth over the past decade and is projected to continue expanding. The

BS market is expected to grow at a CAGR of 11.8% from 2023 to 2028, from an estimated USD 3.9 billion in 2023 to USD 6.8 billion by that time (Marketsandmarkets, 2023a). Factors such as increasing demand for sustainable agricultural practices, rising need for higher crop yields to feed the growing global population, and environmental concerns are driving this growth. The BS market is divided into microbial and non-microbial categories based on the source. The non-microbial BS group includes biopolymers, vitamins, inorganic chemicals, humic substances, PH (or amino acids), and seaweed extracts as active ingredients (Global Market Insights, 2023). These non-microbial ingredients support plant growth without endangering human or environmental health, whereas microbial-based biostimulant products comprise free-living fungi or bacteria that form a symbiotic relationship with the crops, thereby enhancing the overall yield and alleviating the impacts of climate change. Consumer interest in non-microbial and microbial-based biostimulant products has increased recently due to the declining supply of high-quality agricultural land, a trend that is expected to continue in the years to come. These plant growth promoters lessen the effects of climate change on crops and increase agricultural yields when minimal input conditions are met (Fortune Business Insights, 2023).

Globally, the market for BS is expected to be dominated by the European market. Europe has one of the largest vegan populations in the world, mostly in the UK and Germany, because of increased consumer awareness of environmental safety and sustainability. In a mere four years (2016–2020), the number of vegans worldwide is expected to double, according to a survey by Veganz, a vegan grocery chain with headquarters in Berlin (Marketsandmarkets, 2023a). The region's need for high-value commodities, especially fruits and vegetables, is rising as a result of this trend. Several companies operate in the PBs and growth regulators market globally. These include multinational corporations as well as smaller specialized firms. Major players often invest in research and development to introduce innovative products and technologies to cater to evolving agricultural needs. Leading companies in the European BS market are Tradecorp International (Spain), AgriTecno (Spain), Koppert Biological Systems (Netherlands), Biovert S.L. (Spain), Valagro (Italy), and BASF SE (Germany). At the moment, the European market has a limited or elementary regulatory framework for items containing BS. As a result, there has been a noticeable surge in companies, especially in Spain and Italy. On the other hand, indications point to the region's impending establishment of a thorough regulatory structure to ensure the quality of the biostimulant products sold there (Global Market Insights, 2023; Marketsandmarkets, 2023a).

The global market for PGRs or PHs was experiencing steady growth due to increased demand for high-quality agricultural produce, advancements in farming techniques, and a growing focus on sustainable agriculture. The market encompasses various types of growth regulators, including auxins, cytokinins, gibberellins, ABA, and ethylene, among others. PGRs were worth \$2.2 billion globally in 2021; from 2022 to 2031, the market is expected to increase at a CAGR of 13.9% to reach \$7.4 billion (Allied Market Research, 2023b). The main factors behind the expansion of the PGR sector are the increasing demand for organic food by consumers, the shrinking amount of arable land, and the world population. In terms of global revenue, Europe is anticipated to lead the market by the end of 2031, with Asia-Pacific, North America, and LAMEA (Brazil, Saudi Arabia, South Africa, the United Arab Emirates, and the rest of LAMEA) following closely behind (Allied Market Research, 2023b; Marketsandmarkets, 2023b). Europe and Asia-Pacific hold the top spots in the worldwide PGR market, primarily because of the increase in demand for organic food items and the expansion of the organic food sector (Allied Market Research, 2023b). Some of the major companies in the growth regulator market are Xinyi (H.K.) Industrial Co. Ltd., Sumitomo Chemical Australia Pty Ltd., NuFarm Ltd., Redox Industries Ltd., BASF SE, Bayer Crop Science, Corteva Agriscience, Crop Care Australasia Pty Ltd., Barclay Crop Protection, and Sichuan Guoguang Agrochemical Co. Ltd. Based on type, the global market segment for PGRs is divided into submarkets for auxins, gibberellins, cytokinins, ethylene, and ABA (Allied Market Research, 2023b; Marketsandmarkets, 2023b).

1.5 MARKET AVAILABILITY OF COMMERCIAL PLANT BIOSTIMULANTS AND GROWTH REGULATORS

The market for commercial PBs and growth regulators was growing globally, driven by increasing demand for sustainable agricultural practices, the need to improve crop yields, and regulatory support for eco-friendly agricultural inputs. These products are gaining popularity due to their ability to enhance plant growth, improve nutrient uptake, and increase stress tolerance (Markets and Markets, 2023). Some key players in the global market for BS and growth regulators include BASF SE, Bayer CropScience AG, Syngenta AG, Marrone Bio Innovations, Inc., and Valagro SpA, among others. These businesses provide an extensive variety of products derived from various sources, such as seaweed extracts, microbial extracts, amino acids, and humic substances (Allied Market Research, 2023b; Marketsandmarkets, 2023b). The availability of these products varies by region, with more developed markets such as North America and Europe having a wider range of options compared to emerging markets in Asia, Africa, and Latin America (Figure 1.1). However, with increasing awareness and adoption of sustainable agricultural practices worldwide, the availability of BS and growth regulators is expected to continue expanding globally (Allied Market Research, 2023b; Global Market Insights, 2023; Marketsandmarkets, 2023b). Similarly, India's agricultural sector has also seen increased interest in adopting biostimulant and growth regulator-based products due to their potential to improve crop yield, quality, and resilience while reducing reliance on synthetic chemicals. Several companies in India are actively involved in the production and distribution of these products. Some well-known companies in this sector include Bayer CropScience, Syngenta, UPL Limited, FMC Corporation, Indofil Industries Limited, Tata Chemicals, and Gujarat State Fertilizers & Chemicals Ltd. (GSFC).

FIGURE 1.1 Some of the commercial plant biostimulants and growth regulators available in the market.

Apart from these larger companies, there are also numerous smaller manufacturers and distributors of BS and growth regulators catering to specific regions or crops. The availability of these products can vary depending on the region and distribution network. They are commonly found in agricultural input supply stores, agricultural cooperatives, and through online platforms. Additionally, agricultural extension services and government agencies often promote the use of BS and growth regulators as part of sustainable farming practices.

1.6 REGULATORY FRAMEWORK FOR PLANT BIOSTIMULANTS AND GROWTH REGULATORS

Regulations surrounding PBs and growth regulators vary across different countries and regions. However, there has been a growing recognition of the importance of regulating these products to ensure their safety, efficacy, and proper labeling.

1.6.1 PLANT BIOSTIMULANTS REGULATION

The regulatory framework for PBs varies depending on the country or region. However, there were efforts underway in various jurisdictions to establish clearer regulations for these products due to their increasing popularity and importance in agriculture. In many countries, regulatory oversight of BS falls under agricultural or environmental regulatory agencies. These agencies may include the U.S. Environmental Protection Agency (EPA), the European Commission (EC) or its member state agencies, and Fertilizer Control (Order) 1985, India, and so forth. Moreover, one of the primary challenges in regulating BS is defining what they are. Different regions have proposed various definitions, but generally, "*Biostimulants are substances or microorganisms that are administered to soil or plants to improve nutrient uptake, nutrient utilization efficiency, abiotic stress tolerance, and crop quality attributes*." Depending on the jurisdiction, manufacturers may be required to register or obtain approval for their biostimulant products before they can be marketed and sold. The registration/approval process typically involves submitting data on product composition, efficacy, safety, and environmental impact (Fertilizer Association of India, 2024; Sanjuán et al., 2023; SEIPASA, 2022).

1.6.1.1 EUROPEAN UNION

The utilization of PBs has gained significant traction in modern agriculture due to their potential to enhance crop productivity, resilience, and sustainability. However, the regulatory landscape governing these products in the European Union (EU) has been complex and evolving. A comprehensive overview of the regulatory framework for PBs in the EU, focusing on recent developments and future prospects, is as follows (Sanjuán et al., 2023; SEIPASA, 2022):

1. **Definition of Plant Biostimulants:** The European Union defined PBs in 2019 as substances or microorganisms that, when added to plants or the rhizosphere, attempt to accelerate the body's natural processes in order to improve crop quality, endurance to abiotic stress, and nutrient uptake.
2. **Regulation (EU) 2019/1009:** The regulatory framework for PBs in the EU is primarily governed by Regulation (EU) 2019/1009, which establishes harmonized rules for placing these products on the market. This regulation, which came into effect on July 16, 2019, established guidelines for putting EU fertilizer products on the market, including BS. It established requirements for labelling, marketing, and the registration process for placing such products on the EU market. According to Regulation (EU) 2019/1009, manufacturers of PBs need to register their products with the competent authorities of EU member states before placing them on the market. This registration process involves demonstrating the safety, efficacy, and quality of the product. Moreover, the regulation specifies data requirements for the registration of PBs, including information on the identity and properties of the product, its effect on plants and soil, and its safety for humans, animals, and the environment (https://www.legislation.gov.uk/eur/2019/1009/contents).
3. **EU Fertilizing Product Regulation Implementation**: Member states were given a transition period to implement Regulation (EU) 2019/1009 into their national legislation. During this period, stakeholders were expected to adapt to the new requirements, and competent authorities were expected to establish the necessary infrastructure for implementing the regulation (EUR-Lex, 2022).

4. **European Biostimulant Industry Council (EBIC)**: EBIC is an organization representing the interests of the plant biostimulant industry in Europe. It has been actively engaged in discussions with policymakers and stakeholders to ensure that the regulatory framework for PBs in the EU is conducive to innovation and market growth while ensuring product safety and efficacy (https://biostimulants.eu/).

Despite the strides made in harmonizing regulations, challenges remain in implementing and interpreting the EU framework effectively. Key areas of ongoing debate include the classification of BS versus PPPs, the determination of efficacy criteria, and the establishment of standardized testing protocols. Additionally, the incorporation of emerging technologies, such as biostimulant-based BS, poses novel regulatory considerations.

1.6.1.2 UNITED STATES

The regulatory framework for PBs in the United States was undergoing significant changes. Historically, there wasn't a specific category for PBs in the regulatory framework, which led to some ambiguity regarding their classification and oversight. However, in recent years, there have been efforts to establish clearer regulations for these products. The U.S. Environmental Protection Agency (EPA) oversees the regulation of pesticides, including those used in agricultural settings. Historically, PBs often fell into a grey area where they were not clearly defined as pesticides but could have pesticidal properties or effects. This ambiguity led to challenges in regulation and marketing. To address this issue, the 2018 Farm Bill directed the USDA to create a definition for PBs and to establish a regulatory framework for their oversight. This effort aimed to provide clarity for manufacturers, growers, and regulators regarding the classification and regulation of these products (https://www.congress.gov/115/bills/hr2/BILLS-115hr2enr.pdf).

In 2019, the USDA issued a memorandum outlining its approach to regulating PBs. The memorandum indicated that the USDA intended to regulate PBs as agricultural fertilizers under the authority of the Federal Fertilizer Act (FFA), administered by the USDA's Agricultural Marketing Service (AMS) (https://www.ams.usda.gov/). This approach focused on ensuring product safety and efficacy while providing a pathway for product registration and labeling requirements. Additionally, the EPA has been exploring ways to streamline the regulation of PBs under the Federal Insecticide, Fungicide, and Rodenticide Act (FIFRA), which governs the registration and use of pesticides. In 2019, the EPA released a draft guidance document outlining its proposed approach to regulating PBs under FIFRA. The guidance aimed to clarify when PBs would be considered pesticides and provided recommendations for registration and labeling (https://www.epa.gov/).

1.6.1.3 INDIA

In India, all fertilizers are regulated under the Fertilizer (Inorganic, Organic, or Mixed) (Control) Order, 1985, issued by the government of India under the Essential Commodities Act, 1955. This order aims to regulate the production, distribution, and sale of fertilizer in India. It categorizes fertilizers into three main types: inorganic, organic, or mixed, and establishes standards and regulations for each type. The order sets forth guidelines regarding the quality, packaging, labeling, and pricing of fertilizers to ensure their efficacy and safety for agricultural use. It also mandates licensing requirements for manufacturers, distributors, and sellers of fertilizers to maintain quality standards and prevent malpractice in the industry. Overall, the Fertilizer (Inorganic, Organic, or Mixed) (Control) Order, 1985, plays a crucial role in regulating the fertilizer industry in India, ensuring the availability of quality fertilizers to farmers, and promoting agricultural productivity and sustainability (Fertilizer Association of India, 2024; https://upagripardarshi.gov.in/MediaGallery/FCO_Act.pdf).

The Fertilizer (Inorganic, Organic, or Mixed) (Control) Amendment Order, 2021, establishes the legal framework for biostimulant registration in India. India is currently one of the few countries worldwide with distinct laws controlling BS (Press Trust of India, 2023). Schedule VI of the Fertilizer (Inorganic, Organic, or Mixed) (Control) Order (FCO), 1985, regulates the sale of BS. According to the Indian Fertilizer Order (FCO) Amendment Order, 2021, a "biostimulant" is defined as a material, microbes, or both in combination whose main purpose, when applied to plants, seeds, or the rhizosphere, is to stimulate physiological processes in plants and to improve their uptake of nutrients and growth, yield, nutrition efficiency, crop quality, and stress tolerance, irrespective of the amount of nutrients they contain. However, this order does not include PGRs or pesticides, as these are governed by the Insecticide Act, 1968

(Fertilizer Association of India, 2024; https://www.fertcon.net/wp-content/uploads/2021/06/6-AUXI.pdf).

1.6.2 PLANT GROWTH REGULATORS REGULATION

Regulation of PGRs varies by country and region. In many jurisdictions, including the United States, the European Union, and India, plant growth regulators are subject to regulation to ensure their safety for human health, environmental safety, and efficacy in achieving their intended purposes.

1.6.2.1 EUROPEAN UNION

PGRs in the European Union are governed by Regulation (EC) No. 1107/2009 of the European Union, which addresses the marketing of PPPs. PGRs are considered PPPs if they have a direct or indirect action in protecting plants or plant products against pests or if they are used to control plant growth (EU Monitor, 2020; FAOLEX Database, 2009). There are some key aspects of the regulatory framework for PGRs in the European Union (https://www.legislation.gov.uk/eur/2009/1107):

1. **Approval Process**: Before a PGR can be placed on the market in the EU, it must undergo a rigorous approval process. This process involves the submission of comprehensive data on the product's efficacy, safety for humans, animals, and the environment, and its effects on non-target organisms.
2. **Active Substances**: Like other PPPs, PGRs must contain approved active substances. These active substances are evaluated at the EU level and included in Annex I of Regulation (EC) No. 1107/2009 if they meet the necessary criteria regarding safety and efficacy (https://www.legislation.gov.uk/eur/2009/1107).
3. **Authorization**: After an active ingredient receives EU approval, individual PGR products containing that active substance must be authorized by Member States before they can be placed on their markets. Member States evaluate these products based on factors such as intended use, application rates, and potential risks.
4. **Maximum Residue Levels (MRLs)**: MRLs are established for PGRs to ensure that residues on treated crops do not pose

unacceptable risks to consumers. Compliance with MRLs is a prerequisite for the authorization of PPPs.

5. **Monitoring and Enforcement**: EU member states are responsible for monitoring the proper use of PGRs and enforcing regulations to ensure compliance with safety standards. This includes conducting inspections, sampling, and testing of products and crops.
6. **Review and Renewal**: Authorizations for PPPs, including PGRs, are subject to periodic review and renewal. This process ensures that products on the market continue to meet the latest safety and efficacy standards.
7. **Restrictions and Conditions**: Certain restrictions or conditions may be imposed on the use of PGRs to minimize risks to human health, animal health, or the environment. These may include limitations on application rates, application methods, or use in specific crops.

In conclusion, the regulatory framework for PGRs in the EU exemplifies a balanced approach to promoting agricultural productivity while safeguarding human health and the environment. By adhering to stringent scientific standards and continuous monitoring, the EU strives to ensure the responsible use of PGRs, contributing to sustainable and resilient food systems for future generations.

1.6.2.2 UNITED STATES

In the United States, PGRs are primarily regulated by the Environmental Protection Agency (EPA) and the U.S. Department of Agriculture (USDA). The regulatory framework for PGRs is designed to ensure their safe use, effectiveness, and minimal environmental impact (Cohrssen, 2022; Lake et al., 2002). Along with the Environmental Protection Agency (EPA), other agencies are also involved in the regulatory framework for PGRs in the United States.

1. **Environmental Protection Agency (EPA)**: The EPA regulates PGRs under the Federal Insecticide, Fungicide, and Rodenticide Act (FIFRA). PGRs are considered pesticides under this act. Before a PGR can be marketed and sold in the U.S., it must undergo a rigorous registration process with the EPA. This process involves

submitting extensive data on the product's safety, efficacy, and environmental impact. The EPA evaluates the data to determine whether the PGR meets the safety standards outlined in FIFRA. If the product meets these standards, it is granted registration, allowing it to be sold and used in accordance with approved label instructions. The EPA also establishes tolerance levels for pesticide residues on food crops to ensure that PGR residues do not pose unacceptable risks to human health (United States-Environmental Protection Agency, 2023; https://www.epa.gov/).

2. **U.S. Department of Agriculture (USDA)**: The USDA regulates PGRs in the context of their use on crops and in agriculture. The USDA's Animal and Plant Health Inspection Service (APHIS) oversees the regulation of genetically modified crops, including those engineered to produce or respond to PGRs. Moreover, APHIS evaluates the environmental impact of genetically modified crops, including their potential to cross-pollinate with wild relatives and the spread of modified traits to unintended recipients (https://www.usda.gov/).
3. **State Regulation**: In addition to federal regulation, states may have their own regulations governing the sale and use of PGRs. States may impose additional requirements or restrictions on PGRs based on local environmental conditions, agricultural practices, and public health concerns. Some states may require PGR products to be registered with state agencies before they can be sold or used within the state (United States-Environmental Protection Agency, 2023; https://www.epa.gov/).
4. **Labelling and Use Restrictions**: PGR products approved for sale in the U.S. must have labeling that provides instructions for safe and effective use. The labeling typically includes information on application rates, timing, target crops, safety precautions, and environmental considerations. Moreover, users are legally obligated to follow the instructions on the label, and using the product in a manner inconsistent with the label is a violation of federal law (Cohrssen, 2022; Lake et al., 2002).

Overall, the regulatory framework for PGRs in the United States is designed to balance the benefits of these products for agriculture with the need to safeguard both the environment and human health. Compliance with regulatory requirements ensures that PGRs are used safely and responsibly across the agricultural industry.

1.6.2.3 INDIA

The regulatory framework for PGRs in India primarily falls under the purview of the Central Insecticides Board & Registration Committee (CIB&RC), which operates under the Ministry of Agriculture and Farmers' Welfare (Mishra et al., 2020). The regulatory framework aims to ensure the safety, efficacy, and proper usage of PGRs to protect human health, the environment, and agricultural productivity. These are the key components of the regulatory framework for PGRs in India (https://ppqs.gov.in/divisions/central-insecticides-board-registration-committee):

1. **Central Insecticides Board (CIB)**: The CIB is responsible for the overall regulation and registration of insecticides, fungicides, herbicides, and PGRs in India. It oversees the approval process and sets standards for the manufacturing, sale, distribution, and use of these products.

2. **Registration Committee (RC)**: The Registration Committee is a statutory body constituted under the Insecticides Act, 1968. It evaluates applications for the registration of pesticides, including PGRs, and decides whether to grant registration based on safety, efficacy, and other relevant factors. Additionally, manufacturers are required to adhere to Good Manufacturing Practices (GMP) to ensure product quality and safety standards.

3. **The Insecticides Act, 1968, and Rules, 1971**: This legislation provides the legal framework for the regulation of pesticides in India, including PGRs. It outlines the procedures for registration, licensing, import, manufacture, sale, transport, and use of pesticides. The act also specifies penalties for violations. The Insecticides Rules of 1971 elaborate on the provisions of the Insecticides Act of 1968, providing more detailed guidelines and procedures for various aspects of pesticide regulation, including registration, packaging, labeling, and quality control.

The registration process for PGRs typically involves submitting detailed technical data, including efficacy studies, toxicological assessments, and environmental impact studies. Once registered, PGR products are subject to periodic review and monitoring to ensure continued compliance with regulatory standards. Furthermore, the regulatory framework includes provisions for labeling requirements, dosage recommendations, and restrictions on the sale and distribution of PGRs to safeguard human health and the environment (Mishra et al., 2020; https://ppqs.gov.in/divisions/central-insecticides-board-registration-committee).

1.7 FUTURE PROSPECTS

The utilization of PBs and growth regulators has gained significant attention in agricultural practices due to their potential to enhance crop productivity, improve quality, and mitigate environmental stress factors. BS are substances that promote the use of small quantities to promote plant development, while growth regulators are chemicals that influence plant physiological processes, such as hormone levels and development. As global agricultural challenges mount, including climate change, soil degradation, and water scarcity, there is a growing imperative to explore sustainable and innovative solutions to optimize plant growth and ensure food security. The future prospects of PBs and growth regulators are promising, driven by several factors:

1. **Increasing Demand for Sustainable Agriculture:** As the global population continues to grow, there is a heightened demand for food production. Moreover, with increasing concerns over environmental sustainability, there is a shift towards eco-friendly agricultural practices. PBs and growth regulators offer promising alternatives to conventional synthetic fertilizers and pesticides. Their ability to enhance nutrient uptake, promote root development, and improve stress tolerance aligns with the goals of sustainable agriculture by reducing chemical inputs and minimizing environmental impact (Magnabosco et al., 2023; Mannino, 2023).

2. **Enhanced Crop Yield and Quality:** In a world facing increasing food demand, enhancing crop yield and quality is paramount. PBs and growth regulators offer the potential to boost both yield

quantity and nutritional value. By stimulating physiological processes such as flowering, fruiting, and nutrient assimilation, they optimize plant growth and development, resulting in higher yields and improved crop characteristics (Castro-Camba et al., 2020; Drobek et al., 2019).

3. **Compatibility with Organic Farming:** The rising consumer demand for organic produce has fueled interest in organic farming methods. PBs, derived from natural sources such as seaweed extracts, amino acids, and microbial formulations, align well with organic farming principles (Kumar et al., 2023a, b; Rodrigues et al., 2020). Their organic certification and minimal environmental impact make them valuable tools for organic growers seeking to enhance soil fertility and crop productivity without compromising their commitment to sustainability (Kumar et al., 2023c).

4. **Adaptation to Climate Change:** Climate change poses significant challenges to global agriculture, including unpredictable weather patterns, temperature extremes, and water scarcity. PBs and growth regulators can play a crucial role in helping crops adapt to these changing conditions. By improving resilience to unfavorable environmental conditions like heat, salinity, and drought, they enable plants to maintain productivity and optimize resource utilization in a variable climate (Asif et al., 2023; Calvo et al., 2014; Van Oosten et al., 2017).

5. **Technological Advancements:** Ongoing research and technological advancements are driving innovation in the field of PBs and growth regulators. From novel formulations and delivery systems to precision agriculture techniques, there is a wealth of opportunity to improve the efficacy and application of these products. Biotechnological approaches, such as genetic engineering and microbiome manipulation, hold promise for developing customized solutions tailored to specific crops and environmental conditions (Kumar et al., 2023).

6. **Regulatory Landscape:** As awareness of PBs and growth regulators grows, so too does the need for regulatory oversight. Governments and regulatory bodies are working to establish clear guidelines and standards for the registration, labeling, and use of

these products. A transparent regulatory framework will provide clarity for manufacturers, growers, and consumers while ensuring product safety and efficacy (Caradonia et al., 2019; Du Jardin, 2015; Kumar et al., 2023).

In a nutshell, the future prospects of PBs and growth regulators are bright, driven by their potential to address key challenges facing global agriculture. As sustainable, environmentally friendly, and effective solutions, they offer a pathway towards resilient and productive agricultural systems in the face of an uncertain future. Continued research, innovation, and collaboration will be essential to unlocking their full potential and supporting the transition towards a more sustainable food production system.

1.8 CONCLUSION

PBs and growth regulator-based products offer promising avenues for enhancing crop productivity, resilience, and sustainability in agricultural systems. Through their ability to stimulate physiological processes, improve nutrient uptake, and mitigate abiotic stress, these products hold significant potential for optimizing plant growth and development. However, their efficacy can vary depending on factors such as crop species, environmental conditions, and application methods. Moreover, while BS and growth regulators offer alternatives to traditional agricultural inputs, their widespread adoption requires further research to elucidate their mechanisms of action, refine application strategies, and ensure compatibility with existing farming practices. Additionally, considerations regarding regulatory frameworks, product safety, and environmental impact are essential for promoting their responsible use and integration into integrated crop management approaches. The global landscape of PBs and growth regulator-based products is complex and evolving. This chapter overview has shed light on the diverse regulatory frameworks that govern these products across different regions, highlighting both similarities and disparities. Despite efforts to harmonize regulations and establish clear definitions, discrepancies persist, posing challenges for manufacturers, regulators, and farmers alike. Moreover, the increasing demand for sustainable agricultural practices and the growing interest in BS and growth regulators as alternatives to traditional agrochemicals underscore the importance of robust regulatory frameworks. Harmonizing standards, enhancing transparency, and fostering collaboration between stakeholders will be crucial in facilitating market access, ensuring product safety, and promoting innovation in this burgeoning sector. Moving forward, continued dialogue, research, and regulatory refinement will be essential to address emerging issues, capitalize on opportunities, and promote the responsible use of BS and growth regulator-based products in agriculture. By working together, stakeholders can navigate regulatory complexities, unlock the full potential of these products, and contribute to a more sustainable and resilient global food system.

KEYWORDS

- **nutrient use efficiency**
- **phytohemagglucosides**
- **plant biostimulants**
- **plant hormones**
- **plant protection products**
- **relative water content**
- **seaweed extracts**

REFERENCES

Abbas, T., Balal, R. M., Shahid, M. A., Pervez, M. A., Ayyub, C. M., Aqueel, M. A., & Javaid, M. M. (2015). Silicon-induced alleviation of NaCl toxicity in okra (Abelmoschus esculentus) is associated with enhanced photosynthesis, Osmo protectants and antioxidant metabolism. *Acta Physiologies Plantarum, 37*(6). https://doi.org/10.1007/s11738-014-1768-5 (accessed on 6th November 2025).

Abdelaal, K. A. A., Attia, K. A., Alamery, S. F., El-Afry, M. M., Ghazy, A. I., Tantawy, D. S., Al-Doss, A. A., El-Shawy, E. S. E. M., Abu-Elsaoud, A., & Hafez, Y. M. (2020). Exogenous application of proline and salicylic acid can mitigate the injurious impacts of drought stress on barley plants associated with physiological and histological characters. *Sustainability, 12*(5), 1736. https://doi.org/10.3390/su12051736 (accessed on 6th November 2025).

Ahammed, G. J., & Yang, Y. (2021). Mechanisms of silicon-induced fungal disease resistance in plants. *Plant Physiology and Biochemistry, 165*, 200–206. https://doi.org/10.1016/j.plaphy.2021.05.031 (accessed on 6th November 2025).

Ahmad, I., Zhu, G., Zhou, G., Song, X., Hussein Ibrahim, M. E., Ibrahim Salih, E. G., Hussain, S., & Younas, M. U. (2022). Pivotal role of phytohormones and their responsive genes in plant growth and their signaling and transduction pathway under salt stress in cotton. *International Journal of Molecular Sciences, 23*(13), 7339. https://doi.org/10.3390/ijms23137339 (accessed on 6th November 2025).

Ahmad, P., Rasool, S., Gul, A., Sheikh, S. A., Akram, N. A., Ashraf, M., Kazi, A. M., & Gucel, S. (2016). Jasmonates: Multifunctional roles in stress tolerance. *Frontiers in Plant Science, 7*, 813. https://doi.org/10.3389/fpls.2016.00813 (accessed on 6th November 2025).

Ahmad, S. S., Tahir, I., Wani, A. S., Dar, R. A., & Nisar, S. (2018). Adenine type and diphenyl urea derived cytokinin's improve the postharvest performance of Iris germanica L. cut scapes. *Physiology and Molecular Biology of Plants, 24*(6), 1127–1137.

Akhtar, S. S., Mekureyaw, M. F., Pandey, C., & Roitsch, T. (2020). Role of cytokinins for interactions of plants with microbial pathogens and pest insects. *Frontiers in Plant Science, 10*, 1777.

Alfosea-Simón, M., Simón-Grao, S., Zavala-Gonzalez, E. A., Cámara-Zapata, J. M., Simón, I., Martínez-Nicolás, J. J., Lidón, V., Rodríguez-Ortega, W. M., & García-Sánchez, F. (2020). Application of bio stimulants containing amino acids to tomatoes could favor sustainable cultivation: Implications for tyrosine, lysine, and methionine. *Sustainability, 12*(22), 9729. https://doi.org/10.3390/su12229729 (accessed on 6th November 2025).

Alharbi, B. M., Abdulmajeed, A. M., & Hassan, H. (2021). Biochemical and molecular effects induced by triacontanol in acquired tolerance of rice to drought stress. *Genes, 12*(8), 1119. https://doi.org/10.3390/genes12081119 (accessed on 6th November 2025).

Ali, E., Hussain, S., Jalal, F., Khan, M. A., Imtiaz, M., Said, F., Ismail, M., Khan, S., Ali, H. M., Hatamleh, A. A., Al-Dosary, M. A., Mosa, W. F. A., & Shah, F. (2023). Salicylic acid-mitigates abiotic stress tolerance via altering defense mechanisms in Brassica napus (L.). *Frontiers in Plant Science, 14*, 1187260. https://doi.org/10.3389/fpls.2023.1187260 (accessed on 6th November 2025).

Ali, H. M. M., & Perveen, S. (2020). Effect of foliar applied triacontanol on wheat (Triticum aestivum L.) under arsenic stress: A study of changes in growth, yield and photosynthetic characteristics. *Physiology and Molecular Biology of Plants, 26*, 1215–1224. https://doi.org/10.1007/s12298-020-00831-0 (accessed on 6th November 2025).

Amin, A. A., Rashad EL-Sh, M., & EL-Abagy, H. M. H. (2007). Physiological effect of indole-3-butyric acid and salicylic acid on growth, yield and chemical constituents of onion plants. *Journal of Applied Sciences Research, 3*, 1554–1563.

Ampong, K., Thilakaranthna, M. S., & Gorim, L. Y. (2022). Understanding the role of humic acids on crop performance and soil health. *Frontiers in Agronomy, 4*, 848621. https://doi.org/10.3389/fagro.2022.848621 (accessed on 6th November 2025).

Aremu, A. O., Makhaye, G., Tesfay, S. Z., Gerrano, A. S., Du Plooy, C. P., & Amoo, S. O. (2022). Influence of commercial seaweed extract and microbial bio stimulant on growth, yield, phytochemical content, and nutritional quality of five Abelmoschus esculentus genotypes. *Agronomy, 12*(2), 428. https://doi.org/10.3390/agronomy12020428 (accessed on 6th November 2025).

Arkhipova, T., Martynenko, E., Sharipova, G., Kuzmina, L., Ivanov, I., Garipova, M., & Kudoyarova, G. (2020). Effects of plant growth promoting rhizobacteria on the content of abscisic acid and salt resistance of wheat plants. *Plants, 9*(11), 1429. https://doi.org/10.3390/plants9111429 (accessed on 6th November 2025).

Asadi Karam, E., & Keramat, B. (2017). Foliar spray of triacontanol improves growth by alleviating oxidative damage in coriander under salinity. *Indian Journal of Plant Physiology, 22*, 120–124. https://doi.org/10.1007/s40502-017-0286-z (accessed on 6th November 2025).

Asano, K., Takashi, T., Miura, K., Qian, Q., Kitano, H., Matsuoka, M., & Ashikari, M. (2007). Genetic and molecular analysis of utility of sd1 alleles in rice breeding. *Breeding Science, 57*, 53–58.

Asif, A., Ali, M., Qadir, M., Karthikeyan, R., Singh, Z., Khangura, R., Di Gioia, F., & Ahmed, Z. F. R. (2023). Enhancing crop resilience by harnessing the synergistic effects of bio stimulants against abiotic stress. *Frontiers in Plant Science, 14*, 1276117. https://doi.org/10.3389/fpls.2023.1276117 (accessed on 6th November 2025).

Avanci, N. C., Luche, D. D., Goldman, G. H., & Goldman, M. H. S. (2010). Jasmonates are phytohormones with multiple functions, including plant defense and reproduction. *Genetics and Molecular Research, 9*, 484–505.

Awan, F. K., Khurshid, M. Y., & Mehmood, A. J. I. J. I. R. B. (2017). Plant growth regulators and their role in abiotic stress management. *International Journal of Innovative Research in Biosciences, 1*, 9–21.

Ayed, S., Bouhaouel, I., Jebari, H., & Hamada, W. (2022). Use of bio stimulants: Towards sustainable approach to enhance durum wheat performances. *Plants, 11*(1), 133. https://doi.org/10.3390/plants11010133 (accessed on 6th November 2025).

Azad, M. O. K., Park, B. S., Adnan, M., Germ, M., Kreft, I., Woo, S. H., & Park, C. H. (2021). Silicon bio stimulant enhances the growth characteristics and fortifies the bioactive compounds in common and Tartary buckwheat plant. *Journal of Crop Science and Biotechnology, 24*, 51–59. https://doi.org/10.1007/s12892-020-00058-1 (accessed on 6th November 2025).

Azhar, B. J., Abbas, S., Aman, S., Yamburenko, M. V., Chen, W., Müller, L., Uzun, B., Jewell, D. A., Dong, J., Shakeel, S. N., Groth, G., Binder, B. M., Grigoryan, G., & Schaller, G. E. (2023). Basis for high-affinity ethylene binding by the ethylene receptor ETR1 of Arabidopsis. *Proceedings of the National Academy of Sciences, 120*(23), e2215195120. https://doi.org/10.1073/pnas.2215195120 (accessed on 6th November 2025).

Backer, R., Rokem, J. S., Ilangumaran, G., Lamont, J., Praslickova, D., Ricci, E., Subramanian, S., & Smith, D. L. (2018). Plant growth-promoting rhizobacteria: Context, mechanisms of action, and roadmap to commercialization of bio stimulants for sustainable agriculture. *Frontiers in Plant Science, 9*, 1473. https://doi.org/10.3389/fpls.2018.01473 (accessed on 6th November 2025).

Bagautdinova, Z. Z., Omelyanchuk, N., Tyapkin, A. V., Kovrizhnykh, V. V., Lavrekha, V. V., & Zemlyanskaya, E. V. (2022). Salicylic acid in root growth and development. *International Journal of Molecular Sciences, 23*(4), 2228. https://doi.org/10.3390/ijms23042228 (accessed on 6th November 2025).

Bakhat, H. F., Bibi, N., Zia, Z., Abbas, S., Hammad, H. M., Fahad, S., Ashraf, R., Shah, G. M., Rabbani, F., & Saeed, S. (2018). Silicon mitigates biotic stresses in crop plants: a review. *Crop Protection, 104*, 21–34.

Baltazar, M., Correia, S., Guinan, K. J., Sujeeth, N., Bragança, R., & Gonçalves, B. (2021). Recent advances in the molecular effects of bio stimulants in plants: An overview. *Biomolecules, 11*(8), 1096. https://doi.org/10.3390/biom11081096 (accessed on 6th November 2025).

Balzan, S., Johal, G. S., & Carraro, N. (2014). The role of auxin transporters in monocots development. *Frontiers in Plant Science, 5*, 393. https://doi.org/10.3389/fpls.2014.00393 (accessed on 6th November 2025).

Bari, M. A., Prity, S. A., Das, U., Akther, M. S., Sajib, S. A., Reza, M. A., & Kabir, A. H. (2020). Silicon induces phytochelatin and ROS scavengers facilitating cadmium detoxification in rice. *Plant Biology, 22*(3), 472–479. https://doi.org/10.1111/plb.13090 (accessed on 6th November 2025).

Baroccio, F., Barilaro, N., Tolomei, P., & Mascini, M. (2017). Classification of bio stimulants origin using amino acids composition of hydrolyzed proteins. *Journal of Horticultural Science and Research, 1*, 30–35.

Bartucca, M. L., Cerri, M., Del Buono, D., & Forni, C. (2022). Use of bio stimulants as a new approach for the improvement of phytoremediation performance–A review. *Plants, 11*(15), 1946. https://doi.org/10.3390/plants11151946 (accessed on 6th November 2025).

Battacharyya, D., Babgohari, M. Z., Rathor, P., & Prithiviraj, B. (2015). Seaweed extracts as bio stimulants in horticulture. *Scientia Horticulture, 196*, 39–48.

Benjamin, G., Pandharikar, G., & Frendo, P. (2022). Salicylic acid in plant symbioses: Beyond plant pathogen interactions. *Biology, 11*(6), 861. https://doi.org/10.3390/biology11060861 (accessed on 6th November 2025).

Bhandari, S., Bhandari, A., & Shrestha, J. (2021). Effect of different doses of triacontanol on growth and yield of kohlrabi (Brassica oleracea L. var. gongylodes). *Helion, 7*(10), e08242. https://doi.org/10.1016/j.heliyon.2021.e08242 (accessed on 6th November 2025).

Bharath, P., Gahir, S., & Raghavendra, A. S. (2021). Abscisic acid-induced stomatal closure: An important component of plant defense against abiotic and biotic stress. *Frontiers in Plant Science, 12*, 615114. https://doi.org/10.3389/fpls.2021.615114 (accessed on 6th November 2025).

Binder, B. M. (2020). Ethylene signaling in plants. *Journal of Biological Chemistry, 295*(22), 7710–7725.

Binenbaum, J., Weinstain, R., & Shani, E. (2018). Gibberellin localization and transport in plants. *Trends in Plant Science, 23*(5), 410–421. https://doi.org/10.1016/j.tplants.2018.02.005 (accessed on 6th November 2025).

Botton, A., Tonutti, P., & Ruperti, B. (2019). Biology and biochemistry of ethylene. In E. M. Yahia (Ed.), *Postharvest Physiology and Biochemistry of Fruits and Vegetables* (pp. 93–112). Woodhead Publishing. https://doi.org/10.1016/B978-0-12-813278-4.00005-1 (accessed on 6th November 2025).

Brookbank, B. P., Patel, J., Gazzarrini, S., & Nambara, E. (2021). Role of basal ABA in plant growth and development. *Genes, 12*(12), 1936. https://doi.org/10.3390/genes12121936 (accessed on 6th November 2025).

Bulgari, R., Cocetta, G., Trivellini, A., Vernieri, P., & Ferrante, A. (2015). Bio stimulants and crop responses: A review. *Biological Agriculture & Horticulture, 31*(1), 1–17.

Business Research Insights. (2023). *Methyl DihydroJasmonate Market Size, Share, Growth, And Industry Analysis, By Type (Synthesis Methyl Dihydrojasmonate, Natural*

Dihydrojasmonate), By Application (Flavor and Fragrance, Household Chemicals, Others), Regional Insights, And Forecast to 2031. Report ID: BRI105271, Pages: 106. https://www.businessresearchinsights.com/market-reports/methyl-dihydro-jasmonate-market-105271 (accessed on 6th November 2025).

Calvo, P., Nelson, L., & Kloepper, J. W. (2014). Agricultural uses of plant bio stimulants. *Plant and Soil, 383*, 3–41. https://doi.org/10.1007/s11104-014-2131-8 (accessed on 6th November 2025).

Caradonia, F., Battaglia, V., Righi, L., Pascali, G., & La Torre, A. (2019). Plant bio stimulant regulatory framework: Prospects in Europe and current situation at international level. *Journal of Plant Growth Regulation, 38*, 438–448. https://doi.org/10.1007/s00344-018-9853-4 (accessed on 6th November 2025).

Carillo, P., De Micco, V., Ciriello, M., Formisano, L., El-Nakhel, C., Giordano, M., Colla, G., & Rouphael, Y. (2022). Morpho-anatomical, physiological, and mineral composition responses induced by a vegetal-based bio stimulant at three rates of foliar application in greenhouse lettuce. *Plants, 11*(15), 2030. https://doi.org/10.3390/plants11152030 (accessed on 6th November 2025).

Castiglione, A. M., Mannino, G., Contartese, V., Bertea, C. M., & Ertani, A. (2021). Microbial bio stimulants as response to modern agriculture needs: Composition, role and application of these innovative products. *Plants, 10*(8), 1533. https://doi.org/10.3390/plants10081533 (accessed on 6th November 2025).

Castro-Camba, R., Sánchez, C., Vidal, N., & Vielba, J. M. (2020). Plant development and crop yield: The role of gibberellins. *Plants, 11*(19), 2650. https://doi.org/10.3390/plants11192650 (accessed on 6th November 2025).

Caulier, S., Nannan, C., Gillis, A., Licciardi, F., Bragard, C., & Mahillon, J. (2019). Overview of the antimicrobial compounds produced by members of the Bacillus subtilis group. *Frontiers in Microbiology, 10*, 302. https://doi.org/10.3389/fmicb.2019.00302 (accessed on 6th November 2025).

Chakraborty, M., Hasanuzzaman, M., Rahman, M., Khan, M. A. R., Bhowmik, P., Mahmud, N. U., Tanveer, M., & Islam, T. (2020). Mechanism of plant growth promotion and disease suppression by chitosan biopolymer. *Agriculture, 10*(12), 624. https://doi.org/10.3390/agriculture10120624 (accessed on 6th November 2025).

Chandler, P. M., Marion-Poll, A., Ellis, M., & Gubler, F. (2002). Mutants at the Slender1 locus of barley cv Himalaya. Molecular and physiological characterization. *Plant Physiology, 129*, 181–190.

Chang, C. (2016). Q&A: How do plants respond to ethylene and what is its importance? *BMC Biology, 14*, 7. https://doi.org/10.1186/s12915-016-0230-0 (accessed on 6th November 2025).

Chaudhary, P., Singh, S., Chaudhary, A., Sharma, A., & Kumar, G. (2022). Overview of biofertilizers in crop production and stress management for sustainable agriculture. *Frontiers in Plant Science, 13*, 930340. https://doi.org/10.3389/fpls.2022.930340 (accessed on 6th November 2025).

Chaudhuri, A., Halder, K., Abdin, M. Z., Majee, M., & Datta, A. (2022). Abiotic stress tolerance in plants: Brassinosteroids navigate competently. *International Journal of Molecular Sciences, 23*(23), 14577. https://doi.org/10.3390/ijms232314577 (accessed on 6th November 2025).

Chen, D., Zhou, W., Yang, J., Ao, J., Huang, Y., Shen, D., Jiang, Y., Huang, Z., & Shen, H. (2021). Effects of seaweed extracts on the growth, physiological activity, cane yield and sucrose content of sugarcane in China. *Frontiers in Plant Science, 12*, 659130. https://doi.org/10.3389/fpls.2021.659130 (accessed on 6th November 2025).

Chen, X., Yuan, H., Chen, R., Zhu, L., Du, B., Weng, Q., & He, G. (2022). Isolation and characterization of triacontanol-regulated genes in rice (*Oryza sativa* L.): Possible role of triacontanol as a plant growth stimulator. *Plant and Cell Physiology, 43*(8), 869–876. https://doi.org/10.1093/pcp/pcf100 (accessed on 6th November 2025).

Cheng, Y., Tian, Q., & Zhang, W. H. (2016). Glutamate receptors are involved in mitigating effects of amino acids on seed germination of *Arabidopsis thaliana* under salt stress. *Environmental and Experimental Botany, 130*, 68–78. https://doi.org/10.1016/j.envexpbot.2016.05.004 (accessed on 6th November 2025).

Çimrin, K. M., Türkmen, Ö., Turan, M., & Tuncer, B. (2010). Phosphorus and humic acid application alleviate salinity stress of pepper seedling. *African Journal of Biotechnology, 9*(36), 5845–5851. https://www.ajol.info/index.php/ajb/article/view/92903 (accessed on 6th November 2025).

Cocetta, G., & Natalini, A. (2022). Ethylene: Management and breeding for postharvest quality in vegetable crops. A review. *Frontiers in Plant Science, 13*, 968315. https://doi.org/10.3389/fpls.2022.968315 (accessed on 6th November 2025).

Cohrssen, J. J. (2022). Regulating the regulator of plant growth regulators. *Cato Institute, Regulation*, WINTER 2022–2023. https://www.cato.org/regulation/winter-2022-2023/regulating-regulator-plant-growth-regulators (accessed on 6th November 2025).

Colla, G. (2020). Protein hydrolysates as plant bio stimulants. https://www.bio stimulant.com/protein-hydrolysates-as-plant-bio stimulants/ (accessed on 6th November 2025).

Colla, G., Nardi, S., Cardarelli, M., Ertani, A., Lucini, L., Canaguier, R., & Rouphael, Y. (2015). Protein hydrolysates as bio stimulants in horticulture. *Scientia Horticulture, 196*, 28–38.

Colla, G., Rouphael, Y., Canaguier, R., Svecova, E., & Cardarelli, M. (2014). Bio stimulant action of a plant-derived protein hydrolysate produced through enzymatic hydrolysis. *Frontiers in Plant Science, 5*, 448. https://doi.org/10.3389/fpls.2014.00448 (accessed on 6th November 2025).

Colla, G., Rouphael, Y., Lucini, L., Canaguier, R., Stefanoni, W., Fiorillo, A., & Cardarelli, M. (2016). Protein hydrolysate-based bio stimulants: Origin, biological activity and application methods. *Acta Horticulture, 1148*, 27–34.

Cooke, J., DeGabriel, J. L., & Hartley, S. E. (2016). The functional ecology of plant silicon: Geoscience to genes. *Functional Ecology, 30*, 1270–1276.

Cornforth, J. W., Milborrow, B. V., Ryback, G., & Wareing, P. F. (1965). Identity of sycamore 'dormin' with abscisin II. *Nature, 205*, 1269–1270.

Coskun, D., Deshmukh, R., Sonah, H., Menzies, J. G., Reynolds, O., Ma, J. F., Kronzucker, H. J., & Bélanger, R. R. (2019). The controversies of silicon's role in plant biology. *New Phytologist, 221*, 67–85. https://doi.org/10.1111/nph.15343 (accessed on 6th November 2025).

Crampon, C., Boutin, O., & Badens, E. (2011). Supercritical carbon dioxide extraction of molecules of interest from microalgae and seaweeds. *Industrial & Engineering Chemistry Research, 50*, 8941–8953.

Cristiano, G., Pallozzi, E., Conversa, G., Tufarelli, V., & De Lucia, B. (2018). Effects of an animal-derived bio stimulant on the growth and physiological parameters of potted snapdragon (*Antirrhinum majus* L.). *Frontiers in Plant Science, 9*, 861. https://doi.org/10.3389/fpls.2018.00861 (accessed on 6th November 2025).

Danilova, M. N., Kudryakova, N. V., Doroshenko, A. S., Zabrodin, D. A., Vinogradov, N. S., & Kuznetsov, V. V. (2016). Molecular and physiological responses of *Arabidopsis thaliana* plants deficient in the genes responsible for ABA and cytokinin reception and metabolism to heat shock. *Russian Journal of Plant Physiology, 63*, 308–318.

Dar, N. A., Amin, I., Wani, W., Wani, S. A., Shikari, A. B., Wani, S. H., & Masoodi, K. Z. (2017). Abscisic acid: A key regulator of abiotic stress tolerance in plants. *Plant Gene, 11*(B), 106–111. https://doi.org/10.1016/j.plgene.2017.07.003 (accessed on 6th November 2025).

del Orozco-Mosqueda, M. C., Glick, B. R., & Santoyo, G. (2020). ACC deaminase in plant growth-promoting bacteria (PGPB): An efficient mechanism to counter salt stress in crops. *Microbiological Research, 235*, 126439. https://doi.org/10.1016/j.micres.2020.126439 (accessed on 6th November 2025).

Demole, E., Lederer, E., & Mercier, D. (1962). Islemen et determination de la structure du jasmone de methyl, constituent odorant caractéristique de essence de Jasmin. *Helvetica Chemical Acta, 45*, 675–685.

Dempsey, D. A., & Klessig, D. F. (2017). How does the multifaceted plant hormone salicylic acid combat disease in plants and are similar mechanisms utilized in humans? *BMC Biology, 15*, 23. https://doi.org/10.1186/s12915-017-0364-8 (accessed on 6th November 2025).

Detmann, K. C., Araújo, W. L., Martins, S. C., Fernie, A. R., & Damatta, F. M. (2013). Metabolic alterations triggered by silicon nutrition: Is there a signaling role for silicon? *Plant Signaling & Behavior, 8*, e22523. https://doi.org/10.4161/psb.22523 (accessed on 6th November 2025).

Detmann, K. C., Araújo, W. L., Martins, S. C., Sanglard, L. M., Reis, J. V., Detmann, E., Rodrigues, F. Á., Nunes-Nesi, A., Fernie, A. R., & DaMatta, F. M. (2012). Silicon nutrition increases grain yield, which, in turn, exerts a feed-forward stimulation of photosynthetic rates via enhanced mesophyll conductance and alters primary metabolism in rice. *New Phytologist, 196*(3), 752–762.

Dewang, S. P., & Devi, C. U. (2022). Efficacy of organic bio stimulant (fish protein hydrolysate) on the growth and yield of tomato (*Solanum lycopersicum*). *Agricultural Science Digest–A Research Journal, 42*, 20–25.

Drobek, M., Frąc, M., & Cybulska, J. (2019). Plant bio stimulants: Importance of the quality and yield of horticultural crops and the improvement of plant tolerance to abiotic stress—A review. *Agronomy, 9*(6), 335. https://doi.org/10.3390/agronomy9060335 (accessed on 6th November 2025).

Du Jardin, P. (2015). Plant bio stimulants: Definition, concept, main categories and regulation. *Scientia Horticulture, 196*, 3–14. https://doi.org/10.1016/j.scienta.2015.09.021 (accessed on 6th November 2025).

Dubois, M., Van den Broeck, L., & Inzé, D. (2018). The pivotal role of ethylene in plant growth. *Trends in Plant Science, 23*(4), 311–323.

Durrant, W. E., & Dong, X. (2004). Systemic acquired resistance. *Annual Review of Phytopathology, 42*, 185–209.

Ebinezer, L. B., Franchin, C., Trentin, A. R., Carletti, P., Trevisan, S., Agrawal, G. K., Rakwal, R., Quaggiotti, S., Arrigoni, G., & Masi, A. (2020). Quantitative proteomics of maize roots treated with a protein hydrolysate: A comparative study with transcriptomics highlights the molecular mechanisms responsive to bio stimulants. *Journal of Agricultural and Food Chemistry, 68*, 7541–7553.

El-Beltagi, H. S., Ismail, S. A., Ibrahim, N. M., Shehata, W. F., Alkhateeb, A. A., Ghazzawy, H. S., El-Mogy, M. M., & Sayed, E. G. (2022). Unravelling the effect of triacontanol in combating drought stress by improving growth, productivity, and physiological performance in strawberry plants. *Plants, 11*(15), 1913. https://doi.org/10.3390/plants11151913 (accessed on 6th November 2025).

El-Sabagh, A., Islam, M. S., Hossain, A., Iqbal, M. A., Mubeen, M., Waleed, M., et al. (2022). Phytohormones as growth regulators during abiotic stress tolerance in plants. *Frontiers in Agronomy, 4*, 765068. https://doi.org/10.3389/fagro.2022.765068 (accessed on 6th November 2025).

El-Sherbeny, M., & da Teixeira Silva, J. A. (2013). Foliar treatment with proline and tyrosine affect the growth and yield of beetroot and some pigments in beetroot leaves. *Journal of Horticultural Research, 21*, 95–99.

Epstein, E. (1994). The anomaly of silicon in plant biology. *Proceedings of the National Academy of Sciences of the United States of America, 91*, 11–17.

Epstein, E. (1999). Silicon. *Annual Review of Plant Physiology and Plant Molecular Biology, 50*, 641–664.

Ertani, A., Francioso, O., Tinti, A., Schiavon, M., Pizzeghello, D., & Nardi, S. (2018). Evaluation of seaweed extracts from *Laminaria* and *Ascophyllum nodosum* spp. as bio stimulants in *Zea mays* L. using a combination of chemical, biochemical and morphological approaches. *Frontiers in Plant Science, 9*, 428. https://doi.org/10.3389/fpls.2018.00428 (accessed on 6th November 2025).

Ertani, A., Nardi, S., Francioso, O., Sanchez-Cortes, S., Di Foggia, M., & Schiavon, M. (2019). Effects of two protein hydrolysates obtained from chickpea (*Cicer arietinum* L.) and *Spirulina platensis* on *Zea mays* (L.) plants. *Frontiers in Plant Science, 10*, 954. https://doi.org/10.3389/fpls.2019.00954 (accessed on 6th November 2025).

EU Monitor. (2020). Regulation 2009/1107–Placing of plant protection products on the market. https://www.eumonitor.eu/9353000/1/j4nvk6yhcbpeywk_j9vvik7m1c3gyxp/vitgbgiu7xzc (accessed on 6th November 2025).

EUR-Lex. (2022). Safe and effective fertilizing products on the EU market. https://eur-lex.europa.eu/EN/legal-content/summary/safe-and-effective-fertilising-products-on-the-eu-market.html (accessed on 6th November 2025).

Evenson, R. E., & Gollin, D. (2003). Assessing the impact of the green revolution, 1960 to 2000. *Science, 300*, 758–762.

Fadiji, A. E., Babalola, O. O., Santoyo, G., & Perazzolli, M. (2022). The potential role of microbial bio stimulants in the amelioration of climate change-associated abiotic stresses on crops. *Frontiers in Microbiology, 12*, 829099. https://doi.org/10.3389/fmicb.2021.829099 (accessed on 6th November 2025).

Faiza, H., Khan, O., Ali, C., Hussain, T., Haider, S. T., Siddique, T., Liaquat, M., Noor, A., Khan, R. W., Ashraf, S., Rashid, S., Noreen, A., Asghar, S., & Anjum, Q. S. (2024). Foliar application of triacontanol ameliorates heat stress through regulation of the antioxidant defense system and improves yield of eggplant. *Brazilian Journal of Biology, 84*, e253696. https://doi.org/10.1590/1519-6984.253696 (accessed on 6th November 2025).

Fariduddin, Q., Yusuf, M., Ahmad, I., & Ahmad, A. (2014). Brassinosteroids and their role in response of plants to abiotic stresses. *Biologic Plantarum, 58*, 9–17.

Fathy, M., Saad Eldin, S. M., Naseem, M., Dandekar, T., & Othman, E. M. (2022). Cytokinin's: Wide-spread signaling hormones from plants to humans with high medical potential. *Nutrients, 14*(7), 1495. https://doi.org/10.3390/nu14071495 (accessed on 6th November 2025).

Fatma, M., Asgher, M., Iqbal, N., Rasheed, F., Sehar, Z., Sofo, A., & Khan, N. A. (2022). Ethylene signaling under stressful environments: Analyzing collaborative knowledge. *Plants, 11*(17), 2211. https://doi.org/10.3390/plants11172211 (accessed on 6th November 2025).

Fauteux, F., Rémus-Borel, W., Menzies, J. G., & Bélanger, R. R. (2005). Silicon and plant disease resistance against pathogenic fungi. *FEMS Microbiology Letters, 249*(1), 1–6. https://doi.org/10.1016/j.femsle.2005.06.034 (accessed on 6th November 2025).

Franzoni, G., Cocetta, G., Prinsi, B., Ferrante, A., & Espen, L. (2022). Bio stimulants on crops: Their impact under abiotic stress conditions. *Horticulture, 8*(3), 189. https://doi.org/10.3390/horticulturae8030189 (accessed on 6th November 2025).

Gao, S., & Chu, C. (2020). Gibberellin metabolism and signaling: Targets for improving agronomic performance of crops. *Plant and Cell Physiology, 61*(11), 1902–1911. https://doi.org/10.1093/pcp/pcaa104 (accessed on 6th November 2025).

Ghosh, S., & Halder, S. (2018). Effect of different kinds of gibberellin on temperate fruit crops: A review. *The Pharma Innovation Journal, 7*(3), 315–319.

Gianfagna, T. (1995). Natural and synthetic growth regulators and their use in horticultural and agronomic crops. In P. J. Davies (Ed.), *Plant Hormones* (pp. 751–773). Springer, Dordrecht. https://doi.org/10.1007/978-94-011-0473-9_34 (accessed on 6th November 2025).

González, A., Castro, J., Vera, J., & Moenne, A. (2013). Seaweed oligosaccharides stimulate plant growth by enhancing carbon and nitrogen assimilation, basal metabolism, and cell division. *Journal of Plant Growth Regulation, 32*, 443–448.

Growth Market Report. (2023). *Microbial Bio Stimulants Market*. Report ID: MC-5016, Page: 173. https://growthmarketreports.com/request-for-customization/5016 (accessed on 6th November 2025).

Guo, Y., Ren, G., Zhang, K., Li, Z., Miao, Y., & Guo, H. (2021). Leaf senescence: Progression, regulation, and application. *Molecular Horticulture, 1*, 5. https://doi.org/10.1186/s43897-021-00006-9 (accessed on 6th November 2025).

Gupta, K., Wani, S. H., Razzaq, A., Skalicky, M., Samantara, K., Gupta, S., Pandita, D., Goel, S., Grewal, S., Hejnak, V., Shiv, A., El-Sabrout, A. M., Elansary, H. O., Alaklabi, A., & Brestic, M. (2022). Abscisic acid: Role in fruit development and ripening. *Frontiers in Plant Science, 13*, 817500. https://doi.org/10.3389/fpls.2022.817500 (accessed on 6th November 2025).

Gupta, R., & Chakrabarty, S. K. (2013). Gibberellic acid in plant: Still a mystery unresolved. *Plant Signaling & Behavior, 8*(9), e25504.

Gupta, S., Kulkarni, M. G., White, J. F., Strick, W. A., Papenfus, H. B., Doležal, K., Ördög, V., Norrie, J., Critchley, H. T., & Van Staden, J. (2021). Categories of various plant bio stimulants–Mode of application and shelf-life. In S. Gupta & J. Van Staden (Eds.),

Bio Stimulants for Crops from Seed Germination to Plant Development: A Practical Approach (pp. 1–60). Academic Press. https://doi.org/10.1016/B978-0-12-823048-0.00018-6 (accessed on 6th November 2025).

Gurav, R., Nalavade, V., Aware, C., Vyavahare, G., Bhatia, S. K., Yang, Y. H., & Jadhav, J. (2020). Microbial degradation of poultry feather biomass in a constructed bioreactor and application of hydrolysate as bioenhancer to vegetable crops. *Environmental Science and Pollution Research, 27*, 2027–2035. https://doi.org/10.1007/s11356-019-06536-6 (accessed on 6th November 2025).

Hamid, B., Zaman, M., Farooq, S., Fatima, S., Sayyed, R. Z., Baba, Z. A., Sheikh, T. A., Reddy, M. S., El Enshasy, H., & Gafur, A., & Suriani, N. L. (2021). Bacterial plant bio stimulants: A sustainable way towards improving growth, productivity, and health of crops. *Sustainability, 13*(5), 2856. https://doi.org/10.3390/su13052856 (accessed on 6th November 2025).

He, M., He, C. Q., & Ding, N. Z. (2018). Abiotic stresses: General defenses of land plants and chances for engineering multistress tolerance. *Frontiers in Plant Science, 9*, 1771. https://doi.org/10.3389/fpls.2018.01771 (accessed on 6th November 2025).

Hedden, P., & Sponsel, V. (2015). A century of gibberellin research. *Journal of Plant Growth Regulation*, *34*, 740–760. https://doi.org/10.1007/s00344-015-9546-1 (accessed on 6th November 2025).

Hewedy, O. A., Elsheery, N. I., Karkour, A. M., Elhamouly, N., Arafa, R. A., Mahmoud, G. A. E., Dawood, F. A., Hussein, W. E., Mansour, A., Amin, D. H., Allakhverdiev, S. I., Zivcak, M., & Brestic, M. (2023). Jasmonic acid regulates plant development and orchestrates stress response during tough times. *Environmental and Experimental Botany*, *208*, 105260. https://doi.org/10.1016/j.envexpbot.2023.105260 (accessed on 6th November 2025).

Hidangmayum, A., Dwivedi, P., Katiyar, D., & Hemantaranjan, A. (2019). Application of chitosan on plant responses with special reference to abiotic stress. *Physiology and Molecular Biology of Plants*, *25*(2), 313–326. https://doi.org/10.1007/s12298-018-0633-1 (accessed on 6th November 2025).

Hijri, M. (2023). Microbial-based plant bio stimulants. *Microorganisms*, *11*(3), 686. https://doi.org/10.3390/microorganisms11030686 (accessed on 6th November 2025).

Ho, T. T., Murthy, H. N., & Park, S. Y. (2020). Methyl jasmonate induced oxidative stress and accumulation of secondary metabolites in plant cell and organ cultures. *International Journal of Molecular Sciences*, *21*(3), 716. https://doi.org/10.3390/ijms21030716 (accessed on 6th November 2025).

Hu, J., Mitchum, M. G., Barnaby, N., Ayele, B. T., Ogawa, M., Nam, E., Lai, W. C., Hanada, A., Alonso, J. M., Ecker, J. R., Swain, S. M., Yamaguchi, S., Kamiya, Y., & Sun, T. P. (2008). Potential sites of bioactive gibberellin production during reproductive growth in *Arabidopsis*. *Plant Cell*, *20*(2), 320–336. https://doi.org/10.1105/tpc.107.057752 (accessed on 6th November 2025).

Huang, P., Li, Z., & Guo, H. (2022). New advances in the regulation of leaf senescence by classical and peptide hormones. *Frontiers in Plant Science*, *13*, 923136. https://doi.org/10.3389/fpls.2022.923136 (accessed on 6th November 2025).

Hurtado, A. Q., Neish, I. C., Ali, M. K. M., Norrie, J., Pereira, L., Michalak, I., Shukla, P. S., & Critchley, A. T. (2021). Extracts of seaweeds used as bio stimulants on land and sea crops: An efficacious, phyconomic, circular blue economy: With special reference to *Ascophyllum* (brown) and *Kappaphycus* (red) seaweeds. In S. Gupta & J. Van Staden (Eds.), *Bio Stimulants for Crops from Seed Germination to Plant Development* (pp. 263–288). Academic Press. https://doi.org/10.1016/B978-0-12-823048-0.00017-4 (accessed on 6th November 2025).

Hussain, M. A., Fahad, S., Sharif, R., Jan, M. F., Mujtaba, M., Ali, Q., Ahmad, A., Ahmad, H., Amin, N., & Ajayo, B. S. (2020). Multifunctional role of brassinosteroid and its analogues in plants. *Plant Growth Regulation*, *92*, 141–156.

IndustryARC. (2023). *Auxins Market–Forecast (2024–2030)* (Report Code: CMR 24091). https://www.industryarc.com/Research/Global-Auxins-Industry-Market-Research-511818 (accessed on 6th November 2025).

Iqbal, M. A. (2015). Improving the growth and yield of canola (*Brassica napus* L.) with seed treatment and foliar sprays of *Brassica* (*Brassica napus* L.) and *Moringa* (*Moringa oleifera* L.) leaf extracts. *American-Eurasian Journal of Agricultural & Environmental Sciences*, *14*, 1067–1073.

Iqbal, N., Khan, N. A., Ferrante, A., Trivellini, A., Francini, A., & Khan, M. I. R. (2017). Ethylene role in plant growth, development and senescence: Interaction with other phytohormones. *Frontiers in Plant Science*, *8*, 475. https://doi.org/10.3389/fpls.2017.00475 (accessed on 6th November 2025).

Islam, M. S., Hongxin, W., Admassu, H., Noman, A., Ma, C., & Wei, F. A. (2021). Degree of hydrolysis, functional and antioxidant properties of protein hydrolysates from grass turtle (*Chinemys reevesii*) as influenced by enzymatic hydrolysis conditions. *Food Science & Nutrition*, *9*(8), 4031–4047. https://doi.org/10.1002/fsn3.1903 (accessed on 6th November 2025).

Islam, S., & Mohammad, F. (2020). Triacontanol as a dynamic growth regulator for plants under diverse environmental conditions. *Physiology and Molecular Biology of Plants*, *26*(5), 871–883.

Jain, M., & Khurana, J. P. (2009). Transcript profiling reveals diverse roles of auxin-responsive genes during reproductive development and abiotic stress in rice. *FEBS Journal*, *276*, 3148–3162.

Jan, S., Singh, R., Bhardwaj, R., Ahmad, P., & Kapoor, D. (2020). Plant growth regulators: A sustainable approach to combat pesticide toxicity. *3 Biotech*, *10*(11), 466. https://doi.org/10.1007/s13205-020-02454-4 (accessed on 6th November 2025).

Jarukas, L., Ivanauskas, L., Kasparaviciene, G., Baranauskaite, J., Marksa, M., & Bernatoniene, J. (2021). Determination of organic compounds, fulvic acid, humic acid, and human in peat and sapropel alkaline extracts. *Molecules*, *26*(10), 2995. https://doi.org/10.3390/molecules26102995 (accessed on 6th November 2025).

Jurado, S., Abraham, Z., Manzano, C., López-Torrejón, G., Pacios, L. F., & Del Pozo, J. C. (2010). The *Arabidopsis* cell cycle F-box protein SKP2A binds to auxin. *Plant Cell*, *22*, 3891–3904.

Kaushal, P., Ali, N., Saini, S., Pati, P. K., & Pati, A. M. (2023). Physiological and molecular insight of microbial bio stimulants for sustainable agriculture. *Frontiers in Plant Science*, *14*, 1041413. https://doi.org/10.3389/fpls.2023.1041413 (accessed on 6th November 2025).

Kaya, C., Ugurlar, F., Ashraf, M., & Ahmad, P. (2023). Salicylic acid interacts with other plant growth regulators and signal molecules in response to stressful environments in plants. *Plant Physiology and Biochemistry*, *196*, 431–443. https://doi.org/10.1016/j.plaphy.2023.02.006 (accessed on 6th November 2025).

Khan, A. S., Ahmad, B., Jaskani, M. J., Ahmad, R., & Malik, A. U. (2012). Foliar application of mixture of amino acids and seaweed (*Ascophyllum nodosum*) extract improve growth and physicochemical properties of grapes. *International Journal of Agriculture and Biology*, *14*, 383–388.

Khan, M. I. R., Fatma, M., Per, T. S., Anjum, N. A., & Khan, N. A. (2015). Salicylic acid-induced abiotic stress tolerance and underlying mechanisms in plants. *Frontiers in Plant Science*, *6*, 462. https://doi.org/10.3389/fpls.2015.00462 (accessed on 6th November 2025).

Khan, S., Alvi, A. F., & Khan, N. A. (2024). Role of ethylene in the regulation of plant developmental processes. *Stresses*, *4*(1), 28–53. https://doi.org/10.3390/stresses4010003 (accessed on 6th November 2025).

Khan, T. A., Kappachery, S., Karumannil, S., AlHosani, M., Almansoori, N., Almansoori, H., Yusuf, M., Tran, L. S. P., & Gururani, M. A. (2023). Brassinosteroid signaling pathways: Insights into plant responses under abiotic stress. *International Journal of Molecular Sciences*, *24*(24), 17246. https://doi.org/10.3390/ijms242417246 (accessed on 6th November 2025).

Khan, W., Rayirath, U. P., Subramanian, S., Jithesh, M. N., Rayorath, P., Hodges, D. M., Critchley, A. T., Craigie, J. S., Norrie, J., & Prithiviraj, B. (2009). Seaweed extracts as bio stimulants of plant growth and development. *Journal of Plant Growth Regulation*, *28*, 386–399. https://doi.org/10.1007/s00344-009-9103-x (accessed on 6th November 2025).

Kim, E. J., & Russinova, E. (2020). Brassinosteroid signaling. *Current Biology*, *30*, R294–R298.

Kim, H., Seomun, S., Yoon, Y., & Jang, G. (2021). Jasmonic acid in plant abiotic stress tolerance and interaction with abscisic acid. *Agronomy*, *11*(9), 1886. https://doi.org/10.3390/agronomy11091886 (accessed on 6th November 2025).

Kisvarga, S., Farkas, D., Boronkay, G., Neményi, A., & Orlóci, L. (2022). Effects of bio stimulants in horticulture, with emphasis on ornamental plant production. *Agronomy*, *12*(5), 1043. https://doi.org/10.3390/agronomy12051043 (accessed on 6th November 2025).

Klessig, D. F., Choi, H. W., & Dempsey, D. A. (2018). Systemic acquired resistance and salicylic acid: Past, present, and future. *Molecular Plant-Microbe Interactions*, *31*(9), 871–888. https://doi.org/10.1094/MPMI-03-18-0067-CR (accessed on 6th November 2025).

Koo, Y. M., Heo, A. Y., & Choi, H. W. (2020). Salicylic acid as a safe plant protector and growth regulator. *Plant Pathology Journal*, *36*(1), 1–10. https://doi.org/10.5423/PPJ.RW.12.2019.0295 (accessed on 6th November 2025).

Kulikov, S. N., Chirkov, S. N., Il'Ina, A. V., Lopatin, S. A., & Varlamov, V. P. (2006). Effect of the molecular weight of chitosan on its antiviral activity in plants. *Applied Biochemistry and Microbiology*, *42*, 200–203.

Kumar, D., Menon, M. K., Balasubramani, N., Sadalaxmi, A., & Banerjee, M. (2023c). Certification of organic agriculture and trade issues. In D. Kumar, N. Ravisankar, & A. Panghal (Eds.), *Transforming Organic Agri-Produce into Processed Food Products:*

Post-COVID-19 Challenges and Opportunities (pp. 113–151). Apple Academic Press. https://doi.org/10.1201/9781003329770 (accessed on 6th November 2025).

Kumar, D., Singh, S. K., Arya, S. K., Srivastava, D., Rajput, V. D., & Husain, R. (2023b). Multifunctional growth-promoting microbial consortium-based biofertilizers and their techno-commercial feasibility for sustainable agriculture. In J. A. Parray, N. Shameem, D. Egamberdieva, & R. Z. Sayyed (Eds.), *Rhizobiome: Ecology, Management and Application, Microbiome Research in Plants and Soil* (pp. 167–208). https://doi.org/10.1016/B978-0-443-16030-1.00010-9 (accessed on 6th November 2025).

Kumar, D., Willer, H., Ravisankar, N., Chandrabhanu, & Kumar, A. (2023a). Current scenario of organic farming. In D. Kumar, N. Ravisankar, & A. Panghal (Eds.), *Transforming Organic Agri-Produce into Processed Food Products: Post-COVID-19 Challenges and Opportunities* (pp. 1–24). Apple Academic Press. https://doi.org/10.1201/9781003329770 (accessed on 6th November 2025).

Kumar, M. S., & Sharma, S. A. (2021). Toxicological effects of marine seaweeds: A cautions insight for human consumption. *Critical Reviews in Food Science and Nutrition*, *61*, 500–521.

Kumar, S. M., Zeng, X., Wang, Y., Su, S., Soothar, P., Bai, L., Kumar, M., Zhang, Y., Mustafa, A., & Ye, N. (2020). The short-term effects of mineral- and plant-derived fulvic acids on some selected soil properties: Improvement in the growth, yield, and mineral nutritional status of wheat (*Triticum aestivum* L.) under soils of contrasting textures. *Plants*, *9*(2), 205. https://doi.org/10.3390/plants9020205 (accessed on 6th November 2025).

Lake, L. K., Shafer, W. E., Reilly, S. K., & Jones, R. S. (2002). Regulation of biochemical plant growth regulators at the U.S. Environmental Protection Agency. *Hort Technology*, *12*(1), 55–58. https://doi.org/10.21273/HORTTECH.12.1.55 (accessed on 6th November 2025).

Lamar, R. T. (2020). Possible role for electron shuttling capacity in elicitation of PB activity of humic substances on plant growth enhancement. In *Chemical Biology of Plant Bio Stimulants* (pp. 97–121).

Lenka, S. K., Katiyar, A., Chinnusamy, V., & Bansal, K. C. (2011). Comparative analysis of drought-responsive transcriptome in Indica rice genotypes with contrasting drought tolerance. *Plant Biotechnology Journal*, *9*, 315–327. https://doi.org/10.1111/j.1467-7652.2010.00560.x (accessed on 6th November 2025).

Lephatsi, M. M., Meyer, V., Piater, L. A., Dubery, I. A., & Tugizimana, F. (2021). Plant responses to abiotic stresses and rhizobacterial bio stimulants: Metabolomics and epigenetics perspectives. *Metabolites*, *11*(7), 457. https://doi.org/10.3390/metabo11070457 (accessed on 6th November 2025).

Li, J., & Chory, J. (1997). A putative leucine-rich repeat receptor kinase involved in brassinosteroid signal transduction. *Cell*, *90*, 929–938.

Li, J., Van Gerrewey, T., & Geelen, D. (2022). A meta-analysis of bio stimulant yield effectiveness in field trials. *Frontiers in Plant Science*, *13*, 836702. https://doi.org/10.3389/fpls.2022.836702 (accessed on 6th November 2025).

Lima, J. V., & Lobato, A. K. S. (2017). Brassinosteroids improve photosystem II efficiency, gas exchange, antioxidant enzymes and growth of cowpea plants exposed to water deficit. *Physiology and Molecular Biology of Plants*, *23*(1), 59–72.

Lin, Y., & Jones, M. L. (2022). Evaluating the growth-promoting effects of microbial bio stimulants on greenhouse floriculture crops. *Hort Science*, *57*, 97–109.

Liu, J., Qiu, G., Liu, C., Li, H., Chen, X., Fu, Q., Lin, Y., & Guo, B. (2022). Salicylic acid, a multifaceted hormone, combats abiotic stresses in plants. *Life (Basel)*, *12*(6), 886. https://doi.org/10.3390/life12060886 (accessed on 6th November 2025).

Lu, H. (2009). Dissection of salicylic acid-mediated defense signaling networks. *Plant Signaling & Behavior*, *4*(8), 713–717.

Ma, J. F., & Yamaji, N. (2006). Silicon uptake and accumulation in higher plants. *Trends in Plant Science*, *11*, 392–397.

Ma, N., Hu, C., Wan, L., Hu, Q., Xiong, J., & Zhang, C. (2017). Strigulates improve plant growth, photosynthesis, and alleviate oxidative stress under salinity in rapeseed (*Brassica napus* L.) by regulating gene expression. *Frontiers in Plant Science*, *8*, 1671. https://doi.org/10.3389/fpls.2017.01671 (accessed on 6th November 2025).

Magnabosco, P., Masi, A., Shukla, R., Bansal, V., & Carletti, P. (2023). Advancing the impact of plant bio stimulants to sustainable agriculture through nanotechnologies. *Chemical and Biological Technologies in Agriculture*, *10*, 117. https://doi.org/10.1186/s40538-023-00491-8 (accessed on 6th November 2025).

Majeed Zargar, S., Ahmad Macha, M., Nazir, M., Kumar Agrawal, G., & Rakwal, R. (2012). Silicon: A multitalented micronutrient in OMICS perspective–An update. *Current Proteomics*, *9*, 245–254.

Malécange, M., Sergheraert, R., Teulat, B., Mounier, E., Lothier, J., & Sakr, S. (2023). Bio stimulant properties of protein hydrolysates: Recent advances and future challenges. *International Journal of Molecular Sciences*, *24*(11), 9714. https://doi.org/10.3390/ijms24119714 (accessed on 6th November 2025).

Mandal, S., Ghorai, M., Anand, U., Samanta, D., Kant, N., Mishra, T., et al. (2022). Cytokinin and abiotic stress tolerance–What has been accomplished and the way forward? *Frontiers in Genetics*, *13*, 943025. https://doi.org/10.3389/fgene.2022.943025 (accessed on 6th November 2025).

Manghwar, H., Hussain, A., Ali, Q., & Liu, F. (2022). Brassinosteroids (BRs) role in plant development and coping with different stresses. *International Journal of Molecular Sciences*, *23*(3), 1012. https://doi.org/10.3390/ijms23031012 (accessed on 6th November 2025).

Manjunatha, G., Gupta, K. J., Lokesh, V., Mur, L. A., & Neelwarne, B. (2012). Nitric oxide counters ethylene effects on ripening fruits. *Plant Signaling & Behavior*, *7*(4), 476–483.

Mannino, G. (2023). A new era of sustainability: Plant bio stimulants. *International Journal of Molecular Sciences*, *24*(22), 16329. https://doi.org/10.3390/ijms242216329 (accessed on 6th November 2025).

MarketsandMarkets. (2023a). *Bio Stimulants Market by Crop Type (Cereals & Grains, Oilseeds & Pulses, Fruits & Vegetables, Flowers & Ornamentals), Active Ingredients (Humic Substances, Seaweed Extracts), Form (Dry, Liquid), Mode of Application and Region–Global Forecast to 2028* (Report Code AGI: 2920). https://www.marketsandmarkets.com/Market-Reports/bio stimulant-market-1081.html (accessed on 6th November 2025).

MarketsandMarkets. (2023b). *Plant Growth Regulators Market by Type (Auxins, Cytokinin's, Gibberellins, Ethylene), Crop Type (Cereals & Grains, Fruits & Vegetables, Oilseeds & Pulses, Turf & Ornamentals), Formulation, Function and Region–Global*

Forecast to 2028 (Report Code: AGI 3934). https://www.marketsandmarkets.com/Market-Reports/plant-growth-regulators-market-94580110.htm (accessed on 6th November 2025).

Martins, E. M., Pillajo, J. Q., & Jones, M. L. (2024). Humic and fulvic acids promote growth and flowering in petunias at low and optimal fertility. *Hort Science*, *59*(2), 235–244. https://doi.org/10.21273/HORTSCI17554-23 (accessed on 6th November 2025).

Meshram, J. H., Singh, S. B., Raghavendra, K. P., & Waghmare, V. N. (2022). Drought stress tolerance in cotton: Progress and perspectives. In A. K. Shanker, C. Shanker, A. Anand, & M. Maheswari (Eds.), *Climate Change and Crop Stress: Molecules to Ecosystems* (pp. 135–169). Academic Press. https://doi.org/10.1016/B978-0-12-816091-6.00005-5 (accessed on 6th November 2025).

Miceli, A., Moncada, A., & Vetrano, F. (2021). Use of microbial bio stimulants to increase the salinity tolerance of vegetable transplants. *Agronomy*, *11*(6), 1143. https://doi.org/10.3390/agronomy11061143 (accessed on 6th November 2025).

Minh, N., van Hoa, N., & Trung, T. (2020). Preparation, properties, and application of low-molecular-weight chitosan. In *Handbook of Chitin and Chitosan* (Vol. 1, pp. 453–471). Elsevier.

Mir, R. A., Bhat, B. A., Yousuf, H., Islam, S. T., Raza, A., Rizvi, M. A., Charagh, S., Albaqami, M., Sofi, P. A., & Zargar, S. M. (2022). Multidimensional role of silicon to activate resilient plant growth and to mitigate abiotic stress. *Frontiers in Plant Science*, *13*, 819658. https://doi.org/10.3389/fpls.2022.819658 (accessed on 6th November 2025).

Mir, R. A., Bhat, K. A., Shah, A. A., & Zargar, S. M. (2020). Role of silicon in abiotic stress tolerance of plants. In M. I. R. Khan, A. Singh, & P. Poór (Eds.), *Improving Abiotic Stress Tolerance in Plants* (pp. 271–290). CRC Press. https://doi.org/10.1201/9780429027505 (accessed on 6th November 2025).

Mishra, J., Dutta, V., & Arora, N. K. (2020). Biopesticides in India: Technology and sustainability linkages. *3 Biotech*, *10*(5), 210. https://doi.org/10.1007/s13205-020-02192-7 (accessed on 6th November 2025).

Mitani, N., Yamaji, N., Ago, Y., Iwasaki, K., & Ma, J. F. (2011). Isolation and functional characterization of an influx silicon transporter in two pumpkin cultivars contrasting *in silicon* accumulation. *Plant Journal*, *66*, 231–240.

Mitchell, J., Mandava, N., Worley, J., Plimmer, J., & Smith, M. (1970). Brassies—A new family of plant hormones from rape pollen. *Nature*, *225*, 1065–1066.

Möller, M., & Smith, M. L. (1998). The significance of the mineral component of seaweed suspensions on *Lactuca sativa* L. seedling growth. *Journal of Plant Physiology*, *153*(5–6), 658–663.

Moreno-Hernández, J. M., Benítez-García, I., Mazorra-Manzano, M. A., Ramírez-Suárez, J. C., & Sánchez, E. (2020). Strategies for production, characterization and application of protein-based bio stimulants in agriculture: A review. *Chilean Journal of Agricultural Research*, *80*, 274–289.

Mukarram, M., Ali, J., Dadkhah-Aghdash, H., Kurjak, D., Kačík, F., & Ďurkovič, J. (2023). Chitosan-induced biotic stress tolerance and crosstalk with phytohormones, antioxidants, and other signaling molecules. *Frontiers in Plant Science*, *14*, 1217822. https://doi.org/10.3389/fpls.2023.1217822 (accessed on 6th November 2025).

N/A. (2009). Regulation (EC) No. 1107/2009 of the European Parliament and of the Council concerning the placing of plant protection products on the market and repealing Council

Directives 79/117/EEC and 91/414/EEC. *Official Journal of the European Union L 309*. https://www.fao.org/faolex/results/details/en/c/LEX-FAOC103352/ (accessed on 6th November 2025).

N/A. (2020). *Ethylene Market Size, Share & COVID-19 Impact Analysis, By Application (High-Density Polyethylene, Low Density Polyethylene, Ethylene Oxide, Ethyl Benzene, And Others), And Regional Forecast, 2020–2027*. Report ID: FBI104532, Pages: 100. https://www.fortunebusinessinsights.com/ethylene-market-104532 (accessed on 6th November 2025).

N/A. (2021). *The Fertilizer (Inorganic, Organic, Mixed) (Control) Order, 1985 (amended up to July, 2021) along with The Fertilizer (Movement Control) Order, 1973, and The Essential Commodities Act, 1955*. https://upagripardarshi.gov.in/MediaGallery/FCO_Act.pdf (accessed on 6th November 2025).

N/A. (2022). *Salicylic Acid Market Size, Share, Competitive Landscape and Trend Analysis Report by Application (Pharmaceutical, Skin Care, Haircare, and Food Preservatives & Others): Global Opportunity Analysis and Industry Forecast, 2021–2030*. Report code: A00314, Page: 253. https://www.alliedmarketresearch.com/salicylic-acid-market (accessed on 6th November 2025).

N/A. (2022). The European fertilizing products regulation: What you need to know to understand the new EU 2019/1009 regulatory framework. https://www.seipasa.com/en/blog/the-european-fertilizing-products-regulation/ (accessed on 6th November 2025).

N/A. (2023). Agriculture Ministry issues draft guidelines for registration of bio-stimulants. *Business Standard*. https://www.business-standard.com/industry/agriculture/agri-min-issues-draft-guidelines-for-registration-of-bio-stimulants-123061400625_1.html (accessed on 6th November 2025).

N/A. (2023). *Bio Stimulants Market Size, Share & COVID-19 Impact Analysis, By Source (Microbial and Non-Microbial), By Active Ingredient (Seaweed Extracts, Humic Substances, Vitamins & Amino Acids, Microbial Amendments, And Others), By Application (Foliar Application, Soil Treatment, And Seed Treatment), By Crop (Row Crops, Fruit and Vegetables, Turf & Ornamentals, And Others), And Regional Forecast, 2023–2030*. Report ID: FBI100414. https://www.fortunebusinessinsights.com/industry-reports/bio stimulants-market-100414 (accessed on 6th November 2025).

N/A. (2023). *Bio Stimulants Market Size–By Active Ingredients (Acid-Based [Humic Acid, Fulvic Acid, Amino Acids], Extract-Based [Protein Hydrolysates, Seaweed Extracts]), By Crop (Row Crops, Turf & Ornamentals, Fruit & Vegetables), Application, Form & Forecast 2023–2032*. Report ID: GMI1346. https://www.gminsights.com/industry-analysis/bio stimulants-market (accessed on 6th November 2025).

N/A. (2023). *Chitosan Market Size, Share & Trends Analysis Report by Application (Pharmaceutical, Water Treatment, Cosmetics, Biomedical, Food & Beverage), By Region, And Segment Forecasts, 2023–2030*. Report ID: 978-1-68038-798-8, Pages: 126. https://www.grandviewresearch.com/industry-analysis/global-chitosan-market (accessed on 6th November 2025).

N/A. (2023). *Global Gibberellins Market Size, Share, Growth Analysis, By Type (Gibberellic Acid (GA3), Gibberellins A4/A7), by Crop Type (Fruits and Vegetables, Cereals and Grains), by Application (Seed Treatment, Foliar Spray)–Industry Forecast 2023–2030* (Report ID: SQMIG15C2036, pp. 157). https://www.skyquestt.com/report/gibberellins-market# (accessed on 6th November 2025).

N/A. (2023). *Global Triacontanol Market by Type (Solid, Liquor), by Application (Bactericide, Growth Regulator), by Geographic Scope and Forecast* (Report ID: 250274, pp. 220). https://www.verifiedmarketreports.com/product/triacontanol-market/ (accessed on 6th November 2025).

N/A. (2023a). *Seaweed Extracts Market Size, Share, Competitive Landscape and Trend Analysis Report by Source (Red Seaweed, Brown Seaweed, Green Seaweed), By Form (Liquid, Powder, Flakes), By Application (Food and Beverages, Agriculture, Pharmaceuticals, Others): Global Opportunity Analysis and Industry Forecast, 2023–2032*. Report Code: A12569. https://www.alliedmarketresearch.com/seaweed-extracts-market-A12569 (accessed on 6th November 2025).

N/A. (2023b). *Plant Growth Regulators Market Size, Share, Competitive Landscape and Trend Analysis Report by Type (Auxins, Gibberellins, Cytokinin's, Abscisic Acid, Ethylene), By Crop Type (Cereals, Oilseeds and Pulses, Fruits and Vegetables, Turf and Ornamentals, Others): Global Opportunity Analysis and Industry Forecast, 2021–2031*. Report Code: A00228. https://www.alliedmarketresearch.com/plant-growth-regulators-market (accessed on 6th November 2025).

N/A. (2023c). *Humic-Based Bio Stimulants Market by Type (Humic Acid, Fulvic Acid, Potassium Humate), Application (Seed Treatment, Soil Treatment, Foliar Spray), Form, Crop Type, And Region (North America, Europe, APAC, South America, RoW)–Global Forecast to 2027* (Report Code: AGI 7771). https://www.marketsandmarkets.com/Market-Reports/humic-based-bio stimulants-market-45726696.html (accessed on 6th November 2025).

N/A. (2024). *Cytokinin's Market: By Type, Function, End Use & Region–Forecast 2022–2032*. https://www.futuremarketinsights.com/reports/cytokinins-market (accessed on 6th November 2025).

N/A. (2024). *Global 28-Homobrassinolide Market by Type (Type I, Type II), by Application (Application I, Application II), by Geographic Scope and Forecast* (Report ID: 72791, pp. 220). https://www.verifiedmarketreports.com/product/28-homobrassinolide-market-insights-2019-global-and-chinese-analysis-and-forecast-to-2024/ (accessed on 6th November 2025).

N/A. (2024). *Global Protein Hydrolysates Market Size by Source, by Form, by Application, by Geographic Scope, and Forecast* (Report ID: 22541, pp. 202). https://www.verifiedmarketresearch.com/product/protein-hydrolysates-market/ (accessed on 6th November 2025).

Naeem, M., Ansari, A. A., Aftab, T., Shabbir, A., Alam, M. M., Khan, M. M. A., & Uddin, M. (2019). Application of triacontanol modulates plant growth and physiological activities of *Catharanthus roseus* (L.). *International Journal of Botany Studies*, *4*, 131–135.

Naeem, M., Khan, M. M. A., & Moinuddin. (2012). Triacontanol: A potent plant growth regulator in agriculture. *Journal of Plant Interactions*, *7*(2), 129–142. https://doi.org/10.1080/17429145.2011.619281 (accessed on 6th November 2025).

Naeem, M., Khan, M. M. A., & Siddiqui, M. H. (2009). Triacontanol stimulates nitrogen fixation, enzyme activities, photosynthesis, crop productivity and quality of hyacinth bean (*Lablab purpureus* L.). *Scientia Horticulture*, *121*, 389–396.

Nafie, E., Hathout, T., Al, S., & Al, M. (2011). Jasmonic acid elicits oxidative defense and detoxification systems in *Cucumis melo* L. cells. *Brazilian Journal of Plant Physiology*, *23*, 161–174.

Naroozlo, Y. A., Souri, K. M., & Mojtaba, D. (2019). Stimulation effects of foliar applied glycine and glutamine amino acids on lettuce growth. *Open Agriculture*, *4*, 164. https://doi.org/10.1515/opag-2019-0016 (accessed on 6th November 2025).

Nasri, M. (2017). Protein hydrolysates and biopeptides: Production, biological activities, and applications in foods and health benefits. A review. *Advances in Food and Nutrition Research*, *81*, 109–159. https://doi.org/10.1016/bs.afnr.2016.10.003 (accessed on 6th November 2025).

Nolan, T. M., Vukašinović, N., Liu, D., Russinova, E., & Yin, Y. (2020). Brassinosteroids: Multidimensional regulators of plant growth, development, and stress responses. *Plant Cell*, *32*(2), 295–318.

Ordonio, R. L., Ito, Y., Hatakeyama, A., Ohmae-Shinohara, K., Kasuga, S., Tokunaga, T., Mizuno, H., Kitano, H., Matsuoka, M., & Sazuka, T. (2014). Gibberellin deficiency pleiotropically induces culm bending in sorghum: An insight into sorghum semi-dwarf breeding. *Scientific Reports*, *4*, 5287. https://doi.org/10.1038/srep05287 (accessed on 6th November 2025).

Othibeng, K., Nephali, L., Myoli, A., Buthelezi, N., Jonker, W., Huyser, J., & Tugizimana, F. (2022). Metabolic circuits in sap extracts reflect the effects of a microbial bio stimulant on maize metabolism under drought conditions. *Plants (Basel)*, *11*(4), 510. https://doi.org/10.3390/plants11040510 (accessed on 6th November 2025).

Ouellette, S., Goyette, M. H., Labbé, C., Laur, J., Gaudreau, L., Gosselin, A., Dorais, M., Deshmukh, R. K., & Bélanger, R. R. (2017). Silicon transporters and effects of silicon amendments in strawberry under high tunnel and field conditions. *Frontiers in Plant Science*, *8*, 1664–1462.

Padmanabhan, P., & Paliyath, G. (2018). Ethylene signal transduction during fruit ripening and senescence. In G. Paliyath, J. Subramanian, L. T. Lim, K. S. Subramanian, A. K. Handa, & A. K. Mattoo (Eds.), *Postharvest Biology and Nanotechnology* (pp. 53–57). John Wiley & Sons, Inc. https://doi.org/10.1002/9781119289470.ch3 (accessed on 6th November 2025).

Pang, Q., Chen, X., Lv, J., Li, T., Fang, J., & Jia, H. (2020). Triacontanol promotes the fruit development and retards fruit senescence in strawberry: A transcriptome analysis. *Plants*, *9*(4), 488. https://doi.org/10.3390/plants9040488 (accessed on 6th November 2025).

Parthier, B. (1991). Jasmonates, new regulators of plant growth and development: Many facts and few hypotheses on their actions. *Botanica Acta*, *104*(6), 446–454. https://doi.org/10.1111/j.1438-8677.1991.tb00257.x (accessed on 6th November 2025).

Peng, J., Richards, D. E., Hartley, N. M., Murphy, G. P., Devos, K. M., Flintham, J. E., Beales, J., Fish, L. J., Worland, A. J., Pelica, F., Sudhakar, D., Christou, P., Snape, J. W., Gale, M. D., & Harberd, N. P. (1999). 'Green revolution' genes encode mutant gibberellin response modulators. *Nature*, *400*(6741), 256–261. https://doi.org/10.1038/22307 (accessed on 6th November 2025).

Pereira, T. S., da Silva Lobato, A. K., Tan, D. K. Y., da Costa, D. V., Uchoa, E. B., do Nascimento, Pereira, E. S., Ávila, F. W., Marques, D. J., & Guedes, E. M. S. (2013). Positive interference of silicon on water relations, nitrogen metabolism, and osmotic adjustment in two pepper (*Capsicum annuum*) cultivars under water deficit. *Australian Journal of Crop Science*, *7*(8), 1064–1071.

Perminova, I. V., Frimmel, F. H., Kudryavtsev, A. V., Kulikova, N. A., Abbt-Braun, G., Hesse, S., & Petrosyant, V. S. (2003). Molecular weight characteristics of humic substances from different environments as determined by size exclusion chromatography and their statistical evaluation. *Environmental Science & Technology*, *37*(11), 2477–2485.

Pichyangkura, R., & Chadchawan, S. (2015). Bio stimulant activity of chitosan in horticulture. *Scientia Horticulture*, *196*, 49–65.

Polo, J., & Mata, P. (2018). Evaluation of a bio stimulant (Pepto) based in enzymatic hydrolyzed animal protein in comparison to seaweed extracts on root development, vegetative growth, flowering, and yield of gold cherry tomatoes grown under low stress ambient field conditions. *Frontiers in Plant Science*, *8*, 2261. https://doi.org/10.3389/fpls.2017.02261 (accessed on 6th November 2025).

Pospíšilová, J. (2003). Interaction of cytokinin's and abscisic acid during regulation of stomatal opening in bean leaves. *Photosynthetic*, *41*, 49–56.

Povero, G., Mejia, J. F., Di Tommaso, D., Piaggesi, A., & Warrior, P. (2016). A systematic approach to discover and characterize natural plant bio stimulants. *Frontiers in Plant Science*, *7*, 435. https://doi.org/10.3389/fpls.2016.00435 (accessed on 6th November 2025).

Preston, C. A., Laue, G., & Baldwin, I. T. (2004). Plant–plant signaling: Application of trans- or cis-methyl jasmonates equivalent to sagebrush releases does not elicit direct defenses in native tobacco. *Journal of Chemical Ecology*, *30*, 2193–2214.

Rademacher, W. (2015). Plant growth regulators: Backgrounds and uses in plant production. *Journal of Plant Growth Regulation*, *34*, 845–872. https://doi.org/10.1007/s00344-015-9541-6 (accessed on 6th November 2025).

Rashid, N., Wahid, A., Ibrar, D., Irshad, S., Hasnain, Z., Al-Hashimi, A., Elshikh, M. S., Jacobsen, S. E., & Khan, S. (2022). Application of natural and synthetic growth promoters improves the productivity and quality of quinoa crop through enhanced photosynthetic and antioxidant activities. *Plant Physiology and Biochemistry*, *182*, 1–10. https://doi.org/10.1016/j.plaphy.2022.04.012 (accessed on 6th November 2025).

Ratcliffe, S., Jugdaohsingh, R., Vivancos, J., Marron, A., Deshmukh, R., Ma, J. F., Mitani-Ueno, N., Robertson, J., Wills, J., Boekschoten, M. V., Müller, M., Mawhinney, R. C., Kinrade, S. D., Isenring, P., Bélanger, R. R., & Powell, J. J. (2017). *American Journal of Physiology–Cell Physiology*, *312*, C550–C561.

Reyes-Díaz, M., Lobos, T., Cardemil, L., Nunes-Nesi, A., Retamales, J., Jaakola, L., Alberdi, M., & Ribera-Fonseca, A. (2016). Methyl jasmonate: An alternative for improving the quality and health properties of fresh fruits. *Molecules*, *21*(6), 567. https://doi.org/10.3390/molecules21060567 (accessed on 6th November 2025).

Ries, S. K., Wert, V., Sweeley, C. C., & Leavitt, R. A. (1997). Triacontanol: A new naturally occurring plant growth regulator. *Science*, *195*, 1339–1341.

Rodrigues, M., Baptistella, J. L. C., Horz, D. C., Bortolato, L. M., & Mazzafera, P. (2020). Organic plant bio stimulants and fruit quality—A review. *Agronomy*, *10*(7), 988. https://doi.org/10.3390/agronomy10070988 (accessed on 6th November 2025).

Román-Doval, R., Torres-Arellanes, S. P., Tenorio-Barajas, A. Y., Gómez-Sánchez, A., & Valencia-Lazcano, A. A. (2023). Chitosan: Properties and its application in agriculture in context of molecular weight. *Polymers*, *15*(13), 2867. https://doi.org/10.3390/polym15132867 (accessed on 6th November 2025).

Rouphael, Y., Carillo, P., Cristofano, F., Cardarelli, M., & Colla, G. (2021). Effects of vegetal-versus animal-derived protein hydrolysate on sweet basil morpho-physiological and metabolic traits. *Scientia Horticulture*, *284*, 110123. https://doi.org/10.1016/j.scienta.2021.110123 (accessed on 6th November 2025).

Ruan, J., Zhou, Y., Zhou, M., Yan, J., Khurshid, M., Weng, W., Cheng, J., & Zhang, K. (2019). Jasmonic acid signaling pathway in plants. *International Journal of Molecular Sciences*, *20*(10), 2479. https://doi.org/10.3390/ijms20102479 (accessed on 6th November 2025).

Sabagh, A. E., Mbarki, S., Hossain, A., Iqbal, M. A., Islam, M. S., Raza, A., et al. (2021). Potential role of plant growth regulators in administering crucial processes against abiotic stresses. *Frontiers in Agronomy*, *3*, 648694. https://doi.org/10.3389/fagro.2021.648694 (accessed on 6th November 2025).

Sah, S. K., Reddy, K. R., & Li, J. (2016). Abscisic acid and abiotic stress tolerance in crop plants. *Frontiers in Plant Science, 7*, 571. https://www.frontiersin.org/articles/10.3389/fpls.2016.00571/full (accessed on 6th November 2025).

Sanjuán, J., Nápoles, M. C., Pérez-Mendoza, D., Lorite, M. J., & Rodríguez-Navarro, D. N. (2023). Microbials for agriculture: Why do they call them bio stimulants when they mean probiotics? *Microorganisms*, *11*(1), 153. https://doi.org/10.3390/microorganisms11010153 (accessed on 6th November 2025).

Saradadevi, R., Palta, J. A., & Siddique, K. H. M. (2017). ABA-mediated stomatal response in regulating water uses during the development of terminal drought in wheat. *Frontiers in Plant Science*, *8*, 1251.

Sarwar, M., Anjum, S., Alam, M. W., Ali, Q., Ayyub, C. M., Haider, M. S., Ashraf, M. I., & Mahboob, W. (2022). Triacontanol regulates morphological traits and enzymatic activities of salinity affected hot pepper plants. *Scientific Reports*, *12*, 3736. https://doi.org/10.1038/s41598-022-06516-w (accessed on 6th November 2025).

Sarwar, M., Anjum, S., Ali, Q., Alam, M. W., Haider, M. S., & Mehboob, W. (2021). Triacontanol modulates salt stress tolerance in cucumber by altering the physiological and biochemical status of plant cells. *Scientific Reports*, *11*(1), 24504. https://doi.org/10.1038/s41598-021-04174-y (accessed on 6th November 2025).

Sasidharan, R., & Voesenek, L. A. C. J. (2015). Ethylene-mediated acclimations to flooding stress. *Plant Physiology, 169*(1), 3–12.

Savvas, D., & Ntatsi, G. (2015). Bio stimulant activity of silicon in horticulture. *Scientia Horticulture*, *196*, 66–81.

Schaller, G. E., & Binder, B. M. (2017). Inhibitors of ethylene biosynthesis and signaling. *Methods in Molecular Biology*, *1573*, 223–235.

Sestili, F., Rouphael, Y., Cardarelli, M., Pucci, A., Bonini, P., Canaguier, R., & Colla, G. (2018). Protein hydrolysate stimulates growth in tomato coupled with N-dependent gene expression involved in N assimilation. *Frontiers in Plant Science*, *9*, 1233. https://doi.org/10.3389/fpls.2018.01233 (accessed on 6th November 2025).

Shahrajabian, M. H., & Sun, W. (2022b). Sustainable approaches to boost yield and chemical constituents of aromatic and medicinal plants by application of bio stimulants. *Recent Patents on Food, Nutrition & Agriculture*, *13*(2), 72–92. https://www.eurekaselect.com/article/126745 (accessed on 6th November 2025).

Shahrajabian, M. H., Chaski, C., Polyzos, N., Tzortzakis, N., & Petropoulos, S. A. (2021b). Sustainable agriculture systems in vegetable production using chitin and chitosan as plant bio stimulants. *Biomolecules*, *11*(6), 819. https://doi.org/10.3390/biom11060819 (accessed on 6th November 2025).

Shahrajabian, M. H., Cheng, Q., & Sun, W. (2022a). The effects of amino acids, phenols and protein hydrolysates as bio stimulants on sustainable crop production and alleviated stress. *Recent Patents on Biotechnology*, *16*(4), 319–328. https://www.eurekaselect.com/article/122498 (accessed on 6th November 2025).

Shahrajabian, M. H., Sun, W., Soleymani, A., & Cheng, Q. (2021). Traditional herbal medicines to overcome stress, anxiety and improve mental health in outbreaks of human coronaviruses. *Phytotherapy Research*, *35*(3), 1237–1247. https://doi.org/10.1002/ptr.6888 (accessed on 6th November 2025).

Sharif, R., Mujtaba, M., Ur Rahman, M., Shalmani, A., Ahmad, H., Anwar, T., Tianchan, D., & Wang, X. (2018). The multifunctional role of chitosan in horticultural crops; A review. *Molecules*, *23*(4), 872. https://doi.org/10.3390/molecules23040872 (accessed on 6th November 2025).

Sharma, H. S. S., Fleming, C., Selby, C., Rao, J. R., & Martin, T. (2014). Plant bio stimulants: A review on the processing of macroalgae and use of extracts for crop management to reduce abiotic and biotic stresses. *Journal of Applied Phycology*, *26*, 465–490.

Sharma, I., Kaur, N., & Pati, P. K. (2017). Brassinosteroids: A promising option in deciphering remedial strategies for abiotic stress tolerance in rice. *Frontiers in Plant Science*, *8*, 2151. https://doi.org/10.3389/fpls.2017.02151 (accessed on 6th November 2025).

Sharma, L., Priya, M., Kaushal, N., Bhandhari, K., Chaudhary, S., Dhankher, O. P., Prasad, P. V. V., Siddique, K. H. M., & Nayyar, H. (2020). Plant growth-regulating molecules as thermoprotectants: Functional relevance and prospects for improving heat tolerance in food crops. *Journal of Experimental Botany*, *71*, 569–594. https://doi.org/10.1093/jxb/erz333 (accessed on 6th November 2025).

Shaukat, M., Ahmad, A., Khaliq, T., Afzal, I., Muhammad, S., Safdar, B., & Shah, S. H. (2021). Foliar spray of natural and synthetic plant growth promoters accelerates growth and yield of cotton by modulating photosynthetic pigments. *International Journal of Plant Production*, *15*, 615–624. https://doi.org/10.1007/s42106-021-00158-0 (accessed on 6th November 2025).

Shi, Y., Tian, S., Hou, L., Huang, X., Zhang, X., Guo, H., & Yang, S. (2012). Ethylene signaling negatively regulates freezing tolerance by repressing expression of CBF and type-A ARR genes in *Arabidopsis*. *Plant Cell*, *24*(6), 2578–2595.

Sible, C. N., Seebauer, J. R., & Below, F. E. (2021). Plant bio stimulants: A categorical review, their implications for row crop production, and relation to soil health indicators. *Agronomy*, *11*(7), 1297. https://doi.org/10.3390/agronomy11071297 (accessed on 6th November 2025).

Siddiqui, H., Hayat, S., & Bajguz, A. (2018). Regulation of photosynthesis by brassinosteroids in plants. *Acta Physiologiae Plantarum*, *40*, 59. https://doi.org/10.1007/s11738-018-2639-2 (accessed on 6th November 2025).

Sierras, N., Botta, A., Staasing, L., Martinez, M. J., & Bru, R. (2016). Understanding the effect of amino acids-based bio stimulant by an enantiomeric analysis of their active principles and a proteomic profiling approach. *Acta Horticulture*, *1148*, 93–100.

Sohn, S. I., Pandian, S., Rakkammal, K., Largia, M. J. V., Thamilarasan, S. K., Balaji, S., Zoclanclounon, Y. A. B., Shilpha, J., & Ramesh, M. (2022). Jasmonates in plant growth and development and elicitation of secondary metabolites: An updated overview. *Frontiers in Plant Science*, *13*, 942789. https://doi.org/10.3389/fpls.2022.942789 (accessed on 6th November 2025).

Song, A., Li, P., Fan, F., Li, Z., & Liang, Y. (2014). The effect of silicon on photosynthesis and expression of its relevant genes in rice (*Oryza sativa* L.) under high-zinc stress. *PLoS ONE*, *9*, e113782. https://doi.org/10.1371/journal.pone.0113782 (accessed on 6th November 2025).

Sood, M. (2023). Jasmonates: "The master switch" for regulation of developmental and stress responses in plants. *Journal of Plant Growth Regulation*, *42*, 5247–5265. https://doi.org/10.1007/s00344-023-11047-3 (accessed on 6th November 2025).

Souri, Z., Khanna, K., Karimi, N., & Ahmad, P. (2021). Silicon and plants: Current knowledge and future prospects. *Journal of Plant Growth Regulation*, *40*, 906–925.

Sparzak-Stefanowska, B., Krauze-Baranowska, M., & Hałasa, R. (2019). Influence of plant growth regulators on the shoot culture of *Phyllanthus glaucus* and accumulation of indolizidine alkaloids with evaluation of antimicrobial activity. *Acta Physiologiae Plantarum*, *41*, 6. https://doi.org/10.1007/s11738-018-2796-3 (accessed on 6th November 2025).

Sugawara, S., Mashiguchi, K., Tanaka, K., Hishiyama, S., Sakai, T., Hanada, K., Kinoshita-Tsujimura, K., Yu, H., Dai, X., Takebayashi, Y., Takeda-Kamiya, N., Kakimoto, T., Kawaide, H., Natsume, M., Estelle, M., Zhao, Y., Hayashi, K., Kamiya, Y., & Kasahara, H. (2015). Distinct characteristics of indole-3-acetic acid and phenylacetic acid, two common auxins in plants. *Plant and Cell Physiology*, *56*(8), 1641–1654. https://doi.org/10.1093/pcp/pcv088 (accessed on 6th November 2025).

Sun, W., Shahrajabian, M. H., Kuang, Y., & Wang, N. (2024). Amino acids bio stimulants and protein hydrolysates in agricultural sciences. *Plants*, *13*, 210. https://doi.org/10.3390/plants13020210 (accessed on 6th November 2025).

Taiz, L., & Zeiger, E. (2010). *Plant Physiology* (5th ed.). Sinauer Associates Inc.

Taiz, L., Zeiger, E., Møller, I. M., & Murphy, A. (2015). *Plant Physiology and Development* (6th ed.). Sinauer Associates Inc.

Tan, C., Li, S., Song, J., Zheng, X., Zheng, H., Xu, W., Wan, C., Zhang, T., Bian, Q., & Men, S. (2024). 3,4-Dichlorophenylacetic acid acts as an auxin analog and induces beneficial effects in various crops. *Communications Biology*, *7*, 161. https://doi.org/10.1038/s42003-024-05848-9 (accessed on 6th November 2025).

Tang, J., Han, Z., & Chai, J. (2016). Q&A: What are brassinosteroids and how do they act in plants? *BMC Biology*, *14*, 113. https://doi.org/10.1186/s12915-016-0340-8 (accessed on 6th November 2025).

Tanou, G., Ziogas, V., & Molassiotis, A. (2017). Foliar nutrition, bio stimulants and prime-like dynamics in fruit tree physiology: New insights on an old topic. *Frontiers in Plant Science*, *8*, 75. https://doi.org/10.3389/fpls.2017.00075 (accessed on 6th November 2025).

Tewari, S., & Arora, N. K. (2016). *Fluorescent Pseudomonas* sp. PF17 as an efficient plant growth regulator and biocontrol agent for sunflower crop under saline conditions. *Symbiosis*, *68*(1), 99–108.

Traon, D., Amat, L., Zotz, F., & du Jardin, P. (2014). *A Legal Framework for Plant Bio Stimulants and Agronomic Fertilizer Additives in the EU: Report for the European Commission Enterprise & Industry Directorate-General*. Arcadia International, 1–115.

Trevisan, S., Manoli, A., Ravazzolo, L., Franceschi, C., & Quaggiotti, S. (2017). mRNA-sequencing analysis reveals transcriptional changes in root of maize seedlings treated with two increasing concentrations of a new bio stimulant. *Journal of Agricultural and Food Chemistry*, *65*, 9956–9969.

Tromas, A., Braun, N., Muller, P., Khodus, T., Paponov, I. A., Palme, K., Ljung, K., Lee, J. Y., Benfey, P., Murray, J. A., Scheres, B., & Perrot-Rechenmann, C. (2009). The AUXIN BINDING PROTEIN 1 is required for differential auxin responses mediating root growth. *PLoS ONE*, *4*(9), e6648. https://doi.org/10.1371/journal.pone.0006648 (accessed on 6th November 2025).

Tu, C., & Wang, F. (2019). Sixty years' devotion to green revolution and a lifetime commitment to food security—Review on the academic achievements of Huang Yaoxiang, father of semi-dwarf rice breeding. *Guangdong Agricultural Sciences*, *46*, 1–7.

Tuteja, N. (2007). Abscisic acid and abiotic stress signaling. *Plant Signaling & Behavior*, *2*(3), 135–138.

Ubeda-Tomás, S., Beemster, G. T., & Bennett, M. J. (2012). Hormonal regulation of root growth: Integrating local activities into global behaviour. *Trends in Plant Science*, *17*, 326–331.

Uchiyama, T., & Ogasawara, N. (1981). Constituents of plant leaf wax contained in rice callus tissues. *Agricultural and Biological Chemistry*, *5*, 1261–1263.

United States Environmental Protection Agency (US EPA). (2023). Draft guidance for plant regulators and claims, including plant bio stimulants. https://www.epa.gov/pesticides/draft-guidance-plant-regulators-and-claims-including-plant-bio stimulants (accessed on 6th November 2025).

Urmi, T. A., Islam, M. M., Zumur, K. N., Abedin, M. A., Haque, M. M., Siddiqui, M. H., Murata, Y., & Hoque, M. A. (2023). Combined effect of salicylic acid and proline mitigates drought stress in rice (*Oryza sativa* L.) through the modulation of physiological attributes and antioxidant enzymes. *Antioxidants*, *12*(7), 1438. https://doi.org/10.3390/antiox12071438 (accessed on 6th November 2025).

Valverde, S., Williams, P. L., Mayans, B., Lucena, J. J., & Hernández-Apaolaza, L. (2022). Comparative study of the chemical composition and antifungal activity of commercial brown seaweed extracts. *Frontiers in Plant Science*, *13*, 1017925. https://doi.org/10.3389/fpls.2022.1017925 (accessed on 6th November 2025).

Van Oosten, M. J., Pepe, O., De Pascale, S., Silletti, S., & Maggio, A. (2017). The role of bio stimulants and bio effectors as alleviators of abiotic stress in crop plants. *Chemical and Biological Technologies in Agriculture*, *4*, 5. https://doi.org/10.1186/s40538-017-0089-5 (accessed on 6th November 2025).

Verma, T., Bhardwaj, S., Singh, J., Kapoor, D., & Prasad, R. (2022). Triacontanol as a versatile plant growth regulator in overcoming negative effects of salt stress. *Journal of Agriculture and Food Research*, *10*, 100351. https://doi.org/10.1016/j.jafr.2022.100351 (accessed on 6th November 2025).

Vishwakarma, K., Upadhyay, N., Kumar, N., Yadav, G., Singh, J., Mishra, R. K., Kumar, V., Verma, R., Upadhyay, R. G., Pandey, M., & Sharma, S. (2017). Abscisic acid signaling and abiotic stress tolerance in plants: A review on current knowledge and future prospects. *Frontiers in Plant Science*, *8*, 161. https://doi.org/10.3389/fpls.2017.00161 (accessed on 6th November 2025).

Vocciante, M., Grifoni, M., Fusini, D., Petruzzelli, G., & Franchi, E. (2022). The role of plant growth-promoting rhizobacteria (PGPR) in mitigating plant's environmental stresses. *Applied Sciences*, *12*(3), 1231. https://doi.org/10.3390/app12031231 (accessed on 6th November 2025).

Wang, M., Wang, R., Mur, L. A. J., Ruan, J., Shen, Q., & Guo, S. (2021). Functions of silicon in plant drought stress responses. *Horticulture Research*, *8*, 254. https://doi.org/10.1038/s41438-021-00681-1 (accessed on 6th November 2025).

Wang, S., Liu, P., Chen, D., Yin, L., Li, H., & Deng, X. (2015). Silicon enhanced salt tolerance by improving the root water uptake and decreasing the ion toxicity in cucumber. *Frontiers in Plant Science*, *6*, 759. https://doi.org/10.3389/fpls.2015.00759 (accessed on 6th November 2025).

Waqas, M. A., Kaya, C., Riaz, A., Farooq, M., Nawaz, I., Wilkes, A., & Li, Y. (2019). Potential mechanisms of abiotic stress tolerance in crop plants induced by thiourea. *Frontiers in Plant Science*, *10*, 1336. https://doi.org/10.3389/fpls.2019.01336 (accessed on 6th November 2025).

Waqas, M., Shahzad, R., Khan, A. L., Asaf, S., Kim, Y. H., Kang, S. M., & Lee, I. J. (2016). Salvaging effect of triacontanol on plant growth, thermotolerance, macro-nutrient content, amino acid concentration and modulation of defense hormonal levels under heat stress. *Plant Physiology and Biochemistry*, *99*, 118–125.

War, A. R., Paulraj, M. G., War, M. Y., & Ignacimuthu, S. (2011). Role of salicylic acid in induction of plant defense system in chickpea (*Cicer arietinum* L.). *Plant Signaling & Behavior*, *6*(11), 1787–1792.

Wei, Z., & Li, J. (2020). Regulation of brassinosteroid homeostasis in higher plants. *Frontiers in Plant Science*, *11*, 583622. https://doi.org/10.3389/fpls.2020.583622 (accessed on 6th November 2025).

Weremczuk-Jeżyna, I., Hnatuszko-Konka, K., Lebelt, L., Piotrowska, D. G., & Grzegorczyk-Karolak, I. (2022). The effect of the stress-signaling mediator triacontanol on biochemical and physiological modifications in *Dracocephalum forrestii* culture. *International Journal of Molecular Sciences*, *23*(23), 15147. https://doi.org/10.3390/ijms232315147 (accessed on 6th November 2025).

Windisch, S., Walter, A., Moradtalab, N., Walker, F., Höglinger, B., El-Hasan, A., Ludewig, U., Neumann, G., & Grosch, R. (2021). Role of benzoic acid and lettuce in A in the defense response of lettuce against soil-borne pathogens. *Plants (Basel)*, *10*(11), 2336. https://doi.org/10.3390/plants10112336 (accessed on 6th November 2025).

Woodward, A. W., & Bartel, B. (2005). Auxin: regulation, action, and interaction. *Annals of Botany*, *95*(5), 707–735. https://doi.org/10.1093/aob/mci083 (accessed on 6th November 2025).

Wu, W., Du, K., Kang, X., & Wei, H. (2021). The diverse roles of cytokinin's in regulating leaf development. *Horticulture Research*, *8*, 118. https://doi.org/10.1038/s41438-021-00558-3 (accessed on 6th November 2025).

Xu, C., & Leskovar, D. I. (2015). Effects of *Ascophyllum nodosum* seaweed extracts on spinach growth, physiology and nutrition value under drought stress. *Scientia Horticulture*, *183*, 39–47.

Yan, Y., Borrego, E., & Kolomiets, M. V. (2013). Jasmonate biosynthesis, perception and function in plant development and stress response. In R. V. Baez (Ed.), *Lipid Metabolism* (pp. 393–442). InTech.

Yang, C., Li, W., Cao, J., Meng, F., Yu, Y., Huang, J., Jiang, L., Liu, M., Zhang, Z., Chen, X., Miyamoto, K., Yamane, H., Zhang, J., Chen, S., & Liu, J. (2017). Activation of ethylene signaling pathways enhances disease resistance by regulating ROS and phytoalexin production in rice. *The Plant Journal*, *89*(2), 338–353.

Yin, C. C., Zhao, H., Ma, B., Chen, S. Y., & Zhang, J. S. (2017). Diverse roles of ethylene in regulating agronomic traits in rice. *Frontiers in Plant Science*, *8*, 1676. https://doi.org/10.3389/fpls.2017.01676 (accessed on 6th November 2025).

Zaid, A., Asghar, M., Wani, I. A., & Wani, S. H. (2020). Role of triacontanol in overcoming environmental stresses. In A. Roychoudhury & D. K. Tripathi (Eds.), *Protective Chemical Agents in The Amelioration of Plant Abiotic Stress: Biochemical and Molecular Perspectives* (pp. 491–509). John Wiley & Sons Ltd. https://doi.org/10.1002/9781119552154.ch25 (accessed on 6th November 2025).

Zaid, A., Mohammad, F., & Fariduddin, Q. (2019). Plant growth regulators improve growth, photosynthesis, mineral nutrient and antioxidant system under cadmium stress in menthol mint (*Mentha arvensis* L.). *Physiology and Molecular Biology of Plants*, *26*(1), 25–39. https://doi.org/10.1007/s12298-019-00715-y (accessed on 6th November 2025).

Zambounis, A. G., Kalamaki, M. S., Tani, E. E., Paplomatas, E. J., & Tsaftaris, A. S. (2012). Expression analysis of defense-related genes in cotton (*Gossypium hirsutum*) after *Fusarium oxysporum* f. sp. *vasinfectum* infection and following chemical elicitation using a salicylic acid analog and methyl jasmonate. *Plant Molecular Biology Reporter*, *30*, 225–234.

Zargar, S. M., Mahajan, R., Bhat, J. A., Nazir, M., & Deshmukh, R. (2019). Role of silicon in plant stress tolerance: Opportunities to achieve a sustainable cropping system. *3 Biotech*, *9*(3), 73. https://doi.org/10.1007/s13205-019-1613-z (accessed on 6th November 2025).

Zhang, H., Sun, X., & Dai, M. (2021). Improving crop drought resistance with plant growth regulators and rhizobacteria: Mechanisms, applications, and perspectives. *Plant Communications*, *3*(1), 100228. https://doi.org/10.1016/j.xplc.2021.100228 (accessed on 6th November 2025).

Zhang, P., Zhang, H., Wu, G., Chen, X., Gruda, N., Li, X., Dong, J., & Duan, Z. (2021). Dose-dependent application of straw-derived fulvic acid on yield and quality of tomato plants grown in a greenhouse. *Frontiers in Plant Science*, *12*, 736613. https://doi.org/10.3389/fpls.2021.736613 (accessed on 6th November 2025).

Zhang, Q., Gong, M., Xu, X., Li, H., & Deng, W. (2022). Roles of auxin in the growth, development, and stress tolerance of horticultural plants. *Cells*, *11*(17), 2761. https://doi.org/10.3390/cells11172761 (accessed on 6th November 2025).

Zhang, S., Gottschalk, C., & van Nocker, S. (2019). Genetic mechanisms in the repression of flowering by gibberellins in apple (*Malus* × *domestica* Borkh.). *BMC Genomics*, *20*, 747. https://doi.org/10.1186/s12864-019-6090-6 (accessed on 6th November 2025).

Zhang, X., & Ervin, E. H. (2008). Impact of seaweed extract-based cytokinin's and zeatin riboside on creeping bentgrass heat tolerance. *Crop Science*, *48*(1), 364–370.

Zhang, X., Schmidt, R. E., Ervin, E. H., & Doak, S. (2002). Creeping bentgrass physiological responses to natural plant growth regulators and iron under two regimes. *Hort Science*, *37*(6), 898–902.

Zhao, M., Liu, W., Xia, X., Wang, T., & Zhang, W. H. (2014). Cold acclimation-induced freezing tolerance of *Medicago truncatula* seedlings is negatively regulated by ethylene. *Physiologic Plantarum, 152*, 115–129.

Žižková, E., Kubeš, M., Dobrev, P. I., Přibyl, P., Šimura, J., Zahajská, L., Drábková, L. Z., Novák, O., & Motyka, V. (2017). Control of cytokinin and auxin homeostasis in cyanobacteria and algae. *Annals of Botany, 119*(1), 151–166. https://doi.org/10.1093/aob/mcw194.

CHAPTER 2

Usage of Fish Byproducts as Biostimulants: An Overview of Market Potential, Relevant Legislation, and Regulations

ASHA KUMARI and NUTAN KAUSHIK*

Amity Food and Agriculture Foundation, Amity University, Noida, Uttar Pradesh, India

*Corresponding author

ABSTRACT

The growing population of the world increases the burden on agriculture to produce more food without altering the quality of the crop. With the current pattern of agricultural practices, the global goal of zero hunger by 2030 cannot be achieved. A cost-effective, sustainable, and environmentally friendly method needs to be developed to improve crop yield and quality, even in stressful conditions such as disease and drought. Therefore, the biostimulants (BS) substance can be a potential alternative that increases the growth of plants through different modes of action, such as stimulating soil phosphorus release or nitrogen metabolism, stimulating the growth of roots, increasing absorption and transportation of nutrients, and so forth. The global market for BS is increasing at a very fast rate, which is estimated to reach USD 6.2 billion by 2027. Plant biostimulants (PBs) are derived from various organic or inorganic materials, including humic compounds, seaweed extracts, protein hydrolysates (PHs), and leftovers from agricultural or industrial byproducts that promote the growth and productivity of plants. Fish processing byproducts have potential

Plant Biostimulants and Growth Regulators in Sustainable Agriculture.
Deepak Kumar, Sanjay K. Singh, Nutan Kaushik, and Mahaveer P. Sharma (Eds.)

DOI: 10.1201/9781779641458-3

applications as PBs, which may aid in the promotion of plant growth in stressful conditions.

2.1 INTRODUCTION

The seafood sector and the fish market produce a total of 100 million metric tons of byproducts worldwide, with more than 4 million metric tons produced in India. India is one of the world's top fish producers, exporting 13,77,244 MT of fish. The growing need for fish worldwide has increased the demand for processed seafood. As one of the most significant animal protein sources, fish is generally acknowledged to be an excellent source of protein and other nutrients needed to keep the body healthy (Ranvichandran et al., 2012). Therefore, consumption of fish and seafood is the most preferred food worldwide due to its advantages for health. Fish waste is being produced in massive quantities at the same time, primarily as a result of industrial fish processing. This large quantity of fish waste contributes 50–60% of the total fish captured, which has not been utilized effectively, and disposing of aquatic waste may additionally have a significant detrimental influence on the environment (Jafarpour et al., 2020). Therefore, there is an urgent requirement to find ecologically acceptable means for efficiently reutilizing fish waste.

The growing populations of the world are facing challenges to meet their food needs now and in the future due to rising hunger, malnutrition, climate change, and overexploitation of natural resources. The increasing demand for food exerted immense pressure on agriculture to produce high-quality crops. The productivity of agriculture is declining in many parts of the world due to chemical-based agriculture practices and climate change, which may be a potential factor in the food crisis and a major challenge to meeting the global goal of zero hunger by 2030 (Tilman et al., 2011). Agricultural producers need to find an efficient and effective way to produce crops in a sustainable way. Biostimulants (BS) can be a potential and sustainable alternative to protect plants against stress and promote their growth. The global market for BS may increase in the future due to increased consumer awareness about the advantages of organic food. The global market for BS is estimated to be USD 3.5 billion in 2022, which is supposed to increase to USD 6.2 billion by 2027 (CAGR of 11.785) (Marketsandmarkets, 2023). The annual growth rate of plant biostimulants (PBs) was 10.4% from 2016 to 2021. In the year 2021, BS will grow to USD 2.91 billion (Yakhin et al., 2017).

The market for BS in India was approximately USD 230.9 million in 2021, which is expected to increase to USD 734.1 million by 2029 with a CAGR of 15.6% (Fortune Business Insights, 2023). Irrespective of the nutrient contents, a BS is any material or microorganism that is administered to plants with the intention of improving nutritional efficiency, tolerance to abiotic stress, and/or crop quality attributes.

BS are not considered fertilizers because they only facilitate nutrient acquisition by developing new pathways or mechanisms for nutrient uptake. Organisms are applied to plants with the aim of enhancing nutrition efficiency, abiotic stress tolerance, and/or crop quality traits, regardless of their nutrient contents. BS was first explained by Kauffman et al. (2007) as a substance that promotes the growth of plants when applied in low quantities. BS are also termed metabolic enhancers (Du Jardin, 2015). The BS are formulated by varying the ingredients. Kauffman et al. (2007) classified the BS into three major groups: humic substances, hormone-containing products, and amino acid-containing products (Figure 2.1).

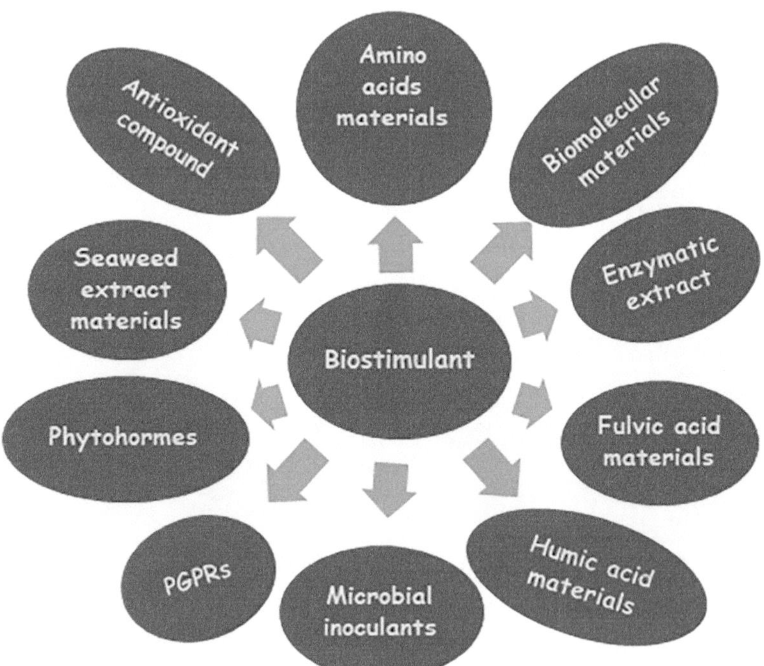

FIGURE 2.1 Different active compounds show biostimulant properties.

BS promote plant growth without being nutrients, soil improvers, or pesticides (Du Jardin, 2015). A biostimulants acts as a growth promoter, modulates quality traits, and enhances tolerance to environmental stress. The advantages of being used as a biostimulant are environmental friendliness, stress mitigation, high yields, and crop protection (García-García et al., 2020). Many industries are key players in the promotion of biostimulant concepts, such as the European Biostimulant Industry Council (EBIC) in Europe and the Biostimulant Coalition in the USA. Thcsc organizations are discussing BS with different regulators, stakeholders, and scientists. The first international symposium was organized in Strasbourg in November 2012 on the topic "The use of BS in agriculture." The Indian government has taken several initiatives (such as the start-up India Seed Fund) to promote BS, for which the Indian government will grant a budget of approximately INR 260 million during the years 2021–22. Different awareness initiatives regarding the use of BS and their benefits are given to farmers at the national and grass-roots levels (Meena and Kumar, 2023; https://www.biostimulant.com/) (Figure 2.2).

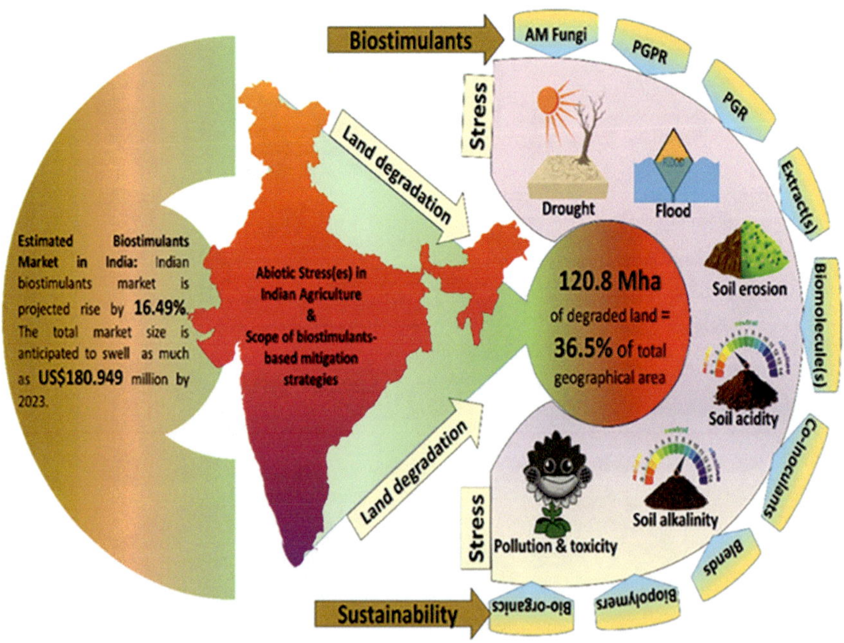

FIGURE 2.2 Scope of biostimulants in India.

Source: Fortune Business Insights (2023); and Meena and Kumar (2023).

2.2 DIFFERENT MATERIALS THAT ARE USED AS BIOSTIMULANT SUBSTANCES

2.2.1 HUMIC SUBSTANCES

Humic substances are organic matter in the soil that is developed after the natural decomposition of plants, animals, and microbial residues. The humic substances increase plant growth, respiration, and crop yield through the interaction of organic matter, microbes, and plant roots. The effectiveness of humic substances depends on the source, environmental conditions, receiving plants, dose, and manner of humic substance application (Rose et al., 2014). Rose et al. (2014) state that overall growth of plant shoots and roots increases by 22% and 21% on a dry weight basis, respectively, using random-effect meta-analysis. The BS effect of humic substances is due to the amelioration of root nutrition. The mechanisms by which humic substances act as BS are:

1. **Cation Exchange of Soil:** The uptake of micronutrients and macronutrients by the roots of plants increases because the cation exchange capacity (CEC) of soil increases due to the presence of polyanionic humic substances in the soil. The availability of phosphorus increases with humic substances by interfering with calcium phosphate precipitation.
2. **Stimulation of the Plasma Membrane H^+ ATPase:** H^+-ATPase helps in the transportation of nitrate and other nutrients by converting free energy released by ATP hydrolysis into a trans-membrane electrochemical potential.

Olivares et al. (2015) reported that humic substances having high molecular mass enhance the growth of hydroponically grown maize seedlings through stress response modulation by humic substances.

2.2.2 PROTEIN HYDROLYSATES AND OTHER NITROGEN-CONTAINING COMPOUNDS

Protein hydrolysates (PHs) are obtained after the hydrolysis of plant and animal byproducts using different techniques such as chemical, enzymatic, and ultrasonic hydrolysis (Halpern et al., 2015). BS containing PH can be used as single or combination compounds. The nitrogenous molecules are betaines, polyamines, and non-protein amino acids. Glycine betaine is an amino acid derivative with anti-stress properties (Chen and Murata, 2011). The mechanism by which PH-containing BS promote plant growth is:

- Modulation of nitrogen uptake
- Regulation of enzymes responsible for nitrogen assimilation
- Act on the signaling pathways of nitrogen acquisition in roots.
- Regulating enzymes in TCA cycles
- Chelating properties

PHs contain fat-free amino acids, including serine, glutamic acid, aspartic acid, hydroxyproline, threonine, proline, glycine, tyrosine, isoleucine, leucine, alanine, methionine, melatonin, free amino acids, peptides, proteins, and organic matter (Chalamaiah et al., 2012). PH containing a biostimulant containing proline aids in mobility and the acquisition of micronutrients and protects plants against heavy metals. The application of PH in the soil increases soil fertility, respiration, and microbial biomass. Corte et al. (2014) tested the hydrolyzed proteins from animal sources on yeast and plants and observed that hydrolyzed proteins do not have genotoxicity, phytotoxicity, or ecotoxicity. Whereas the EU Commission's Implementing Regulation No. 354/2014 in respect to organic production, labelling, and control banned the use of hydrolyzed protein from animal sources on edible parts of organic crops.

2.2.3 SEAWEED EXTRACT

The use of seaweed extract as a biostimulant has been reported recently. The growth-promoting compounds present in seaweeds are polysaccharides, laminarin, alginates, and carrageenans and their derivatives, sterols, betaines, and hormones (Craigie, 2011). Seaweed extract can be applied either to soil, plants, hydroponic solutions, or foliar treatments. Polysaccharides present in seaweed contribute to the formation of gel and increase water retention and soil aeration. Seaweed containing polyanionic compounds contributes to the fixation and exchange of cations, the fixation of heavy metals, and soil remediation. Calvo et al. (2014) reported that seaweed extract shows antistress effect properties because it contains antioxidants and regulates endogenous stress-responsive genes.

2.2.4 CHITOSAN

Chitosan is produced naturally or industrially by the deacetylation of biopolymer chitin. Application of chitosan as a biostimulant helps to protect the plants against abiotic stress (drought, salinity, and cold stress). Chitosan increases the binding of polycationic compounds to cellular components such as DNA, plasma membrane, and cell wall constituents, as well as specific receptors involved in defense gene activation (Katiyar et al., 2015).

2.3 MECHANISM OF ACTION OF THE BIOSTIMULANTS

The guideline or regulation on commercialization of the BS products was not cleared due to a lack of knowledge about modes of action (Colla et al., 2017). The mechanism through which BS increase plant growth and protection is by activating nitrogen metabolism or phosphorus release from soils, which enhances plant establishment and stimulates root growth (Du Yakhin et al., 2017) (Figure 2.3). BS affect the solubility, assimilation, absorption, and transportation of micronutrients through soil by altering the root structure and boosting soil activity on an enzymatic and microbiological level (Colla et al., 2015). BS increase the availability of nutrients from the soil to the plant by increasing cation exchange through the supply of nitrogen and improving the solubility of minerals in the soil (De Pascale et al., 2017). The BS increases the availability of insoluble elements such as iron (Fe) to plants by forming complexes with insoluble materials.

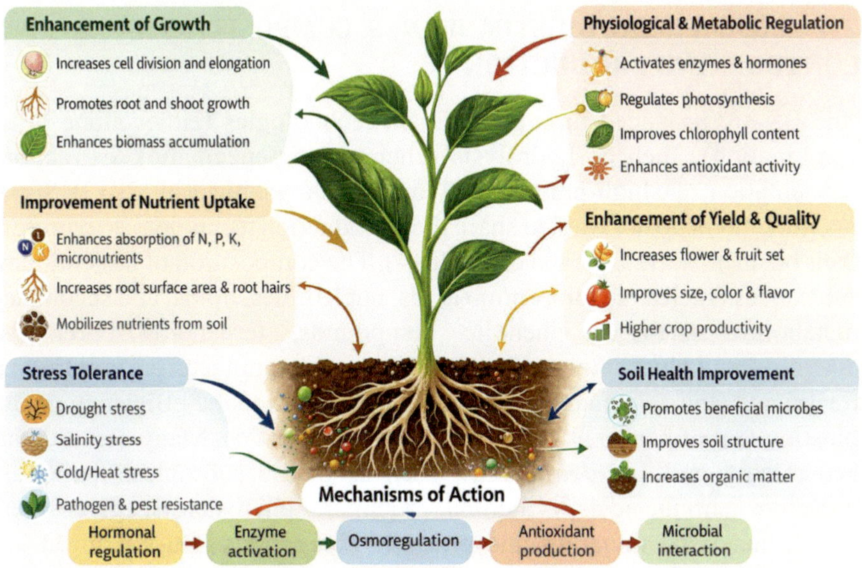

FIGURE 2.3 Functions of plant biostimulants

Source: Created with BioRender.com.

The BS, in the form of humic materials, increases the availability and uptake of nutrients by lowering the pH of the soil and root surface through an increase in plasma membrane H^+-ATPase activity, which releases H^+ into the soil (Canellas and Olivares, 2014). As discussed in the earlier sections, PHs are an important category of BS that increase plant growth. Fish PHs are a combination of amino acids, oligopeptides, and polypeptides that can be applied near the root of a plant in the soil or used as foliar sprays (Colla et al., 2015). The PHs also stimulated the growth of soil microorganisms that can utilize PHs as sources of nitrogen and carbon. PHs also play a role in the chelation of minerals in the soil so that they can be more easily accessible to plants (Paradikovi et al., 2011). Trevisan et al. (2017) reported a significant increase in the dry weight of the root and the ratio of root to shoot due to transcriptional changes in the root of maize seedlings. Trevisan et al. (2019) also discussed the biostimulant APR® mode of action on maize seedlings under abiotic stress (such as hypoxia, salt, and nutrient deprivation stresses). Similarly, Wilson et al. (2015) examined the impact of adding gelatine capsules to the soil close to cucumber seeds. They observed a significant increase in leaf area, weight biomass, and nitrogen content in the seedlings of cucumbers.

2.4 COMMERCIAL BIOSTIMULANTS: COMPOSITION, QUALITY CONTROL, AND PRODUCTION

The BS are produced through different technologies (either alone or in combination), such as hydrolysis (chemical or enzymatic), extraction, fermentation, and high-pressure cell rupture (Du Yakhin et al., 2017). Some manufacturers do not want to share their production technologies to protect their business secrets (Traon et al., 2014). The composition of biostimulant substances (such as sugars, amino acids, nucleotides, lipids, and secondary metabolites such as phenolic compounds, terpenoids/isoprenoids, alkaloids, and glucosinolates) may differ due to manufacturing conditions, treatments, and raw materials, which also determine the efficiency of the biostimulant (Aharoni et al., 2011; Lötze et al., 2015). Many commercial BS contain multiple compounds, such as humic substances, peptides, bioactive amino acids, vitamins, betaines, phytosterols, glycosides, lipids, beneficial minerals, and phenolic compounds (Du Yakhin et al., 2017). The efficiency and quality of BS are analyzed using different methods like SDS-PAGE, GC/MS, FTIR, bioassays, amino acid analysis, high-performance liquid chromatography and mass spectrometry (HPLC/MS), and ELISA (Sharma et al., 2012).

The PBs derived from fish PH are either produced in liquid form (90% moisture content) or in dry or solid form. The solid form of fish PHs as BS is more preferred due to long-term storage, easy transportation, and storage (Petrova et al., 2018). The procedure for the manufacture of fish PHs as BS starts with the collection of raw materials (fish or byproducts) from the processing industry (Figure 2.4), followed by the hydrolysis of the raw materials to extract low-molecular-weight peptides or individual amino acids. The temperature and time of hydrolysis depend on the approach of hydrolysis, the raw material, the nature of the final product, and its applications. The PH fraction is purified by centrifugation and filtration, and the oil fraction is removed or reduced to 0.5% to decrease the oxidation due to fat in the final products (He et al., 2013). The bioactive component from

FIGURE 2.4 Flow diagram of the production of protein hydrolysate from fish and fish byproducts.

fish PH was obtained through membrane filtration techniques followed by drying using different drying methods such as spray drying, freeze drying, and roller drum drying (Abejón et al., 2016; Petrova et al., 2018). The choice of drying method depends on the availability of equipment, energy, and costs. The fish PH is packed and stored at 4°C before use.

2.5 FISH HYDROLYSATES AS PLANT BIOSTIMULANTS

The PHs from fish byproducts are potential sources of proteins, omega-3 fatty acids (EPA and DHA), minerals, and vitamins. Currently, fish byproducts are used for the production of low-value products such as fish feed and silage, depending on the storage and handling methods of the byproducts. The PHs from fish byproducts show antioxidant activity, antimicrobial activity, antihypertensive activity, hemolytic activity, and metal chelating activity (Rajendran et al., 2018; Tang et al., 2015; Vázquez et al., 2019) (Table 2.1). Due to these properties, fish PHs as BS can be used in organic production systems like other commercially available PHs. The PH derived from fish acts as a biostimulant to increase the number of leaves, stem diameter, mass of shoots and roots, and the variety of plant' succulence. Xu et al. (2017) reported the use of fish-derived PH as a biostimulant to increase the photosynthesis and chlorophyll content for sustainable production of lettuce (*Lactuca sativa* L.) and improve the soil properties. He also observed that soil content was higher in C/N, increased in K and Fe, increased water holding capacity, increased CEC, increased electrical conductivity, a lower pH of the soil, and lower P, Mg, Cu, Mn, B, and Na in the soil after the harvesting of lettuce treated with biostimulant.

The biostimulant comprises PHs, an increased rate of photosynthesis, the content of chlorophyll in the leaf, stomatal conductance, and transpiration (Xu et al., 2017). In another study, Kumari et al. (2022) reported that PH obtained from pink perch head and viscera hydrolysis using the alcalase enzyme contains a higher amount of arginine, asparagine/aspartate, glutamine/glutamate, glycine, alanine, and proline. According to previous literature, PH was used to increase the growth of several plants, such as lettuce, winter wheat, tomatoes, corn, and peppers (Colla et al., 2014; Ertani et al., 2014; Popko et al., 2018). The biostimulant in the form of PHs increases the uptake of amino acids by plants, which aids in saving energy for amino acid synthesis and increases the rate of reconstruction during transplantation, flowering periods, plant diseases, and times of climatic stress (Bulgari et al., 2015).

TABLE 2.1 Protein Hydrolysate Extracted from Fish and Fish by Products

Name of Fish	Method of Hydrolysis	Constituent and Functions	References
Pollock	Enzymatic hydrolysis (Alcalase and Flavourzyme)	Peptides showing hormone stimulated growth	Zheng et al. (2013)
Barbel	Enzymatic hydrolysis (Alcalase)	Peptides having antimicrobials or antibiotic adjuvants protect plants	Sila et al. (2014)
Anchovy	Enzymatic hydrolysis	Peptides having antibacterial activity to protect plants	Wang et al. (2019)
Pollock	Chemical hydrolysis (Formic acids)	Peptides having immune-modulatory activity to increases the growth of plants	Liang et al. (2006)
Pacific hake	Enzymatic hydrolysis (Flavourzyme)	Peptides and amino acids having cryoprotectant activity	Jenkelunas and Li-Chan (2018)
Catshark	Enzymatic hydrolysis (Flavourzyme)	Peptides having emulsifying properties	García-Moreno et al. (2016)
Shark Capelin	Enzymatic hydrolysis (Alcalase)	Peptides having foaming properties	Shahidi et al. (1995)
Salmon	Enzymatic hydrolysis (Alcalase, Flavourzyme, Corolase)	Peptides having water binding properties	Kristinsson and Rasco (2000)
Sardine, Tuna	Enzymatic hydrolysis (Alcalase, Neutrase, Papain and Pepsin)	Peptides and amino acids having antioxidant activity	Pezeshk et al. (2017)
Tuna	Enzymatic hydrolysis (Alcalase, Neutrase, Protamex)	Peptides having antihypertensive ACE-inhibitory activity	Je et al. (2009)
Yellow fin tuna	Enzymatic hydrolysis (Protamex)	Peptides having antimicrobial properties	Pezeshk et al. (2019)
Lizard Fish	Enzymatic hydrolysis (Papain)	Peptides having antianemic activity	Dong et al. (2005)
Cod and Saithe fish	Enzymatic hydrolysis (Protamex)	Peptides having ACE inhibitory activity	Ween et al. (2017)

Some amino acids in PHs act as chelating agents for minerals, which increase metal ion absorption and transportation within the plant and increase plant growth. BS increase plant growth by improving nutrient uptake, increasing soil microbial metabolism, increasing root length, and increasing nutrient metabolism due to enzyme activity (Lucini et al., 2015). Lisiecka et al. (2011) studied and found that animal-derived PH used as a biostimulant may not have a beneficial effect on the growth of strawberry plants, such as a decrease in strawberry number, length, diameter, crown diameter, leaf number, and weight. The growth of plants due to PH used as a biostimulant depends on the methods of application, source of PH, and condition of growth.

Many animal-based PHs have been commercially available on the market. In Ireland, PH from fish is commercially used as a PBs (http://www.cfish.ie/plant-biostimulants/). But on contradictory grounds, Corte et al. (2014) assessed the safety of PHs obtained after hydrolysis of animal sources on plant and yeast models and reported that PHs are safe to use because they do not show any ecotoxicity, genotoxicity, or phytotoxicity. The bioactive properties of fish PHs have been shown to have a positive effect on human health and are used as feed supplements for animals. Therefore, these bioactive PHs, when used as biostimulants, may have a positive effect on the growth of plants. The fish PH can be used in agriculture as a plant biostimulant. According to EU regulation No. 354/2014 with regards to organic production, labelling, and control (http://data.europa.eu/eli/reg_impl/2014/354/oj), PHs from animal sources cannot be used on edible parts of crops, but it can be equally applied to non-edible parts of crops such as seeds, roots, and leaves.

2.6 LEGISLATION AND REGULATION OF BIOSTIMULANTS

2.6.1 PLANT BIOSTIMULANTS REGULATION IN EUROPE AND THE UNITED STATES

There are no specific regulations or frameworks available for BS due to a lack of formal definitions and concepts of acceptance. The legal and regulatory definition of BS was not defined anywhere in the world, including in the European Union and the United States. According to European REACH Regulation (EC) No. 1907/2006 concerning registration, evaluation, authorization, and restriction of chemicals, BS are categorized under substances of variable composition. In February 2014, the American Association of Plant Food Control Officials (AAPFCO) agreed to modify the existing lists of beneficial substances to include BS. In Europe, BS are launched on the market on the basis of national regulations on fertilizer and pesticides. In the European market, all forms of BS are covered under Plant Protection Products (PPPs) of European Commission Regulation No. 1107/2009. All the plant growth regulations and herbicide safety have been registered under PPP regulations. Many companies producing BS introduce their products to the European market by an alternative route, namely the fertilizer route, instead of PPPS due to the lengthy and costly procedure. Regulation (EC) No. 2003/2003 on fertilizer is not applicable to BS because Article 2 of this regulation says that fertilizer's main function is to provide nutrients (such as nitrogen, phosphorous, and potassium), secondary nutrients (such as calcium, magnesium, sodium, and sulfur), and micronutrients (boron, cobalt, copper, iron, manganese, molybdenum, and zinc) to plants. But later, this regulation includes other compounds used as fertilizer additives, that is, nitrification and urease inhibitors. This led to modifications in the European fertilizer regulation, which was amended by EC No. 1107/2008 to introduce products that do not provide nutrients but are additives to fertilizers, enhancing fertilizer performance. The BS (especially humic acids, seaweed extract, and PH) introduced in the European market according to national fertilizer law must be analyzed for assessment of efficacy, toxicity, and ecotoxicity (La Torre et al., 2015). The need for reformation of fertilizer regulation for harmonization of legislation on BS, fertilizing materials, and additives such as nutrient performance enhancers, organic and organo-mineral fertilizers, soil improvers, and growing media. Similarly, the USA also does not have an approved definition of BS that are launched on the market as per the fertilizer law. Different regulatory bodies and federal agencies, such as the American Association of Plant Food Control officials, the EPA, and the USDA, will work on improving regulations on biostimulant products.

2.6.2 REGULATION OF PLANT BIOSTIMULANTS IN INDIA

In India, regulations on BS are as per the FCO, 1985. BS containing pesticides or similar chemicals are being sold on the market and are regulated under the Insecticides Act, 1968. In India, regulations on BS were initiated in December 2016. According to legislation, the first provisional registration is done to ensure the availability of similar products on the market. A two-year buffer period was given before the grant of inclusion in Schedule VI of FCO, 1985.

The state government grants permission to sell BS substances according to the FCO, 1985, and provides a certificate in Form G-2. BS do not include pesticides or plant growth regulators (PGRs), which are regulated under the

Insecticides Act, 1968 (46 of 1968). The state government grants certificates for the following substances as BS: botanical extracts, biochemicals, protein hydrolysates, vitamins, cell-free microbial products, antioxidants, antitranspirants, humic and fulvic acids, and their derivatives, according to the Gazette of India, issued on February 23, 2021. BS and similar substances are tested for bioefficacy, acute toxicity (oral toxicity in rats, dermal toxicity in rats, inhalation toxicity in rats, primary skin irritation in rabbits, and eye irritation in rabbits), and ecotoxicity (e.g., effects on birds, freshwater fish, honeybees, and earthworms). According to the Gazette of India issued by the Government of India on February 23, 2021, biostimulants shall not contain heavy metals above the prescribed limits (Table 2.2).

TABLE 2.2 Permissible Limit of Heavy Metal Contents in Biostimulants in India

SL. No.	Mineral	In mg/kg (maximum)
1	Cadmium (Cd)	5.00
2	Chromium (VI) (Cr)	50.0
3	Copper (Cu)	300.0
4	Zinc (Zn)	1000.0
5	Lead (Pb)	100.0
6	Arsenic (As_2O_3)	10.0

Manufacturers or importers need to show the 10-rupee affidavit on non-judicial stamp paper stating that the biostimulant substance does not contain pesticides beyond the permissible limit of 0.01 ppm. The Central Biostimulant Committee in India is:

- Agriculture commissioner, DoA & FW: chairperson, ex officio
- Director General of Health Services or his authorized representative; member, ex officio
- Deputy Director General (NRM), ICAR, or his authorized representative: member, ex officio
- Deputy director general (crop) of ICAR or his authorized representative: member, ex-officio
- Joint secretary (INM), DoA, & FW: member ex officio.
- Director, Central Fertilizer Quality Control and Training Institute (FQC & TI) FBD: member, ex officio
- Authorized representative of the Indian Toxicology Research Centre (ITRC), Lucknow: member, ex officio
- Deputy Secretary (INM), DoA & FW: member secretary, ex officio.

The issues on which the central biostimulant committee shall advise the central government are:

- Inclusion of a new biostimulant
- Specification of various biostimulants
- Methods of drawing samples and their analysis
- Minimum requirements of the laboratory
- Method of testing biostimulants
- Any other matter referred to the committee by the central government

Andhra Pradesh is the first state to issue guidelines on BS before the notified legislation. The provisional registrations are filed after submission of the G-1 form to the controller of fertilizers (Joint Secretary (INM), DAC & FW, Krish Bhavan); after that, the state government issues the G-2 form within 15–20 days of receiving the application from the manufacturer. Then, the manufacturer submits the completed Form G-2, which is thoroughly scrutinized by the Controller of Fertilizers following the standard procedures, and the certificate of provisional registration under Clause 20C of the FCO is issued to the manufacturer in Form G-3. The procedure and type of form filled out for provisional registration (Figures 2.5–2.8).

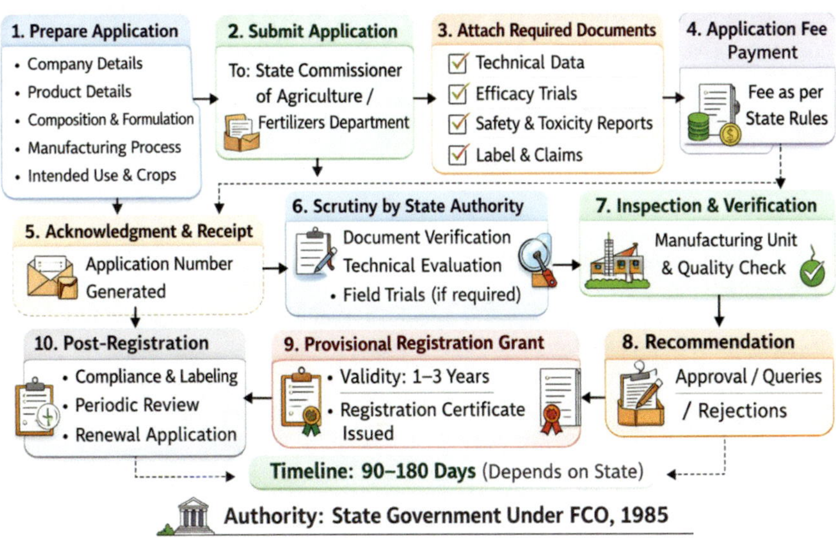

FIGURE 2.5 Procedure to file the provisional registration for biostimulants with the state government under the Fertilizer (Control) Order, 1985.

Source: Created with BioRender.com.

FORM- G

[See clause 20C(3)]

APPLICATION FOR INCLUSION OF BIOSTIMULANTS IN SCHEDULE VI

To,

The Controller of Fertiliser

(i) Name of manufacturer along with address of unit where it is located:

(ii) Address of the Registered office of the company:

(iii) Location and full address of storage / godown(s):

(iv) Name and address of the applicant along with his designation, contact number and email ID:

(v) PAN and Adhaar number of the authorized person:

(vi) Name of biostimulant (indicate the category of biostimulant as specified in sub-clause (2) of clause 20C):

(vii) Brand Name:

(viii) In case of importer, following additional information shall be provided:

(a) country of origin:

(b) is biostimulant registered under country of origin (provide the copy):

(c) whether the agronomic efficacy trials and toxicity trials are conducted in the country of origin:

(ix) Technical details of product:

(x) Labels and leaflet details:

(xi) Quantity manufactured and sold during last three years (attach sale document):

Verification:-

I do hereby solemnly verify that to the best of my knowledge and belief the statement given above is correct and complete.

I clearly understand that the certificate of registration is liable to be canceled if any information or data, submitted with application is found incorrect or false at any stage.

Signature of the applicant

FIGURE 2.6 Form for application for inclusion of biostimulant in Schedule VI of Fertilizer Control Order (1985)

Source: https://acfiindia.com/home/central-notice/25.pdf.

FORM- G-1

[See clause 20C(5)]

APPLICATION FOR PROVISIONAL REGISTRATION OF BIOSTIMULANTS

To,

The Controller of Fertiliser

(i) Name of manufacturer along with address of unit where it is located:

(ii) Address of the Registered office of the company:

(iii) Location and full address of storage / godown(s):

(iv) Name and address of the applicant along with his designation, contact number and email ID:

(v) PAN and Adhaar number of the authorized person:

(vi) Name of biostimulant (indicate the category of biostimulant as specified in sub-clause (2) of clause 20C):

(vii) Brand Name:

(viii) In case of importer, following additional information shall be provided:

(a) country of origin:

(b) is biostimulant registered under country of origin (provide the copy):

(c) whether the agronomic efficacy trials and toxicity trials are conducted in the country of origin:

(ix) Technical details of product:

(x) Labels and leaflet details:

(xi) Quantity manufactured and sold during last three years (attach sale document):

(xii) Date and year when introduced in market:

(xiii) State(s) in which product is sold:

Verification:-

I do hereby solemnly verify that to the best of my knowledge and belief the statement given above is correct and complete.

I clearly understand that the certificate of registration is liable to be canceled if any information or data, submitted with application is found incorrect or false at any stage.

Signature of the applicant

FIGURE 2.7 Form for application for provisional registration of biostimulant.

FORM- G-2

[See clause 20C(5)(b)]

(To be issued by the State Government(s))

This is to certify that M/s is selling biostimulant (Name) brand name prior to the (date of publication of notification) and during last three years, ----- metric ton/ litre of biostimulant were sold in the state.

This is further certified that no complaint has so far been reported of ill effects or hazardous effects of the said biostimulant on humans as well as on animals or environment.

Director of Agriculture

FORM G-3

[See clause 20C(6)]

(To be issued by Controller)

Certificate of Provisional Registration

M/s....... (name) and (address) is hereby given the Certificate of Provisional Registration to manufacture the biostimulant (Name) under sub-clause (6) of clause 20C of the Fertiliser (Inorganic, Organic or Mixed) (Control) Order, 1985, details of which are as under:—

1. Name of the bio-stimulant:
2. Brand Name of the bio-stimulant:
3. Location of manufacturing unit:
4. Place of storage:
5. Certificate Registration Number:

Book No. and Page No.

The provisional registration shall be valid upto.....day of August, 2022.

The provisional registration shall be subject to the following terms and conditions:—

1. The holder shall manufacture only the biostimulant for which the registration is provided.
2. The heavy metal content and Pesticide content shall not exceed the limit prescribed under the Order.
3. The holder shall comply with the provisions of the said Order and notification issued thereunder.
4. For selling of biostimulant, the authorization letter shall be obtained from the Notified Authority of the concerned State Government.
5. Any information provided in Form G-1 if found incorrect, or on violation of any terms and conditions, the Controller of Fertiliser shall cancel the provisional registration.

(Controller)

[F.No.3-14/2017 Fert Law]

NEERAJA ADIDAM, Jt. Secy.

FIGURE 2.8 Form issued by the state government for final provisional registration of the biostimulant.

2.7 CONCLUSION

The current trends in agricultural practices are unable to fulfill the demand for food for the growing population of the world. The strict regulation and consumer awareness of the use of chemical fertilizers in agriculture increase the need to find alternative methods for sustainable plant production that are economical and environmentally friendly. Fish byproducts can be used as BS to promote the growth of plants by enhancing nutrient uptake, yields, and product quality. BS is a possible substitute for traditional chemical fertilizers, and demand for biostimulant will increase exponentially in the market in the next few years. The fish-producing industries generate a huge amount of byproduct (approximately 60% of fish weight), which can be used to produce high-value products such as PBs. Fish byproducts can be hydrolyzed chemically or enzymatically to produce PHs, which can be used in agriculture to improve the growth of plants. The regulations on the use of BS are not available due to a lack of knowledge on the mode of action of the BS. In Europe, the USA, and India, the framework for the regulation of BS is provided by the EC, USDA, and FCO, respectively. The BS developed from fish PH can be launched on the market under the Insecticides Act of 1968 in India and fertilizer and pesticide laws in the USA and Europe.

KEYWORDS

- **biostimulants**
- **high-performance liquid chromatography**
- **mass spectrometry**
- **protein hydrolysates**
- **regulatory framework**
- **sustainable agriculture**

REFERENCES

Abejón, R., Abejón, A., Belleville, M. P., Sánchez-Marcano, J., Garea, A., & Irabien, Á. (2016). Multiobjective optimization of membrane networks for fractionation of protein hydrolysate from fish by-products. In *Computer Aided Chemical Engineering, 38*, 415–420.

Aharoni, A., & Galili, G. (2011). Metabolic engineering of the plant primary–secondary metabolism interface. *Current Opinion in Biotechnology, 22*(2), 239–244.

Bulgari, R., Franzoni, G., & Ferrante, A. (2019). Bio stimulants application in horticultural crops under abiotic stress conditions. *Agronomy, 9*(6), 306. https://doi.org/10.3390/agronomy9060306 (accessed on 6th November 2025).

Calvo, P., Nelson, L., & Kloepper, J. W. (2014). Agricultural uses of plant bio stimulants. *Plant and Soil, 383*(1), 3–41.

Canellas, L. P., & Olivares, F. L. (2014). Physiological responses to humic substances as plant growth promoter. *Chemical and Biological Technologies in Agriculture, 1*(1), 1–11.

Chalamaiah, M., Hemalatha, R., & Jyothirmayi, T. (2012). Fish protein hydrolysates: Proximate composition, amino acid composition, antioxidant activities and applications: A review. *Food Chemistry, 135*(4), 3020–3038.

Chen, T. H., & Murata, N. (2011). Glycine betaine protects plants against abiotic stress: Mechanisms and biotechnological applications. *Plant, Cell & Environment, 34*(1), 1–20.

Colla, G., Hoagland, L., Ruzzi, M., Cardarelli, M., Bonini, P., Canaguier, R., & Rouphael, Y. (2017). Bio stimulant action of protein hydrolysates: Unraveling their effects on plant physiology and microbiome. *Frontiers in Plant Science, 8*, 2202. https://doi.org/10.3389/fpls.2017.02202 (accessed on 6th November 2025).

Colla, G., Nardi, S., Cardarelli, M., Ertani, A., Lucini, L., Canaguier, R., & Rouphael, Y. (2015). Protein hydrolysates as bio stimulants in horticulture. *Scientia Horticulture, 196*, 28–38.

Colla, G., Rouphael, Y., Canaguier, R., Svecova, E., & Cardarelli, M. (2014). Bio stimulant action of a plant-derived protein hydrolysate produced through enzymatic hydrolysis. *Frontiers in Plant Science, 5*, 448. https://doi.org/10.3389/fpls.2014.00448 (accessed on 6th November 2025).

Corte, L., Dell'Abate, M. T., Magini, A., Migliore, M., Felici, B., Roscini, L., Sardella, R., Tancini, B., Emiliani, C., Cardinali, G., & Benedetti, A. (2014). Assessment of safety and efficiency of nitrogen organic fertilizers from animal-based protein hydrolysates–a laboratory multidisciplinary approach. *Journal of the Science of Food and Agriculture, 94*, 235–245.

Craigie, J. S. (2011). Seaweed extract stimuli in plant science and agriculture. *Journal of Applied Phycology, 23*(3), 371–393.

De Pascale, S., Rouphael, Y., & Colla, G. (2017). Plant bio stimulants: Innovative tool for enhancing plant nutrition in organic farming. *European Journal of Horticultural Science, 82*(6), 277–285.

Dong, Y. L., Sheng, G. Y., Fu, J. M., & Wen, K. W. (2005). Chemical characterization and anti-anemia activity of fish protein hydrolysate from *Saurida elongata*. *Journal of the Science of Food and Agriculture, 85*(12), 2033–2039.

Du Jardin, P. (2015). Plant bio stimulants: Definition, concept, main categories and regulation. *Scientia Horticulture, 196*, 3–14.

Ertani, A., Pizzeghello, D., Francioso, O., Sambo, P., Sanchez-Cortes, S., & Nardi, S. (2014). *Capsicum chinensis* L. growth and nutraceutical properties are enhanced by bio stimulants in a long-term period: Chemical and metabolomic approaches. *Frontiers in Plant Science, 5*, 375.

García-García, A. L., García-Machado, F. J., Borges, A. A., Morales-Sierra, S., Boto, A., & Jiménez-Arias, D. (2020). Pure organic active compounds against abiotic stress: A bio stimulant overview. *Frontiers in Plant Science, 11*, 575829. https://doi.org/10.3389/fpls.2020.575829 (accessed on 6th November 2025).

García-Moreno, P. J., Stephansen, K., van der Kruijs, J., Guadix, A., Guadix, E. M., Chronakis, I. S., & Jacobsen, C. (2016). Encapsulation of fish oil in nanofibers by emulsion electrospinning: Physical characterization and oxidative stability. *Journal of Food Engineering, 183*, 39–49.

Halpern, M., Bar-Tal, A., Ofek, M., Minz, D., Muller, T., & Yermiyahu, U. (2015). The use of bio stimulants for enhancing nutrient uptake. *Advances in Agronomy, 130*, 141–174.

He, S., Franco, C., & Zhang, W. (2013). Functions, applications and production of protein hydrolysates from fish processing co-products (FPCP). *Food Research International, 50*(1), 289–297.

Jafarpour, A., Gregersen, S., Marciel Gomes, R., Marcatili, P., Hegelund Olsen, T., Jacobsen, C., & Sørensen, A. D. M. (2020). Bio functionality of enzymatically derived peptides from codfish (*Gadus morhua*) frame: Bulk *in vitro* properties, quantitative proteomics, and bioinformatic prediction. *Marine Drugs, 18*(12), 599.

Je, J. Y., Lee, K. H., Lee, M. H., & Ahn, C. B. (2009). Antioxidant and antihypertensive protein hydrolysates produced from tuna liver by enzymatic hydrolysis. *Food Research International, 42*(9), 1266–1272.

Jenkelunas, P. J., & Li-Chan, E. C. (2018). Production and assessment of Pacific hake (*Merluccius productus*) hydrolysates as cryoprotectants for frozen fish mince. *Food Chemistry, 239*, 535–543.

Katiyar, D., Hemantaranjan, A., & Singh, B. (2015). Chitosan as a promising natural compound to enhance potential physiological responses in plant: A review. *Indian Journal of Plant Physiology, 20*(1), 1–9.

Kauffman, G. L., Kneivel, D. P., & Watschke, T. L. (2007). Effects of a bio stimulant on the heat tolerance associated with photosynthetic capacity, membrane thermostability, and polyphenol production of perennial ryegrass. *Crop Science, 47*(1), 261–267.

Kumari, A., Kaushik, N., & Slizyte, R. (2022). Production and microencapsulation of protein hydrolysate of pink perch (*Nemipterus japonicus*) by-products obtained from surimi industry for its sustainable utilization. *Waste and Biomass Valorization*, 1–18.

La Torre, A., Battaglia, V., & Caradonia, F. (2016). An overview of the current plant bio stimulant legislations in different European Member States. *Journal of the Science of Food and Agriculture, 96*(3), 727–734.

Liang, M., Wang, J., & Chang, Q., & Mai, K. (2006). Effects of different levels of fish protein hydrolysate in the diet on the nonspecific immunity of Japanese sea bass, *Lateolabrax japonicus* (Cuvier et Valenciennes, 1828). *Aquaculture Research, 37*(1), 102–106.

Lisiecka, J., Knaflewski, M., Spizewski, T., Fraszczak, B., Kaluzewicz, A., & Krzesinski, W. (2011). The effect of animal protein hydrolysate on quantity and quality of strawberry daughter plants cv. 'Elsanta.' *Acta Scientiarum Polonorum Hortorum Cultus, 10*(1), 31–40.

Lötze, E., & Hoffman, E. W. (2016). Nutrient composition and content of various biological active compounds of three South African-based commercial seaweed bio stimulants. *Journal of Applied Phycology, 28*(2), 1379–1386.

Lucini, L., Rouphael, Y., Cardarelli, M., Canaguier, R., Kumar, P., & Colla, G. (2015). The effect of a plant-derived bio stimulant on metabolic profiling and crop performance of lettuce grown under saline conditions. *Scientia Horticulture, 182*, 124–133.

Madende, M., & Hayes, M. (2020). Fish by-product uses as bio stimulants: An overview of the current state of the art, including relevant legislation and regulations within the EU and USA. *Molecules, 25*(5), 1122. https://doi.org/10.3390/molecules25051122 (accessed on 6th November 2025).

MarketsandMarkets. (2023). Bio stimulants market by crop type (cereals & grains, oilseeds & pulses, fruits & vegetables, flowers & ornamentals), active ingredients (humic substances, seaweed extracts), form (dry, liquid), mode of application and region – Global forecast to 2028. Report Code: AGI 2920, Pages: 334. https://www.marketsandmarkets.com/Market-Reports/bio stimulant-market-1081.html (accessed on 6th November 2025).

Meena, K. K., & Kumar, P. (2023). Bio stimulants in agriculture: An Indian perspective. https://www.bio stimulant.com/bio stimulants-in-agriculture-an-indian-perspective/ (accessed on 6th November 2025).

N/A. (2023). India bio stimulants market size, share & COVID-19 impact analysis, by source (microbial and nonmicrobial), by active ingredient (seaweed extracts, humic substances, vitamins & amino acids, microbial amendments, and others), by application (foliar treatment, soil treatment, and seed treatment), by crop type (row crops, fruits & vegetables, turf & ornamentals, and others), and forecast, 2023–2030. Report ID: FBI106785, Pages: 108. https://www.fortunebusinessinsights.com/india-bio stimulants-market-106785 (accessed on 6th November 2025).

Olivares, F. L., Aguiar, N. O., Rosa, R. C. C., & Canellas, L. P. (2015). Substrate biofortification in combination with foliar sprays of plant growth promoting bacteria and humic substances boosts production of organic tomatoes. *Scientia Horticulture, 183*, 100–108.

Parađiković, N., Vinković, T., Vinković Vrček, I., Žuntar, I., Bojić, M., & Medić-Šarić, M. (2011). Effect of natural bio stimulants on yield and nutritional quality: An example of sweet yellow pepper (*Capsicum annuum* L.) plants. *Journal of the Science of Food and Agriculture, 91*(12), 2146–2152.

Petrova, I., Tolstorebrov, I., & Eikevik, T. M. (2018). Production of fish protein hydrolysates step by step: Technological aspects, equipment used, major energy costs and methods of their minimizing. *International Aquatic Research, 10*(3), 223–241.

Pezeshk, S., Ojagh, S. M., Rezaei, M., & Shabanpour, B. (2017). Antioxidant and antibacterial effect of protein hydrolysis of yellowfin tuna waste on flesh quality parameters of minced silver carp. *Journal of Genetic Resources, 3*(2), 103–112.

Popko, M., Michalak, I., Wilk, R., Gramza, M., Chojnacka, K., & Górecki, H. (2018). Effect of the new plant growth bio stimulants based on amino acids on yield and grain quality of winter wheat. *Molecules, 23*(2), 470. https://doi.org/10.3390/molecules23020470 (accessed on 6th November 2025).

Rajendran, C. K., Mohan, S. R., Khiari, A., Udenigwe, Z. C. C., & Mason, B. (2018). Yield, physicochemical, and antioxidant properties of Atlantic salmon visceral hydrolysate: Comparison of lactic acid bacterial fermentation with Flavoenzyme proteolysis and formic acid treatment. *Journal of Food Processing and Preservation, 42*(6), e13620. https://doi.org/10.1111/jfpp.13620 (accessed on 6th November 2025).

Ravichandran, S., Joseph, F. S., Kanagalakshmi, R., & Ramya, M. S. (2012). Variation in nutritive composition of two commercially important. *International Journal of Zoological Research, 8*(1), 43–51.

Rose, M. T., Patti, A. F., Little, K. R., Brown, A. L., Jackson, W. R., & Cavagnaro, T. R. (2014). A meta-analysis and review of plant-growth response to humic substances: Practical implications for agriculture. *Advances in Agronomy, 124*, 37–89.

Shahidi, F., & Onodenalore, A. C. (1995). Water dispersions of myofibrillar proteins from capelin (*Mallotus villosus*). *Food Chemistry, 53*(1), 51–54.

Sharma, S. H. S., Lyons, G., McRoberts, C., McCall, D., Carmichael, E., Andrews, F., & McCormack, R. (2012). Brown seaweed species from Strangford Lough: Compositional analyses of seaweed species and bio stimulant formulations by rapid instrumental methods. *Journal of Applied Phycology, 24*, 1141–1157.

Sila, A., Nedjar-Arroume, N., Hedhili, K., Chataigné, G., Balti, R., Nasri, M., & Bougatef, A. (2014). Antibacterial peptides from barbel muscle protein hydrolysates: Activity against some pathogenic bacteria. *LWT–Food Science and Technology, 55*(1), 183–188.

Tang, W., Zhang, H., Wang, L., Qian, H., & Qi, X. (2015). Targeted separation of antibacterial peptide from protein hydrolysate of anchovy cooking wastewater by equilibrium dialysis. *Food Chemistry, 168*, 115–123.

Tejada, M., Rodríguez-Morgado, B., Gómez, I., Franco-Andreu, L., Benítez, C., & Parrado, J. (2016). Use of biofertilizers obtained from sewage sludges on maize yield. *European Journal of Agronomy, 78*, 13–19.

Tilman, D., Balzer, C., Hill, J., & Befort, B. L. (2011). Global food demand and the sustainable intensification of agriculture. *Proceedings of the National Academy of Sciences, 108*(50), 20260–20264.

Traon, D., Amat, L., Zotz, F., & du Jardin, P. (2014). A legal framework for plant bio stimulants and agronomic fertilizer additives in the EU–Report to the European Commission, DG Enterprise & Industry. https://op.europa.eu/en/publication-detail/-/publication/dbeffd43-98a5-4e39-a930-7dfa21816f8c (accessed on 6th November 2025).

Trevisan, S., Manoli, A., & Quaggiotti, S. (2019). A novel bio stimulant, belonging to protein hydrolysates, mitigates abiotic stress effects on maize seedlings grown in hydroponics. *Agronomy, 9*(1), 28. https://doi.org/10.3390/agronomy9010028 (accessed on 6th November 2025).

Trevisan, S., Manoli, A., Ravazzolo, L., Franceschi, C., & Quaggiotti, S. (2017). mRNA-sequencing analysis reveals transcriptional changes in root of maize seedlings treated with two increasing concentrations of a new bio stimulant. *Journal of Agricultural and Food Chemistry, 65*(46), 9956–9969.

Vázquez, J. A., Meduíña, A., Durán, A. I., Nogueira, M., Fernández-Compás, A., Pérez-Martín, R. I., & Rodríguez-Amado, I. (2019). Production of valuable compounds and bioactive metabolites from by-products of fish discards using chemical processing, enzymatic hydrolysis, and bacterial fermentation. *Marine Drugs, 17*(3), 139. https://doi.org/10.3390/md17030139 (accessed on 6th November 2025).

Wang, J., Wei, R., & Song, R. (2019). Novel antibacterial peptides isolated from the Maillard reaction products of half-fin anchovy (*Setipinna taty*) hydrolysates/glucose and their mode of action in *Escherichia coli*. *Marine Drugs, 17*(1), 47. https://doi.org/10.3390/md17010047 (accessed on 6th November 2025).

Ween, O., Stangeland, J. K., Fylling, T. S., & Aas, G. H. (2017). Nutritional and functional properties of fishmeal produced from fresh by-products of cod (*Gadus morhua* L.) and saithe (*Pollachius virens*). *Helion, 3*(7), e00343. https://doi.org/10.1016/j.heliyon.2017.e00343 (accessed on 6th November 2025).

Wilson, H. T., Xu, K., & Taylor, A. G. (2015). Transcriptome analysis of gelatin seed treatment as a bio stimulant of cucumber plant growth. *The Scientific World Journal, 2015*, 391234. https://doi.org/10.1155/2015/391234 (accessed on 6th November 2025).

Xu, C., & Mou, B. (2017). Drench application of fish-derived protein hydrolysates affects lettuce growth, chlorophyll content, and gas exchange. *Hort Technology, 27*(4), 539–543.

Xu, L., & Geelen, D. (2018). Developing bio stimulants from agro-food and industrial by-products. *Frontiers in Plant Science, 9*, 1567. https://doi.org/10.3389/fpls.2018.01567 (accessed on 6th November 2025).

Yakhin, O. I., Lubyanov, A. A., Yakhin, I. A., & Brown, P. H. (2017). Bio stimulants in plant science: A global perspective. *Frontiers in Plant Science, 7*, 2049. https://doi.org/10.3389/fpls.2016.02049 (accessed on 6th November 2025).

Zheng, K., Liang, M., Yao, H., Wang, J., & Chang, Q. (2013). Effect of size-fractionated fish protein hydrolysate on growth and feed utilization of turbot (*Scophthalmus maximus* L.). *Aquaculture Research, 44*(6), 895–902. https://doi.org/10.1111/j.1365-2109.2012.03094.x (accessed on 6th November 2025).

PART II

Plant Biostimulants

CHAPTER 3

Production and Application of Biochar in Carbon Sequestration and Soil Fertility

UMESH PATLE,[1,*] DEEPAK KUMAR,[2,3,*] HARSH GUPTA,[1] EKTA PANDRE,[1] TULIKA PANIGRAHI,[4] VEERENDRA K. PATEL,[1] and VIVEK KUMAR SINGH[1]

[1]*Faculty of Agriculture, Mahatma Gandhi Chitrakoot Vishwavidyalaya, Chitrakoot, Satna, Madhya Pradesh, India*

[2]*R&D Division, Nextnode Bioscience Pvt. Ltd., Gujarat, India*

[3]*Shri Ram Solvent Extraction Pvt. Ltd., Jaspur, Uttarakhand, India*

[4]*Faculty of Agriculture Science and Technology, AKS University, Satna, Madhya Pradesh, India*

*Corresponding author

ABSTRACT

By turning agricultural waste into biochar, a sustainable alternative to burning biomass was offered. Biochar is made by an anaerobic process. When biochar was applied to soil as opposed to parent husks or stover, improvements were seen in the soil's ability to hold water, total organic carbon, cation exchange capacity (CEC), and CO_2 emissions, increase holding on to soil nutrients, and recycle trash, resulting in longer-term carbon sequestration, and the biochar increased soil fertility and crop development. The ways in which biochar improves soil carbon sequestration are examined, including how it functions as a physical

Plant Biostimulants and Growth Regulators in Sustainable Agriculture.
Deepak Kumar, Sanjay K. Singh, Nutan Kaushik, and Mahaveer P. Sharma (Eds.)

DOI: 10.1201/9781779641458-5

barrier to prevent carbon loss, encourages stable soil aggregates, and affects soil microorganisms. Strategies for optimizing the efficiency of biochar through amendments are discussed, along with obstacles and constraints such as changes in the characteristics of biochar and ideal application rates. Biochar can be applied to soil as an amendment or for sequestering carbon. When it comes to using biochar to improve soil fertility and productivity, burning the plants after harvest was a frequent practice among farmers a few decades ago. Despite their ignorance of the benefits of biochar, the farmers believed that they were irradiating insects as well as many illnesses and diseases. By increasing the crop's yield and the fertility and texture of the soil, farmers are thereby indirectly benefiting themselves. All that biochar is carbon in a stable form that may be left in the soil for many years.

3.1 INTRODUCTION

The term "char" is commonly used to describe the solid material left behind after burning carbon-containing substances (Amin et al., 2016). A familiar example is charcoal, which humans have been making since the early days of creating fire. Another type of char is biochar, made from organic materials like wood, plants, or animal waste but produced with limited oxygen, unlike charcoal. Biochar is created through biomass burning at relatively low temperatures, usually below 700°C. Biochar, a term derived from "biological charcoal," refers to a form of carbon-rich, porous material produced through the pyrolysis of organic biomass under controlled conditions. This sustainable and environmentally friendly practice has gained attention due to its potential applications in agriculture, specifically in enhancing soil fertility and mitigating climate change through carbon sequestration (Borel et al., 2020).

Carbon sequestration is a vital process aimed at capturing and storing atmospheric carbon dioxide (CO_2) to mitigate its impact on global warming (Lal et al., 2015). Biochar consists of carbon that is resistant to microbial breakdown, making it a valuable contributor to environmental sustainability by aiding in carbon sequestration. Utilizing crop residue for biochar production provides an eco-friendly alternative to the environmentally harmful practice of open-field crop residue burning (Sharma, 2018). The importance of carbon sequestration cannot be overstated, as excessive carbon emissions contribute significantly to climate change. Agricultural practices, often associated with soil degradation and carbon loss, can be transformed through the incorporation of biochar. This process not only aids in climate change mitigation by sequestering carbon but also enhances soil fertility, offering a sustainable solution to address challenges faced by modern agriculture (Hussain et al., 2021; Rodrigues et al., 2023).

The significance of biochar and carbon sequestration in soil fertility lies in their multifaceted impact on agricultural ecosystems. Biochar acts as a soil amendment, improving water retention, nutrient availability, and microbial activity. Simultaneously, carbon sequestration promotes a healthier environment by reducing greenhouse gas emissions. This integration of biochar and carbon sequestration represents a promising approach to sustainable agriculture, emphasizing the interconnected roles of soil health and environmental stewardship (Khan et al., 2024; Nepal et al., 2023).

3.2 PRODUCTION TECHNIQUES FOR BIOCHAR

The growing interest in using biochar for various purposes has led to an increased focus on its production. Thermochemical conversion, a widely employed technology for biochar production, encompasses various techniques such as pyrolysis, hydrothermal carbonization (HTC), gasification, torrefaction, and hydrothermal liquefaction (Yaashikaa et al., 2020).

3.2.1 PYROLYSIS

Pyrolysis is a thermochemical method that produces biochar, bio-oil, and syngas from biomass (Bridgwater and Peacocke, 2000). It involves heating and breaking down biomass under low oxygen conditions at temperatures between 400°C and 1200°C (Borel et al., 2020). The absence of oxygen ensures biomass heating beyond its stability limit, resulting in robust products like solid residues. The effective temperature range for pyrolysis is typically between 300 and 700°C (Kwaku Armah et al., 2023). The process involves complex reactions in the reacting zone, favoring char formation at lower temperatures and longer residence times (Balogun et al., 2017). Operating variables, such as temperature, significantly impact product efficiency. Biochar, bio-oil, and syngas (collectively CO_2, CO, and H_2) are produced as solid, liquid, and gaseous byproducts, respectively. The choice of pyrolysis type, whether slow or fast, determines the primary output:

biochar or bio-oil. Slow pyrolysis is optimal for biochar production, while fast pyrolysis is more suitable for bio-oil (Oziegbe and Olawole, 2019).

In essence, pyrolysis offers an alternative method to convert biomass into diverse products, with the process being adaptable to achieve specific desired outcomes. Different reactors, such as paddle kilns and fluidized beds, and operating parameters like temperature play crucial roles in determining the yield and composition of biochar during pyrolysis (Cantrell et al., 2012; Wei et al., 2019) (Figure 3.1).

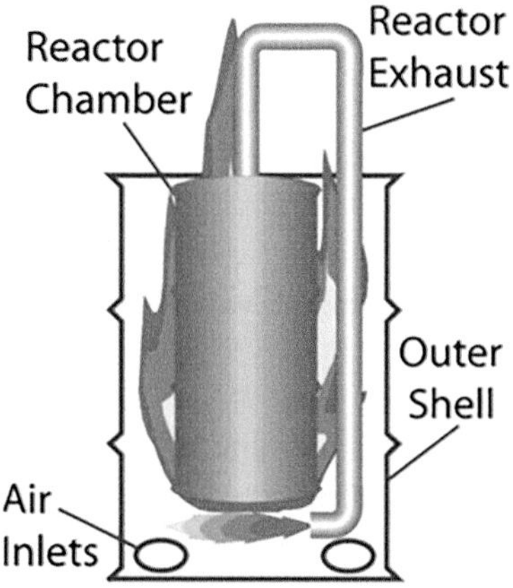

FIGURE 3.1 Schematic representation of pyrolysis process.

3.2.2 DIFFERENT TYPES OF PYROLYSIS PROCESSES

Pyrolysis is a process of thermal decomposition of organic materials at elevated temperatures in the absence of oxygen. This process can be categorized into various types based on different parameters such as the feedstock used, the operating conditions, and the end products obtained. Each type of pyrolysis process has its advantages and limitations, and the choice of method depends on factors such as feedstock availability, desired end products, energy requirements, and economic feasibility (Pahnila et al., 2023; Shah et al., 2023).

3.2.2.1 SLOW PYROLYSIS

In slow pyrolysis, the process takes its time, with a residence period exceeding one hour, and the primary product is biochar (Liu et al., 2015).

This conventional method involves heating biomass between 300°C and 600°C at a moderate rate of 5–7°C/min (Gabhane et al., 2020). The extended residence time allows secondary reactions to occur, leading to the creation of solid carbonaceous biochar. While slow pyrolysis predominantly favors char development, it also produces liquid and gaseous byproducts in moderate quantities, with biochar constituting 35–45%, bio-oil 25–35%, and syngas 20–30% of the final products (Al Arni, 2018).

3.2.2.2 FAST PYROLYSIS

Fast pyrolysis is a swift thermochemical process that converts solid biomass into high-energy liquid bio-oil. It operates without oxygen at temperatures above 500°C and has a rapid heating rate of over 300°C/min, completing biochar generation in just a few seconds. This process yields 60% bio-oil, 20% biochar, and 20% syngas (Dai et al., 2017). Higher temperatures (850–1250°C) with a short residence time (1–10 seconds) have also been reported, resulting in 60%–75% liquid products, 15%–25% biochar, and 10–20% non-condensable gaseous products (Bridgwater et al., 2003).

3.2.2.3 FLASH PYROLYSIS

Flash pyrolysis is an enhanced version of fast pyrolysis, where biomass rapidly decomposes at extremely high temperatures (900–1200°C) within less than one second (Demirbas and Arin, 2002). Despite yielding a high amount of bio-oil due to a combination of high temperature, high heating rate, and short vapor residence time, the biochar yield is reduced compared to fast pyrolysis (Tripathi et al., 2020). This technique, performed in fluidized bed reactors, has limited industrial applicability due to its high temperature.

3.2.2.4 VACUUM PYROLYSIS

Vacuum pyrolysis involves thermal biomass decomposition under low pressure in an isolated oxygen environment (0.5–2 bar, 450–600°C). Similar to slow pyrolysis, it maintains low heating rates. The efficient discharge of vapors through condensation minimizes vapor residence time, reducing secondary reactions. Vacuum or low-pressure extraction plays a crucial role in product quality and yield (Tripathi et al., 2016).

3.2.2.5 INTERMEDIATE PYROLYSIS

Intermediate pyrolysis combines elements of both slow and fast pyrolysis, striking a balance between solid and liquid product yields. Operating at 1 bar pressure, temperatures range from 500 to 650°C, with heating rates between 0.1 and 10°C/min and residence times of 5 to 17 minutes (Yazdani et al., 2019). The finished products typically consist of 40–60% liquid, 20–30% non-condensable gases, and 15–25% biochar.

3.2.2.6 HYDROPYROLYSIS

Hydropyrolysis, a novel technique, involves injecting hydrogen into a high-pressure reactor during biomass conversion (50–200 bar, 350–600°C) (Tripathi et al., 2016). It shares similarities with fast pyrolysis but is not ideal for biochar production as hydrogen acts as a reducing agent, reducing oxygen content in bio-oils. Catalysts are often employed to eliminate oxygen, water, and COX, but developing catalysts for this purpose poses a challenge.

3.2.3 GASIFICATION

The process breaks down carbon-rich materials into gaseous products, including CO, CO_2, CH_4, H_2, and trace hydrocarbons collectively called syngas (Dasappa et al., 2022). Carried out at high temperatures between 700 and 900°C and in an environment with limited oxygen supply, such as nitrogen, steam, carbon dioxide, or a mix of these gases, gasification yields syngas as the primary product, while char becomes a byproduct with a lower yield, accompanied by ash, tar, and some oil (Gao et al., 2023; Maitlo et al., 2022).

Unlike combustion, where energy is released, partial oxidation of biomass during gasification converts the intrinsic chemical energy of carbon into combustible fuel gases, making it a more efficient and convenient utilization of biomass (Maitlo et al., 2022). Commercially, gasification surpasses traditional techniques like pyrolysis, combustion, and fermentation due to lower levelized emissions and higher syngas volume. Achieving high gasification efficiency depends on the O/C ratio, and using biomass with a low O/C ratio, achieved through torrefaction as a pre-treatment, enhances efficiency (Glushkov et al., 2021; Lu et al., 2019).

Gasification and pyrolysis, closely related processes, show no clear separation when combined. The small amount of oxygen used in gasification leads to partial oxidation, altering the final product's characteristics. Gasification results in approximately 85% gaseous products, 10% solid char, and 5% liquid products (Rabea et al., 2023; Tripathi et al., 2016). The gasification process includes several steps, such as drying, pyrolysis, oxidation/combustion, reduction, and cracking. Drying removes biomass moisture, pyrolysis heats biomass with restricted oxygen, and oxidation/combustion reactions create CO_2, CO, and water. Reduction reactions with char and volatiles produce CO, H_2, and other combustible gases. Lastly, cracking during gasification transforms tar gases into non-condensable gases, light hydrocarbons, and unconverted tar (Brar et al., 2012; Kushwah et al., 2022; Nguyen et al., 2021; Shah et al., 2023) (Table 3.1).

TABLE 3.1 Comparison of Typical Operating Conditions and Product Properties of Various Biochar Production Processes

Process	Process Temperature	Residence Time	Solid Product Yield on a Dry Wood Feedstock Basis (mass %)	Carbon Content of the Solid Product (mass %)	Carbon Yield (Mass Carbon, Product/Mass Carbon Feedstock)
Slow pyrolysis	≈400	Minute to days	≈30	95%	≈0.58
Fast pyrolysis	≈500	≈1 s	12–26	74%	0.2–0.26
Gasification	≈800	≈10–20 s	≈10	—	—
HTC	≈180–250	1–12 h	<66%	<70%	≈0.88
Flash carbonization	≈300–600	<30 min	37	≈85%	≈0.65
Torrefaction	≈290	10–60 min	61–84	51–55%	0.67–0.85

Source: Adapted from Li et al. (2023) and Meyer et al. (2011).

3.3 APPLICATIONS OF BIOCHAR IN SOIL HEALTH AND WASTE MANAGEMENT

Biochar, along with bio-oil and gases, is a byproduct derived from the pyrolysis of biomass. The growing adoption of biochar is attributed to its ability to mitigate the adverse environmental effects associated with biomass (Laird et al., 2008; Li et al., 2023; Novotný et al., 2023). Its unique properties make it a valuable soil amendment, offering benefits to both agricultural productivity and environmental sustainability (Zhang et al., 2022). Some of the benefits of applying biochar to soil include (Figure 3.2):

1. **Carbon Sequestration:** Carbon sequestration refers to the process of capturing and storing carbon dioxide (CO_2) from the atmosphere to mitigate its concentration and thereby reduce global warming and climate change (Rodrigues et al., 2023). Carbon sequestration is seen as one of the important strategies to complement efforts to reduce CO_2 emissions and mitigate climate change (Don et al., 2024). Biochar plays a crucial role in carbon sequestration as its stable carbon structure resists microbial degradation. By incorporating biochar into the soil, it helps capture and store carbon for an extended period, mitigating greenhouse gas emissions and contributing to climate change mitigation (Kabir et al., 2023; Wang et al., 2023).
2. **Improved Soil Fertility:** Biochar enhances soil fertility by improving nutrient availability and retention. Its porous structure provides a habitat for beneficial microorganisms, promoting nutrient cycling and increasing the soil's CEC. This results in better availability of essential nutrients for plant growth (Li and Tasnady, 2023). Biochar has a porous structure that can adsorb and retain nutrients such as nitrogen, phosphorus, and potassium, preventing them from leaching away with water runoff. This retention can enhance nutrient availability to plants over time (Rasse et al., 2022). The porous structure of biochar also enhances soil water retention. This can be especially advantageous in drought-prone regions or soils with poor water-holding capacity, as it helps to maintain adequate soil moisture levels for plant growth (Ndede et al., 2022). Moreover, biochar provides a habitat for beneficial microorganisms in the soil, including bacteria and fungi. These microorganisms contribute to nutrient cycling, decomposition of organic matter, and overall soil health, which can lead to improved fertility (Ding et al., 2016; Mukherjee et al., 2022).
3. **Increase of Cation Exchange Capacity (CEC):** Biochar has a high surface area with numerous negatively charged sites. This property increases the soil's CEC, allowing it to retain more essential nutrients like potassium, calcium, and magnesium. Higher CEC enhances nutrient availability for plants (Alkharabsheh et

al., 2021; Kabir et al., 2023). Moreover, biochar plays a significant role in increasing CEC in soils. CEC is a measure of a soil's ability to retain and exchange positively charged ions (cations) such as calcium (Ca^{2+}), magnesium (Mg^{2+}), potassium (K^+), and ammonium (NH_4^+). Biochar contains functional groups such as carboxyl, hydroxyl, and phenolic groups on its surface. These functional groups have a negative charge and can attract and retain positively charged ions, thereby increasing CEC. Overall, the addition of biochar to soils can significantly improve their CEC, thereby enhancing nutrient retention, water holding capacity, and overall soil fertility. This makes biochar a valuable tool in sustainable agriculture and soil management practices (Ambaye et al., 2021; Zhang et al., 2022).

4. **Minimization of Nutrient Leaching:** Nutrient leaching refers to the process by which essential nutrients, such as nitrogen, phosphorus, and potassium, are washed away from the soil due to water movement, eventually reaching groundwater or surface water bodies. This can lead to nutrient depletion in the soil, water pollution, and ecosystem degradation. Biochar helps reduce nutrient leaching by binding nutrients to its surface, preventing them from being washed away by rainwater. This contributes to improved nutrient retention in the root zone, making nutrients more accessible to plants (Sharma, 2024). Biochar has a high surface area and a porous structure, which enable it to adsorb and retain nutrients within its matrix. When nutrients are applied to soil amended with biochar, they can bind to its surface, reducing the risk of leaching (Kabir et al., 2023; Wang et al., 2022, 2023).

5. **Fostering Beneficial Microbes:** The porous structure of biochar provides a habitat for beneficial soil microorganisms. These microbes contribute to nutrient cycling, disease suppression, and overall soil health. Biochar promotes a thriving microbial community, enhancing the soil's biological activity (Dai et al., 2021). Biochar provides a porous and stable physical habitat for microbes to colonize. Its high surface area and pore structure create a favorable environment for microbial attachment and growth (Bolan et al., 2023). Biochar serves as a stable carbon source for soil microorganisms. Microbes utilize carbon as an

energy and nutrient source, and the slow-release nature of carbon from biochar can support microbial growth over extended periods. Studies have shown that biochar-amended soils often exhibit increased microbial diversity (Giagnoni and Renella, 2022). This diversity is essential for the functioning of soil ecosystems, as different microbial species perform various functions such as nutrient cycling, disease suppression, and organic matter decomposition. This can lead to increased nutrient availability through processes like mineralization and organic matter decomposition (Osman et al., 2022; Xiang et al., 2022).

6. **Rise in Soil Organic Carbon (SOC) Stock:** Biochar has gained attention for its potential role in enhancing soil organic carbon (SOC) stocks. When incorporated into soil, biochar can persist for hundreds to thousands of years, acting as a stable carbon sink (Kabir et al., 2023). Biochar is a stable form of carbon that persists in the soil for an extended period. Its addition increases the soil's organic carbon content, contributing to improved soil fertility and long-term carbon sequestration (Kabir et al., 2023; Wang et al., 2023). However, the effectiveness of biochar in enhancing SOC stocks can vary depending on factors such as feedstock type, pyrolysis conditions, soil properties, and climate. Additionally, the sustainability of biochar production and its potential impacts on land use and biodiversity need to be carefully considered (Bo et al., 2023; Tsolis and Barouchas, 2023).

7. **Crop Yield Enhancement:** The positive effects of biochar on soil structure, nutrient availability, and microbial activity translate into increased crop yields. Biochar-amended soils often show improved plant growth, resilience to stress, and higher productivity (Frimpong et al., 2021; Knoblauch et al., 2021). The impact of various biochar kinds and rates on the yield, growth, and yield components of garden peas in Bule Weda, southern Ethiopia, was investigated by Berihun et al. (2017). Four biochar rates (0, 6, 12, and 18 t ha^{-1}) and two forms of biochar (corncobs and *Lantana camara*) make up the treatments. The findings demonstrated the considerable effects of biochar application on soil bulk density, pH, porosity, and exchangeable acidity. Similarly, the results of Knoblauch et al. (2021) indicated that the application of relatively small amounts of biochar could improve crop yields and the nutrients available to plants in marginal

arable soils. It may also reduce the need for mineral fertilizers by increasing the soil's ability to sequester organic carbon. Application rates of 5–20 t ha^{-1} of biochar have been found to increase the vines and increase the number of leaves, tubers, and weight of the tubers. The incorporation of biochar indicates its capacity to regulate plant-parasitic nematodes in a manner that is dependent on rate (Edussuriya et al., 2023). It is imperative to examine the enduring impacts of biochar on augmenting crop yields in order to ensure food security and foster sustainable agricultural growth. Jiang et al. (2024) examined the structural equation model (SEM) and meta-analysis in order to address this problem. The findings showed that biochar increased crop output considerably and sustainably. Over the course of a two-year monitoring period, biochar also increased SOC on average by 36.2%. On a long-term basis, crop yields grew by an average of 16% following the application of biochar; however, the rise differed among crop types, and the application of biochar worked better on corn and wheat than on paddy rice production. Overall, the application of biochar to agricultural soils offers a range of benefits that can lead to significant enhancements in crop yield, soil fertility, and environmental sustainability. However, the effectiveness of biochar may vary depending on factors such as feedstock, production method, soil type, climate, and crop species, highlighting the importance of site-specific considerations and further research in optimizing its use (Vijay et al., 2021; Ye et al., 2020).

8. **Waste Management:** Waste management through biochar production is an innovative approach that offers both environmental and agricultural benefits. Biochar is obtained from the pyrolysis, or heating, of biomass materials such as agricultural waste, wood chips, or organic residues (Amalina et al., 2022; Tomczyk et al., 2020). This process converts organic waste into a stable form of carbon that can be used to improve soil quality and sequester carbon dioxide from the atmosphere. Biochar offers a sustainable solution for managing agricultural and organic waste. By converting biomass residues into biochar, it not only reduces waste disposal issues but also transforms the waste into a valuable resource for soil improvement (Allohverdi et al., 2021; Mulabagal et al., 2021). Apart from waste management and carbon sequestration, biochar

production can also generate renewable energy in the form of syngas and bio-oil, which can be used for heat and power generation. Overall, waste management through biochar production provides a sustainable solution to both waste disposal and soil degradation issues while contributing to climate change mitigation efforts (Kabir et al., 2023; Li and Tasnady, 2023; Wang et al., 2023).

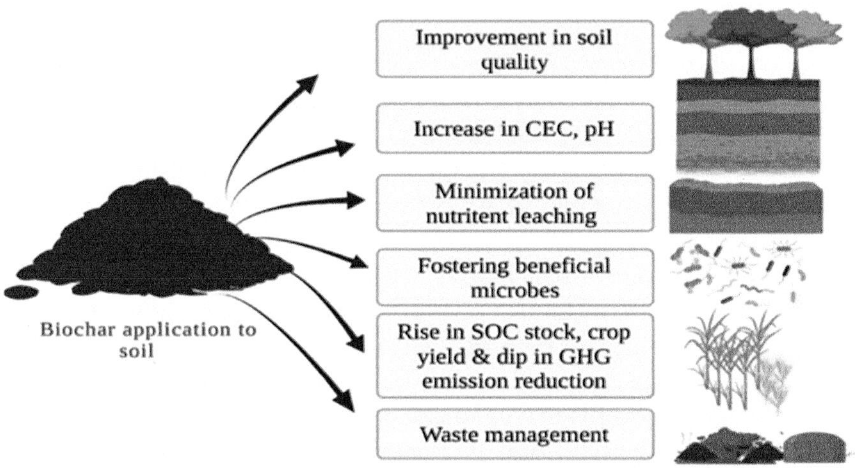

FIGURE 3.2 Applications of biochar in soil health and waste management.

3.4 GLOBAL BIOCHAR MARKET AND REGULATORY FRAMEWORK

As the utilization of biochar expands across various applications, a parallel increase in the number of manufacturers is observed. This growth underscores the necessity for established standards and guidelines to regulate the production of biochar. The estimated value of the global biochar market was USD 220.27 million in 2022, and it is expected to rise at a compound annual growth rate (CAGR) of 11.14% from 2023 to 2032, reaching around USD 633.31 million. The demand for biochar is expected to rise in the global market due to its growing significance in enhancing soil porosity and decreasing soil water loss (Precedence Research, 2023).

Over the projected period of 2023–2032, the worldwide biochar market is expected to have a significant increase. The main factor propelling the biochar market's expansion is thought to be the increase in government spending in the organic agriculture sector. Another factor fueling the market's expansion is growing consumer knowledge of globally available organic yet eco-friendly products as a result of growing environmental concerns. Moreover, the use of biochar in the energy production industries has increased due to the growing significance of renewable energy; in the upcoming years, there is anticipated to be an increase in demand. Biochar is another sustainable method for addressing climate change, energy security, and food security. The biochar business will expand because many countries are concentrating on enacting regulations to lower carbon emissions from agricultural activities (Fortune Business Insights, 2022; Precedence Research, 2023).

Asia Pacific currently holds a dominant geographical position in the global biochar market, and growth in this market is anticipated to be substantial during the predicted period. The biochar industry in the region is expanding thanks to the swift expansion of the agriculture sector in nations like China and India. One of the main factors driving the growth of the biochar market in the region is the increasing amount of research and development being done to create products made of biochar to increase crop production. The second-biggest market for biochar is North America. The North American biochar market is expanding as a result of consumers' growing apprehensions about organic food products. Throughout the projection period, Europe is anticipated to see a change. The region's need for biochar has increased as a result of strict environmental regulations governing agricultural operations (Fortune Business Insights, 2022; Precedence Research, 2023) (Table 3.2). There are many commercial biochar products available in the global markets that are very famous among farmers and used for soil applications and nutrient management (Figure 3.3).

TABLE 3.2 Biochar Market Scope

Biochar Market Dynamics	Details
Base year	2022
Forecast period	2023–2032
Largest market	Asia Pacific
Second largest market	North America
Market size in 2023	USD 244.81 Million
Market size in 2032	USD 633.31 Million
Growth rate from 2023 to 2032	CAGR of 11.14%

TABLE 3.2 *(Continued)*

Biochar Market Dynamics	Details
Biochar market companies	Airex Energy Inc. (Canada); Biochar Supreme (US); Pacific Biochar Corporation (US); Phoenix Energy (US); ProActive Agriculture (US); Novocarbo *(India);* Carbonis GmbH & Co. KG (Germany); Farm2Energy (India); Chardust Ltd. (Kenya); Safi Organics (Kenya); Charline Gmbh (Austria); Cookswell Jikos (Kenya)

Source: Precedence Research (2023).

The regulatory framework surrounding biochar varies from country to country and is often a blend of environmental, agricultural, and waste management regulations. There are key aspects commonly found in regulatory frameworks for biochar:

1. **Environmental Regulations**: These regulations focus on the environmental impact of biochar production, including air and water quality standards, waste disposal regulations, and emissions controls. They may govern aspects such as the handling and treatment of feedstock materials, emissions from pyrolysis or gasification processes, and the disposal or beneficial use of biochar byproducts (Rodriguez et al., 2024).
2. **Agricultural Regulations**: Agriculture-related regulations often address the use of biochar as a soil amendment or fertilizer. They may include guidelines on application rates, nutrient content, and labeling requirements for biochar products intended for agricultural use. These regulations aim to ensure that biochar application practices support soil health, crop productivity, and environmental sustainability (Major, 2010; Rodriguez et al., 2024).
3. **Waste Management Regulations**: Biochar production generates byproducts and residues that may be subject to waste management regulations. These regulations govern the handling, transport, storage, and disposal of biochar byproducts, such as ash and char fines, to prevent environmental contamination and public health risks (Ndirangu et al., 2019; Rodriguez et al., 2024; Saletnik et al., 2019).
4. **Product Standards and Labelling**: Some jurisdictions have established product standards and labeling requirements for biochar to ensure quality, consistency, and safety. These standards may specify physical and chemical properties, such as particle

size, surface area, pH, and contaminant levels, as well as labeling requirements for product identification and consumer information (Alexander Associates, 2019).

5. **Certification and Quality Assurance Programs**: Certification programs and quality assurance schemes may be implemented to verify compliance with regulatory requirements and industry standards. These programs may involve third-party certification bodies or government agencies responsible for monitoring and enforcing compliance with biochar regulations (https://indocert.org/biochar/).

FIGURE 3.3 Some of the selected commercial biochar products available in the global market.

3.5 NEGATIVE IMPACTS AND LIMITATIONS OF BIOCHAR APPLICATIONS

Despite the numerous advantages associated with the application of biochar in agriculture, such as carbon sequestration, reduction of greenhouse gas emissions, and enhanced crop yields, it is essential to acknowledge potential negative impacts and limitations (Hussain et al., 2017). Biochar application can introduce specific adverse effects, including toxicity, retention of heavy metals, and other limitations (Chen et al., 2018; Van Zwieten et al., 2010). The pyrolysis process of biochar production results in the formation of organic pollutants such as PAHs, dioxins, and furans, indicating varying degrees of these contaminants in biochar (Hale et al., 2012). Additionally, as biochar stimulates soil microbial activity, it may contribute to the degradation of naturally occurring PAHs, potentially contaminating the soil and posing a risk to human health (Quilliam et al., 2013).

Beyond organic contaminants, biochar may carry heavy metals like Cu, Cd, Cr, Ni, and Zn, impacting crop growth and human health (Hospido et al., 2005). Although biochar can reduce the bioavailability of heavy metals through processes like adsorption, plant uptake, and food chain transfer, certain instances show an increase in Cu and As levels in soil after biochar application (Ahmad et al., 2014; Beesley et al., 2010). Moreover, biochar-induced pH changes in acidic soils may affect Cu mobility and availability, with potential long-term impacts on soil health (Chen et al., 2018). The aging of biochar due to biotic and abiotic factors may diminish its ability to adsorb toxic compounds like heavy metals over time, raising concerns about potential soil contamination and negative impacts on ecosystem services. It is crucial to note that these negative impacts are often associated with the pyrolysis process, biochar feedstock, and environmental conditions during application. Further research in these aspects is necessary for a more informed and cautious approach to biochar application.

The positive effects of biochar are not uniform across all soil types, with its benefits more pronounced in degraded and less fertile soils, aiding in increasing crop yields (Lehmann et al., 2014). The impact of biochar on agricultural productivity is also influenced by the specific plant target site. For instance, while applying 14 t ha^{-1} of biochar increased the asexual growth of tomato plants, it did not significantly affect fruit yield (Vaccari et al., 2015). Biochar application might even lead to delayed plant flowering (Hol et al., 2017). In soils dominated by nitrification processes, biochar tends to increase soil N_2O emissions due to enhanced nitrification, but in cases where denitrification is the primary pathway for controlling N_2O, biochar application usually reduces N_2O emissions (Liu et al., 2018).

Concerning soil biology, biochar application may interfere with organic matter decomposition, resulting in a reduced abundance of fungal species like *Ascomycetes* and *Basidiomycetes* (Zheng et al., 2016). Weed-related issues may also arise, with biochar applications leading to increased weed growth during lentil cultivation, and repeated applications potentially posing challenges for weed control (Safaei et al., 2018). In conclusion, these identified negative impacts and limitations underscore the need for a cautious and context-specific approach when employing biochar, recognizing that it may not universally serve as a cure-all for achieving sustainable agricultural development.

3.6 FUTURE PROSPECTS OF BIOCHAR APPLICATION

Looking ahead, the future prospects of biochar hinge significantly on understanding how its properties are influenced by the production process. It is crucial to recognize that the type of biochar used in water treatment may differ substantially from the one employed in energy or agriculture applications. The existing body of literature presents diverse findings on the impact of biochar on agriculture, particularly on crop production, as soils exhibit significant variability. The addition of biochar can lead to either an increase or decrease in crop yields, contingent upon factors such as soil type and fertilizer management (Gaskin et al., 2010). Similarly, the chemical interactions between biochar and heavy metal ions exhibit inconsistencies (Abdelhafez et al., 2014). Clearly, the intricate mechanisms governing the interplay among biochar, soil, and plants are critical and yet not fully understood. Therefore, there is a pressing need for more comprehensive research efforts focused on unraveling the properties of biochar and its reciprocal effects on soil and crop responses. This investigation should extend across both field conditions and climate-controlled environments to provide a more nuanced understanding of biochar's applications in diverse settings. Moreover, there are some key areas where biochar could make a significant impact in the future:

1. **Renewable Energy**: The production of biochar often involves the pyrolysis or gasification of biomass, which can also generate bioenergy in the form of syngas or bio-oil. This renewable energy can be used for heat and power generation, further contributing to sustainability and reducing reliance on fossil fuels (Osman et al., 2023).

2. **Climate-Resilient Agriculture**: Biochar-amended soils have been shown to improve crop resilience to drought, heat stress, and other climate-related factors. By enhancing soil water retention and nutrient availability, biochar can help buffer crops against

climate variability and contribute to food security in a changing climate (Ayaz et al., 2021; Huang et al., 2023).

3. **Biodiversity Conservation**: Healthy soils enriched with biochar can support diverse microbial communities and beneficial soil organisms, fostering ecosystem resilience and biodiversity conservation. By promoting soil biodiversity, biochar application can contribute to the restoration of degraded lands and ecosystems (Murtaza et al., 2023).

Overall, the future prospects of biochar application are promising, with potential benefits for agriculture, environmental sustainability, and climate change mitigation. However, further research, technological advancements, and widespread adoption are needed to fully realize its potential and address challenges such as scalability, cost-effectiveness, and environmental impacts.

3.7 CONCLUSION

In conclusion, the utilization of biochar, derived from the anaerobic pyrolysis of agricultural waste, presents a sustainable alternative to biomass burning, offering numerous benefits for soil improvement and environmental sustainability. As a byproduct of the thermochemical conversion process, biochar has gained prominence for its ability to enhance soil fertility, sequester carbon, and mitigate climate change impacts. Production techniques such as pyrolysis and gasification contribute to the creation of biochar with distinct properties, emphasizing the importance of understanding the interplay between biochar, soil, and plants. The multifaceted advantages of biochar application in soil include carbon sequestration, improved fertility, increased CEC, reduced nutrient leaching, and fostering beneficial microbes. Additionally, the rise in SOC stock and enhanced crop yield underscores the potential of biochar as a valuable soil amendment. However, it is crucial to acknowledge the negative impacts and limitations associated with biochar application, including the introduction of organic pollutants and the potential retention of heavy metals.

Looking forward, the future prospects of biochar rely on comprehensive research efforts to unravel its properties and understand the intricate mechanisms influencing soil and crop responses. The varying effects of biochar on different soil types and crops necessitate context-specific approaches. As biochar applications expand across diverse sectors, including agriculture, water treatment, and energy, the establishment of standardized guidelines becomes imperative to regulate its production. This careful and informed approach is essential to harnessing the full potential of biochar while mitigating any potential adverse effects, ensuring its role as a sustainable solution for agricultural development and environmental stewardship.

KEYWORDS

- **carbon sequestration**
- **cation exchange capacity**
- **hydrothermal carbonization**
- **soil organic carbon**
- **structural equation model**

REFERENCES

Abdelhafez, A. A., Li, J., & Abbas, M. H. (2014). Feasibility of biochar manufactured from organic wastes on the stabilization of heavy metals in a metal smelter contaminated soil. *Chemosphere, 117*, 66–71.

Ahmad, M., Rajapaksha, A. U., Lim, J. E., Zhang, M., Bolan, N., Mohan, D., Vithanage, M., Lee, S. S., & Ok, Y. S. (2014). Biochar as a sorbent for contaminant management in soil and water: A review. *Chemosphere, 99*, 19–33. https://doi.org/10.1016/j.chemosphere.2013.10.071 (accessed on 6th November 2025).

Al Arni, S. (2018). Comparison of slow and fast pyrolysis for converting biomass into fuel. *Renewable Energy, 124*, 197–201.

Alkharabsheh, H. M., Seleiman, M. F., Battaglia, M. L., Shami, A., Jalal, R. S., Alhammad, B. A., Almutairi, K. F., & Al-Saif, A. M. (2021). Biochar and its broad impacts in soil quality and fertility, nutrient leaching and crop productivity: A review. *Agronomy, 11*, 993. https://doi.org/10.3390/agronomy11050993 (accessed on 6th November 2025).

Allohverdi, T., Mohanty, A. K., Roy, P., & Misra, M. (2021). A review on current status of biochar uses in agriculture. *Molecules, 26*(18), 5584. https://doi.org/10.3390/molecules26185584 (accessed on 6th November 2025).

Amalina, F., Syukor, A., Razak, A., Krishnan, S., Sulaiman, H., Zularisam, A. W., & Nasrullah, M. (2022). Biochar production techniques utilizing biomass waste-derived materials and environmental applications–A review. *Journal of Hazardous Materials Advances, 7*, 100134. https://doi.org/10.1016/j.hazadv.2022.100134 (accessed on 6th November 2025).

Ambaye, T. G., Vaccari, M., van Hullebusch, E. D., Amrane, A., & Rtimi, S. (2021). Mechanisms and adsorption capacities of biochar for the removal of organic and inorganic pollutants from industrial wastewater. *International Journal of Environmental Science and Technology, 18*, 3273–3294. https://doi.org/10.1007/s13762-020-03060-w (accessed on 6th November 2025).

Amin, F. R., Huang, Y., He, Y., Zhang, R., Liu, G., & Chen, C. (2016). Biochar applications and modern techniques for characterization. *Clean Technologies and Environmental Policy, 18*(5), 1457–1473.

Ayaz, M., Feizienė, D., Tilvikienė, V., Akhtar, K., Stulpinaitė, U., & Iqbal, R. (2021). Biochar role in the sustainability of agriculture and environment. *Sustainability, 13*(3), 1330. https://doi.org/10.3390/su13031330 (accessed on 6th November 2025).

Balogun, A. O., Sotoudehniakarani, F., & McDonald, A. G. (2017). Thermo-kinetic, spectroscopic study of brewer's spent grains and characterization of their pyrolysis products. *Journal of Analytical and Applied Pyrolysis, 127*, 8–16.

Beesley, L., Moreno-Jiménez, E., & Gomez-Eyles, J. L. (2010). Effects of biochar and green waste compost amendments on mobility, bioavailability and toxicity of inorganic and organic contaminants in a multi-element polluted soil. *Environmental Pollution, 158*, 2282–2287. https://doi.org/10.1016/j.envpol.2010.02.003 (accessed on 6th November 2025).

Berihun, T., Tolosa, S., Tadele, M., & Kebede, F. (2017). Effect of biochar application on growth of garden pea (*Pisum sativum* L.) in acidic soils of Bule Woreda Gedeo Zone Southern Ethiopia. *2017*, 6827323. https://doi.org/10.1155/2017/6827323 (accessed on 6th November 2025).

Bo, X., Zhang, Z., Wang, J., Guo, S., Li, Z., Lin, H., Huang, Y., Han, Z., Kuzyakov, Y., & Zou, J. (2023). Benefits and limitations of biochar for climate-smart agriculture: A review and case study from China. *Biochar, 5*, 77. https://doi.org/10.1007/s42773-023-00279-x (accessed on 6th November 2025).

Bolan, S., Hou, D., Wang, L., Hale, L., Egamberdieva, D., Tammeorg, P., Li, R., Wang, B., Xu, J., Wang, T., Sun, H., Padhye, L. P., Wang, H., Siddique, H. M., Rinklebe, J., Kirkham, M. B., & Bolan, N. (2023). The potential of biochar as a microbial carrier for agricultural and environmental applications. *Science of the Total Environment, 886*, 163968. https://doi.org/10.1016/j.scitotenv.2023.163968 (accessed on 6th November 2025).

Borel, L. D., Reis Filho, A. M., Xavier, T. P., Lira, T. S., & Barrozo, M. A. (2020). An investigation on the pyrolysis of the main residue of the brewing industry. *Biomass and Bioenergy, 140*, 105698. https://doi.org/10.1016/j.biombioe.2020.105698 (accessed on 6th November 2025).

Brar, J. S., Singh, K., Wang, J., & Kumar, S. (2012). Cogasification of coal and biomass: A review. *International Journal of Forestry Research*, Article ID: 363058. https://doi.org/10.1155/2012/363058 (accessed on 6th November 2025).

Bridgwater, A. V. (2003). Renewable fuels and chemicals by thermal processing of biomass. *Chemical Engineering Journal, 91*(2–3), 87–102.

Bridgwater, A., & Peacocke, G. (2000). Fast pyrolysis processes for biomass. *Renewable and Sustainable Energy Reviews, 4*(1), 1–73.

Cantrell, K. B., Hunt, P. G., Uchimiya, M., Novak, J. M., & Ro, K. S. (2012). Impact of pyrolysis temperature and manure source on physicochemical characteristics of biochar. *Bioresource Technology, 107*, 419–428.

Chen, D., Liu, X., Bian, R., Cheng, K., Zhang, X., Zheng, J., Joseph, S., Crowley, D., Pan, G., & Li, L. (2018). Effects of biochar on availability and plant uptake of heavy metals—A meta-analysis. *Journal of Environmental Management, 222*, 76–85. https://doi.org/10.1016/j.jenvman.2018.05.004 (accessed on 6th November 2025).

Dai, L., Fan, L., Liu, Y., Ruan, R., Wang, Y., Zhou, Y., Zhao, Y., & Yu, Z. (2017). Production of bio-oil and biochar from soap stock via microwave-assisted co-catalytic fast pyrolysis. *Bioresource Technology, 225*, 1–8. https://doi.org/10.1016/j.biortech.2016.11.017 (accessed on 6th November 2025).

Dai, Z., Xiong, X., Zhu, H., Xu, H., Leng, P., Li, J., Tang, C., & Xu, J. (2021). Association of biochar properties with changes in soil bacterial, fungal and fauna communities and nutrient cycling processes. *Biochar, 3*, 239–254. https://doi.org/10.1007/s42773-021-00099-x (accessed on 6th November 2025).

Dasappa, S., & Shivapuji, A. M. (2022). Biomass gasification: Thermochemical route to energetic bio-chemicals. *Advanced Biofuel Technologies, 12*, 305–332.

Demirbas, A., & Arin, G. (2002). An overview of biomass pyrolysis. *Energy Sources, 24*(5), 471–482.

Ding, Y., Liu, Y., Liu, S., Li, Z., Tan, X., Huang, X., Zeng, G., Zhou, L., & Zheng, B. (2016). Biochar to improve soil fertility: A review. *Agronomy for Sustainable Development, 36*, 36. https://doi.org/10.1007/s13593-016-0372-z (accessed on 6th November 2025).

Don, A., Leifeld, J., Kätterer, T., Martin, M., Pellerin, S., Emde, D., Seitz, D., & Chenu, C. (2024). Carbon sequestration in soils and climate change mitigation—Definitions and pitfalls. *Global Change Biology, 30*(1), e16983. https://doi.org/10.1111/gcb.16983 (accessed on 6th November 2025).

Edussuriya, R., Rajapaksha, A. U., Jayasinghe, C., Pathirana, C., & Vithanage, M. (2023). Influence of biochar on growth performances, yield of root and tuber crops and controlling plant-parasitic nematodes. *Biochar, 5*, 68. https://doi.org/10.1007/s42773-023-00261-7 (accessed on 6th November 2025).

Fortune Business Insights. (2022). Biochar market size, share & COVID-19 impact analysis, by technology (pyrolysis and gasification), by application (farming, livestock, power generation, and others) and regional forecasts, 2023–2030. Report ID: FBI100750, Pages: 170. https://www.fortunebusinessinsights.com/industry-reports/biochar-market-100750 (accessed on 6th November 2025).

Frimpong, K. A., Phares, C. A., Boateng, I., Abban-Baidoo, E., & Apuri, L. (2021). One-time application of biochar influenced crop yield across three cropping cycles on tropical sandy loam soil in Ghana. *Helion, 7*(2), e06267. https://doi.org/10.1016/j.heliyon.2021.e06267 (accessed on 6th November 2025).

Gabhane, J. W., Bhange, V. P., Patil, P. D., Bankar, S. T., & Kumar, S. (2020). Recent trends in biochar production methods and its application as a soil health conditioner: A review. *SN Applied Sciences, 2*(7), 1–21.

Gao, Y., Wang, M., Raheem, A., Wang, F., Wei, J., Xu, D., Song, X., Bao, W., Huang, A., Zhang, S., & Zhang, H. (2023). Syngas production from biomass gasification: Influences of feedstock properties, reactor type, and reaction parameters. *ACS Omega, 8*(35), 31620–31631.

Gaskin, J. W., Speir, R. A., Harris, K., Das, K. C., Lee, R. D., Morris, L. A., & Fisher, D. S. (2010). Effect of peanut hull and pine chip biochar on soil nutrients, corn nutrient status, and yield. *Agronomy Journal, 102*(2), 623–633. https://doi.org/10.2134/agronj2009.0083 (accessed on 6th November 2025).

Giagnoni, L., & Renella, G. (2022). Effects of biochar on the C use efficiency of soil microbial communities: Components and mechanisms. *Environments, 9*(11), 138. https://doi.org/10.3390/environments9110138 (accessed on 6th November 2025).

Glushkov, D., Nyashina, G., Shvets, A., Pereira, A., & Ramanathan, A. (2021). Current status of the pyrolysis and gasification mechanism of biomass. *Energies, 14*(22), 7541. https://doi.org/10.3390/en14227541 (accessed on 6th November 2025).

Hale, S. E., Lehmann, J., Rutherford, D., Zimmerman, A. R., Bachmann, R. T., Shitumbanuma, V., O'Toole, A., Sundqvist, K. L., Arp, H. P. H., & Cornelissen, G. (2012). Quantifying the total and bioavailable polycyclic aromatic hydrocarbons and dioxins in biochar's. *Environmental Science & Technology, 46*, 2830–2838. https://doi.org/10.1021/es203984k (accessed on 6th November 2025).

Hol, W. H. G., Vestergård, M., ten Hooven, F., Duyts, H., van de Voorde, T. F. J., & Bezemer, T. M. (2017). Transient negative biochar effects on plant growth are strongest after microbial species loss. *Soil Biology and Biochemistry, 115*, 442–451. https://doi.org/10.1016/j.soilbio.2017.09.016 (accessed on 6th November 2025).

Hospido, A., Moreira, T., Martín, M., Rigola, M., & Feijoo, G. (2005). Environmental evaluation of different treatment processes for sludge from urban wastewater treatments: Anaerobic digestion versus thermal processes. *International Journal of Life Cycle Assessment, 10*, 336–345. https://doi.org/10.1065/lca2005.05.210 (accessed on 6th November 2025).

Huang, Y., Tao, B., Lal, R., Lorenz, K., Jacinthe, P. A., Shrestha, R. K., Bai, X., Singh, M. P., Lindsey, J. E., & Ren, W. (2023). A global synthesis of biochar's sustainability in climate-smart agriculture–Evidence from field and laboratory experiments. *Renewable and Sustainable Energy Reviews, 172*, 113042. https://doi.org/10.1016/j.rser.2022.113042 (accessed on 6th November 2025).

Hussain, M., Farooq, M., Nawaz, A., Al-Sadi, A. M., Solaiman, Z. M., Alghamdi, S. S., Ammara, U., Ok, Y. S., & Siddique, K. H. M. (2017). Biochar for crop production: Potential benefits and risks. *Journal of Soils and Sediments, 17*, 685–716. https://doi.org/10.1007/s11368-016-1360-2 (accessed on 6th November 2025).

Hussain, S., Hussain, S., Guo, R., Sarwar, M., Ren, X., Krstic, D., Aslam, Z., Zulifqar, U., Rauf, A., Hano, C., & El-Esawi, M. A. (2021). Carbon sequestration to avoid soil degradation: A review on the role of conservation tillage. *Plants (Basel), 10*(10), 2001. https://doi.org/10.3390/plants10102001 (accessed on 6th November 2025).

Jiang, Y., Li, T., Xu, X., Sun, J., Pan, G., & Cheng, K. (2024). A global assessment of the long-term effects of biochar application on crop yield. *Current Research in Environmental Sustainability, 7*, 100247. https://doi.org/10.1016/j.crsust.2024.100247 (accessed on 6th November 2025).

Kabir, E., Kim, K. H., & Kwon, E. E. (2023). Biochar as a tool for the improvement of soil and environment. *Frontiers in Environmental Science, 11*, 1324533. https://doi.org/10.3389/fenvs.2023.1324533 (accessed on 6th November 2025).

Khan, S., Irshad, S., Mehmood, K., Hasnain, Z., Nawaz, M., Rais, A., Gul, S., Wahid, M. A., Hashem, A., Abd_Allah, E. F., & Ibrar, D. (2024). Biochar production and characteristics, its impacts on soil health, crop production, and yield enhancement: A review. *Plants, 3*, 166. https://doi.org/10.3390/plants13020166 (accessed on 6th November 2025).

Knoblauch, C., Priyadarshani, S. H. R., Haefele, S. M., Schröder, N., & Pfeiffer, E. M. (2021). Impact of biochar on nutrient supply, crop yield and microbial respiration on sandy soils of northern Germany. *European Journal of Soil Science, 72*(4), 1885–1901. https://doi.org/10.1111/ejss.13088 (accessed on 6th November 2025).

Kushwah, A., Reina, T. R., & Short, M. (2022). Modelling approaches for biomass gasifiers: A comprehensive overview. *Science of the Total Environment, 834*, 155243. https://doi.org/10.1016/j.scitotenv.2022.155243 (accessed on 6th November 2025).

Kwaku Armah, E., Chetty, M., Adebisi Adedeji, J., Erwin Estrice, D., Mutsvene, B., Singh, N., & Tshemese, Z. (2023). Biochar: Production, application and the future. *IntechOpen*, 1–26. https://www.intechopen.com/chapters/82129 (accessed on 6th November 2025).

Laird, D. A. (2008). The charcoal vision: A win–win–win scenario for simultaneously producing bioenergy, permanently sequestering carbon, while improving soil and water quality. *Agronomy Journal, 100*(1), 178–181.

Lal, R., Wakene, N., & Klaus, L. (2015). Carbon sequestration in soil. *Current Opinion in Environmental Sustainability, 15*, 79–86.

Lehmann, J., Abiven, S., Kleber, M., Pan, G., Singh, B. P., Sohi, S. P., & Zimmerman, A. R. (2014). Persistence of biochar in soil. In *Biochar for Environmental Management: Science, Technology and Implementation* (2nd ed., pp. 233–280).

Li, S., & Tasnady, D. (2023). Biochar for soil carbon sequestration: Current knowledge, mechanisms, and future perspectives. *C, 9*, 67. https://doi.org/10.3390/c9030067 (accessed on 6th November 2025).

Li, Y., Gupta, R., Zhang, Q., & You, S. (2023). Review of biochar production via crop residue pyrolysis: Development and perspectives. *Bioresource Technology, 369*, 128423. https://doi.org/10.1016/j.biortech.2022.128423 (accessed on 6th November 2025).

Liu, Q., Zhang, Y., Liu, B., Amonette, J. E., Lin, Z., Liu, G., Ambus, P., & Xie, Z. (2018). How does biochar influence soil N cycle? A meta-analysis. *Plant and Soil, 426*, 211–225. https://doi.org/10.1007/s11104-018-3619-4 (accessed on 6th November 2025).

Liu, W. J., Jiang, H., & Yu, H. Q. (2015). Development of biochar-based functional materials: Toward a sustainable platform carbon material. *Chemical Reviews, 115*(22), 12251–12285.

Lu, X., Cao, L., Wang, H., Peng, W., Xing, J., Wang, S., Cai, S., Shen, B., Yang, Q., Nielsen, C. P., & McElroy, M. B. (2019). Gasification of coal and biomass as a net carbon-negative power source for environment-friendly electricity generation in China. *PNAS, 116*(17), 8206–8213. https://doi.org/10.1073/pnas.1812239116 (accessed on 6th November 2025).

Maitlo, G., Ali, I., Mangi, K. H., Ali, S., Maitlo, H. A., Unar, I. N., & Pirzada, A. M. (2022). Thermochemical conversion of biomass for syngas production: Current status and future trends. *Sustainability, 14*(5), 2596. https://doi.org/10.3390/su14052596 (accessed on 6th November 2025).

Major, J. (2010). Guidelines on practical aspects of biochar application to field soil in various soil management systems. *International Biochar Initiative*, 1–23. https://

biochar-international.org/wp-content/uploads/2018/04/IBI_Biochar_Application.pdf (accessed on 6th November 2025).

Meyer, S., Glaser, B., & Quicker, P. (2011). Technical, economical, and climate-related aspects of biochar production technologies: A literature review. *Environmental Science & Technology, 45*(22), 9473–9483.

Mukherjee, S., Sarkar, B., Aralappanavar, V. K., Mukhopadhyay, R., Basak, B. B., Srivastava, P., Marchut-Mikołajczyk, O., Bhatnagar, A., Semple, K. T., & Bolan, N. (2022). Biochar–microorganism interactions for organic pollutant remediation: Challenges and perspectives. *Environmental Pollution, 308*, 119609. https://doi.org/10.1016/j.envpol.2022.119609 (accessed on 6th November 2025).

Mulabagal, V., Baah, D. A., Egiebor, N. O., Sajjadi, B., Chen, W. Y., Viticoski, R. L., & Hayworth, J. S. (2021). Biochar from biomass: A strategy for carbon dioxide sequestration, soil amendment, power generation, CO_2 utilization, and removal of perfluoroalkyl and polyfluoroalkyl substances (PFAS) in the environment. In Lackner, M., Sajjadi, B., & Chen, W. Y. (Eds.), *Handbook of Climate Change Mitigation and Adaptation* (pp. 1–64). Springer, New York, NY. https://doi.org/10.1007/978-1-4614-6431-0_80-2 (accessed on 6th November 2025).

Murtaza, G., Ahmed, Z., Eldin, S. M., Ali, B., Bawazeer, S., Usman, M., Iqbal, R., Neupane, D., Ullah, A., Khan, A., Hassan, M. U., Ali, I., & Tariq, A. (2023). Biochar–soil–plant interactions: A cross talk for sustainable agriculture under changing climate. *Frontiers in Environmental Science, 11*, 1059449. https://doi.org/10.3389/fenvs.2023.1059449 (accessed on 6th November 2025).

N/A. (2019). Biochar labelling guidance document. *US Biochar Initiative*, USA, 1–10. https://biochar-us.org/sites/default/files/Biochar%20labeling%20guidance%20 5-23-19%20final.pdf (accessed on 6th November 2025).

N/A. (2023). Biochar market (By technology: Pyrolysis, gasification, others; by feedstock: Woody biomass, agricultural waste, animal manure, others; by application: Agriculture, forestry, electricity generation, others)–Global industry analysis, size, share, growth, trends, regional outlook, and forecast 2023–2032. Report Code: 2613, Pages: 150. https://www.precedenceresearch.com/biochar-market# (accessed on 6th November 2025).

Ndede, E. O., Kurebito, S., Idowu, O., & Tokunari, T., & Jindo, K. (2022). The potential of biochar to enhance the water retention properties of sandy agricultural soils. *Agronomy, 12*(2), 311. https://doi.org/10.3390/agronomy12020311 (accessed on 6th November 2025).

Ndirangu, S. M., Liu, Y., Xu, K., & Song, S. (2019). Risk evaluation of pyrolyzed biochar from multiple wastes. *Journal of Chemistry, 2019*, 4506314. https://doi.org/10.1155/2019/4506314 (accessed on 6th November 2025).

Nepal, J., Ahmad, W., Munsif, F., Khan, A., & Zou, Z. (2023). Advances and prospects of biochar in improving soil fertility, biochemical quality, and environmental applications. *Frontiers in Environmental Science, 11*, 1114752. https://doi.org/10.3389/fenvs.2023.1114752 (accessed on 6th November 2025).

Nguyen, N. M., Alobaid, F., Dieringer, P., & Epple, B. (2021). Biomass-based chemical looping gasification: Overview and recent developments. *Applied Sciences, 11*(15), 7069. https://doi.org/10.3390/app11157069 (accessed on 6th November 2025).

Novotný, M., Marković, M., Raček, J., Šipka, M., Chorazy, T., Tošić, I., & Hlavínek, P. (2023). The use of biochar made from biomass and biosolids as a substrate for green infrastructure: A review. *Sustainable Chemistry and Pharmacy, 32*, 100999. https://doi.org/10.1016/j.scp.2023.100999 (accessed on 6th November 2025).

Oni, A., Oziegbe, O., & Olawole, O. O. (2019). Significance of biochar application to the environment and economy. *Annals of Agricultural Sciences, 64*(2), 222–236.

Osman, A. I., Fawzy, S., Farghali, M., El-Azazy, M., Elgarahy, A. M., Fahim, R. A., Maksoud, M. I. A. A., Ajlan, A. A., Yousry, M., Saleem, Y., & Rooney, D. W. (2022). Biochar for agronomy, animal farming, anaerobic digestion, composting, water treatment, soil remediation, construction, energy storage, and carbon sequestration: A review. *Environmental Chemistry Letters, 20*, 2385–2485. https://doi.org/10.1007/s10311-022-01424-x (accessed on 6th November 2025).

Osman, A., Lai, Z. Y., Farghali, M., Yiin, C. L., Elgarahy, A. M., Hammad, A., Ihara, I., Al-Fatesh, A. S., Rooney, D. W., & Yap, P. S. (2023). Optimizing biomass pathways to bioenergy and biochar application in electricity generation, biodiesel production, and biohydrogen production. *Environmental Chemistry Letters, 21*, 2639–2705. https://doi.org/10.1007/s10311-023-01613-2 (accessed on 6th November 2025).

Oziegbe, O., & Olawole, O. O. (2019). Significance of biochar application to the environment and economy. *Annals of Agricultural Sciences, 64*(2), 222–236. https://doi.org/10.1016/j.aoas.2019.12.006 (accessed on 6th November 2025).

Pahnila, M., Koskela, A., Sulasalmi, P., & Fabritius, T. (2023). A review of pyrolysis technologies and the effect of process parameters on biocarbon properties. *Energies, 16*(19), 6936. https://doi.org/10.3390/en16196936 (accessed on 6th November 2025).

Quilliam, R. S., DeLuca, T. H., & Jones, D. L. (2013). Biochar application reduces nodulation but increases nitrogenase activity in clover. *Plant and Soil, 366*, 83–92. https://doi.org/10.1007/s11104-012-1411-4 (accessed on 6th November 2025).

Rabea, K., Michailos, S., Hughes, K. J., Ingham, D., Ingham, M., & Pourkashanian, M. (2023). Comprehensive process simulation of a biomass-based hydrogen production system through gasification within the BECCS concept in a commercial two-stage fixed bed gasifier. *Energy Conversion and Management, 298*, 117812. https://doi.org/10.1016/j.enconman.2023.117812 (accessed on 6th November 2025).

Rasse, D. P., Weldon, S., Joner, E. J., Joseph, S., Kammann, C. I., Liu, X., O'Toole, A., Pan, G., & Kocatürk-Schumacher, N. P. (2022). Enhancing plant N uptake with biochar-based fertilizers: Limitation of sorption and prospects. *Plant and Soil, 475*, 213–236. https://doi.org/10.1007/s11104-022-05365-w (accessed on 6th November 2025).

Rodrigues, C. I. D., Brito, L. M., & Nunes, L. J. R. (2023). Soil carbon sequestration in the context of climate change mitigation: A review. *Soil Systems, 7*, 64. https://doi.org/10.3390/soilsystems7030064 (accessed on 6th November 2025).

Rodriguez Franco, C., Page-Dumroese, D. S., Pierson, D., Miller, M., & Miles, T. (2024). Policy and regulations for mobile biochar production in the United States of America. *Forests, 15*(1), 192. https://doi.org/10.3390/f15010192 (accessed on 6th November 2025).

Safaei Khorram, M., Fatemi, A., Khan, M. A., Kiefer, R., & Jafarnia, S. (2018). Potential risk of weed outbreak by increasing biochar's application rates in slow-growth legume, lentil (*Lens culinaris* Medik.). *Journal of the Science of Food and Agriculture, 98*, 2080–2088. https://doi.org/10.1002/jsfa.8689 (accessed on 6th November 2025).

Saletnik, B., Zaguła, G., Bajcar, M., Tarapatskyy, M., Bobula, G., & Puchalski, C. (2019). Biochar as a multifunctional component of the environment—A review. *Applied*

Sciences, 9(6), 1139. https://doi.org/10.3390/app9061139 (accessed on 6th November 2025).

Shah, H. H., Amin, M., Iqbal, A., Nadeem, I., Kalin, M., Soomar, A. M., & Galal, A. M. (2023). A review on gasification and pyrolysis of waste plastics. *Frontiers in Chemistry, 10*, 960894. https://doi.org/10.3389/fchem.2022.960894 (accessed on 6th November 2025).

Sharma, P. (2024). Biochar application for sustainable soil erosion control: A review of current research and future perspectives. *Frontiers in Environmental Science, 12*, 1373287. https://doi.org/10.3389/fenvs.2024.1373287 (accessed on 6th November 2025).

Sharma, S. P. (2018). Biochar for carbon sequestration: Bioengineering for sustainable environment. In Azevedo, V., & Barh, D. (Eds.), *Omics Technologies and Bio-Engineering* (pp. 365–385). Academic Press, Elsevier Inc.

Tomczyk, A., Sokołowska, Z., & Boguta, P. (2020). Biochar physicochemical properties: Pyrolysis temperature and feedstock kind effects. *Reviews in Environmental Science and Biotechnology, 19*, 191–215. https://doi.org/10.1007/s11157-020-09523-3 (accessed on 6th November 2025).

Tripathi, M., Bhatnagar, A., Mubarak, N. M., Sahu, J. N., & Ganesan, P. (2020). RSM optimization of microwave pyrolysis parameters to produce OPS char with high yield and large BET surface area. *Fuel, 277*, 118184. https://doi.org/10.1016/j.fuel.2020.118184 (accessed on 6th November 2025).

Tripathi, M., Sahu, J. N., & Ganesan, P. (2016). Effect of process parameters on production of biochar from biomass waste through pyrolysis: A review. *Renewable and Sustainable Energy Reviews, 55*, 467–481.

Tsolis, V., & Barouchas, P. (2023). Biochar as soil amendment: The effect of biochar on soil properties using VIS-NIR diffuse reflectance spectroscopy, biochar aging and soil microbiology—A review. *Land, 12*(8), 1580. https://doi.org/10.3390/land12081580 (accessed on 6th November 2025).

Vaccari, F. P., Maienza, A., Miglietta, F., Baronti, S., Di Lonardo, S., Giagnoni, L., Lagomarsino, A., Pozzi, A., Pusceddu, E., Ranieri, R., Valboa, G., & Genesio, L. (2015). Biochar stimulates plant growth but not fruit yield of processing tomato in a fertile soil. *Agriculture, Ecosystems & Environment, 207*, 163–170. https://doi.org/10.1016/j.agee.2015.04.015 (accessed on 6th November 2025).

Van Zwieten, L., Kimber, S., Morris, S., Chan, K. Y., Downie, A., Rust, J., Joseph, S., & Cowie, A. (2010). Effects of biochar from slow pyrolysis of papermill waste on agronomic performance and soil fertility. *Plant and Soil, 327*, 235–246. https://doi.org/10.1007/s11104-009-0050-x (accessed on 6th November 2025).

Vijay, V., Shreedhar, S., Adlak, K., Payyanad, S., Sreedharan, V., Gopi, G., Sophia van der Voort, T., Malarvizhi, P., Yi, S., Gebert, J., & Aravind, P. (2021). Review of large-scale biochar field-trials for soil amendment and the observed influences on crop yield variations. *Frontiers in Energy Research, 9*, 710766. https://doi.org/10.3389/fenrg.2021.710766 (accessed on 6th November 2025).

Wang, C., Luo, D., Zhang, X., Huang, R., Cao, Y., Liu, G., Zhang, Y., & Wang, H. (2022). Biochar-based slow-release of fertilizers for sustainable agriculture: A mini review. *Environmental Science and Ecotechnology, 10*, 100167. https://doi.org/10.1016/j.ese.2022.100167 (accessed on 6th November 2025).

Wang, L., Deng, J., Yang, X., Hou, R., & Hou, D. (2023). Role of biochar toward carbon neutrality. *Carbon Research, 2*, 2. https://doi.org/10.1007/s44246-023-00035-7 (accessed on 6th November 2025).

Wei, J., Tu, C., Yuan, G., Liu, Y., Bi, D., Xiao, L., Lu, J., Theng, B. K. G., Wang, H., Zhang, L., & Zhang, X. (2019). Assessing the effect of pyrolysis temperature on the molecular properties and copper sorption capacity of a halophyte biochar. *Environmental Pollution, 251*, 56–65.

Xiang, L., Harindintwali, J. D., Wang, F., Redmile-Gordon, M., Chang, S. X., Fu, Y., He, C., Muhoza, B., Brahushi, F., Bolan, N., Jiang, X., Ok, Y. S., Rinklebe, J., Schaeffer, A., Zhu, Y. G., Tiedje, J. M., & Xing, B. (2022). Integrating biochar, bacteria, and plants for sustainable remediation of soils contaminated with organic pollutants. *Environmental Science & Technology, 56*(23), 16546–16566.

Yaashikaa, P., Kumar, P. S., Varjani, S., & Saravanan, A. (2020). A critical review on the biochar production techniques, characterization, stability and applications for circular bioeconomy. *Biotechnology Reports, 28*, e00570. https://doi.org/10.1016/j.btre.2020.e00570 (accessed on 6th November 2025).

Yazdani, E., Hashemabadi, S. H., & Taghizadeh, A. (2019). Study of waste tire pyrolysis in a rotary kiln reactor in a wide range of pyrolysis temperature. *Waste Management, 85*, 195–201. https://doi.org/10.1016/j.wasman.2018.12.020 (accessed on 6th November 2025).

Ye, L., Camps-Arbestain, M., Shen, Q., Lehmann, J., Singh, B., & Sabir, M. (2020). Biochar effects on crop yields with and without fertilizer: A meta-analysis of field studies using separate controls. *Soil Use and Management, 36*(1), 2–18. https://doi.org/10.1111/sum.12546 (accessed on 6th November 2025).

Zhang, P., Duan, W., Peng, H., Pan, B., & Xing, B. (2022). Functional biochar and its balanced design. *ACS Environmental Au, 2*(2), 115–127. https://doi.org/10.1021/acsenvironau.1c00032 (accessed on 6th November 2025).

Zhang, Y., He, M., Wang, L., Yan, J., Zhu, X., Ok, Y. S., Mechtcherine, V., & Tsang, D. C. W. (2022). Biochar as construction materials for achieving carbon neutrality. *Biochar, 4*, 59. https://doi.org/10.1007/s42773-022-00182-x (accessed on 6th November 2025).

Zheng, J., Chen, J., Pan, G., Liu, X., Zhang, X., Li, L., Bian, R., Cheng, K., & Jinwei, Z. (2016). Biochar decreased microbial metabolic quotient and shifted community composition four years after a single incorporation in a slightly acid rice paddy from southwest China. *Science of the Total Environment, 571*, 206–217. https://doi.org/10.1016/j.scitotenv.2016.07.135 (accessed on 6th November 2025).

CHAPTER 4

Valorization of Organic Wastes from Agro-Food and Industrial Byproducts for the Production of Growth-Promoting Biostimulants

AYUSHI SHARMA and NUTAN KAUSHIK*

Amity Food and Agriculture Foundation, Amity University, Noida, Uttar Pradesh, India

*Corresponding author

ABSTRACT

A key objective in agriculture is to find eco-friendly and sustainable techniques to generate fertilizing reagents with biological origins due to the rising demand for food and agricultural yields that are higher quality and more sustainable. Biostimulants (BS) are a class of natural compounds that help to increase plant growth and nutrient uptake while minimizing reliance on chemical fertilizers. Biostimulants derived from byproducts open the way for waste recycling and reduction, benefiting growers, the food sector, registration and distribution firms, and consumers. The following factors must be present for designated byproducts to be chosen for biostimulant valorization: no pesticide residue, low collection and storage costs, adequate supply, and synergy with other valorization pathways. A scientifically oriented biostimulant industry and well-crafted laws pertaining to these substances are predicated on the clarification of the biological underpinnings of biostimulant function. It is suggested that a biostimulant be defined as "a biologically derived formulated product that increases plant productivity due to the unique or emergent qualities of the complex of elements rather than

Plant Biostimulants and Growth Regulators in Sustainable Agriculture.
Deepak Kumar, Sanjay K. Singh, Nutan Kaushik, and Mahaveer P. Sharma (Eds.)

DOI: 10.1201/9781779641458-6

only because known plant growth regulators (PGRs), vital plant nutrients, or plant protection chemicals are present" in order to distinguish them from the existing legislative product categories. In this chapter, all aspects of agro-industrial organic waste valorization for producing plant growth-promoting BS are critically analyzed and presented in regards to raw material source, types, and commercially available BS.

4.1 INTRODUCTION

It is expected that further research and development will increase the number of biostimulants (BS) derived from byproducts as the worldwide market for BS expands. The development of BS for a targeted industry is also necessary due to the global nutritional imbalance. Here, we examine some instances of BS made from agricultural waste and talk about why agricultural biomass is such a useful resource for the development of new agrochemicals. It is important to note that waste streams from the food and agricultural industries are equally significant sources for the creation of BS. Aquaculture waste streams, manures, composts, sewage treatment, vermicompost, and food waste are some of the waste streams that produce BS (Yakhin et al., 2017). Recycling organic waste as biofertilizer has historically been a widespread technique. Around the world, manure is still used to fertilize the land. However, environmental issues arise when local production outpaces local application and storage capabilities, leading to inconsiderate disposal and the ensuing environmental leakage that leads to water contamination, ammonia poisoning, overfertilization, and the infection of human pathogens, amongst other problems (Dominguez and Edwards, 2010). Finding alternate methods of reusing such materials in the agricultural production system is required to prevent environmental damage.

4.2 PRIMARY SOURCES

In this section, examples of BS derived from agricultural byproducts are reviewed, and it is discussed why agricultural biomass is such a useful resource for the development of novel agrochemical products. Large amounts of organic biomass from production or processing processes are a problem for the agricultural and food industries due to the increased demand for nutritious and fresh food. The amount of biomass produced may significantly vary depending on the types of crops and how they are processed.

Additionally, the primary sources and physiological properties of BS are diverse. It also has implications for how BS are categorized. For instance, fish byproducts such as heads, skin, fins, frames, and so forth, account for nearly 60% of the biomass (Chalamaiah et al., 2012). A massive 5.9 Mt of crustacean exoskeletons are produced globally by the fishing industry, with 35–45% of them being abandoned as trash (head and thorax) (Sharp, 2013). Because they come from cells and tissues that are exposed to the outside of the organisms' bodies, which are designed to ward against intruders and diseases, many of these sources are abundant in secondary metabolites. For example, potato peel contains steroidal alkaloids in higher concentrations, which are linked to protection against bacterial, fungal, and insect diseases (reviewed in Fritsch et al., 2017). One metric ton of wet salted hides generates 200 kg of leather and 450–600 kg of solid waste, if not transformed, leaving over 60% of biomass to be disposed of in the environment as waste. This is because to make leather, the tanning industry processes the hides of animals and skins, which produces large quantities of byproducts (Alexander et al., 1992; Verheijen et al., 1996). As a result, the exploitation of these goods is favored by the huge amount of byproduct and the presence of a diverse mix of metabolite profiles. The use of animal byproducts as feedstock has been diminishing over time, nevertheless, as a result of stricter regulations. Additionally, methods for increasing the value of animal feed offer very modest financial rewards. However, PHs are increasingly acknowledged as parts of BS in agriculture that boost crop yield and quality (Colla et al., 2017).

As an illustration, BS have been widely produced from aquatic algae and their extracts (McHugh, 2013). Moreover, BS are formulations made from natural-source materials. Based on their place of origin, BS are divided into microbial and non-microbial categories. Microbial BS differ from non-microbial BS because they are derived from sources like microorganisms, fermented items, organic byproducts, and many more, whereas non-microbial BS are made of plant-derived products, amino acids, protein hydrolysates (PHs), herbal extracts, non-protein substances, humic acid, and fulvic acid, and so forth. Rosemary-based extracts, for example, encourage the growth of tomatoes (Rouphael et al., 2015). Plant and animal materials can be hydrolyzed by enzymes to produce a wide variety of BS (Ugolini et al., 2015). The hydrolytic products are an intricate PH combination (amino acids and peptides). Chemical acid or alkaline hydrolysis has been used to convert raw materials derived from animals, such as fish waste, bone meal, casein, and skin collagen. Enzymatic hydrolysis of substances such as leftover fruit or vegetable material, lentils, alfalfa hay, and so forth, results in the production of BS (Pacholczak et al., 2012; Scaglia et al., 2017). They promote plant growth, require little fertilizer, are environmentally friendly, and are cost-effective (Pacholczak et al., 2012). A microbe-based fermentation technique is the answer that simultaneously allows the production of biostimulating preparations and the minimization of organic waste. They might also be made of materials that have been anaerobically digested. The organic substance that dissolves during fermentation might also exhibit biostimulatory properties. Typically, plant materials, lignin biomass, and animal manure serve as substrates for the dissolved organic matter (Xu et al., 2015). Biopreparations for marine algae may be made using low-molecular-weight amino acids, polypeptides, phytohormones, enzymes, carbohydrates, antioxidants, and vitamins. These elements cause the rhizogenesis mechanism, which in turn causes beneficial morphological and anatomical changes in plants. Despite the fact that brown, green, and red macroalgae are frequently utilized as seaweed extracts (SWE) in agriculture and horticulture and that there are numerous commercial products available on the market, seaweed extracts still represent a significant category of organic non-microbial BS. The chemical composition and quality of macroalgae's raw material are often hampered by the fact that they are typically taken from seas and oceans, which makes it difficult to standardize and obtain reliable performance from the extracted products (Rouphael and Colla, 2020). Higher plant parts, such as seeds, leaves, and roots, as well as exudates from families such as Amaryllidaceae, Ericaceae, Moringaceae, Solanaceae, Brassicaceae, Vitaceae, Fabaceae, Theaceae, Fagaceae, Poaceae, Plantaginaceae, and Rosaceae, are often used as raw materials for BS (Yakhin et al., 2017).

4.3 TYPES OF BIOSTIMULANTS

Based on the source material, the method of action, and other factors, numerous categories of BS have been identified over time by various authors. For instance, du Jardin (2015) categorized BS into 7 types, including seaweed extracts, PHs, organic acids (i.e., fulvic and humic acids), inorganic chemicals, chitosan, helpful fungi, and bacteria (du Jardin, 2015). Over the last two decades, there has been a lot of interest in plant-based chemicals and other bioactive, natural molecules like fulvic and humic acids (Assainar et al., 2018; Rang et al., 2018; Rouphael et al., 2017a, b). Several BS containing active ingredients are commercially available (Table 4.1). Depending on the composition and intended benefits, BS might be applied as a soil or foliar spray (Shah et al., 2018).

TABLE 4.1 List of Commercially Available Biostimulants and Their Active Ingredients

Sl. No.	Biostimulants	Bioactive Components Present
1.	Algreen®	Algae extracts, alginic acid; amino acids, plant hormones, and vitamins
2.	Asahi SL (Atonik)®	0.3% sodium para-nitrophenolate, 0.2% sodium ortho-nitrophenolate, and 0.1% sodium 5-nitroguaiacolate
3.	Benefit	Amino acids, nucleotides, free enzymatic proteins, vitamins S- 92
4.	Benefit®	Free amino acids, vitamins, and nucleotides
5.	Bio- Algeen	Seaweed *Ascophyllum nodosum* (L.) Le Jolis extract
6.	Bio Rhizotonic®	Algae extract, and vitamins
7.	Bio Root®	Rock phosphate, organic acids, yeast, soybean meal, potassium sulphate, and alfalfa
8.	Bio-algeenS-92®	*Ascophyllum nodosum* extract
9.	Biozyme®	GA_3, zeatin, *Ascophyllum nodosum* extract, IAA, and chelated micronutrients,
10.	Biplantol ®®	Micro and macro elements, and uronic acid
11.	Acker-Schachtelhalm extrakt (*Equisetum extract)*	Silicon, glycosides, Flavonoids, Si, and organic acids
12.	Ergonfill ®	Protein hydrolysates, keratin derivatives, and cysteine
13.	Fermented plant extract (Fermentierter Pflanzenextrakt®)	Lactic acid and photosynthetic bacteria, molasses, yeast, pepper extract, garlic and grasses
14.	Goëmar BM 86®	Extract of *Ascophyllum nodosum*
15.	Grow-plex SP®	Liquid humate
16.	Kendal®	Glutathione, protein hydrolysate, oligosaccharides, saponins, urea,
17.	KE-Plantasalva®	Bio-molasses and herbs extracts
18.	Megafol®	Vitamins, cytokines, auxin, amino acids, gibberellins, betaines
19.	Radifarm®	Amino acids, glycosides, micronutrients, organic acids, carbohydrates, saponins, and vitamins
20.	Retrosal®	Zinc and calcium
21.	Root and Shoot Builder	Seaweed extract (*Ascophyllum nodosum*), micronutrients, amino acids, natural chelating agents

TABLE 4.1 *(Continued)*

Sl. No.	Biostimulants	Bioactive Components Present
22.	Root Juice	Seaweed extract, humic acid, and Fulvic acid
23.	Roots 2	Seaweed, vitamins, humic acid
24.	Ruter AA®	Amino acids and Micro- and Macronutrients
25.	Slavol®	Phosphate-solubilizing and nitrogen-fixing bacteria, and auxins
26.	Stimulate®	auxin, cytokinin, and gibberellic acid
27.	Tablet®	*Rhizophagus intraradices* and *Trychoderma atroviride* spores
28.	Terra-Sorb®	Hydrolyzed bioactive amino acids
29.	Tytanit®	Titanium ascorbate
30.	Viva®	Amino acids, polypeptides, polysaccharides, vitamin complexes, proteins, and humic acid

Source: Adapted from Jasmitha and Swamy (2021); Parađiković et al. (2019); Yakhin et al. (2017).

BS are produced by hydrolyzing fruit or vegetable waste, pulses, forages, and other plant materials. Peptides, amino acids, and other non-protein compounds make up PHs. Biopreparations may be derived from phytohormones, low-molecular weight polypeptides, vitamins, amino acids, enzymes, antioxidants, and carbohydrates. Microorganisms are commonly employed in the production of BS and can be produced from bacteria, yeasts, or fungi. These preparations may contain living and/or non-living microbes and their metabolites (Yakhin et al., 2017). Many researchers, including Colla et al. (2015), Matyjaszczyk (2015), Ravensberg (2015), Sofo et al. (2014), and Xavier and Boyetchko (2002), have all documented preparations based on microorganisms as BS.

Animal-derived amino acids and PHs, including wastes and byproducts, can also be used to make BS (Kolomaznk et al., 2012; Maini, 2006; Mladenova et al., 1998). Moreover, Jannin et al. (2011), Kelting et al. (1998), and Sanders et al. (1990) evaluated humate-based raw materials for BS. BS that are formed from food waste extracts or industrial waste streams, composts and manures, compost extracts, aquaculture residues, vermicompost, waste streams, and sewage treatments are among the main groups of BS. It is difficult to pinpoint the mode of action of these compounds due to the variety of source materials and extraction techniques (Yakhin et al., 2017).

4.4 ROLE OF BIOSTIMULANTS IN ABIOTIC STRESSES

BS activate physiological and biochemical processes when they enter the tissues and cells of the host plant. This results in modifications in the hormone regulation, synthesis routes, and signaling pathways required for the development of plants (Figure 4.1). Natural plant biostimulants (BS), which improve nutrient use efficiency (NUE), fruit set, flowering, crop productivity, and plant growth while also strengthening tolerance to a variety of abiotic stresses, are a promising and eco-friendly innovation (Colla and Rouphael, 2015). The term "plant biostimulants" was first used to exclude some functionality, such as plant protection substances and fertilizers. BS are defined as "materials that, in tiny amounts, stimulate plant growth" (Zhang and Schmidt, 1997). The researchers intended to subtly distinguish BS from soil amendments and nutrients, which also encourage plant growth but are visibly administered in higher quantities, by characterizing BS as "minute quantities." Abiotic stress makes cover crops incapable of germination, further cultivation, and successful seeding. Seed priming, conditioning, and coating are examples of seed improvement techniques that are frequently used to improve pest resistance, stand homogeneity, seed performance during sowing, and seedling growth (Harris et al., 2001; Sanpp et al., 2008).

The most often used BS include seaweed extracts, microbial inoculants, humic acids, biopolymers, hormones, nitrogen derivatives, and herbal extracts (Craigie et al., 2011; Khan et al., 2009; Taylor et al., 1998). The primary source of numerous compounds that may be used as growth supplements, including antimicrobials, osmoprotectants, phytohormones, proteins, lipids, carbohydrates, bioactive peptides, and amino acids, is seaweed-based extracts (Carvalho et al., 2013, 2018; Sharma et al., 2013;). Seaweed extract-based BS for horticulture crops increase the vigor of bean seeds (Colla et al., 2015) and stimulate proline production in leaves during drought (Paraikovi et al., 2014). Microbial inoculants and PHs, which can help to promote stress tolerance and may increase nutrient accumulation, mobilization, and distribution, are frequently mentioned in definitions of BS (Malik et al., 2020). In addition, BS promote growth, increase nutrient uptake, raise crop productivity, improve seed vigor and consistency, reduce dormancy, improve root system efficiency, boost photosynthetic rate and activity, increase fruit size and ripening, and monitor flowering (Rouphael et al., 2017).

Table 4.2 summarizes research on the impact of several BS on various crop developments. Sustainable crop quality, growth, and production are the results of all these factors. The primary issue facing manufacturers, developers, legislators, scientists, and everyone else interested in BS is their role in sustainable agriculture. It is still unclear how growing conditions affect the biostimulant's deposition and biochemical activity in a plant. Wheat stands establishment in poor soil types is strengthened by seed priming, which increases germination rate and average seedling size in cold climates (Falloon et al., 1983; Fernández et al., 2018; Vartha et al., 1973). According to Amirkhani et al. (2016) and Paraikovi et al. (2017), the use of fungicides in conjunction with BS may help perennial ryegrass develop a plant stand more quickly. The definition of a PB given by Traon et al. (2017) is: "Any substance or microbe that, when added to seeds, plants, or the root rhizosphere, promotes the NUE's durability to stresses (both biotic and abiotic) and natural processes—regardless of the nutrients they contain—or any blend of these substances or microbes designed for this purpose is referred to as a plant biostimulant."

To reduce reliance on traditional pesticides and fertilizers, which are frequently abused in agricultural crop systems and are less effective, more studies on biostimulant applications in agriculture have been made in recent years (Traon et al., 2014; Wilson et al., 2018). Due to the smaller surface area of treated seeds compared to foliar applications, seed treatments require even lower concentrations of active substances per hectare and promote faster germination and better plant growth than untreated seeds (Yaronskaya et al., 2006). Regardless of the nutritional makeup of the crops, these chemicals are beneficial in low concentrations and support crop quality, resilience to environmental challenges, and nutrition (Cuto et al., 2012). These substances, which are largely cytokinins, auxins, and gibberellins, may work similarly to known growth-regulating hormones when taken exogenously (Qiu et al., 2020). Utilizing BS as seed coating material has a great deal of promise to hasten the growth of seedlings and early stand establishment. These seedlings are referred to as "biostimulant seedlings," which may grow in severe, arid environments as well as in soil with poor nutrient management (Malik et al., 2020). A study showed that BS may increase the growth of epiphytic bacteria (*Pseudomonas sp.*, *Bacillus sp.*, and *Acinetobacter sp.*) with the support of biological pathogen surveillance and PGP, therefore functioning in conjunction with PB organic compounds to enhance fresh lettuce production for sale (Malik et al., 2020).

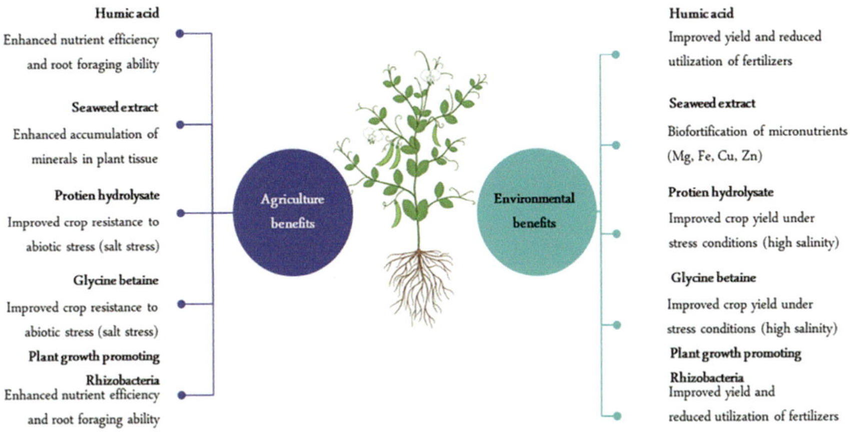

FIGURE 4.1 The advantages of various types of biostimulants in agriculture and the environment.

4.5 GLOBAL PERSPECTIVE AND OPPORTUNITIES FOR AGRICULTURAL WASTE REUSE IN SUSTAINABLE AGRICULTURE

The production of agriculture has greatly improved during the past 50 years, reaching 23.7 million metric tons of food every day. As a result of the significant pressure this rise in production has placed on the environment, some aspects of agricultural sustainability are in doubt. Aside from producing food, agriculture also produces a large amount of biomass that is waste because it cannot be used to produce food (Duque-Acevedo et al., 2020). Promising approaches have been studied over the years, such as waste thermal revalorization by the generation of bioenergy. A method that takes into account all agro-industrial waste as a starting material for the diverse manufacture of compounds to be employed in agriculture but with varied goals is completely lacking. Furthermore, it should be noted that over the coming years, there is going to be an increasing necessity to boost agriculture production in order to sustain the world's expanding population, which has nearly doubled since 1960 and is expected to rise much faster in the upcoming years. Similarly, according to some estimates, 9.1 billion people will inhabit the planet by 2050 (Leisner, 2020; Puglia et al., 2021). Recently, more explicit efforts have been made to change legislative frameworks in favor of an approach that is more mindful and encourages a resource-conserving and eco-friendly culture (Rhozyel et al., 2018). For

TABLE 4.2 Effect of Different Biostimulants on Plant Growth and Development

Sl. No.	Biostimulants	Plant/Crop	Activity	Dosage	References
1.	Asahi SL (Atonik)®	Bean (*Phaseolus vulgaris* L.)	Increased the size and seeds' weight as well as the number of pods, which had an impact on bean yield.	0.30%	Kocira et al. (2015)
2.	Biozyme®	Wheat (*Triticum aestivum*)	Biostimulant application enhanced fruit yield parameters	400 mL/ha	Singh and Chandel (2005)
3.	Biplantol®	Barley (*Hordeum vulgare*), Rye (*Secale cereale*) and Wheat (*Triticum aestivum*)	Germination rates increased with increasing coat size	NA	Gorim and Asch (2012)
4.	Ergonfill®	Cherry tomato seedlings (*Lycopersicon esculentum* Mill. cv. Sakura)	Better adaptation of the seedlings to drought	NA	Murtic et al. (2019)
5.	Goëmar BM 86®	Broccoli (*Brassica oleracea* var. italica)	Broccoli cultivation improved the marketable and superior production.	1 L/ha	Gajc-Wolska et al. (2012)
6.	Grow-plex SP®	Tomato (*Solanum lycopersicum*)	Increased the number of leaves, fresh and dry weights of the plants	90 gm/ 100 L	Abdel-Mawgoud et al. (2007)
7.	Kendal®	French marigold (*Tagetes patula*)	Plant quality was improved by boosting their ability to withstand temperature stress and nitrogen uptake from the soil	300–400 mL/100 ml	Zeljkovic et al. (2023)
8.	KE-Plantasalva®	Garden cress (*Lepidium sativum* L.).	Increase in fresh weight of seedlings	NA	Lisjak et al. (2015)
9.	Megafol®	Corn (*Zea mays* L.)	Increase in germination, biomass production, vigor index	NA	Panfili et al. (2019)
10.	Radifarm®	Scarlet sage (*Salvia splendens* L.)	Enhanced roots growth and aboveground mass	0.25%	Zeljkovic et al. (2010)

TABLE 4.2 *(Continued)*

Sl. No.	Biostimulants	Plant/Crop	Activity	Dosage	References
11.	Retrosal®	Romaine lettuce (*Lactuca sativa* 'Longifolia')	Increase in chlorophyll content measured *in vivo* (+45% in the highest dose) and that a general positive effect was observable on net photosynthesis rate. Improved gas exchange ability under our experimental conditions	0.2 mL/plant dose	Bulgari et al. (2019)
12.	Ruter AA®	Spring barley (*Hordeum vulgare*)	Increased yields by 20–21% on average; maintaining higher transpiration and photosynthesis resulted in higher yield	NA	Gavelienė et al. (2018)
13.	Slavol®	Cauliflower (*Brassica oleracea*)	Increased average leaf length and width in plants	NA	Dimovska et al. (2017)
14.	Terra-Sorb®	Sunflower (*Helianthus annuus*)	Elevated growth, increased content of amino acids, and elevated activity of acetolactate synthase (ALS)	8 mL/L	Navarro-León et al. (2022)
15.	Tytanit®	Tomato (*Solanum lycopersicum*)	Increase in the yield of fruits with the greatest diameters as well as an increase in nitrogen content in fruits	960 g ha^{-1}	Kleiber and Markiewicz (2013)
16.	Viva®	Tomato (*Solanum lycopersicum*)	Improvement in fruit quality parameters (total soluble solids, total acidity, ascorbic acid and lycopene content) and yield were also maintained in reduced macronutrient fertilization when biostimulant was added	10.5 mL/plant (dissolved in 0.5 L H_2O in the ratio 1:50 v/v) (standard nutrition with biostimulant – SV)	Koleška et al. (2017)

instance, the UN's "Paris Agreement," which was signed in 2015, referred to a number of steps to limit GHG emissions and climate change while also emphasizing the significance of preserving food supply and security. The urgent need to increase the ecological systems' resilience has received special attention in this publication, along with the necessity to develop new materials and technologies that have little to no environmental impact. The European Commission gave a strategy for making the EU economy sustainable special priority and attention in December 2019. In order to combat climate change, achieve more efficient resource use, and advance towards a clean and circular economy, the EU has selected agriculture as one of its major priorities. The Green Deal's suggestions were made with the intention of developing agricultural systems based on sustainable practices, such as better management of fertilizers and water as well as lower greenhouse gas emissions (European Commission, 2019). For the transition to a circular economy based on renewable energy and bio-based products made from biomass feedstock, recycling agricultural waste is of utmost strategic importance in this context (Kapoor et al., 2020; Sherwood, 2020). The core concept of the circular economy is to use renewable resources instead of the conventional linear economy, which uses the "take-make-dispose" cycle. Since it consumes natural resources and has an adverse effect on the environment, the production method based on a linear economy is no longer seen as sustainable (Mathews and Tan, 2016).

4.6 CONCLUSION

The valorization of organic wastes from agro-food and industrial byproducts for the production of growth-promoting BS presents a promising avenue for sustainable agricultural practices. Through innovative processes, these organic wastes can be converted into valuable BS that enhance plant growth, productivity, and resilience while simultaneously reducing the environmental impacts associated with waste disposal. In conclusion, the valorization of organic wastes for the production of growth-promoting BS represents a win-win solution for agriculture, the environment, and the economy. By harnessing the potential of waste resources, stakeholders can foster a more resilient and sustainable agricultural sector while addressing pressing challenges such as soil degradation, resource depletion, and climate change mitigation. Continued research, innovation, and collaboration are vital to realizing the full potential of this approach and maximizing its benefits for future generations.

KEYWORDS

- **abiotic stresses**
- **biostimulants**
- **plant growth regulators**
- **protein hydrolysates**
- **seaweed extracts**
- **sustainable agriculture**

REFERENCES

Abdel-Mawgoud, A. M., El-Greadly, N. H., Helmy, Y. I., & Singer, S. M. (2007). Responses of tomato plants to different rates of humic-based fertilizer and NPK fertilization. *Journal of Applied Sciences Research, 3*(2), 169–174.

Amirkhani, M., Netravali, A. N., Huang, W., & Taylor, A. G. (2016). Investigation of soy protein–based bio stimulant seed coating for broccoli seedling and plant growth enhancement. *Hort Science, 51*, 1121–1126.

Assainar, S. K., Abbott, L. K., Mickan, B. S., Whiteley, A. S., Siddique, K. H. M., & Solaiman, Z. M. (2018). Response of wheat to a multiple species microbial inoculant compared to fertilizer application. *Frontiers in Plant Science, 9*, 1601. https://doi.org/10.3389/fpls.2018.01601 (accessed on 6th November 2025).

Battacharyya, D., Babgohari, M. Z., Rathor, P., & Prithiviraj, B. (2015). Seaweed extracts as bio stimulants in horticulture. *Scientia Horticulture, 196*, 39–48.

Bulgari, R., Trivellini, A., & Ferrante, A. (2019). Effects of two doses of organic extract-based bio stimulant on greenhouse lettuce grown under increasing NaCl concentrations. *Frontiers in Plant Science, 9*, 1870. https://doi.org/10.3389/fpls.2018.01870 (accessed on 6th November 2025).

Canellas, L. P., Olivares, F. L., Aguiar, N. O., Jones, D. L., Nebbioso, A., Mazzei, P., & Piccolo, A. (2015). Humic and fulvic acids as bio stimulants in horticulture. *Scientia Horticulture, 196*, 15–27.

Carvalho, M. E. A., Castro, P. R. C., Novembre, A. D. C., & Chamma, H. (2013). Seaweed extract improves the vigor and provides the rapid emergence of dry bean seeds. *American Journal of Agricultural and Environmental Sciences, 13*, 1104–1107.

Carvalho, M. E. A., Castro, P. R., Gaziola, S. A., & Azevedo, R. A. (2018). Is seaweed extract an elicitor compound? Changing proline content in drought-stressed bean plants. *Communicate Scientiae, 9*, 292–297.

Chalamaiah, M., Dinesh Kumar, B., Hemalatha, R., & Jyothirmayi, T. (2012). Fish protein hydrolysates: Proximate composition, amino acid composition, antioxidant activities and applications: A review. *Food Chemistry, 135*, 3020–3038.

Colla, G., Hoagland, L., Ruzzi, M., Cardarelli, M., Bonini, P., Canaguier, R., & Rouphael, Y. (2017). Bio stimulant action of protein hydrolysates: Unraveling their effects on plant physiology and microbiome. *Frontiers in Plant Science, 8*, 2202. https://doi.org/10.3389/fpls.2017.02202 (accessed on 6th November 2025).

Colla, G., Nardi, S., Cardarelli, M., Ertani, A., Lucini, L., Canaguier, R., & Rouphael, Y. (2015). Protein hydrolysates as bio stimulants in horticulture. *Scientia Horticulture, 196*, 28–38.

Couto, C. A., Peixoto, C. P., Vieira, E. L., Carvalho, E. V., & Peixoto, V. A. B. (2012). Action of cantina, butyric acid and gibberellic acid on the emergency of sunflower under aluminum stress/Acao da cinetina, acido indolbutirico e acido giberelico na emergencia do girassol sob estresse por aluminio. *Comunicata Scientiae, 3*, 206–210.

Craigie, J. S. (2011). Seaweed extract stimuli in plant science and agriculture. *Journal of Applied Phycology, 23*, 371–393.

Dimovska, D., Bogevska, Z., Iljovski, I., Zdravkovska, M., Kungulovski, D., & Atanasova-Pancevska, N. (2017). Impact of Salvo microbiological fertilizer on soil microorganisms during cauliflower (*Brassica oleracea* L. var. *botrytis*) growth. In *VIII International Scientific Agriculture Symposium "AGROSYM 2017,"* Jahorina, October 05, 2017, 340–345.

Dominguez, J., & Edwards, C. A. (2010). Relationships between composting and vermicomposting. In Edwards, C. A., Arancon, N. Q., & Sherman, R. (Eds.), *Vermiculture Technology: Earthworms, Organic Wastes, and Environmental Management* (pp. 11–26). Boca Raton, FL: CRC Press.

Duque-Acevedo, M., Belmonte-Ureña, L. J., Cortés-García, F. J., & Camacho-Ferre, F. (2020). Agricultural waste: Review of the evolution, approaches and perspectives on alternative uses. *Global Ecology and Conservation, 22*, e00902. https://doi.org/10.1016/j.gecco.2020.e00902 (accessed on 6th November 2025).

Falloon, R., & Fletcher, R. (1983). Increased herbage production from perennial ryegrass following fungicide seed treatment. *New Zealand Journal of Agricultural Research, 26*, 1–5.

Fernández, J. A., Ceglie, F., Tuzel, Y., Koller, M., Koren, A., Hitchings, R., & Tittarelli, F. (2018). Organic substrate for transplant production in organic nurseries: A review. *Agronomy for Sustainable Development, 38*, 35. https://doi.org/10.1007/s13593-018-0508-4 (accessed on 6th November 2025).

Gajc-Wolska, J., Spiżewski, T., & Grabowska, A. (2012). The effect of seaweed extracts on the yield and quality parameters of broccoli (*Brassica oleracea* var. *cymosa* L.) in open field production. In *I World Congress on the Use of Bio Stimulants in Agriculture, 1009*, 83–89.

Gavelienė, V., Pakalniškytė, L., Novickienė, L., & Balčiauskas, L. (2018). Effect of bio stimulants on cold resistance and productivity formation in winter rapeseed and winter wheat. *Irish Journal of Agricultural and Food Research, 57*, 71–83.

Gorim, L., & Asch, F. (2012). Effects of composition and share of seed coatings on the mobilization efficiency of cereal seeds during germination. *Journal of Agronomy and Crop Science, 198*(2), 81–91.

Harris, D., Raghuwanshi, B. S., Gangwar, J. S., Singh, S. C., Joshi, K. D., Rashid, A., & Hollington, P. A. (2001). Participatory evaluation by farmers of on-farm seed priming in wheat in India, Nepal and Pakistan. *Experimental Agriculture, 37*, 403–415.

Jasmitha, B. G., & Swamy, G. S. (2021). Bio stimulants use in horticulture–A review. *Agrico's e-Newsletter, 3*, 9–12.

Kapoor, R., Ghosh, P., Kumar, M., Sengupta, S., Gupta, A., Kumar, S. S., Vijay, V., Kumar, V., Kumar Vijay, V., & Pant, D. (2020). Valorization of agricultural waste for biogas based circular economy in India: A research outlook. *Bioresource Technology, 304*, 123036. https://doi.org/10.1016/j.biortech.2020.123036 (accessed on 6th November 2025).

Khan, W., Rayirath, U. P., Subramanian, S., Jithesh, M. N., Rayorath, P., Hodges, D. M., Critchley, A. T., Craigie, J. S., Norrie, J., & Prithiviraj, B. (2009). Seaweed extracts as bio stimulants of plant growth and development. *Journal of Plant Growth Regulation, 28*, 386–399.

Kleiber, T., & Markiewicz, B. (2013). Application of "Tytanit" in greenhouse tomato growing. *Acta Scientiarum Polonorum Hortorum Cultus, 12*(3), 117–126.

Kocira, A., Kocira, S., & Stryjecka, M. (2015). Effect of Asahi SL application on common bean yield. *Agriculture and Agricultural Science Procedia, 7*, 103–107.

Koleška, I., Hasanagić, D., Todorović, V., Murtić, S., Klokić, I., Parađiković, N., & Kukavica, B. (2017). Bio stimulant prevents yield loss and reduces oxidative damage in tomato plants grown on reduced NPK nutrition. *Journal of Plant Interactions, 12*(1), 209–218.

Leisner, C. P. (2020). Review: Climate change impacts on food security – Focus on perennial cropping systems and nutritional value. *Plant Science, 293*, 110412. https://doi.org/10.1016/j.plantsci.2020.110412 (accessed on 6th November 2025).

Lisjak, M., Tomić, O., Špoljarević, M., Teklić, T., Stanisavljević, A., & Balas, J. (2015). Garden cress germinability and seedling vigor after treatment with plant extracts. *Poljoprivreda, 21*(2), 41–46.

Malik, A., Mor, V. S., Tokas, J., Punia, H., Malik, S., Malik, K., Sangwan, S., Tomar, S., Singh, P., Singh, N., & Singh, G. (2020). Bio stimulant-treated seedlings under sustainable agriculture: A global perspective facing climate change. *Agronomy, 11*(1), 14. https://doi.org/10.3390/agronomy11010014 (accessed on 6th November 2025).

Mathews, J. A., & Tan, H. (2016). Circular economy: Lessons from China. *Nature, 531*, 440–442.

Murtic, S., Oljaca, R., Murtic, M. S., Koleska, I., & Muhic, A. (2019). Enzymatic antioxidant responses to bio stimulants in cherry tomato subjected to drought. *Journal of Animal and Plant Sciences, 29*(6), 1664–1672.

N/A. (2015). The Paris Agreement. https://unfccc.int/process-and-meetings/the-paris-agreement (accessed on 6th November 2025).

N/A. (2019). The European Green Deal. https://commission.europa.eu/strategy-and-policy/priorities-2019-2024/european-green-deal_en (accessed on 6th November 2025).

Nabti, E., Jha, B., & Hartmann, A. (2016). Impact of seaweeds on agricultural crop production as biofertilizer. *International Journal of Environmental Science and Technology, 14*, 1119–1134.

Navarro-León, E., Borda, E., Marín, C., Sierras, N., Blasco, B., & Ruiz, J. M. (2022). Application of an enzymatic hydrolyzed L-α-amino acid based bio stimulant to improve sunflower tolerance to Imazamox. *Plants, 11*(20), 2761. https://doi.org/10.3390/plants11202761 (accessed on 6th November 2025).

Pacholczak, A., Szydło, W., Jacygrad, E., & Federowicz, M. (2012). Effect of auxins and the biostimulator Alamino Plant on rhizogenesis in stem cuttings of two dogwood cultivars (*Cornus alba* 'Aurea' and 'Elegantissima'). *Acta Scientiarum Polonorum Hortorum Cultus, 11*, 93–103.

Panfili, I., Bartucca, M. L., Marrollo, G., Povero, G., & Del Buono, D. (2019). Application of a plant bio stimulant to improve maize (*Zea mays*) tolerance to metolachlor. *Journal of Agricultural and Food Chemistry, 67*(44), 12164–12171.

Parađiković, N., Teklić, T., Zeljković, S., Lisjak, M., & Špoljarević, M. (2019). Bio stimulants research in some horticultural plant species–A review. *Food and Energy Security, 8*(2), e00162. https://doi.org/10.1002/fes3.162 (accessed on 6th November 2025).

Parađiković, N., Tkalec, M., Zeljković, S., & Vinković, T. (2014). Bio stimulant application in transplants production of *Allium sativum* L. and *Rosa canina* L. In *Proceedings of the Fifth International Scientific Agricultural Symposium "Agrosym 2014,"* Jahorina, Bosnia and Herzegovina, 23–26 October, p. 694.

Parađiković, N., Zeljković, S., Tkalec, M., Vinković, T., Maksimović, I., & Haramija, J. (2017). Influence of bio stimulant application on growth, nutrient status and proline concentration of *Begonia* transplants. *Biological Agriculture & Horticulture, 33*, 89–96.

Puglia, D., Pezzolla, D., Gigliotti, G., Torre, L., Bartucca, M. L., & Del Buono, D. (2021). The opportunity of valorizing agricultural waste, through its conversion into bio stimulants, biofertilizers, and biopolymers. *Sustainability, 13*(5), 2710. https://doi.org/10.3390/su13052710 (accessed on 6th November 2025).

Qiu, Y., Amirkhani, M., Mayton, H., Chen, Z., & Taylor, A. (2020). Bio stimulant seed coating treatments to improve cover crop germination and seedling growth. *Agronomy, 10*(2), 154. https://doi.org/10.3390/agronomy10020154 (accessed on 6th November 2025).

Rang-Jin, X., Kyriacou, M., & Colla, G. (2018). Vegetable grafting: A toolbox for securing yield stability under multiple stress conditions. *Frontiers in Plant Science, 8*, 2255. https://doi.org/10.3389/fpls.2017.02255 (accessed on 6th November 2025).

Rhozyel, M. S., & Žalpyte, J. (2018). A macroeconomic perspective on green growth. In W. Leal Filho, D. Pociovălişteanu, P. Borges de Brito, & I. Borges de Lima (Eds.), *Towards a Sustainable Bioeconomy: Principles, Challenges and Perspectives* (pp. 63–73). World Sustainability Series; Springer: Cham, Switzerland.

Rouphael, Y., & Colla, G. (2020). Bio stimulants in agriculture. *Frontiers in Plant Science, 11*, 40. https://doi.org/10.3389/fpls.2020.00040 (accessed on 6th November 2025).

Rouphael, Y., Cardarelli, M., Bonini, P., & Colla, G. (2017). Synergistic action of a microbial-based bio stimulant and a plant-derived protein hydrolysate enhances lettuce tolerance to alkalinity and salinity. *Frontiers in Plant Science, 8*, 131. https://doi.org/10.3389/fpls.2017.00131 (accessed on 6th November 2025).

Rouphael, Y., Colla, G., Graziani, G., Ritieni, A., Cardarelli, M., & De Pascale, S. (2017b). Phenolic composition, antioxidant activity and mineral profile in two seed-propagated artichoke cultivars as affected by microbial inoculants and planting time. *Food Chemistry, 234*, 10–19.

Rouphael, Y., De Micco, V., Arena, C., Raimondi, G., Colla, G., & De Pascale, S. (2017a). Effect of *Ecklonia maxima* seaweed extract on yield, mineral composition, gas exchange, and leaf anatomy of zucchini squash grown under saline conditions. *Journal of Applied Phycology*, *29*, 459–470.

Rouphael, Y., Franken, P., Schneider, C., Schwarz, D., Giovannetti, M., Agnolucci, M., De Pascale, S., Bonini, P., & Colla, G. (2015). Arbuscular mycorrhizal fungi act as bio stimulants in horticultural crops. *Scientia Horticulture*, *196*, 91–108.

Scaglia, B., Pognani, M., & Adani, F. (2017). The anaerobic digestion process capability to produce bio stimulant: The case study of the dissolved organic matter (DOM) vs. auxin-like property. *Science of the Total Environment*, *589*, 36–45.

Schmitt, A., Koch, E., Stephan, D., Kromphardt, C., Jahn, M., Krauthausen, H. J., Forsberg, G., Werner, S., Amein, T., Wright, S. A. I., Tinivella, F., & van der Wolf, J. S. P. (2009). Evaluation of non-chemical seed treatment methods for the control of *Phoma valerianellae* on lamb's lettuce seeds. *Journal of Plant Diseases and Protection*, *116*, 200–207. https://doi.org/10.1007/BF03356311 (accessed on 6th November 2025).

Shah, Z. H., Rehman, H. M., Akhtar, T., Alsamadany, H., Hamooh, B. T., Mujtaba, T., Daur, I., Al Zahrani, Y., Alzahrani, H. A. S., Ali, S., Yang, S. H., & Chung, G. (2018). Humic substances: Determining potential molecular regulatory processes in plants. *Frontiers in Plant Science*, *9*, 263. https://doi.org/10.3389/fpls.2018.00263 (accessed on 6th November 2025).

Sharma, H. S. S., Fleming, C. C., Selby, C., Rao, J. R., & Martin, T. (2013). Plant bio stimulants: A review on the processing of macroalgae and use of extracts for crop management to reduce abiotic and biotic stresses. *Journal of Applied Phycology*, *26*, 465–490.

Sherwood, J. (2020). The significance of biomass in a circular economy. *Bioresource Technology*, *30*, 122755. https://doi.org/10.1016/j.biortech.2020.122755 (accessed on 6th November 2025).

Singh, P. K., & Chandel, A. S. (2005). Effect of biozyme on yield and quality of wheat (*Triticum aestivum*). *Indian Journal of Agronomy*, *50*(1), 58–60.

Singh, U. B., Malviya, D., Khan, W., Singh, S., Karthikeyan, N., Imran, M., Rai, J. P., Sarma, B. K., Manna, M. C., Chaurasia, R., Sharma, A. K., Paul, D., & Oh, J. W. (2018). Earthworm grazed-*Trichoderma harzianum* biofortified spent mushroom substrates modulate accumulation of natural antioxidants and bio-fortification of mineral nutrients in tomato. *Frontiers in Plant Science*, *9*, 1017. https://doi.org/10.3389/fpls.2018.01017 (accessed on 6th November 2025).

Snapp, S., Price, R., & Morton, M. (2008). Seed priming of winter annual cover crops improve germination and emergence. *Agronomy Journal*, *100*, 1506–1510.

Taylor, A. G., Allen, P. S., Bennett, M. A., Bradford, K. J., Burris, J. S., & Misra, M. K. (1998). Seed enhancements. *Seed Science Research*, *8*, 245–256.

Traon, D., Amat, L., Zotz, F., & du Jardin, P. (2014). A legal framework for plant bio stimulants and agronomic fertilizer additives in the EU–Report to the European Commission. *DG Enterprise & Industry: Copenhagen, Denmark*. https://op.europa.eu/en/publication-detail/-/publication/dbeffd43-98a5-4e39-a930-7dfa21816f8c (accessed on 6th November 2025).

Trejo-Téllez, L. I., Estrada-Ortiz, E., Gómez-Merino, F. C., Becker, C., Krumbein, A., & Schwarz, D. (2019). Flavonoid, nitrate and glucosinolate concentrations in *Brassica* species are differentially affected by photosynthetically active radiation, phosphate and phosphite. *Frontiers in Plant Science*, *10*, 371. https://doi.org/10.3389/fpls.2019.00371 (accessed on 6th November 2025).

Ugolini, L., Cinti, S., Righetti, L., Stefan, A., Matteo, R., D'Avino, L., & Lazzeri, L. (2015). Production of an enzymatic protein hydrolysate from defatted sunflower seed meal for potential application as a plant bio stimulant. *Industrial Crops and Products*, *75*, 15–23.

Vartha, E. W., & Clifford, P. T. P. (1973). Effects of seed coating on establishment and survival of grasses, surface-sown on tussock grasslands. *New Zealand Journal of Experimental Agriculture*, *1*, 39–43.

Xu, C., & Leskovar, D. I. (2015). Effects of *Ascophyllum nodosum* seaweed extracts on spinach growth, physiology and nutrition value under drought stress. *Scientia Horticulture*, *183*, 39–47.

Yakhin, O. I., Lubyanov, A. A., Yakhin, I. A., & Brown, P. H. (2017). Bio stimulants in plant science: A global perspective. *Frontiers in Plant Science*, *7*, 2049. https://doi.org/10.3389/fpls.2016.02049 (accessed on 6th November 2025).

Yaronskaya, E., Vershilovskaya, I., Poers, Y., Alawady, A. E., Averina, N., & Grimm, B. (2006). Cytokinin effects on tetrapyrrole biosynthesis and photosynthetic activity in barley seedlings. *Planta*, *224*, 700–709.

Zeljković, S. B., Parađiković, N. A., Babić, T. S., Đurić, G. D., Oljača, R. M., Vinković, T. M., & Tkalec, M. B. (2010). Influence of bio stimulant and substrate volume on root growth and development of scarlet sage (*Salvia splendens* L.) transplants. *Journal of Agricultural Sciences, Belgrade*, *55*, 29–36.

Zeljković, S., Parađiković, N., Maksimović, I., Teklić, T., & Tkalec Kojić, M. (2023). Growth and nutrient status of French marigold (*Tagetes patula* L.) under bio stimulant application. *New Zealand Journal of Crop and Horticultural Science*, *51*(1), 614–624. https://doi.org/10.1080/01140671.2022.2058024 (accessed on 6th November 2025).

Zhang, X., & Schmidt, R. E. (1997). The impact of growth regulators on the α-tocopherol status in water-stressed *Poa pratensis*. *International Turfgrass Society Research Journal*, *8*, 1364–1371.

CHAPTER 5

Humic and Fulvic Acids as Biostimulants in Agriculture

AYESHA ANSARI,[1] PALLAVI GAUR,[2] TARA CHAND YADAV,[3] PANKAJ GAUTAM,[4] NAVIN KUMAR,[5] and PAYAL GUPTA[5,*]

[1]*Department of Biosciences, Graphic Era Deemed to be University, Dehradun, Uttarakhand, India*

[2]*Department of Microbiology, SGRR University, Dehradun, Uttarakhand, India*

[3]*Department of Computer Science and Biosciences, Marwadi University, Rajkot, Gujarat, India*

[4]*Department of Life Sciences, Graphic Era University, Dehradun, Uttarakhand, India*

[5]*Department of Biotechnology, Graphic Era University, Dehradun, Uttarakhand, India*

*Corresponding author

ABSTRACT

Increased food production is required to feed the growing population, and to meet this goal, synthetic fertilizers have been widely used in the agriculture sector. This unsupervised practice feeds the global population a nutrition-compromised and chemical-loaded diet. Also, unregulated chemical practices have questioned the sustainable development goal of humans, as their agricultural practices are impacting soil and plant health, too. Therefore, in pursuit of eco-friendly and healthy alternatives in agriculture, biostimulants (BS) have been favored, and their application

Plant Biostimulants and Growth Regulators in Sustainable Agriculture.
Deepak Kumar, Sanjay K. Singh, Nutan Kaushik, and Mahaveer P. Sharma (Eds.)

DOI: 10.1201/9781779641458-7

is encouraged. These compounds stimulate and improve the growth of plants and enhance soil quality. In modern agriculture, the use of BS is an eco-friendly approach that promotes crop productivity and improves agricultural practices. Humic acid and fulvic acid are components of soil humic substances with potential biostimulant properties. This chapter provides detailed information on BS, their types, and omics-based characterization, with a focus on humic acid and fulvic acid as BS.

5.1 INTRODUCTION

From the beginning of human civilization, agricultural practices have been the only way to feed the ever-expanding population globally. With advancements in modern agricultural practices like the implementation of modern machines and the use of fertilizers, pesticides, and transgenic varieties, though fertilizers and pesticides play an important role in promoting both the quality and productivity of plant produce, they are troublesome from an environmental and health viewpoint (Alewu and Nosiri, 2011; Gunnell et al., 2007). Moreover, the continuous exposure of microbes, pests, and insects has made them resistant to these chemicals but proved hazardous to the food chain (Yassin et al., 2021). A number of regulatory norms and rules have been formulated globally by international and national organizations, but they have also proved ineffective in controlling or restricting their use completely (Chuan et al., 2019; Ju et al., 2016; Shuqin et al., 2018). In another pursuit of environmentally and health-safe options, some natural, biological substances that would address both issues simultaneously have been explored. The development of transgenic varieties is one of the most acceptable and stable ways to increase plant productivity, as well as restrict the use of chemicals, but the option is time-consuming, costly, and regulated by several norms. Another safe, cost-effective, and non-binding use of natural substances includes the use of biofertilizers, biopesticides, biostimulants (BS), and microbial inoculants (Ajmal et al., 2018; Patil et al., 2016).

BS are organic substances responsible for better plant growth and development when applied in small amounts. The benefits of BS on plant health are incomparable to those attributed to traditional plant nutrients (Drobek et al., 2019). The term "biostimulants" was first defined by Kauffman in 2007. According to him, BS are substances other than fertilizers that enhance plant growth even in low quantities. Previously, BS were also called metabolic enhancers and products with hormone-like activities. These are the most commonly used stimulants in the agriculture sector because they can be used in the form of soil preparations (powder, granules, or solutions) (Drobek et al., 2019). More precisely, BS are defined as organic products of complex constituents with unique properties that enable enhanced plant productivity and stress-mitigating responses, as opposed to plant nutrients, regulators, and other substances (Yakhin et al., 2017). This distinction from other growth-promoting additives is made with technical precision. To highlight the importance of these agrochemical products, the term "BS" first appeared in Regulation (European Union) 2019/1009, where BS were classified as microbial and non-microbial. As per Business Wire, Dublin, the global market for BS in 2021 was nearly 3.2 billion USD, and it is believed to increase to 5.6 billion USD till 2026 at a CAGR of 12.1%. This level of growth in the biostimulant market is due to its nurturing impact on plant growth development via numerous points of action throughout the crop lifecycle, regardless of its application to soil, seed, or plant. BS improve the nutrient uptake ability of the crops, and this shows the way to sustainable agriculture by discouraging the application of synthetic chemicals.

The sources of BS are diverse, where microalgae and seaweed (macroalgae) extracts contribute a huge portion to meeting the demand for BS globally. Seaweed-based biostimulants (SWB) are rich sources of phytohormones and trace elements essential for crop growth. The SWB improves the growth and productivity of crops, even under stressful conditions, by improving the aeration and structure of the soil, increasing seed germination, and promoting root growth, flowering, and fruit production. Microalgae is also a rich source of BS with considerable impact on plant growth and crop yield. Although both macroalgae and microalgae are rich sources of BS, macroalgae-based BS are most explored and have a well-established market (Kapoore et al., 2021). The commercial potential and market of microalgae-based BS are lagging due to insufficient research. Conventionally, blue-green algae have been used for fixing nitrogen in paddy fields, which is beneficial for crops as it satisfies the nitrogen need (Chakdar et al., 2012). The microalgae biomass, when applied to soil, slowly releases nutrients and conditions the soil (Garcia-Gonzalez et al., 2016).

BS have long been farmers' first choice for a variety of purposes, including fostering plant growth, development, and performance, enhancing crop tolerance to biotic and abiotic stress, and promoting translocation and assimilation of nutrients. Crop metabolism is naturally stimulated in terms of genomics, flowering, and fruiting (Ali et al., 2021; Carmody et al., 2020; Rouphael et al., 2018; Shukla et al., 2019). Many studies on improved crop yield and quality using BS suggested that BS increase the amount of chlorophyll in leaves, enhance photosynthesis, and thus boost the yield of crop production with good nutritional efficiency, as well as increase the water holding capacity of the soil (Calvo et al., 2014; Rouphael et al., 2018).

5.2 TYPES OF BIOSTIMULANTS AND THEIR IMPORTANCE

The criteria for defining and characterizing BS for plants are continuously evolving. Up until now, several definitions have been proposed and recognized by the scientific community, regulatory authorities, and stakeholders, but no legal or concrete definition has been proposed for PBs anywhere in the world, including in the European Union and the United States. For general reference, BS are categorized into nine groups depending upon the nature of the substance: (a) organic substances; (b) humic materials; (c) chemical elements; (d) inorganic salts; (e) antiperspirants; (f) chitin/chitosan and their derivatives; (g) PGPR/arbuscular mycorrhizal fungi/*Trichoderma* spp.; (h) N-containing/amino acid substances; and (i) algae extracts (Carletti et al., 2021; García-Sánchez et al., 2022) (Figures 5.1 and 5.2). The major categories of BS that are widely used in agriculture are described in subsections.

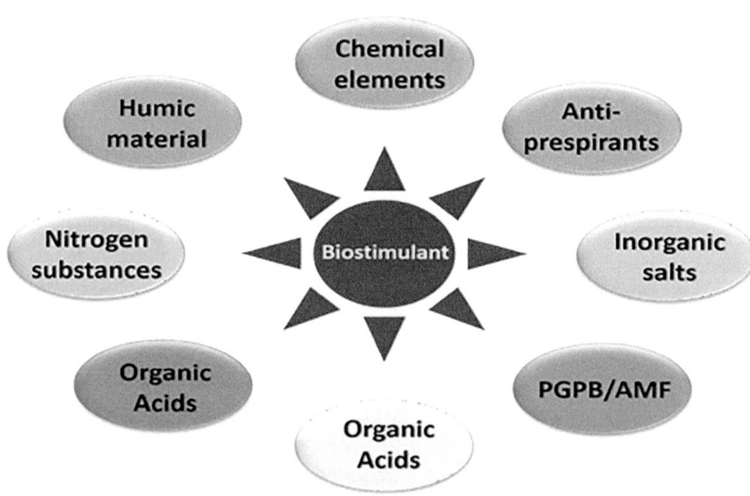

FIGURE 5.1 Types of biostimulants based on their source.

5.2.1 AMINO ACID SUBSTANCES

Nitrogen-rich compounds are well known to stimulate plant development. This class of BS is composed of hydrolyzed proteins and amino acids (AAs) derived from either animals or vegetables (Yakhin et al., 2017). The chemical and enzymatic protein hydrolysis of agro-industrial leftovers, extracted from both plant and animal byproducts, results in peptide mixtures and amino acids, which are often used as PBs to enhance their growth and development (Navarro-León et al., 2022). Protein-rich BS contain non-protein compounds, amino acids, and peptides. It falls into the category of non-microbial BS, which comprise signaling peptides and free amino acids. Other nitrogen-rich compounds include polyamines and betaines. Glycine betaine is one of the nitrogen-rich compounds that helps in abiotic stress tolerance in plants (Chen et al., 2011). Animal-origin AA-based BS enhanced the heat tolerance of *Diplotaxis tenuifolia* and *Arabidopsis thaliana* plants (García-Sánchez et al., 2022). The mode of action of these BS is broad and includes nitrogen metabolism, the release of phosphorus, stimulating microbial activity in soil, and root growth. The AA-based BS work in favor of plant growth by improving the ion transport, stress response, photosynthesis, and assimilation of essential nutrients (N, S, and C) (Di Mola et al., 2020). They are responsible for regulating the enzymes of the TCA cycle, which result in increased nitrogen metabolism.

5.2.2 CHITIN/CHITOSAN-DERIVED SUBSTANCES

Chitin is a biopolymer with multiple applications in agriculture, food, paper, and cosmetics. The deacetylated form of chitin is known as chitosan (Gupta et al., 2021). Like other BS, chitosan activates genes involved in plants' defense against pathogens. They help improve plant resistance to abiotic and biotic stressors. Chitosan has a specific role in the agriculture industry; it increases protein assimilation and photosynthesis, alters plant metabolism to increase root and shoot biomass, improves fruit quality, and enhances antioxidant activities (Dias et al., 2022). Chitosan is also responsible for the increased content of flavonoids, carotenoids, and anthocyanins in fruits. Chitosan has defensive properties; it can be used as a coating agent for a variety of fruits to protect against pathogens, which improves the shelf life of food products (Stasińska-Jakubas et al., 2022).

5.2.3 MICROBIAL BIOSTIMULANTS

Microbial-based BS are innovative technologies that can ensure agricultural yields with high nutritional values. They contain fungi like AMF (arbuscular mycorrhizal fungi), including *Trichoderma* species, *Basidiomycota*, *Atroviride, Heteroconium chaetospira,* and plant growth-promoting bacteria (PGPB), including *Azospirillum, Rhodococcus, Pseudomonas, and Streptomyces*. *Enterobacter* and *Pseudomonas* species are two bacterial species that have reportedly been used as microbial inoculates (Colla et al., 2015). The microbial BS help in mitigating the adverse effect of agricultural activity on ecosystems by improving soil biodiversity, as the microbes play a vital role in the rhizosphere, endosphere, and phyllosphere of the crop and thus increase nutrient availability and absorption (Castiglione et al., 2021; Lin et al., 2022). PGPB releases antioxidant enzymes, which help in scavenging reactive oxygen species (ROS) and thus helping in coping with abiotic stress tolerance. They help provide nutrients to plants, therefore promoting plant growth. AMF promotes the synthesis of chlorophyll and regulates the opening and closing of the stomatal aperture. They increase nutrient uptake by plants, promote cellulose synthesis, and help to strengthen the plant cell wall (Hasanuzzaman et al., 2021).

5.2.4 ORGANIC ACID BIOSTIMULANTS

A variety of organic acids are known to be present in the close vicinity of plants and thus impact plant growth and development (Hadavi et al., 2022). Humic acid and fulvic acid are two important organic acids that have potential biostimulant properties. The degradation of plants and animals by soil microorganisms results in the formation of humic substances. They improve the quality and economy of agriculture and animal husbandry. Humic compounds are of great importance and have several benefits as BS because they increase soil fertility, act as natural constituents of organic matter, and increase nutrient uptake. They enhance oxygen diffusion at the soil level and compensate for nutritional deficiency in the roots and shoots of plants (Canellas et al., 2015; Nardi et al., 2002). They also modify plant physiology and metabolism in a desirable manner and protect plants from abiotic stress like droughts (Hasanuzzaman et al., 2021).

5.2.5 INORGANIC COMPOUNDS

Aluminum, cobalt, sodium, silicon, and selenium are the inorganic compounds found in soil that enhance plant growth and development. These elements are present in soils and are not essential for all plants, known as beneficial elements (Pilon-Smits et al., 2009). Selenium, silicon, cobalt, aluminum, and sodium are beneficial elements that are present in the form of inorganic salts in the soil as well as plants (Azad et al., 2021; Savvas et al., 2015). Silica deposits help to harden the plant cell wall. They are reported to have specific activities in osmoregulation, increasing plant enzymatic activities, protection against oxidative stress, protection against pathogenic microbe resistance, and heavy metal toxicity. Silica protects plants against diseases and enhances the mechanical properties of leaves, like leaf stiffness and vigor. Silica shows a controlled effect on water mobility in plants. Inorganic compounds also delay plant senescence and aging by reducing the translocation of toxic metals from soil to plants. They improved the absorption and efficiency of nutrients (Dias et al., 2022).

5.2.6 ALGAE BIOSTIMULANTS

Seaweeds are green, brown, and red marine macroalgae and are an excellent source of organic matter that is an effective replacement for fertilizers in the agriculture sector (Kapoore et al., 2021). Seaweed extracts are utilized as food products and have botanical applications. Algal biomass can be directly applied to soils or as foliar for soil treatment in terms of bioremediation. They are an excellent soil conditioner and promote the growth and resistivity of plants against abiotic stress (González-Pérez et al., 2022). They are reported to have antioxidant properties and enhance nutrient uptake from the soil, thus increasing plant biomass and crop yield. They also improve soil quality by showing root-microbe interactions, which help in root growth and elongation. The following species are reported to act as microalgal extracts: *Chlorella ellipsoidea, Spirulina maxima, Calothrix elenkinii*, and *Spirulina platensis* (Chiaiese et al., 2018). Such species act as microalgal-based BS in the agriculture industry. Seaweed extracts are also reported to stimulate seed germination. They enhance the health of soil by increasing its moisture-holding capacity (Dias et al., 2022).

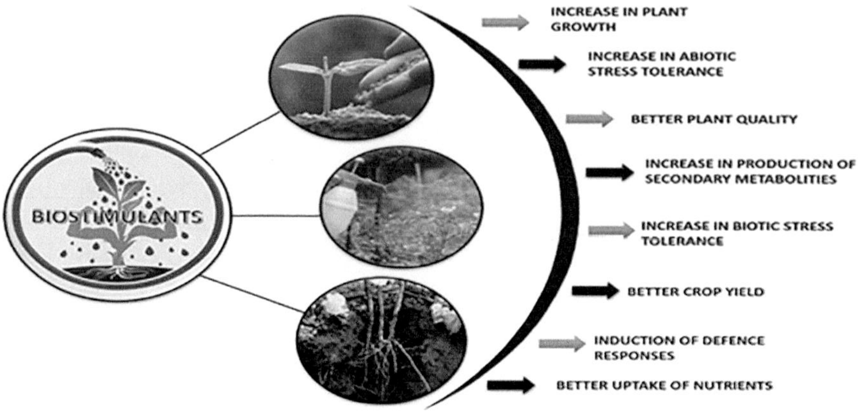

FIGURE 5.2 Beneficial effects of biostimulants applied to roots, soil, and plant shoots.

5.3 HUMIC ACID

The degradation of plants and animals results in the formation of humic substances, which constitute a major organic carbon pool on the earth's surface (Piccolo et al., 1996). Based upon their solubility in acidic and alkaline pH, they can be categorized into two groups, including humic acid and fulvic acid. Humic acid shows greater solubility in alkaline pH and is insoluble in acidic pH, while fulvic acid is soluble at both acidic and alkaline pH (Gaffney et al., 1996).

The major part of the humic substance is composed of acids, including humic acid and fulvic acid. Humic acid is the major part of humic substances with a large molecular structure and is biochemically the most active material found in soil. Due to its abundance, it is used in the agriculture sector to improve the economy and ecology of production. These are often described as the natural polymers that are derived from the biological, chemical, and microbial decomposition of organic matter in plants and animals, which is further humidified to form humic acid. It is a dark-colored substance; its color may range from dark brown to black, and it is soluble at higher pH values. The major constituent of humic acid is carbon, followed by nitrogen and phosphorous. It can also be obtained from lignite coals. Low-rank coal can be used as a rich source of humic acids in agriculture. In the category of coal, leonardite is the best source of humic acid, which is the sedimented soft brown coal (Kurniati et al., 2018). Even in small quantities, humic acid has the potential to protect living cells against the toxic environment of natural sources (Duary, 2020; Md. Shaiful Islam et al., 2005). In the agriculture sector, farmers have employed humates for the improvement of rhizome growth (Canellas et al., 2008). It is reported to have antioxidant, anti-fungal, and anti-bacterial properties. Humic acid can be applied to potted plants to improve the texture and structure of the soil. A humic substance is a heterogeneous mixture of organic acids formed by a simple process of humification (Figure 5.3).

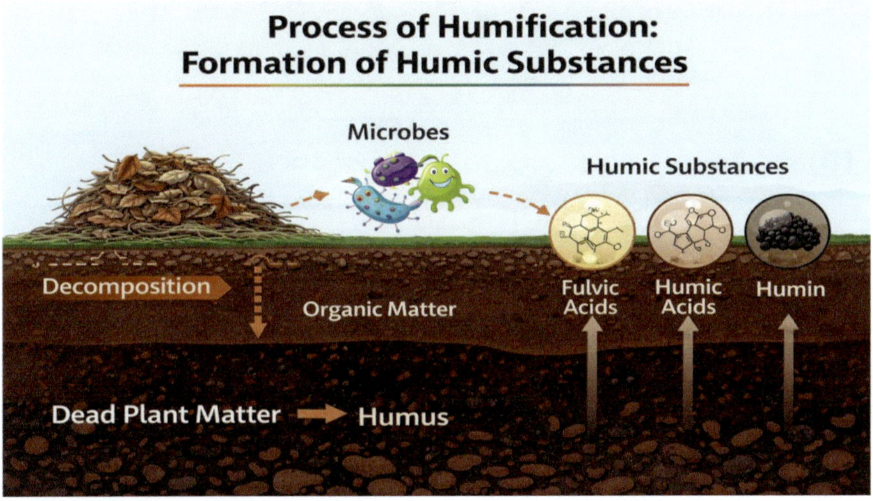

FIGURE 5.3 A schematic representation of the process of humification through which humic substances form.

Source: Created with BioRender.com.

5.3.1 *STRUCTURE OF HUMIC ACID*

Humic acid is highly aromatic and turns insoluble when the carboxylate groups are protonated at low pH values. Humic acid is made by the association of hydrophobic compounds that are stabilized by van der Waal forces, π-π bonds, and CH-π bonds (Figure 5.4A). The presence of a phenolate group and a carboxylate group can form complexes with magnesium ions, calcium ions, ferrous ions, and ferric ions. They have two or more groups to form chelating complexes. It contains aromatic and aliphatic structures; both are extensively substituted with oxygen-containing functional groups. It contains free and bound phenolic OH groups, quinone structures, nitrogen and oxygen as bridge units, and carboxyl groups based on aromatic rings (Gaffney et al., 1996).

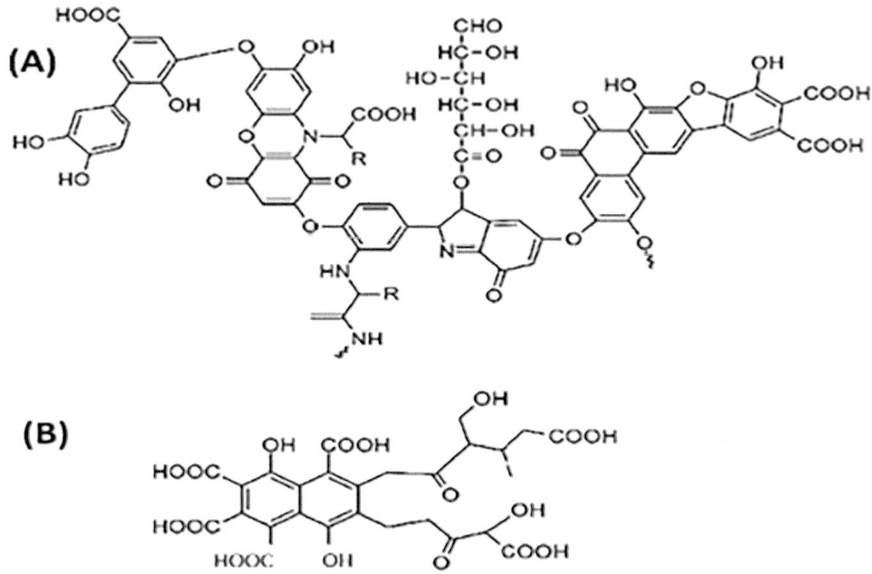

FIGURE 5.4 The chemical structure of (A) humic acid; and (B) fulvic acid.

5.3.2 PHYSICAL AND CHEMICAL PROPERTIES OF HUMIC ACID

Humic acid is highly colloidal and amorphous in nature. The main extractable component of soil humic compounds is humic acid. The color ranges from dark brown to black (Duary, 2020). It is weakly dissociated, determines the buffering capacity of soil, and can be precipitated in dilute acids. The molecular formula of humic acid is $C_9H_9NO_6$, with a molecular weight of 227.1 g/mol. It is a long-chain molecule of moderate molecular weight. Chemically, humic acid is known as nitrobicyclo [2.2.1] hept-5-ene-2,3-dicarboxylic acid. It is a carbon-containing compound with linear polymers having cross-linked bonds. Some humic acid molecules have aromatic carbon rings. They are insoluble in acids and soluble in alkalis. They form salts with inorganic trace elements, and they act as chelating substances (Flaig et al., 1975). Elementally, humic material is rich in C (40–60%), O (30–50%), N (4–5%), S (1–2%), and P; humic acid has more C, H, N, S, and less O (Suffet et al., 1988). Often, humic materials are rich in radicals; humic acid has a high content of semiquinones, which may exist either as permanent or transient species depending on the pH of the soil, solar irradiation, and chemical reduction (Chen et al., 1978).

The most important role of these free radicals is in the oxidation-reduction reaction, which thus mediates the reduction of metals and their migration (Skogerboe et al., 1981).

5.3.3 MODE OF ACTION OF HUMIC ACID

Humic acid, when used as a biostimulant, has a toxic and antagonistic effect against plant parasitic nematodes, including *Rotylenchulus reniformis, Meloidogyne* spp., *Helicotylenchus multicinctus*, and *Radopholus similis*. They work efficiently as bioinoculants with microbial species such as phosphobacterium, *Azospirillum* spp., *Bacillus megaterium*, *Trichoderma viride, Pseudomonas fluorescens*, *Pochonia chlamydosporia, Glomus* spp., *Purpureocillium lilacinum*, and *Trichoderma asperellum* (Baloach et al., 2014; Nagachandrabose and Baidoo, 2021; Olivares et al., 2017). It can be applied to soil at very low concentrations of 0.04%–2%. Nematode control can be achieved through soil application or drenching, seedling root dip therapy, and foliar spray on leaves. Humic acid regulates plant-parasitic nematodes by a variety of methods, including the destruction of juveniles, the inhibition of hatching, the reduction of nematode infectivity and reproduction, and the induction of systemic resistance (Nagachandrabose and Baidoo, 2021). Inhibition of both the classical and alternative pathways of complement activation as well as the degranulation of phagocytes and the production of inflammatory cytokines such as IL-1 beta, IL-6, IL-10, and TNF alpha also contribute to its mode of action (Junek et al., 2009; van Rensburg, 2015). Humic acids increase the availability of P205. It regulates and chelates toxins from the soil, which helps to remove obstacles to growth and provides plants with better resistance against diseases. Humic acid also increases the permeability of plant membranes, and it promotes the uptake of nutrients (Trevisan et al., 2010).

5.3.4 IMPORTANCE OF HUMIC ACID

Humic acid functions as a naturally occurring organic substance in the soil. It increases plant nutrient intake and boosts chlorophyll synthesis. It improves root growth and seed germination, makes the crop disease-free, and hence improves yield. Because of their small size, they help the plants accumulate nutrients from the soil in a desirable manner (Senn et al., 2007).

Moreover, humic acid promotes plant growth by modulating nitrate reductase, glutamate dehydrogenase, and glutamine synthetase, which are key players in C and N metabolism and assimilation pathways (Albuzio et al., 1986; Canellas et al., 2013; Hernandez et al., 2015; Muscolo et al., 1999). The role of humic acid in regulating secondary metabolism in crop plants is comprehensively reviewed in other review papers (Canellas et al., 2015). It increases drought tolerance, seed germination, microbial activity, and vitamin content in the soil. Humic acid also increases the water holding capacity of soil by positively impacting its physical, chemical, and biological properties, and to some extent, it also reduces the levels of toxic minerals in soil (Senn et al., 2007). It alters the primary and secondary metabolisms of plants to enhance the high yield of crop production. They are proven to be effective in plant growth. Humic acid has a beneficial effect on human health, as it improves dietary digestibility as well as calcium and trace element usage in the body (Buschmann et al., 2006).

5.4 FULVIC ACID

Fulvic acid is another example of the humic substances that are also formed by the decomposition of plants and animals. In terms of bioremediation, fulvic acid is the best medicine in nature. Soil scientists have characterized it as the world's most diverse substance (Mahler et al., 2021). It is soluble in water at all pH values, which makes it the component that renders the nutrients accessible. It delivers more than 60 minerals and trace elements to plant root cells. Fulvic acid can be obtained from plants that have deep roots, such as turnips, sweet potatoes, pumpkins, tomatoes, melons, lima beans, blackstrap molasses made from sugarcane, and seawater (Ogner and Schnitzer, 1971).

5.4.1 STRUCTURE OF FULVIC ACID

Fulvic acid is more aliphatic and less aromatic than humic acid. They are richer in carboxylic acid, phenolic, and ketonic groups. Structure is characterized by the loose assembly of aromatic organic polymers with many carboxyl groups (COOH) that release hydrogen ions (Figure 5.4B). Further, the XRD analysis of fulvic acid depicts that its carbon skeleton is made up of a broken array of poorly condensed rings where a considerable number of disordered aliphatic chains or alicyclic structures are present near the edges of aromatic layers (Kodama et al., 1967). It is especially reactive when metals form strong complexes with aluminum ions and copper ions, leading to their increased solubility in natural waters (Canellas et al., 2015).

5.4.2 PHYSICAL AND CHEMICAL PROPERTIES OF FULVIC ACID

The color of fulvic acid ranges from yellow to light brown due to the presence of aromatic and aliphatic rings. The presence of high oxygen content as carboxyl and hydroxyl groups in its structure makes it more chemically reactive than humic acid. In its soluble state, it is the most powerful electrolyte known to exist, which charges, regenerates, regulates, and delivers free electrons and energy to living cells (Canellas et al., 2015). Fulvic acid is necessary for the utilization of vitamins, enzyme production, and hormone structures by plants. It helps in the cellular detoxification process. Fulvic acid is a low-molecular-weight organic substance with a molecular formula of $C_{14}H_{12}O_8$ and a molecular weight of 308.24 g/mol. Chemically, it is known as 3,7,8-trihydroxy-3-methyl-10-oxo-1,4 dihydropyrano[4,3-b] chromene-9-carboxylic acid. The elemental analysis of fulvic acid reveals 50.9% C, 3.35% H, 0.75% N, 0.25% S, and 44.75% O (Gong et al., 2020). It contains 9.1 m equiv COOH, 3.6 m equiv alcoholic OH, 3.3 m equiv phenolic OH, and 3.1 m equiv C=O (Schnitzer, 1969). It reacts with metals and forms complexes that are soluble in water. It occurs in ionic form and contains amino acids, peptides, sugars, phytochemical compounds, and nucleic acids as trace elements (Canellas et al., 2015).

5.4.3 MODE OF ACTION OF FULVIC ACID

Crop output is reportedly increased by the application of fulvic acid, although very little is known about its workings. Fulvic acid works effectively when formulated with a microbial inoculant and directly applied to plants. It increased rhizobium growth and stimulated root nodulation in *Medicago sativa.* The RNA transcript analysis of fulvic acid-treated plants highlighted the upregulation of early nodulation signaling genes (Capstaff et al., 2020). Fulvic acid appears to aid in the tolerance to drought; according to another study, in *Paeonia ostii* under drought conditions, it lowers membrane lipid peroxidation, serves as an antioxidant by removing ROS, and lessens damage to photosynthesis (Fang et al., 2020). In another study, fulvic acid aids in drought tolerance in tea plants by stimulating the expression of genes related to starch and sucrose metabolism, phenylpropanoid biosynthesis, and triterpenoid biosynthesis (Qiu et al., 2021). Fulvic acid is the best-known bioactive molecule that increases the potassium level in leaves. In addition to accelerating plant metabolism and enzymatic reactions like transaminase and invertase, they also speed up crop growth and nutrient uptake (Figure 5.5).

Further, fulvic acid enhances the permeability of membranes. Fulvic acid promoted the growth of high-grade flowers such as *Clivia* and vegetables such as cucumber and pepper (Gong et al., 2020). Fulvic acid improves the soil's ability to hold water and offers the ideal environment for good microbes (Sootahar et al., 2019). Fulvic acid acts as a buffering substance by influencing the primary and secondary metabolisms of plants. It targets the activation of the H^+-ATPase pump, leading to calcium channel activation. It helps in reducing the surface tension of water, which helps in the easy emulsification and dispersion of pesticides (Hayase and Tsubota, 1986).

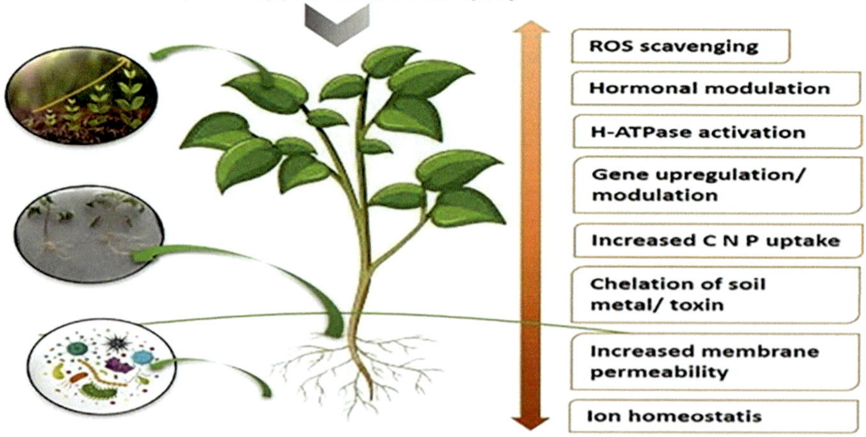

FIGURE 5.5 The mode of action of humic acid and fulvic acid when applied to different plant.

5.5 OMICS-BASED CHARACTERIZATION OF BIOSTIMULANTS

Many unanswered questions regarding the activity and properties of BS are still waiting for researchers' attention. Advances in omics technologies and high-throughput instruments have made it convenient for scientists to comprehensively provide insights into the properties, activities, and workings of BS. The development of omics techniques and access to large datasets have made it possible to study the workings and activities of molecules at the molecular level. Omics is the collection of studies of different biomolecules (lipids, proteins, genes, RNA, and metabolites), where one field is dedicated to the investigation of one molecule only. For instance, metabolomics deals with the study of the whole cellular metabolite pool; transcriptomics for RNA; lipidomics for lipids; epigenomics for modified DNA; and histones in chromosomes (Misra et al., 2019). Of many motivations, the key point of undertaking omics studies is to comprehensively understand the biological system. The application of BS promotes crop growth and development by impacting different cellular and physiological processes. Modulation in the mRNA level in response to BS in crops is vital to investigate using transcriptomics in order to trace the effect at the transcript level. This reflects the expression of which gene gets affected in response to BS and thus gives insights into the workings of BS. Usually, plants respond to BS by altering the expression of genes that act as activators, which in turn activate the cascade of gene expression. Besides, the regulatory pathways involved in crosstalk and signal transmission also get activated. These studies also help in finding the molecular marker responsible for plant stress responses under stress conditions. Moreover, the insights in BS mitigate stress in plants.

Microarray is the most widely adopted way of getting the transcriptome profile of a whole cell so that the complete set of genes can be studied in one go. A similar microarray-based study delineated the mode of action of humic acid and reported differential expression of carbon, nitrogen, and sulfur metabolic pathways. The genes particularly responsible for coding proteins have a role in uptake, transport, and assimilation and were primarily impacted (Ertani et al., 2017; Jannin et al., 2012). BS also changed the transcripts of genes related to secondary metabolism (phenol, terpene) and primary metabolism (photosynthesis, nutrient absorption, and plant development). Microarray has been used to elucidate the metabolic target, antioxidative pathways (SOD enzyme), and signaling processes

(MAPK) of a variety of old and new BS (EXPANDO; CYT31) (Blaszczak et al., 2015; Contartese et al., 2015). Undoubtedly, the mode of action and activity of BS vary with their source, but they also differ with the manufacturing strategy of BS, which is an extremely critical aspect of the biostimulant industry that needs to be regulated (Goni et al., 2016). In order to boost and provide further support to microarray-based studies, RT-PCR, qRT-PCR, and sequencing are some techniques that are employed to get more valid and complementary results (Du Jardin et al., 2020; Franzoni et al., 2019).

Next-genome sequencing (NGS) is another high-throughput, robust, and highly recommended technique used for genomic and transcriptomic studies, even in non-modelled plants. Illumina IG is the most widely used form of NGS where sufficient bioinformatic tools are present to store, analyze, and process large datasets. Super 50 is an *Ascophyllum nodosum* containing biostimulant that positively impacts the growth of *Arabidopsis thaliana.* NGS-based investigation has given insights into drought stress responses. The data showed involvement of the cell cycle (CYCP2 and 1) and plant growth (RD26, BES1) (Rasul et al., 2021). Besides delineating the mode of action of the BS, omics technologies are also employed to design new products as per need. Based on NGS analysis, the ratio of plant extract and seaweed was decided for formulating a biostimulant product for promoting the growth and production of corn and soyabean (Briglia et al., 2019). The combination was found to positively modulate the expression of nitrogen, phosphate, and cytokinin metabolism, along with carbon transport and metabolism genes, in corn crops. In addition to this, metal ion transport and amino acid biosynthesis pathway genes were upregulated in soyabean. Omics has also given a new insight into the role of humic acid in mitigating abiotic stress by activating both enzymatic and non-enzymatic defense pathways in plants (Canellas et al., 2020). In maize seedlings, humic acid enhanced the expression of gibberellic acid, auxin, and abscisic acid (ABA) metabolism genes. Metabolomics is an emerging field of omics that gives both qualitative and quantitative details about the metabolite pool (Dunn et al., 2005). The technique is the most robust among all omics tools, and sample preparation as well as reuse are the most convenient. The plant metabolome is complex as metabolites are categorized into various biochemical compound families with different chemical properties, and thus a combination of different analytical procedures is required to disseminate the metabolite profile.

NMR spectroscopy and mass spectrometry are the most widely used technologies for metabolic fingerprinting. Metabolomics also finds application in biostimulant research in defining the underpinning action of BS on plants (Nephali et al., 2020).

5.6 CONCLUSION

The use of BS is becoming more popular due to the growing desire for the adoption of environmentally friendly technologies that provide fewer health risks. Organic minerals, including humic acid and fulvic acid, are the most abundant minerals found in soil. Due to their abundance and easy accessibility, they can either be formulated or used as individual potent BS for improving crop productivity in terms of both quality and quantity. The benefit of using humic acid and fulvic acid as BS is not only restricted to plants but also to soil. These are the components that are responsible for improving soil health, with the soil microflora especially benefiting. These acids, as BS, are eco-friendly and cost-effective alternatives to reduce the use of synthetic chemicals in the agriculture sector.

KEYWORDS

- **amino acids**
- **biostimulants**
- **next-genome sequencing**
- **plant growth-promoting bacteria**
- **reactive oxygen species**
- **seaweed-based biostimulants**

REFERENCES

Ajmal, M., Ali, H. I., Saeed, R., Akhtar, A., Tahir, M., Mehboob, M. Z., & Ayub, A. (2018). Biofertilizer as an alternative for chemical fertilizers. *Agricultural and Food Sciences, Environmental Science*. https://www.rroij.com/peer-reviewed/biofertilizer-as-an-alternative-for-chemical-fertilizers-86649.html (accessed on 6th November 2025).

Albuzio, A., Ferrari, G., & Nardi, S. (1986). Effects of humic substances on nitrate uptake and assimilation in barley seedlings. *Canadian Journal of Soil Science*, *66*(4), 731–736.

Alewu, B., & Nosiri, C. (2011). Pesticides and human health. In *InTech* (pp. 231–250). https://www.intechopen.com/predownload/19601 (accessed on 6th November 2025).

Ali, O., Ramsubhag, A., & Jayaraman, J. (2021). Bio stimulant properties of seaweed extracts in plants: Implications towards sustainable crop production. *Plants (Basel)*, *10*(3), 531. https://doi.org/10.3390/plants10030531 (accessed on 6th November 2025).

Anjum, S. A., Wang, L. C., Farooq, M., Xue, L., & Ali, S. (2011). Fulvic acid application improves the maize performance under well-watered and drought conditions. *Journal of Agronomy and Crop Science*, *197*(6), 409–417.

Azad, M. O. K., Park, B. S., Adnan, M., Germ, M., Kreft, I., Woo, S. H., & Park, C. H. (2021). Silicon bio stimulant enhances the growth characteristics and fortifies the bioactive compounds in common and Tartary buckwheat plant. *Journal of Crop Science and Biotechnology*, *24*(1), 51–59.

Baloach, N., Yousaf, M., Parvez Akhter, W., Fahad, S., Ullah, B., Qadir, G., & Ahmed, Z. I. (2014). Integrated effect of phosphate solubilizing bacteria and humic acid on physio orphic attributes of maize. *International Journal of Current Microbiology and Applied Sciences*, *3*(6), 549–554.

Blaszczak, A. G., Smith, R., Gutierrez, A., Galbraith, D. W., Janda, J., Vanier, C., & Wozniak, E. M. (2015). Molecular mechanism of action for the novel bio stimulant CYT31 in plants exposed to drought stress. *II World Congress on the Use of Bio stimulants in Agriculture*, *1148*, 85–92.

Briglia, N., Petrozza, A., Hoeberichts, F. A., Verhoef, N., & Povero, G. (2019). Investigating the impact of bio stimulants on the row crops corn and soybean using high-efficiency phenotyping and next generation sequencing. *Agronomy*, *9*(11), 761. https://doi.org/10.3390/agronomy9110761 (accessed on 6th November 2025).

Buschmann, J., Kappeler, A., Lindauer, U., Kistler, D., Berg, M., & Sigg, L. (2006). Arsenate and arsenate binding to dissolved humic acids: Influence of pH, type of humic acid, and aluminum. *Environmental Science & Technology*, *40*(19), 6015–6020.

Calvo, P., Nelson, L., & Kloepper, J. W. (2014). Agricultural uses of plant bio stimulants. *Plant and Soil*, *383*, 3–41. https://doi.org/10.1007/s11104-014-2131-8 (accessed on 6th November 2025).

Canellas, L. P., Balmori, D. M., Médici, L. O., Aguiar, N. O., Campostrini, E., Rosa, R. C. C., Façanha, A. R., & Olivares, F. L. (2013). A combination of humic substances and *Herbaspirillum seropedicae* inoculation enhances the growth of maize (*Zea mays* L.). *Plant and Soil*, *366*(1), 119–132.

Canellas, L. P., Canellas, N. O. A., da S. Irineu, L. E. S., & Olivares, F. L., & Piccolo, A. (2020). Plant chemical priming by humic acids. *Chemical and Biological Technologies in Agriculture*, *7*(1), 1–17.

Canellas, L. P., Olivares, F. L., Aguiar, N. O., Jones, D. L., Nebbioso, A., Mazzei, P., & Piccolo, A. (2015). Humic and fulvic acids as bio stimulants in horticulture. *Scientia Horticulture*, *196*, 15–27.

Canellas, L. P., Teixeira, L. R. L., Dobbss, L. B., Silva, C. A., Medici, L. O., Zandonadi, D. B., & Façanha, A. R. (2008). Humic acids cross interactions with root and organic acids. *Annals of Applied Biology*, *153*(2), 157–166.

Capstaff, N. M., Morrison, F., Cheema, J., Brett, P., Hill, L., Munõz-García, J. C., Khimyak, Y. Z., Domoney, C., & Miller, A. J. (2020). Fulvic acid increases forage legume growth inducing preferential up-regulation of nodulation and signaling-related genes. *Journal of Experimental Botany, 71*(18), 5689–5704.

Carletti, P., García, A. C., & Silva, C. A., & Merchant, A. (2021). Editorial: Towards a functional characterization of plant bio stimulants. *Frontiers in Plant Science, 12*, 677772. https://doi.org/10.3389/fpls.2021.677772 (accessed on 6th November 2025).

Carmody, N., Goñi, O., Łangowski, Ł., & O'Connell, S. (2020). *Ascophyllum nodosum* extract bio stimulant processing and its impact on enhancing heat stress tolerance during tomato fruit set. *Frontiers in Plant Science, 11*, 807. https://doi.org/10.3389/fpls.2020.00807 (accessed on 6th November 2025).

Castiglione, A. M., Mannino, G., Contartese, V., Bertea, C. M., & Ertani, A. (2021). Microbial bio stimulants as response to modern agriculture needs: Composition, role and application of these innovative products. *Plants, 10*(8), 1533. https://doi.org/10.3390/plants10081533 (accessed on 6th November 2025).

Chakdar, H., Jadhav, S. D., Dhar, D. W., & Pabbi, S. (2012). Potential applications of blue green algae. Chen, T. H. H., & Murata, N. (2011). Glycine betaine protects plants against abiotic stress: Mechanisms and biotechnological applications. *Plant, Cell and Environment, 34*(1), 1–20.

Chen, Y., Senesi, N., & Schnitzer, M. (1978). Chemical and physical characteristics of humic and fulvic acids extracted from soils of the Mediterranean region. *Geoderma, 20*(2), 87–104.

Chiaiese, P., Corrado, G., Colla, G., Kyriacou, M. C., & Rouphael, Y. (2018). Renewable sources of plant bio stimulation: Microalgae as a sustainable means to improve crop performance. *Frontiers in Plant Science, 9*, 1782. https://doi.org/10.3389/fpls.2018.01782 (accessed on 6th November 2025).

Chuan, L., Zheng, H., Sun, S., Wang, A., Liu, J., Zhao, T., & Zhao, J. (2019). A sustainable way of fertilizer recommendation based on yield response and agronomic efficiency for Chinese cabbage. *Sustainability, 11*(16), 4368. https://doi.org/10.3390/su11164368 (accessed on 6th November 2025).

Colla, G., Nardi, S., Cardarelli, M., Ertani, A., Lucini, L., Canaguier, R., & Rouphael, Y. (2015). Protein hydrolysates as bio stimulants in horticulture. *Scientia Horticulture, 196*, 28–38. https://doi.org/10.1016/j.scienta.2015.08.037 (accessed on 6th November 2025).

Contartese, V., Garabello, C., Occhipinti, A., Barbero, F., & Bertea, C. M. (2015). Effects of a new bio stimulant on gene expression and metabolic responses of tomato plants. *II World Congress on the Use of Bio stimulants in Agriculture, 1148*, 35–42.

Di Mola, I., Cozzolino, E., Ottaiano, L., Nocerino, S., Rouphael, Y., Colla, G., El-Nakhel, C., & Mori, M. (2020). Nitrogen use and uptake efficiency and crop performance of baby spinach (*Spinacia oleracea* L.) and lamb's lettuce (*Valerianella locusta* L.) grown under variable sub-optimal N regimes combined with plant-based bio stimulant application. *Agronomy, 10*(2), 278. https://doi.org/10.3390/agronomy10020278 (accessed on 6th November 2025).

Dias, M. C., Araújo, M., Silva, S., & Santos, C. (2022). Sustainable olive culture under climate change: The potential of bio stimulants. *Horticulture, 8*(11), 1048. https://doi.org/10.3390/horticulturae8111048 (accessed on 6th November 2025).

Drobek, M., Frąc, M., & Cybulska, J. (2019). Plant bio stimulants: Importance of the quality and yield of horticultural crops and the improvement of plant tolerance to abiotic stress – a review. *Agronomy*, *9*(6), 335. https://doi.org/10.3390/agronomy9060335 (accessed on 6th November 2025).

Du Jardin, P., Xu, L., & Geelen, D. (2020). Agricultural functions and action mechanisms of plant bio stimulants (PBs): An introduction. In D. Geelen & L. Xu (Eds.), *The Chemical Biology of Plant Bio Stimulants* (pp. 1–30). John Wiley & Sons Ltd.

Dunn, W. B., & Ellis, D. I. (2005). Metabolomics: Current analytical platforms and methodologies. *TrAC Trends in Analytical Chemistry*, *24*(4), 285–294.

Ertani, A., Schiavon, M., & Nardi, S. (2017). Transcriptome-wide identification of differentially expressed genes in *Solanum lycopersicon* L. in response to an alfalfa-protein hydrolysate using microarrays. *Frontiers in Plant Science*, *8*, 1159. https://doi.org/10.3389/fpls.2017.01159 (accessed on 6th November 2025).

Fang, Z., Wang, X., Zhang, X., Zhao, D., & Tao, J. (2020). Effects of fulvic acid on the photosynthetic and physiological characteristics of *Paeonia ostii* under drought stress. *Plant Signaling and Behavior*, *15*(7), 1774714. https://doi.org/10.1080/15592324.2020.1774714 (accessed on 6th November 2025).

Flaig, W., Beutelspacher, H., & Rietz, E. (1975). Chemical composition and physical properties of humic substances. In J. E. Gieseking (Ed.), *Soil Components. Volume I. Organic Components* (pp. 1–22). Springer-Verlag.

Franzoni, G., Cocetta, G., Trivellini, A., & Ferrante, A. (2019). Transcriptional regulation in rocket leaves as affected by salinity. *Plants*, *9*(1), 20. https://doi.org/10.3390/plants9010020 (accessed on 6th November 2025).

Gaffney, J. S., Marley, N. A., & Clark, S. B. (1996a). Humic and fulvic acids and organic colloidal materials in the environment. *Humic and Fulvic Acids, ACS Symposium Series*, *652*, 2–16. https://pubs.acs.org/doi/10.1021/bk-1996-0651.ch001 (accessed on 6th November 2025).

Garcia-Gonzalez, J., & Sommerfeld, M. (2016). Biofertilizer and bio stimulant properties of the microalga *Acutodesmus dimorphus*. *Journal of Applied Phycology*, *28*(2), 1051–1061.

García-Sánchez, F., Simón-Grao, S., Navarro-Pérez, V., & Alfosea-Simón, M. (2022). Scientific advances in bio stimulation reported in the 5th Bio stimulant World Congress. *Horticulture*, *8*(7), 665. https://doi.org/10.3390/horticulturae8070665 (accessed on 6th November 2025).

Gong, G. Q., Yuan, X., Zhang, Y. J., Li, Y. J., Liu, W. X., Wang, M., Zhao, Y. F., & Xu, L. W. (2020). Characterization of coal-based fulvic acid and the construction of a fulvic acid molecular model. *RSC Advances*, *10*(9), 5468–5477.

Goni, O., Fort, A., Quille, P., McKeown, P. C., Spillane, C., & O'Connell, S. (2016). Comparative transcriptome analysis of two *Ascophyllum nodosum* extract bio stimulants: Same seaweed but different. *Journal of Agricultural and Food Chemistry*, *64*(14), 2980–2989.

González-Pérez, B. K., Rivas-Castillo, A. M., Valdez-Calderón, A., & Gayosso-Morales, M. A. (2022). Microalgae as bio stimulants: A new approach in agriculture. *World Journal of Microbiology and Biotechnology*, *38*(1), 1–12.

Gunnell, D., Eddleston, M., Phillips, M. R., & Konradsen, F. (2007). The global distribution of fatal pesticide self-poisoning: Systematic review. *BMC Public Health*, *7*, 357. https://doi.org/10.1186/1471-2458-7-357 (accessed on 6th November 2025).

Gupta, P., Gupta, H., & Poluri, K. M. (2021). Biomedical applications of polysaccharide-based nanocomposites from fungal origin. In S. Das & H. R. Das (Eds.), *Microbial and Natural Macromolecules: Synthesis and Application* (pp. 233–272). Academic Press. https://doi.org/10.1016/B978-0-12-820084-1.00010-7 (accessed on 6th November 2025).

Hadavi, E., & Ghazijahani, N. (2022). Simple organic acids as plant bio stimulants. In N. Ramawat & V. Bhardwaj (Eds.), *Bio Stimulants: Exploring Sources and Applications. Plant Life and Environment Dynamics* (pp. 71–105). Springer, Singapore. https://doi.org/10.1007/978-981-16-7080-0_4 (accessed on 6th November 2025).

Hasanuzzaman, M., Parvin, K., Bardhan, K., Nahar, K., Anee, T. I., Masud, A. A. C., & Fotopoulos, V. (2021). Bio stimulants for the regulation of reactive oxygen species metabolism in plants under abiotic stress. *Cells*, *10*(10), 2537. https://doi.org/10.3390/cells10102537 (accessed on 6th November 2025).

Hayase, K., & Tsubota, H. (1983). Sedimentary humic acid and fulvic acid as surface active substances. *Geochimica et Cosmochimica Acta*, *47*(5), 947–952. https://doi.org/10.1016/0016-7037(83)90160-6 (accessed on 6th November 2025).

Hernandez, O. L., Calderín, A., Huelva, R., Martínez-Balmori, D., Guridi, F., Aguiar, N. O., Olivares, F. L., & Canellas, L. P. (2015). Humic substances from vermicompost enhance urban lettuce production. *Agronomy for Sustainable Development*, *35*(1), 225–232. https://doi.org/10.1007/s13593-014-0221-x (accessed on 6th November 2025).

Jannin, L., Arkoun, M., Ourry, A., Laîné, P., Goux, D., Garnica, M., Fuentes, M., Francisco, S. S., Baigorri, R., & Cruz, F. (2012). Microarray analysis of humic acid effects on *Brassica napus* growth: Involvement of N, C and S metabolisms. *Plant and Soil*, *359*(1), 297–319. https://doi.org/10.1007/s11104-012-1191-x (accessed on 6th November 2025).

Ju, X., Gu, B., Wu, Y., & Galloway, J. N. (2016). Reducing China's fertilizer use by increasing farm size. *Global Environmental Change*, *41*, 26–32.

Junek, R., Morrow, R., Schoenherr, J. I., Schubert, R., Kallmeyer, R., Phull, S., & Klöcking, R. (2009). Bimodal effect of humic acids on the LPS-induced TNF-α release from differentiated U937 cells. *Phytomedicine*, *16*(5), 470–476.

Kapoore, R. V., Wood, E. E., & Llewellyn, C. A. (2021). Algae bio stimulants: A critical look at microalgal bio stimulants for sustainable agricultural practices. *Biotechnology Advances*, *49*, 107754. https://doi.org/10.1016/j.biotechadv.2021.107754 (accessed on 6th November 2025).

Kodama, H., & Schnitzer, M. (1967). X-ray studies of fulvic acid, a soil humic compound. *Fuel*, *46*(2), 87–94.

Kurniati, E., Muljani, S., Virgani, D. G., & Neno, B. P. (2018). Humic acid isolations from lignite by ion exchange method. *Journal of Physics: Conference Series*, *953*(1), 012234. https://iopscience.iop.org/article/10.1088/1742-6596/953/1/012234 (accessed on 6th November 2025).

Lin, Y., & Jones, M. L. (2022). Evaluating the growth-promoting effects of microbial bio stimulants on greenhouse floriculture crops. *Hort Science*, *57*(1), 97–109.

Mahler, C. F., Dal Santo Svierzoski, N., & Bernardino, C. A. R. (2021). Chemical characteristics of humic substances in nature. In A. Makan (Ed.), *Humic Substances* (pp. 1–14). IntechOpen. https://www.intechopen.com/chapters/76810 (accessed on 6th November 2025).

Md Shaiful Islam, K., Islam, K., Schuhmacher, A., & Gropp, J. (2005). Humic acid substances in animal agriculture. *Pakistan Journal of Nutrition*, *4*(3), 126–134.

Misra, B. B., Langefeld, C., Olivier, M., & Cox, L. A. (2019). Integrated omics: Tools, advances and future approaches. *Journal of Molecular Endocrinology*, *62*(1), R21–R45.

Muscolo, A., Bovalo, F., Gionfriddo, F., & Nardi, S. (1999). Earthworm humic matter produces auxin-like effects on *Daucus carota* cell growth and nitrate metabolism. *Soil Biology and Biochemistry*, *31*(9), 1303–1311.

Nardi, S., Pizzeghello, D., Muscolo, A., & Vianello, A. (2002). Physiological effects of humic substances on higher plants. *Soil Biology and Biochemistry*, *34*(11), 1527–1536.

Navarro-León, E., López-Moreno, F. J., Borda, E., Marín, C., Sierras, N., Blasco, B., & Ruiz, J. M. (2022). Effect of L-amino acid-based bio stimulants on nitrogen use efficiency (NUE) in lettuce plants. *Journal of the Science of Food and Agriculture*, *102*(15), 7098–7106.

Nephali, L., Piater, L. A., Dubery, I. A., Patterson, V., Huyser, J., Burgess, K., & Tugizimana, F. (2020). Bio stimulants for plant growth and mitigation of abiotic stresses: A metabolomics perspective. *Metabolites*, *10*(12), 505. https://doi.org/10.3390/metabo10120505 (accessed on 6th November 2025).

Ogner, G., & Schnitzer, M. (1971). Chemistry of fulvic acid, a soil humic fraction, and its relation to lignin. *Canadian Journal of Chemistry*, *49*(7), 1053–1063. https://doi.org/10.1139/v71-175 (accessed on 6th November 2025).

Olivares, F. L., Busato, J. G., de Paula, A. M., da Silva Lima, L., Aguiar, N. O., & Canellas, L. P. (2017). Plant growth promoting bacteria and humic substances: Crop promotion and mechanisms of action. *Chemical and Biological Technologies in Agriculture*, *4*(1), 30. https://doi.org/10.1186/s40538-017-0112-x (accessed on 6th November 2025).

Pascual, J. A., García, C., & Hernandez, T. (1999). Comparison of fresh and composted organic waste in their efficacy for the improvement of arid soil quality. *Bioresource Technology*, *68*(3), 255–264. https://doi.org/10.1016/S0960-8524(98)00160-6 (accessed on 6th November 2025).

Patil, H. J., & Solanki, M. K. (2016). Microbial inoculant: Modern era of fertilizers and pesticides. In D. Singh, H. Singh, & R. Prabha (Eds.), *Microbial Inoculants in Sustainable Agricultural Productivity* (pp. 319–343). Springer, New Delhi. https://doi.org/10.1007/978-81-322-2647-5_19 (accessed on 6th November 2025).

Piccolo, A., Nardi, S., & Concheri, G. (1996). Macromolecular changes of humic substances induced by interaction with organic acids. *European Journal of Soil Science*, *47*(3), 319–328. https://doi.org/10.1111/j.1365-2389.1996.tb01405.x (accessed on 6th November 2025).

Pilon-Smits, E. A. H., Quinn, C. F., Tapken, W., Malagoli, M., & Schiavon, M. (2009). Physiological functions of beneficial elements. *Current Opinion in Plant Biology*, *12*(3), 267–274. https://doi.org/10.1016/j.pbi.2009.04.009 (accessed on 6th November 2025).

Qiu, C., Sun, J., Shen, J., Zhang, S., Ding, Y., Gai, Z., Fan, K., Song, L., Chen, B., Ding, Z., & Wang, Y. (2021). Fulvic acid enhances drought resistance in tea plants by regulating the starch and sucrose metabolism and certain secondary metabolism. *Journal of Proteomics*, *247*, 104337. https://doi.org/10.1016/j.jprot.2021.104337 (accessed on 6th November 2025).

Rasul, F., Gupta, S., Olas, J. J., Gechev, T., Sujeeth, N., & Mueller-Roeber, B. (2021). Priming with a seaweed extract strongly improves drought tolerance in *Arabidopsis*.

International Journal of Molecular Sciences, *22*(3), 1469. https://doi.org/10.3390/ijms22031469 (accessed on 6th November 2025).

Rouphael, Y., Giordano, M., Cardarelli, M., Cozzolino, E., Mori, M., Kyriacou, M. C., Bonini, P., & Colla, G. (2018). Plant- and seaweed-based extracts increase yield but differentially modulate nutritional quality of greenhouse spinach through bio stimulant action. *Agronomy*, *8*(7), 126. https://doi.org/10.3390/agronomy8070126 (accessed on 6th November 2025).

Savvas, D., & Ntatsi, G. (2015). Bio stimulant activity of silicon in horticulture. *Scientia Horticulture*, *196*, 66–81.

Schnitzer, M. (1969). Reactions between fulvic acid, a soil humic compound and inorganic soil constituents. *Soil Science Society of America Journal*, *33*(1), 75–81. https://doi.org/10.2136/sssaj1969.03615995003300010022x (accessed on 6th November 2025).

Senn, T. L. (2007). A review of humus and humic acids. *Agricultural and Food Sciences, Environmental Science*. https://www.semanticscholar.org/paper/A-REVIEW-OF-HUMUS-AND-HUMIC-ACIDS-by-T. L. Senn%2C-of-Senn/3bea9de48efb352f25052c700012b8613a45513e (accessed on 6th November 2025).

Shukla, P. S., Mantin, E. G., Adil, M., Bajpai, S., Critchley, A. T., & Prithiviraj, B. (2019). *Ascophyllum nodosum*-based bio stimulants: Sustainable applications in agriculture for the stimulation of plant growth, stress tolerance, and disease management. *Frontiers in Plant Science*, *10*, 655. https://doi.org/10.3389/fpls.2019.00655 (accessed on 6th November 2025).

Shuqin, J., & Fang, Z. (2018). Zero growth of chemical fertilizer and pesticide use: China's objectives, progress and challenges. *Journal of Resources and Ecology*, *9*(sp1), 50–58.

Skogerboe, R. K., & Wilson, S. A. (1981). Reduction of ionic species by fulvic acid. *Analytical Chemistry*, *53*(2), 228–232.

Sootahar, M. K., Zeng, X., Su, S., Wang, Y., Bai, L., Zhang, Y., Li, T., & Zhang, X. (2019). The effect of fulvic acids derived from different materials on changing properties of albic black soil in the Northeast Plain of China. *Molecules*, *24*(8), 1535. https://doi.org/10.3390/molecules24081535 (accessed on 6th November 2025).

Stasińska-Jakubas, M., & Hawrylak-Nowak, B. (2022). Protective, bio stimulating, and eliciting effects of chitosan and its derivatives on crop plants. *Molecules*, *27*(9), 2801. https://doi.org/10.3390/molecules27092801 (accessed on 6th November 2025).

Suffet, I. H., & Patrick, M. (1988). *Aquatic Humic Substances: Influence On Fate and Treatment of Pollutants*. https://www.semanticscholar.org/paper/Aquatic-Humic-Substances%3A-Influence-on-Fate-and-of-Suffet-MacCarthy/95e95c783e93dcc5c6b31e9af98ddd0b70640b37 (accessed on 6th November 2025).

Trevisan, S., Francioso, O., Quaggiotti, S., & Nardi, S. (2010). Humic substances biological activity at the plant-soil interface: From environmental aspects to molecular factors. *Plant Signaling & Behavior*, *5*(6), 635–643.

Van Oosten, M. J., Pepe, O., De Pascale, S., Silletti, S., & Maggio, A. (2017). The role of bio stimulants and bio effectors as alleviators of abiotic stress in crop plants. *Chemical and Biological Technologies in Agriculture*, *4*, 5. https://doi.org/10.1186/s40538-017-0089-5 (accessed on 6th November 2025).

Van Rensburg, C. E. (2015). The anti-inflammatory properties of humic substances: A mini review. *Phytotherapy Research*, *29*(6), 791–795. https://doi.org/10.1002/ptr.5319 (accessed on 6th November 2025).

Yakhin, O. I., Lubyanov, A. A., Yakhin, I. A., & Brown, P. H. (2017). Bio stimulants in plant science: A global perspective. *Frontiers in Plant Science*, *7*, 2049. https://doi.org/10.3389/fpls.2016.02049 (accessed on 6th November 2025).

Yassin, M., Ton, J., Rolfe, S. A., Valentine, T. A., Cromey, M., Holden, N., & Newton, A. C. (2021). The rise, fall and resurrection of chemical-induced resistance agents. *Pest Management Science*, *77*(9), 3900–3909. https://doi.org/10.1002/ps.6370 (accessed on 6th November 2025).

CHAPTER 6

Chitosan and Nanoparticulate Chitosan as Plant Growth Stimulants and Protectants for Sustainable Agriculture

SHIVANGNI SINGH, VAIDEHI BHAGWAT, G. MAHESWARI, DEEPALI CHOUDHARI, and VANDANA GHORMADE*

Nanobioscience Group, Agharkar Research Institute, Pune, Maharashtra, India

*Corresponding author

ABSTRACT

Chitosan is sourced from chitinous crustacean shells or fungal cell walls. Chitosan is a unique natural cationic polymer containing D-glucosamine and *N*-acetyl-D-glucosamine. Chitosan is extracted from crustacean waste by chemical demineralization and deacetylation. Chitosan is characterized by the degree of deacetylation (DD) and the molecular weight of the polymer. Recently, there has been a surge of interest in the agricultural applications of chitosan due to its natural origin, cationic nature, biocompatibility, biodegradability, and muco-adhesiveness. The polymer is reported to play an important role as a plant growth promoter and stimulant for seed germination. The application of chitosan protects plants, fruits, or seeds and reduces the incidence of disease by microbes. Furthermore, chitosan can be formulated as a nanocarrier for the delivery of fertilizers and pesticides. Chitosan and nanoparticulate chitosan have immense benefits for sustainable agriculture.

Plant Biostimulants and Growth Regulators in Sustainable Agriculture.
Deepak Kumar, Sanjay K. Singh, Nutan Kaushik, and Mahaveer P. Sharma (Eds.)

DOI: 10.1201/9781779641458-8

6.1 INTRODUCTION

Chitosan is a glucosamine-containing natural polymer that is often found in the cell walls of fungi or in insect cuticles. Chitosan is obtained by deacetylation of chitin, an abundant natural polysaccharide second to cellulose (Amiri et al., 2022). Chitin and chitosan are natural linear co-polymers of D-glucosamine and *N*-acetyl-D-glucosamine linked by β-1,4-glycosidic bonds. The deacetylation degree (DD) and their solubility in dilute acid determine the nature of the polymer. The deacetylation degree is the ratio of the amino groups ($–NH_2$) to the acetylamino groups ($–NHCOCH_3$) present in chitin and is often expressed as a percentage. The English scientist A. Hachett was the first to report the extraction of the material from crustacean shells and its resistance to ordinary chemicals in 1799. Later, H. Braconnot, a French researcher, isolated chitin from fungi and termed it fungine. Eventually, A. Odier isolated the material from insect cuticles and named it chitin (Greek: chiton—tunic, coat). Later, F. Hoppe-Seiler, a German chemist, termed the deacetylated form of chitin "chitosan" in 1894. Chitosan has a higher deacetylation degree than chitin and is soluble in dilute acids. Chitosan is naturally present in fungal cell walls and insect cuticles. Chitosan is often sourced by deacetylating chitin from abundant marine invertebrate shells generated as seafood processing industry waste. Chitosan is formed either by chemical or enzymic deacetylation. The chemical process involves demineralizing chitin by acid treatment to dissolve calcium carbonate in crustacean shells (shrimps, crabs, lobsters, and krill). It is followed by deproteinization by alkaline solutions like NaOH or KOH. The deacetylation process leading to chitosan formation occurs by the removal of the acetyl groups ($–CH_3CO$) from chitin and depends chiefly on the concentration of the aqueous NaOH solution (40–50%), the reaction temperature (90–120°C), and its duration. Enzymic chitin deacetylation occurs by the enzyme chitin deacetylase, which is present in the fungal cell walls and insect cuticles. Chitin deacetylase deacetylates chitin by removing the acetyl groups ($–CH_3CO$) from chitin, leading to chitosan formation.

6.2 CHITOSAN FROM CRUSTACEAN SOURCES

Waste crustacean shells from the seafood industry are the main source of the commercial production of chitosan. Crustacean shell waste contributes over 10,000 tons of chitin annually (Merzendorfer, 2011). Commonly, chitosan is extracted from crab, shrimp, and krill shells, while lobster and crayfish shells are also used. The chitin content of crustacean shells is 20–30%, while lobsters of *Nephrops* and *Homarus* sp. contain high chitin (60–75%) contents (Kou et al., 2021). In 2015, high crustacean production of 13.7 million metric tons globally generated 6–8 million tons of shell waste (Amiri et al., 2022). The extensive seafood industry generates enormous waste that is discarded, burned, buried, or left to degrade. The extraction of chitin and chitosan can reduce waste and potential environmental hazards and yield valuable polymers with exceptional properties and wide applications.

Commercially, chitosan is produced by exposure of crustacean shell chitin to strong alkalis (30%–50% w/v) and high temperatures (80–150°C) for extended periods to carry out the deacetylation process (Huq et al., 2022). Generally, the material is first demineralized to reduce the presence of calcium carbonate and then deproteinized to remove the protein contents. However, shellfish raw material supplies are limited and dependent on seasonal variations (Huq et al., 2022). Moreover, the chemically processed chitosan displays heterogeneous polymer properties like deacetylation level and chain length (Huq et al., 2022). Grants related to exploring chitin- and chitosan-biopolymers have increased by 55 million euros in Europe. Projects such as value-added production reduce the environmental impact of the production process, and green methods for chitin- and chitosan-biopolymers have been initiated. Academic and industrial collaborators have reported a two-stage green fermentation using microbial organisms like *Serratia marcescens* and *Lactobacillus plantarum*. The successful depolymerization of chitin into chitosan was reported after microbial demineralization and deproteination. The Nano3Bio project, with a grant value of 12 million euros, sought to develop an eco-friendly valorization of marine waste. The agricultural applications of chitin and chitosan are estimated to be US$60 billion. The utilization of these biopolymers for agrarian applications as a formulation component for pesticide delivery and for soil enrichment increased due to their biocompatibility, biodegradability, and eco-friendliness. In addition to these properties, the film-forming capacity of chitosan makes it suitable for application as a food packaging material. The chelating properties of cationic chitosan as a flocculant and coagulant are essential for wastewater treatment.

First-generation chitosan is defined as blended chitosan with batch-to-batch variations. Reproducible and controlled chemical processes produce well-defined second-generation chitosan. Third-generation Chitosan is generated by controlled recombinant production and is suitable for biomedical applications. Although both chemical and biological processes are developed for chitosan preparation, the simpler and quicker chemical process currently provides a cost-effective edge over the biological process for commercial production. The polymer products display optimal physiological characteristics, like low MW and high DD. In comparison, biological processes incorporate microbial degradation and fermentation for deproteination and deacetylation steps that are comparatively mild in specialized equipment and a purification process after fermentation. The chitosan products are generated by biological process refinement and chemical processes to obtain suitable quality. Kuo et al. (2021) suggested alternate methods involving unconventional chemical routes, like KOH deacetylation and the use of microwaves or ultrasounds, for industrial-scale chitosan production. However, further refinement of these routes is suggested before their consideration for industrial applications. In addition to unconventional routes, the integration of enzymic deacetylation with chemical deacetylation is proposed for obtaining sufficiently deacetylated chitosan. Kim et al. (2008) reported the role of *Mortierella* sp. extracellular chitin deacetylase in carrying out the deacetylation of crab shell chitosan. Kolodziejska et al. (2000) integrated chemical and enzymic treatments for the generation of chitosan from krill chitin. The two-stage process involved initial alkali treatment (50% NaOH) at 60–115°C to yield 55–95% DD chitosan. In the second stage, chemically deacetylated chitosan was subjected to extracellular chitin deacetylase from *Mucor rouxii,* which further enhanced the deacetylation degree of chitosan (88–99%). The chemical pretreatment contributed to a reduction in the crystallinity of crustacean chitin and the formation of amorphous material that enhanced enzymic deacetylation. Pareek et al. (2013) reported that chitin deacetylase from *Penicillium oxalicum* SAEM-51 could generate chitosan with ~ 80% degree of deacetylation (DD). Chen et al. (2017) proposed a mechanochemical process for the generation of low-molecular-weight chitosan with chemical treatment during mechanical milling. The process permitted the shearing of long-chain polymers with simultaneous deacetylation and depolymerization, thereby reducing the requirement for an alkaline base.

6.3 CHITOSAN FROM FUNGAL SOURCES

Fungi represent an alternate source of chitosan due to the presence of chitin and chitosan in their cell walls. The demineralization step for the recovery of chitosan is avoided due to the low ash contents of the fungal cell walls. Chitosan production from fungal sources is independent of season variations and provides more product consistency (Abo Elsoud and El Kady, 2019). The extraction of chitosan can provide additional value for the fungal product industry. Furthermore, large quantities of mycelial waste generated by biotechnological industries can be remediated effectively (Pellis et al., 2022). The presence of chitin in fungal cell walls and septa provides the strength required for maintaining cellular integrity and shape. Fungi belonging to different fungal classes like Ascomycetes, Zygomycetes, Basidiomycetes, and Deuteromycetes contain chitin, while chitosan is present in the cell walls of certain Zygomycetes fungi belonging to the Mucorales order. Cultivation of fungi for cell wall chitin/ chitosan recovery is possible with simple nutrients or cost-effective waste materials like soybean milk and mungbean noodles waste (Suntornsuk et al., 2002). Zygomycetous fungi like *Absidia* and *Mucor* are reported for their high production of chitosan. Pochanavanich and Suntornsuk (2002) studied the production and properties of chitosan from four different filamentous fungi: *Lentinus edodes* (Basidiomycetes) and *Pleurotus sajo-caju* (Basidiomycetes), *Rhizopus oryzae* (Zygomycetes), *Aspergillus Niger* (Deuteromycetes), and two Ascomycetes yeast strains, *Zygosaccharomyces rouxii* and *Candida albicans*. The filamentous fungi and yeasts were grown in submerged fermentation (SmF) in PDB and YMB broth, respectively, in a 250-ml flask. The highest chitosan production was by *Aspergillus niger* (107 mg/g) and *Rhizopus oryzae* (138 mg/g). Mushroom strains grew slowly, showing little chitosan production. The yield of chitosan contents from fungi depends on the fungal strains, media composition, cultivation methods like submerged or solid-state, and culture conditions (pH, temperature, etc.). The progress in fermentation technology for fungal cultivation permits the large-scale production of fungi as an alternate source of chitosan. Submerged-state fermentation is advantageous over solid-state fermentation (SSF) due to easier control over the fermentation conditions.

Nwe et al. (2011) compared solid-state and SmF for chitosan production from *G. butleri*. They obtained 12.7% and 9.2% chitosan yields from fungal biomass in SSF and SMF, respectively. While the DDA was > 92% for both methods, the SSF yielded low-molecular-weight chitosan, which was preferable. Cell wall chitosan was extracted from the mycelium of the dimorphic zygomycetous fungus, *Benjaminiella poitrasii,* grown by SmF in YPG medium. The dried mycelial biomass yielded 61 mg/g of chitosan, having 93% DD (Mane et al., 2022). Various biotech industries that use fungi for processes like brewing, baking, food, or the production of antibiotics, organic acids, enzymes, and secondary metabolites generate a large amount of waste fungal biomass. The cell wall polymers of fungi, such as chitin and chitosan, can be extracted from the mycelium and are a renewable source of chitosan production. *Aspergillus Niger* cultivation for citric acid production globally generates ~ 80,000 tons of waste mycelium (Wu et al., 2005). As compared to commercial crustaceans, source chitin has a high crystallinity (1.25), whereas chitin from *A. Niger* has a lower crystallinity (0.37) (Wu et al., 2005). Lv et al. (2016) treated *A. Niger* waste mycelium with *Lactococcus lactis* recombinant chitinase for the hydrolysis of fungal chitin for six days. The production of chitooligosaccharide, N, N′-diacetylchitobiose, was 10% of the waste mycelium with 70% purity. An enhanced chitooligosaccharide yield of 24%, having 78% purity, was reported with a combination of chitinase and snailase. Short-chain chitooligomers and their derivatives display improved water solubility and absorption and are helpful as bacteriostatic agents or food additives. Another potential source of fungal waste arises from the burgeoning mushroom industry. Mushrooms are a popular food and are cultivated worldwide for their consumption. Mushroom waste, consisting of stalks or irregularly sized mushrooms, amounts to 5–20% of production volume. Almost 50,000 metric tons of waste are generated annually, creating issues for disposal and the environment. Kannan et al. (2010) reported the extraction of chitosan from mushrooms like *Agaricus* sp., *Ganoderma* sp., and *Pleurotus* sp., ranging from 5.9–14% of dry weight with 82–89% DD.

The wine industry depends on the yeast-based fermentation of grape juice, generating wine lees as waste. Yeast cell walls are composed of chitin, chitosan, and other polymers. In Australia, the wine industry generates almost 1000 tons of waste. The spent yeast biomass can be effectively utilized as a source of chitosan. The cell wall chitosan extracted from fungi belonging to different wine yeasts such as *Hanseniaspora guilliermondii*, *Issatchenkia orientalis*, *Pichia membranifaciens*, and *Saccharomyces cerevisiae* was in the range of 15–31 mg/g dry biomass, while the DD was 70–89% (Mane et al., 2022). Value-added products can be developed by utilizing food-industry processing waste. *R. oryzae* produced 4.3 g/kg and 1.6 g/kg chitosan from soybean and mungbean residues, respectively (Suntornsuk et al., 2002). Yang et al. (2017) cultivated *Rhizopus oryzae* on xylose-rich lignocellulosic corn stover prehydrolysate and extracted 0.11 g/g of chitosan from the biomass. The chitosan had a high DD (91.27%) with a low viscosity (2.67 mPas) as compared to the chitosan from shellfish (87.25%), which had a high viscosity (22.25 mPas). Chitosan production from *Rhizopus arrhizus* was explored using corn steep liquor and honey as agro-industrial sources (Cardoso et al., 2012). They reported biomass and chitosan yield enhancements of 20.61 g/L and 29.3 mg/g, respectively. Berger et al. (2014) reported 30% higher biomass and chitosan yields in *R. arrhizus* than *C. elegans* in agricultural waste media containing molasses and corn steep liquor. *R. arrhizus* produced 49.31 and 40.67 $mg \cdot g^{-1}$ Chitosan, while *C. elegans* produced 26.29 and 33.13 $mg \cdot g^{-1}$ Chitosan in the molasses and corn steep liquor media, respectively. Both fungi displayed 82% DD with a viscosity of F3.79–3.40 cP and a low molecular weight (LMW) of 4.68–5.08 × 103 g/mol. A newer method for chitosan extraction, like microwave-assisted extraction (MAE), was used at 300 W for 22 min to extract 13.5% (w/w) chitosan from *Rhizopus oryzae* dried fungal biomass as compared to that of the conventional heating process (6.6% w/w) (Sebastian et al., 2019). MAE-extracted chitosan displayed high DD (95%), against conventional autoclaving (91%). MAE is an eco-friendly, efficient process of chitosan extraction that requires lower energy consumption.

Several studies show that chitosan is formed by the enzymatic deacetylation of chitin in the fungal cell walls. The chitin and chitosan pathways in fungi bring together the nitrogen and carbon synthesis pathways by converting fructose to glucosamine 6-phosphate. After acetylation followed by isomerization, *N*-acetylglucosamine-1-P reacts with uridine triphosphate to generate UDP-*N*-acetylglucosamine-1-P, the chitin precursor. Chitin synthase (CS, EC 2.4.1.16) utilized UDP-NAG as a GlNAc donor for the sequential addition of GlNAc to the non-reducing end of the growing chitin chain. In the next step, chitin deacetylase (CDA, EC 3.5.1.41) deacetylates GlNAc residues of chitin to form chitosan.

The synergistic activity of CS, a membrane-integrated enzyme, and CDA deacetylates the nascent chitin chain. Niederhofer and Muller (2004) modified the standard chitosan extraction method to directly extract LMW from fungi, avoiding the step of thermal or chemical depolymerization. Chitin deacetylase is reported from different fungi, such as *Mucor rouxii*, *Rhizopus oryzae*, *Colletotrichum lindemuthianum*, and *Aspergillus nidulans* (Chatterjee et al., 2008). However, the insolubility of the chitin substrate and its crystalline nature inhibit the deacetylase activity (Dhillon et al., 2012). The enzymatic deacetylation process for consistent product quality and an eco-friendly process warrants further exploration. Doiphode et al. (2009) reported that enzymatic deacetylation using extracellularly produced chitin deacetylase from *Metarhizium anisopliae* enhanced chitosan recovery by 15–20% from the zygomycetous fungus *B. poitrasii*. Cardoso et al. (2012) utilized food agribusiness byproducts as a substrate for *Rhizopus arrhizus* and obtained 29.30 mg of chitosan per gramme of dry mass of fungal mycelium. Berger et al. (2014) extracted 57.82 mg of chitosan per gramme of dry biomass of *Cunnigamella ellegans*. Batista et al. (2018) managed to increase yield by working with a medium composed of the byproduct of the shrimp industry, reaching up to 89.95 mg of chitosan per 1 g of biomass of *Rhizopus* spp.

6.4 PROPERTIES OF CHITOSAN

Chitosan is a natural polysaccharide polymer characterized by its viscosity, molecular weight, and chain length. The degree of deacetylation of chitosan often defines the polymer dissolution in aqueous solvents. Several techniques are utilized for the characterization of chitosan. These studies include viscosity determination by viscometry, molecular weight by gel permeation, and degree of deacetylation by infrared spectroscopy, or NMR. The characterization of the molecular weight is related to determining the polymer's intrinsic viscosity. The viscosity of chitosan solutions is determined by the movement of the solution in a temperature-stabilized Ubbelohde capillary. The molecular weight of chitosan is calculated as a function of the Mark-Houwink-Sakurada equation:

$$[\eta] = KM_w^{\ a}$$

$$K = 1.81 \times 10^{-5}\ dl/g;$$

$$a = 0.93$$

'K' and the exponent 'a' are viscometric constants for a particular solvent system

Several studies have described the viscometric constants for chitosan (Kasaai, 2008; Roberts and Domszy, 1982; Wang et al., 1991). Pochanavanich and Suntornsuk (2002) reported the viscosity of fungal chitosan as 3.1–6.2 centipoise (cP). Low-MW chitosan's are soluble at neutral pH, while higher-MW chitosans form highly viscous solutions and have low solubility at neutral pH. Gel permeation chromatography can also determine the molecular weight of the chitosan polymer. Columns are calibrated with pullulan standards; peaks are monitored with a refractive index detector. The molecular weight of fungal chitosan was 2.7×10^4 – 1.9×10^5 Da. Niederhofer and Muller (2004) reported the 4.5×10^4 Da molecular weight of fungal chitosan by the gel permeation method.

El-Sawy et al. (2010) demonstrated that an 80 KGy gamma irradiation dose reduced the molecular weight of crustacean chitosan from 1×10^7 to 3×10^6 Da. Mane et al. (2022) showed that the dimorphic zygomycetous fungus *Benjaminiella poitrasiii* displayed a higher molecular weight of chitosan (46 Kda) as compared to the yeast form (25 Kda), as determined by gel permeation chromatography. In comparison, chitosan from ascomycetous yeasts and basidiomycetous mushrooms showed the presence of lower molecular weight chitosan (~ 6.5 Kda). The deacetylation level of *B. poitrasiii* hyphae was 93% as compared to the ascomycetous yeasts (73–83%). Generally, fungal chitosan displays low-medium MWs (6.4×10^3 to 1.4×10^6 Da), while crustacean chitosan displays a higher molecular weight distribution (~ 1.5×10^6 Da) (Crognale et al., 2022). Low-MW chitosan's are characterized by their solubility at neutral pH and less viscous nature in solutions than high-MW chitosan's. Further, fungal sources of chitosan require less processing with acid or alkaline solutions during extraction and are more environmentally benign than crustacean chitosan. LMWCs and chitooligosaccharide's can be generated by the hydrolysis of chitosan, either by chemical or enzymatic methods.

Mainly, the presence of amino groups in chitosan is responsible for its physico-chemical properties. The presence of the amino group confers properties such as improved solubility and an interactive nature. The presence of amino groups can be determined by IR and NMR methods. FTIR spectroscopy is a user-friendly, quick, and sensitive method to estimate the degree of deacetylation of chitosan. The absorbance of the amide band at approximately 1655 cm^{-1} and the C-H band at 3450 cm^{-1} provided valuable information for measuring the degree of deacetylation of chitosan (Brugnerotto et al., 2001; Roberts and Domszy, 1982). Several studies have described the degree of deacetylation of chitosan by evaluating different absorption ratios such as A1560/A2875, A1655/A2875, A1655/A3450, A1320/A1420, and so forth (Kasaai, 2008; Mane et al., 2022).

Generally, a higher degree of chitosan deacetylation (DD) is associated with enhanced antibacterial activity as compared to lower DD (Kong et al., 2010). Chitosan is a natural polymer that displays non-toxicity, biodegradability, and biocompatibility, making it a suitable alternative to artificial polymers. Pochanavanich and Suntornsuk (2002) studied the deacetylation, mol wt, and viscosity of chitosan from four filamentous fungi and two yeast strains. They showed that the polymer DD% ranged from 83–97% with a viscosity of 3.2–6.2 centipoise. The mol/wt ranged from 0.27 to 1.9×10^5 Da. *Lentinus edodes* had the largest Mol wt polymer with the highest viscosity (1.9×10^5 Da). Low-molecular-weight chitosan is essential for its pharmaceutical and medical applications. Chitosan has applications as a hemostatic dressing, wound healing device, and drug delivery. Niederhofer and Muller (2004) reported directly extracting low-molecular-weight chitosan (average MW of 4.5×10^4 g/mol) from *Absidia coerulea* fungal mycelium, thereby avoiding chemical and thermal processes.

The modification of the chitosan extraction process with microwave-assisted extraction (MAE) is quicker. The MAE process resulted in chitosan with ~95% degree of deacetylation, which was higher than chitosan extracted by conventional heating (90.6%). However, the molecular weight of MAE chitosan (127.58 kDa) was similar to the chitosan from conventional heating (Sebastian et al., 2019). The biological activity of chitosan oligosaccharides depends on their properties, namely, their polymer length and degree and deacetylation pattern. Controlling polymer homogeneity by chemical processes is challenging due to the generation of random acetylation patterns in the chitosan oligosaccharides. Enzymically generated chitosan oligosaccharides by the action of chitin deacetylases of bacterial origin generate chitosan oligosaccharides with specific acetylation patterns (Grifoll-Romero et al., 2018). Recombinant chitin deacetylase from the fungus *Puccinia graminis f. sp. tritici* generates tetrameric to hexameric chitosan oligomers due to its activity on the chitin substrate (Naqvi et al., 2016). Mass spectrometric sequencing of the oligomers revealed the specific acetylation patterns of A-A-D-D,

A-A-D-D-D, and A-A-D-D-D-D, respectively (A, GlcNAc; D, GlcN). These results indicated that the two GlcNAc units at the non-reducing end of the oligomer were not deacetylated by the *P. graminis* CDA. Such enzymatically produced, well-defined chitosan oligomers have potential for a wide variety of applications (Naqvi et al., 2016).

6.5 ROLE OF CHITOSAN IN SEED GERMINATION AND PLANT GROWTH PROMOTION

Chitosan promotes plant growth by influencing the growth rate of roots, shoots, and flowering. Chitosan also helps to reduce stress damage in plants. It increases water uptake and other nutrients by modifying the cell's osmotic pressure and activating plant growth hormones. It also reduces the damage caused by reactive oxygen species (ROS) during seed germination through the activation of antioxidant enzymes (Chakraborty et al., 2020). It has been observed that plant responses to chitosan application depend on the chitosan's molecular weight, the functional groups present, the degree of deacetylation, and so forth. Also, chitosan is insoluble in normal water and, therefore, requires the presence of weak acids for dissolution. The lower molecular-weight derivatives of chitosan induce a similar response in plants, but the main advantage of this over native chitosan is that the former is quickly soluble in water (Ochoa-Meza et al., 2021). Therefore, along with native chitosan, its derivatives like chitosan lactate (Hawrylak-Nowak et al., 2021), oligochitosan (Hai et al., 2019), and gamma-irradiated chitosan (Dar et al., 2015) have also been used to study their effect on plant growth and disease control. The effect of seed priming with chitosan on the germination and seedling growth of wheat (*Triticum aestivum*) was studied by Hameed et al. (2013). Seed priming is a method of seed preparation. Here, the seeds are soaked before sowing. In this study, the seeds were soaked in aerated solutions of 0.1, 0.25, and 0.5% chitosan for 8 hours. Hydropriming (soaking of seeds in water) was taken as the control. Improved seed germination percentage and germination rate were observed in all chitosan priming treatments. An increase in antioxidant enzymes like catalase (CAT), superoxide dismutase (SOD), and peroxidase (POD) was observed with 0.5% chitosan-treated seeds, showing a two-fold increase in CAT activity. A decrease in the total phenolic content of seeds was observed for all the treatments. Increased levels of malondialdehyde were observed in chitosan-primed seeds. Chitosan-treated seeds showed a 37% increase in non-reducing sugar content. Increased seed reserve mobilizing enzyme activity was also observed in chitosan-treated seeds.

Hai et al. (2019) studied the effect of chito oligosaccharide (COS) on soyabean seed germination. COS (100 mg/L) improved the germination energy (3rd germination) and germination percentage to 91% and 93%, respectively. Coating the seed with COS was considered the most suitable method for enhancing seed germination. Liu et al. (2022) studied the effect of chitosan soaking on the seed germination of *Platycodon grandiflorus*, a medicinal plant used in Chinese medicine. The treatment significantly enhanced the germination rate, germination energy, germination index, cotyl length, and radicle length, with 0.155% chitosan showing the best results. The leaf width, length, and area were significantly enhanced by the 0.15–0.25% chitosan treatment. The height overground part weight of 0.15% was superior to all other treatments. The seeds treated with 0.15% chitosan had the highest photosynthetic rate and transpiration rate. Chitosan is reported to enhance the growth of plants and affect the seedling stage, plant height, chlorophyll content, leaf number, leaf area, and total protein content. The effect of foliar application of chitosan on plant growth and curcumin content in turmeric (*Curcuma longa* L.) plants was studied by Anusuya et al. (2016). Chitosan (0.1% w/v) was sprayed on 30-day-old plants with water and fungicide treatment with mancozeb (0.3%) as controls. Chitosan-treated plants showed an increased number of leaves and plants, increased shoot length, and fresh plant weight. In chitosan-treated plants, there is an increase in total protein content and defense-related enzymes like protease inhibitors, β-1,3 glucanases, peroxidases, and polyphenol oxidases. The treated plants showed a 4-fold increase in curcumin content.

Hawrylak-Nowak et al. (2021) studied the effect of foliar application of chitosan lactate (ChL) on basil (*Ocimum basilicum* L.) and lemon balm (*Melissa officinalis* L.) plants. Chitosan treatment (100 and 500 mg/L) to the plants 35 days after sowing was used to evaluate the effects on chlorophyll and carotenoid content. The effect of chitosan lactate was species-specific, with basil plants showing a slight decrease in chlorophyll (12%) and carotenoid (17%) content after 100 mg/L treatment. However, lemon balm plants treated with 500 mg/L showed increased chlorophyll a (6%) and chlorophyll b (9%) content. The 100 mg/L treatment increased lemon balm's total phenolic compound content by 13%. Anthocyanin content was also increased in lemon balm plants for both treatments. The accumulation of rosmarinic acid (RA) in both plants was ChL-dependent, with the 100 mg/L treatment showing a 32% increase in basil plants, and both 100 and 500 mg/L showing a 46% and 32% increase in lemon balm plants, respectively. ChL at a 100 mg/L concentration increased catalase (CAT) activity in basil by 35%. However, 500 mg/L led to a decrease in CAT activity but increased guaiacol peroxidase (GPOX) activity. The GPOX activity in lemon balm was increased dramatically by ChL treatment; no CAT activity was detected. Similar results were obtained by Salachna et al. (2022) in perilla plants and Stasinska-Jakubas et al. (2023) in lemon balm. Similar studies involving the use of chitosan for enhancing the production of secondary metabolites were carried out on peppermint (Ahmad et al., 2019), lemongrass (Mukarram et al., 2022), and *Dracocephalum kotschyi* (Kahromi et al., 2021). Rahman et al. (2018) studied the effect of varying doses of chitosan in the form of foliar application on fruit yield and antioxidant contents in strawberry fruit in field conditions. Four different concentrations of chitosan were used: 125, 250, 500, and 1000 ppm (pH 6.5). It was observed that the application of chitosan had a significant effect on the growth parameters of the strawberry plants. The highest increase in plant height, root length, canopy diameter, and leaf width were observed at 500 ppm, while the highest leaf length and leaf number were observed at 1000 ppm chitosan treatment. The highest fruit weight and fruit per plant were observed at 500 ppm chitosan treatment, and this treatment increased the fruit yield by 42%. The highest content of flavonoids and carotenoids was observed in the 1000 ppm treatment, while the highest content of anthocyanins was observed in the 500-ppm treatment. Chitosan treatment significantly increased the antioxidant activity, with 1000 ppm treatment showing the highest activity. Moreover, applications and advantages of chitosan in crop production for different crops are compiled in Table 6.1 and elaborated in Figure 6.1 (Palacio-Márquez et al., 2022).

6.6 ROLE OF CHITOSAN IN DISEASE CONTROL

In recent years, there has been a surge of interest in chitosan for agricultural applications due to its natural origin, cationic nature, and mucoadhesiveness (El Hadrami et al., 2010). Chitosan also possesses antimicrobial, antifungal, and antiviral properties. Chitosan (a natural biopolymer) is

TABLE 6.1 Applications and Advantages of Chitosan as a Plant Biostimulant in Crop Production

Crops	Modes of Application	Results	References
Capsicum annuum L.	Foliar spray	Growth in the quantity of fruits, leaves, and chlorophyll. Furthermore, it decreased the frequency of invasions by *Phytophthora capsici*.	Esyanti et al. (2019)
Capsicum annuum L. Cv. 'Yolo Wonder'	Foliar spray on leaves and fruits	Gains in fruit output, weight, and diameter	Mahmood et al. (2017)
Solanum lycopersicum L. Cv. 'BINAtomato-6'	Foliar spray	An increase in factors influencing the nitrate reductase enzyme's activity and yield	Mondal et al. (2016)
Solanum lycopersicum L. Cv. 'Marglobe'	Soil treatment/application	Growth in the parasitic fungus *Pochonia chlamydosporia*, which infects different species of Meloidogyne, colonizing roots.	Escudero et al. (2017)
Lycopersicon esculentum L.	Foliar spray	Reduced water stress, increased biomass, yield, and chlorophyll content	Hassnain et al. (2020)
Cucumis sativus L.	Foliar spray	Maintained rates of growth and production despite temperature stress	Ali et al. (2020)
Cucumis sativus L.	As a chitosan protective covering (edible coating) after harvest	Preserved quality and decreased emissions of carbon dioxide	Olawuyi et al. (2019)
Cucumis sativus L. Cv. 'Celebrity F1'	Foliar spray	Improved production, quality, and vegetative development for the two prior seasons	Shehata et al. (2012)
Solanum tuberosum L. Cv. 'Agria'	*In vitro*	Increased yield and weight of fresh tubers	Amini (2015)
Solanum tuberosum L. Cv. 'Spunta'	Both in hydroponic solution and *in vitro*	Significantly decreased the number of *Fusarium* species infestations.	Mejdoub- Trabelsi et al. (2020)
Phaseolus vulgaris L. Cv. 'Giza 3'	Seed priming	Decreased severity of damping-off and root	El-Mohamedy et al. (2017)

TABLE 6.1 *(Continued)*

Crops	**Modes of Application**	**Results**	**References**
Phaseolus vulgaris L. Cv. 'Nebraska'	Foliar spray	Unexpected yield parameter results under temperature stress	Ibrahim and Ramadan (2015)
Citrullus lanatus L. Cv. 'Jubilee'	Chitosan hydrogel coating	Enlarged stoma width and root size	Gonzalez-Gomez et al. (2017)
Zea mays L. Cv. 'Giza 9'	Foliar spray and seed priming	Low salt stress and higher yield and content of chlorophyll.	ALKahtani et al. (2020)
Zea mays L. Cv. 'Pioneer SR03' and '3906'	Foliar spray	Decreased the impact of salinity stress	Al-Tawaha et al. (2018)

Source: Adapted from Palacio-Márquez et al. (2022).

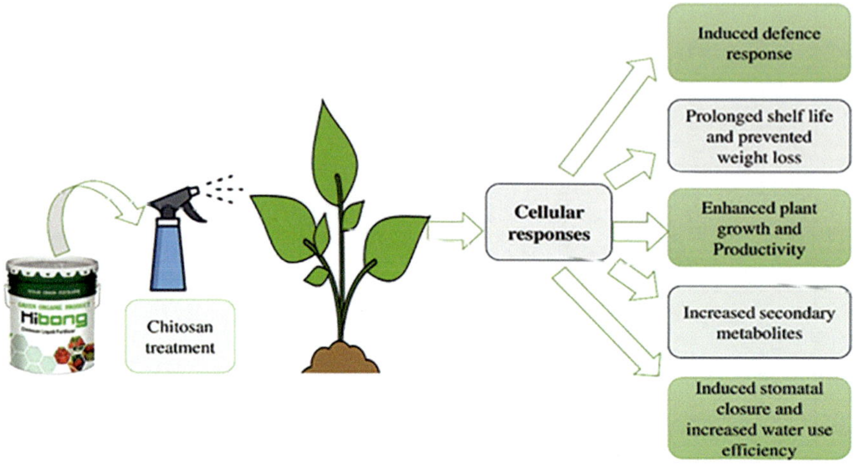

FIGURE 6.1 Roles of chitosan as a plant biostimulant.
Source: Adapted from Pichyangkura and Chadchawan (2015).

biodegradable, non-toxic, and antibacterial, making it suitable for use as an elicitor molecule in various host-pathogen interaction studies (Pimentel et al., 2009) and also serving as an effective elicitor of plant systemic acquired resistance (SAR) to the pathogen (No et al., 2007). Another potential application of chitosan is infection by microbial pathogens. Allan and Hadwiger (1979) described chitosan as an antibacterial agent for the first time. The exact mechanism behind the antimicrobial activity of chitosan is not fully understood, but it may be because of the agglutination of microbial cells, disruption of cells, and cell death. This may be because the chitosan binds to the negatively charged molecules present on the cell surface. Chitosan also induces plant immune responses by activating the genes for the secretion of antimicrobial enzymes like peroxidase, catalase, chitinase, and glucanase (Ma et al., 2013). When chitosan interacts with pectin and pectin dimers, it induces a signal about the degradation of cell walls, which induces a defense response in plants. Chitosan also stimulates the synthesis of fungistatic metabolites like abscisic acid (ABA), methyl jasmonate, chitinase, glucanase, protease, and so forth (Stasinska-Jakubas et al., 2022).

As an effective elicitor, chitosan boosts plant defense systems against diseases, promoting natural defense systems in plants, fruits, and vegetables. Some mechanisms that govern drug release from chitosan nanoparticles (CNPs) are polymer swelling (Liu et al., 2018), absorption drug diffusion, drug diffusion via the polymeric matrix, and polymer erosion or degradation (Singh et al., 2009). Chitosan evokes disease resistance in plant systems. Chitosan stimulates gene expression in defense (Chen et al., 2009). Chitosan stimulates plant defenses in tomato plants (Benhamou et al., 1992), cucumbers (Ben-Shalom et al., 2003), Chilli seeds (Photchanachai et al., 2006), strawberry fruits (El Ghaouth et al., 1992), and other plants. Chitosan also promotes plant innate immunity in rice by stimulating hydrogen peroxide generation (Lin et al., 2005; Yin et al., 2010). It stimulates a defense response mechanism in tobacco by activating nitric oxide pathways (Zhao et al., 2007). The mechanical characteristics of silica/chitosan composites and nanocomposites can be improved by introducing chitosan chains into silica interlayers. Furthermore, combining these two elicitors synergistically generates more substantial and efficient resistance, resulting in a viable method for plant pathogen management (Kiirika et al., 2013). Chitosan stimulates plant defenses against downy mildew and grey mold (Anjali et al., 2012). Chitosan is reported to induce defense responses against *Magnaporthe grisea*, the rice blast pathogen, by stimulating the synthesis of phytoalexins (sakuranetin and monilactone A) in rice seedlings (Rodriguez et al., 2007). Chitosan induced defense responses in the cucumber plants and displayed antifungal activity against *Botrytris cinerea*, the grey mold (Ben-Shalom et al., 2003). Grey mold incidence was reduced by 65%, 82%, and 87% on spraying of chitosan 1, 4, or 24 h, respectively, before inoculation of *B. cinerea* conidia (Ben-Shalom et al., 2003). With chitin and transplant root-dip with chitosan, field trials to determine the efficacy of soil amendments showed that chitosan soil amendment suppresses *Fusarium wilts* (Bell et al., 1998) and chitosan inhibits *Fusarium solani* growth by increasing the synthesis of pisatin phytoalexin in peas (Pérez-de-Luque et al., 2012). Chitosan's ability to suppress infections was boosted when combined with other biocontrol agents (Zhao et al., 2011).

Ma et al. (2013) checked the effect of chitosan and oligochitosan on peach fruit quality and resistance to brown rot caused by *Monilinia fructicola*. The activity and transcript expression of different enzymes after chitosan treatment were also checked. The concentrations used for both chitosan and oligochitosan were 0.5 g/L and 5 g/L. Significant disease control was observed for both 0.5 and 5 g/L chitosan and oligochitosan treatments, with 0.5 g/L chitosan showing a better effect than 0.5 g/L oligochitosan. Both chitosan and oligochitosan enhanced the fruit firmness, with oligochitosan showing better results. Peroxidase, catalase, chitinase, and glucanase activities also increased after chitosan treatment. In a study carried out by Rendina et al. (2019), the efficacy of chitosan in inducing resistance against the cucumber mosaic virus in tomato plants was investigated. They demonstrated that the accumulation of CMV in chitosan-treated plants (1 g/L of chitosan solution) was significantly reduced. An increase in the expression of genes for peroxidase, a non-expressor of pathogenesis-related genes 1, and phytoene synthase 2 was observed. Obianom et al. (2019) showed that the treatment of avocado fruit with a 1.5% chitosan solution reduced the post-harvest stem-end rot (caused by *Lasiodiplodia theobromae*) and anthracnose (*Colletotrichum gloeosporioides*) diseases. An upregulation in the expression of phenylalanine ammonia-lyase (PAL) and a downregulation in the expression of lipoxygenase were observed in fruits inoculated with *C. gloeosporioides* and treated with chitosan, which could help in anthracnose control. Also, an increase in chitinase and SOD activities led to better control of stem-rot disease. Similar results were obtained by Liu et al. (2007) in tomato plants and Chien et al. (2006) in tankan citrus fruit.

6.7 IMPROVEMENT OF SHELF LIFE

Chitosan-based edible coatings are non-toxic, biodegradable, have film-forming qualities, and have antibacterial capabilities. They are an environmentally and biologically acceptable preservation technique for reducing the qualitative deterioration of fresh and ready-to-eat fruits during post-harvest life. Through the interaction of pathways and enzymes involved in ROS production and the scavenging mechanisms that essentially constitute the basic ROS cycle, chitosan-based coatings modulate or control oxidative stress, maintaining the appropriate balance of ROS in fruit cells in different ways (Adiletta et al., 2021; Nair et al., 2020). Chitosan, a naturally occurring biopolymer, is used as a coating to preserve and protect seeds and fruits. The application of chitosan coating has contributed to the extended shelf life of pre-harvest and post-harvest fruits and vegetables. A thin layer of chitosan and other edible coatings forms around fresh food, providing protection and increasing its shelf life. These coatings may also regulate the ripening process and preserve its nutritional qualities (Romanazzi et al., 2022). Chitosan's film-forming abilities, biodegradability, and antibacterial activity make it a desirable substitute for synthetic plastic polymers. However, by including active substances, it prevents the growth of microorganisms and their biochemical and physical damage and boosts their quality and shelf life (Riseh et al., 2023).

Chitosan is a carrier and control release matrix that has numerous advantages, such as shielding biomolecules from extremes in pH, light, and temperature and allowing the active ingredients in its matrix to release gradually, protecting plant cells from the damaging effects of burst release (Mujtaba et al., 2020). Consumption of green leafy vegetables is advised for good health. However, fresh leaves have a short shelf life. Meitha et al. (2022) studied the application of chitosan to spinach leaves for postharvest management. Applying a chitosan solution (0.1% w/v) to post-harvest spinach delayed the decomposition process. Chitosan elicited the accumulation of proline and contributed to the retention of chlorophyll. The upregulation of antioxidant genes (Cu/Zn-SOD and CAT) and the reduction of ROS concentration were reported to play a role in the prolonged shelf life (Meitha et al., 2022). Chitosan was reported to protect mature grape and strawberry grape fruits from *Botrytis cinerea* infection. Coating of chitosan solution (1.0%) inhibited *B. cinerea* spore development (70%) and protected strawberries and two grape cultivars (Shine-Muscat and Kyoho) from infection. Chitosan enhanced the cell wall contents, phenolic compounds, jasmonic acid (JA) formation, and reduced oxidative stress in the fruits (Peian et al., 2021).

6.8 NANOCHITOSAN

Chitosan, formulated in the form of nanoparticles (NPs), has provided advantages as a carrier system for fertilizers, pesticides, and antimicrobial compounds. Chitosan is a unique cationic polymer and can be devised into NPs employing crosslinkers or anionic polymers (Román-Doval et al., 2023). CNPs can be used as a carrier system for encapsulating chemicals, essential oils, phytochemicals, or genetic materials. NPs can protect the encapsulated or entrapped materials and enhance their properties. In the case of genetic materials, the NPs are protected from degradation by nucleases and can help target the cargo to a specific site. Chemicals are protected from degradation, and slow, sustained release from the polymeric NPs contributes to the long-lasting effect and reduction in the dosage concentrations (Ingle et al., 2022).

6.8.1 PLANT PROTECTION FROM PEST

Incorporation of chitosan in emamectin benzoate-cellulose microspheres contributed to enhanced pesticide delivery for effective control of rice leaf folder (*Cnaphalocrocis medinalis*) in rice stems and leaves. The control efficiency was 14-fold better than the commercial emamectin benzoate and was maintained during the rice booting stage (Liu et al., 2023). Chitosan-based NPs have recently been utilized for their sustainable production and role in plant protection worldwide (Sanzari et al., 2019). Gene silencing by rosin-modified polyethylene glycol and chitosan nanoformulation was employed to deliver dsRNA against the *chitin synthase gene* of *Nilaparvata lugens*, the brown plant hopper (BPH). The nanoformulation directly penetrated the body walls of *Nilaparvata lugens*. It reached the hemocoel within 6 hours, resulting in a 54.3% reduction in the relative expression of chitin synthetase A (CHSA) in BPH and a 65.8% death rate. The nanoformulation contributed to an appropriate delivery method for dsRNA application in the BPH, which was helpful for large-scale topical application (Lyu et al., 2023).

CNPs increased plant immunity, and nitric oxide (NO), a key signaling molecule in plant defense, rose after CNP treatment and was beneficial against phytopathogens (Chandra et al., 2015). CNPs operate as strong herbicide carriers, increasing efficiency and controlling release kinetics with minimal toxicity for combinatorial herbicides such as Imazapyr and Imazapic (Maruyama et al., 2016). Chitosan has chelating properties for various organic and inorganic chemicals, making it an ideal biopolymer with improved stability, solubility, and considerable biocidal activity. Furthermore, due to changing physicochemical features such as size, surface area, encapsulation effectiveness, cationic nature, active functional groups, and so on, they demonstrate considerable variety in biological activities when used alone or in combination with other compounds (Du et al., 2009). Several hypotheses have been offered regarding the antibacterial activity of chitosan and its nano-metallic derivatives, but the precise mode of action remains unknown. The most frequently accepted theory is that chitosan's positively charged amine groups bind to the negatively charged components of the microbial cell membrane, triggering intracellular component leakage and cell death (Sudarshan et al., 1992). In pot experiments, water-soluble carboxymethyl chitosan (CMCS)-grafted daphnetin (DA) NPs (DA@CMCS-NPs) effectively suppressed the development of tobacco bacterial wilt, with control efficiencies as high as 74.23%, 67.80%, and 61.67% at 8, 10, and 12 days after inoculation. Furthermore, DA@CMCS-NPs exhibit good biosafety, and the use of these NPs in tobacco manipulation induces defense responses against *R. solanacearum*, which can be linked to systemic resistance (Fan et al., 2023). Pests and mosquitoes were toxicologically affected by the synthesized chitosan-coated cadmium nanoparticles (Ch-CdNps). The toxicological effects of the synthesized Ch-CdNps on pests and mosquitoes were studied. Insecticidal activity influenced the growth and development of *H. armigera* and *S. litura*. *Anopheles stephensi* is the most impacted by mosquitocidal activity, followed by *Aedes aegypti* and *Culex quinquefasciatus* (Suriyakala et al., 2023).

The plant cell wall is the crop's most challenging barrier to gene transfer. Conventional methods like *Agrobacterium* or PEG-mediated gene transfer, electroporation, and biolistic gene guns are expensive, laborious, time-consuming, and have low efficiency (Sivamani et al., 2009). Chitosan has several features that make it an appealing substrate for dsRNA delivery. Chitosan interacts efficiently with RNA, protects it, and can penetrate the cell membrane. CNPs could effectively distribute dsRNA. Chitosan's physical and chemical properties make it relatively easy to handle, with numerous forms available on the market differing in molecular weight and degree of deacetylation (Cooper et al., 2021). Chitosan/dsRNA NPs are effective at inducing RNAi responses in *Caenorhabditis elegans* (Lichtenberg et al., 2019), *Anopheles gambiae* (Zhang et al., 2010), and *Aedes aegypti* (Kumar et al., 2016). Chitosan has recently received much attention for its use in formulations containing small interfering RNAs (siRNA). Chitosan's cationic nature allows it to form complexes with siRNA easily and create NPs. Several studies have shown that CNP-entrapped siRNA can be a carrier for siRNA delivery (Katas et al., 2006; Malmo et al., 2012). Chitosan and DNA NPs can be quickly produced by coating positively charged amine groups on chitosan and negatively charged phosphate groups on DNA. Chitosan polymer, on the other hand, has a low transfection efficiency. Transfection efficiency has been demonstrated to depend on chitosan molecular weight, degree of deacetylation, transfecting medium pH, and cell type (Mao et al., 2010). One recent study showed that CNPs successfully delivered dsRNA (against chitin synthase genes) in a stabilized form to African malaria mosquito larvae (*Anopheles gambiae*) via feeding. Mosquito larvae treated with chitosan/*AgCHS1* dsRNA NPs showed reduced *AgCHS1* gene expression (62.8%) and chitin contents (33.8%) as compared to the chitosan/GFP dsRNA NP-treated control larvae (Zhang et al., 2010).

CNPs play an important role in sustainable farming. Gong et al. (2023) demonstrated that pesticides containing chitosan/siRNA NPs contributed to pesticide reduction and effective control of *S. frugiperda*. Low doses of NPs composed of cyromazine-chitosan (CYR-CS)/siRNA were effective at inhibiting chitin formation in the cuticle and peritrophic membrane of *S. frugiperda*. They displayed high mortality (84%; Gong et al., 2023). It is also demonstrated that CNPs effectively mediate specific dsRNA delivery against Helicoverpa armigera juvenile hormone methyltransferase (JHAMT) and acetylcholinesterase (AChE) target genes and cause 100% insect mortality (Kolge et al., 2021). The *Apolygus lucorum* green plant bug nymph was fed with chitosan/dsRNA, which efficiently inhibited the expression of the G protein-coupled receptor kinase 2 gene (70%), but resulted in much higher mortality (50%), lower weight (26.54%), and a longer developmental time (8.04%) compared with those of the control larvae fed on chitosan. The feeding delivery system had a good gene-silencing impact on *A. lucorum*, suggesting that it could become a practical and effective technique for controlling *A. lucorum* and other insects (Qiao et al., 2023). Delivering NP-conjugated dsRNA to *lepidoptera*n insects may improve RNAi efficiency. Feeding the Cotton Spotted Bollworm *Earias vittella* (*Lepidoptera*: Nolidae) dsCadherin-chitosan, dsAminopeptidase-chitosan, and dsAlpha-amylase-chitosan dramatically reduced gene expression by 78.6%, 83.9%, and 76.6%, respectively (Sandal et al., 2023). These findings clearly show that siRNA formulations based on CNPs may help plant disease and pest management while bypassing the time-consuming procedure of conventional plant transformations.

6.8.2 NANOCHITOSAN AS FERTILIZER AND FERTILIZER CARRIER SYSTEM

Nanofertilizers employ nanoscale structures with the purpose of delivering nutritional components to crops. Such nanofertilizers contribute to increased productivity, targeted delivery, and controlled release of nutrients to crops (Fellet et al., 2021). Kondal et al. (2021) investigated the effect of the slow release of nitrogen nutrients from biodegradable chitosan-urea nanocomposite on soil enzymatic activities and soil microflora. They reported that the urea-nano-polymer enhanced the organic carbon content, soil dehydrogenase activity (DHA), and available potassium with respect to the control (bare urea or conventional urea). The eubacterial and nitrate-reducing bacteria populations were significantly reduced in nanocomposite-treated soil (Kondal et al., 2021). Deshpande et al. (2017) reported that foliar application of CNPs complexed with zinc (Zn/g ~20 mg w/w) enhanced the micronutrient level and improved grain quality in starved wheat plants after anthesis. The zinc concentration was increased (27–42%), which also induced its translocation inside seed tissue and leaves. Dapkekar et al. (2018) reported using zinc-complexed chitosan nanoparticles (ZnCNPs) for zinc fortification (~36%) in wheat plants. The field application of ZnCNPs increased the protein content, kernel weight, yield, and spikelet's per spike of wheat plants (Dapkekar et al., 2018). Dhlamini and co-workers used chitosan nanoformulation with sulphate-supplemented nitrogen, phosphorus, and potassium (NPKS) fertilizer. They found that sulphate-supplemented NPKS nanofertilizers containing 1% CS/NPKS (40 ppm) enhanced the growth of maize plants with respect to control treatments. The tallest maize plant of 37.73 cm with the maximum number of leaves was observed in the treatment of sulphate-supplemented NPKS nanofertilizers. The nanofertilizer also enhanced the plant's chlorophyll content (Dhlamini et al., 2020). In maize, foliar application of Cu-chitosan nanoparticles (CuCNPs) enhanced the plant growth promoter factors, and a significant defense response against *Curvularia* leaf spot (CLS) disease was observed (Choudhary et al., 2017). The Cu-CNPs increased shoot length, root length, root number, stem diameter, and chlorophyll content.

Choudhary et al. (2017) reported that CuCNP-treated maize plants exhibited a significant defense response against CLS disease due to the 4–6 times enhanced activity of SOD with respect to bulk chitosan. SOD aids in growth promotion by converting radicles of toxic superoxide into less toxic components. Elevated activity of peroxidase was recorded post-treatment in leaves. These enhanced activities might aid in the balancing and degeneration of ROS and protect plants from oxidative stress during pathogen invasion. Choudhary et al. (2017) reported the porous structure of CNPs in which Cu was trapped by the ionic gelation method. The copper and chitosan formed a metal-polymer complex by forming C-N bonds. The porous structure aided in the slow release of Cu from the nanocomposite. They observed that during plant-pathogen infection, the acidic pH helps in the faster release of copper, which acts as a strong antifungal agent (Choudhary et al., 2017). Mohammadi et al. (2020) studied the nitrogen release rate from urea granules coated by sepiolite-chitosan nanocomposites. They observed that clay-polymer nanocomposites with more chitosan displayed a slower, sustained release of nitrogen. The chitosan coating aided in the controlled release of nitrogen in the soil. They observed that sepiolite-chitosan nanocomposites showed nitrogen release (75%) in 25 days, while that of urea was in 3 days (Mohammadi et al., 2020).

Namasivayam et al. (2018) encapsulated the isolated insecticidal metabolites from the *Metarhizium anisopliae* fungal strain in chitosan nanocomposites. The nanocomposite showed increased stability (over six months), high entrapment, and loading efficiency (87%). Chitosan nanocomposites treatment in zebrafish models and human peripheral blood cells did not induce any toxic effects. This observation suggested the deployment of chitosan nanocomposites for effective, greener pest management (Namasivayam et al., 2018). Many studies have shown the role of chitosan-based nanocomposites as a greener and more economical pesticide carrier in agriculture (Elsherbiny et al., 2022). Chitosan-based hydrogel has been used in agricultural sectors due to its low price, stability, selectivity, biodegradability, and high affinity for specific ions. These have been effectively used as pesticides and fungicides (Qu et al., 2020). The chitosan nanoformulation binds to the plant membrane and plays a role in activating various pathways responsible for defense-related proteins like SOD, PR proteins, and CAT. Chitosan acts as an elicitor for plant defense mechanisms (Yu et al., 2021). In the study of Van et al. (2013), coffee (*Coffea canephora Piere*) plants subjected to CNPs synthesized from high-MW chitosan (600 kDa) generated with a nanospray dryer showed a 30–60% improvement in net photosynthesis rate and a 30–50% rise in chlorophyll level. Moreover, CNPs increased the nitrogen, phosphorus, and potassium ion uptake by 9.8–27.4%, 17.3–30.4%, and 30–45%, respectively (Van et al., 2013). Sathiyabama et al. (2016) reported that CNP-treated chickpea seeds displayed reduced infection by phytopathogens and exhibited positive morphology effects such as increased germination percentage, seed vigor index, and seedling vegetative biomass.

These composites act as BS by promoting the growth and development of plants, in addition to activating defense pathways during plant-pathogen interaction. By ensuring controlled delivery, using chitosan-based nanoformulations for fertilizer and pesticide delivery can effectively lower their dose and, as a result, lower the environmental damage (Yu et al., 2021). Abdel-Aziz et al. showed the use of CNPs-NPK fertilizers (nitrogen, phosphorous, and potassium) in foliar applications to enhance wheat yield. The CNPs-NPK increased the yield and even reduced the crop life cycle. The nanofertilizers led to increased total saccharide content levels in the wheat grains. Their study suggested the foliar application of 10% nanofertilizers was found to be more effective as, unlike bare NPK fertilizers, the CNPs perhaps aided the NPK components to readily dissolve in water, which helped to increase their uptake and lead to improved wheat qualities (Abdel-Aziz et al., 2018).

6.9 CONCLUSION

Chitosan is a natural polymer with several beneficial agricultural applications due to its natural, biocompatible, and biodegradable nature. Chitosan possesses antimicrobial properties due to its adherence to the cell membrane and possible disruption. It acts as an elicitor molecule to fortify the plant with SAR by activating the genes for secretion of antimicrobial enzymes like peroxidase, catalase, chitinase, and glucanase against the pathogens. Chitosan also stimulates the synthesis of fungistatic metabolites like ABA, methyl jasmonate, chitinase, glucanase, protease, and so forth. These advantageous properties propel the use of chitosan as a seed or fruit coating as a protectant and as a foliar spray on plants for protection and stimulation of the defense mechanisms. The nanoparticulate form of chitosan is advantageous for its applications as a pesticide and fertilizer carrier. These NPs have displayed their utility in reducing the pesticide dosage and increasing the effectiveness and biocompatibility of the pesticide. Chitosan nanoformulation also shows slow-release profiles as fertilizer nanocarriers and contributes to reducing the dosage of the fertilizer. The application of chitosan and chitosan nanocarriers contributes to sustainable agriculture.

ACKNOWLEDGMENTS

Dr. V. Ghormade is grateful for the funding from the DBT Project (BT/PR36430/NNT/28/1713/2020). S. Singh and V. Bhagwat thank DBT for their research studentship. G. Maheswari and Deepali Choudhary are thankful for the fellowship awards to INSPIRE, DST, and UGC, respectively.

KEYWORDS

- **chito oligosaccharide**
- **degree of deacetylation**
- **nanochitosan**
- **slow release**
- **increased stability**

REFERENCES

Abdel-Aziz, H., Hasaneen, M. N., & Omar, A. (2018). Effect of foliar application of nano chitosan NPK fertilizer on the chemical composition of wheat grains. *Egyptian Journal of Botany*, *58*, 87–95.

Abo Elsoud, M. M., & El Kady, E. M. (2019). Current trends in fungal biosynthesis of chitin and chitosan. *Bulletin of the National Research Centre*, *43*, 59. https://doi.org/10.1186/s42269-019-0105-y (accessed on 6th November 2025).

Adiletta, G., Di Matteo, M., & Petriccione, M. (2021). Multifunctional role of chitosan edible coatings on antioxidant systems in fruit crops: A review. *International Journal of Molecular Sciences*, *22*(5), 2633. https://doi.org/10.3390/ijms22052633 (accessed on 6th November 2025).

Ahmad, B., Jaleel, H., Shabbir, A., Khan, M. M. A., & Sadiq, Y. (2019). Concomitant application of depolymerized chitosan and GA_3 modulates photosynthesis, essential oil and menthol production in peppermint (*Mentha piperita* L.). *Scientia Horticulture*, *246*, 371–379.

Ali, M., Ayyub, C. M., Hussain, Z., Hussain, R., & Rashid, S. (2020). Optimization of chitosan level to alleviate the drastic effects of heat stress in cucumber (*Cucumis sativus* L.). *Journal of Pure and Applied Agriculture*, *5*(1), 30–38.

ALKahtani, M. D., Attia, K. A., Hafez, Y. M., Khan, N., Eid, A. M., Ali, M. A., & Abdelaal, K. A. (2020). Chlorophyll fluorescence parameters and antioxidant defense system can display salt tolerance of salt acclimated sweet pepper plants treated with chitosan and plant growth promoting rhizobacteria. *Agronomy*, *10*(8), 1180. https://doi.org/10.3390/agronomy10091325 (accessed on 6th November 2025).

Al-Tawaha, A. R., Turk, M. A., Al-Tawaha, A. R. M., Alu'datt, M. H., Wedyan, M., Al-Ramamneh, E. A. D. M., & Hoang, A. T. (2018). Using chitosan to improve growth of maize cultivars under salinity conditions. *Bulgarian Journal of Agricultural Science*, *24*(3), 437–442.

Amini, J. (2015). Induced resistance in potato plants against *Verticillium* wilt invoked by chitosan and Acibenzolar-S-methyl. *Australian Journal of Crop Science*, *9*(6), 570–576.

Amiri, H., Aghbashlo, M., Sharma, M., Gaffey, J., Manning, L., Moosavi Basri, S. M., Kennedy, J. F., Gupta, V. K., & Tabatabaei, M. (2022). Chitin and chitosan derived from crustacean waste valorization streams can support food systems and the UN Sustainable Development Goals. *Nature Food*, *3*, 822–828.

Anjali, C. H., Sharma, Y., Mukherjee, A., & Chandrasekaran, N. (2012). Neem oil (*Azadirachta indica*) nano emulsion—A potent larvicidal agent against *Culex quinquefasciatus*. *Pest Management Science*, *68*, 158–163.

Anusuya, S., & Sathiyabama, M. (2016). Effect of chitosan on growth, yield and curcumin content in turmeric under field condition. *Biocatalysis and Agricultural Biotechnology*, *6*, 102–106.

Batista, A. C. D. L., Neto, F. E. D. S., & Paiva, W. D. S. (2018). Review of fungal chitosan: Past, present and perspectives in Brazil. *Polymers*, *28*(3), 275–283.

Bell, A. A., Hubbard, J. C., Liu, L., Davis, R. M., & Subbarao, K. V. (1998). Effects of chitin and chitosan on the incidence and severity of *Fusarium* yellows of celery. *Plant Disease*, *82*, 322–328.

Benhamou, N., & Thériault, G. (1992). Treatment with chitosan enhances resistance of tomato plants to the crown and root rot pathogen *Fusarium oxysporum* f. sp. *radicis-lycopersici*. *Physiological and Molecular Plant Pathology*, *41*, 33–52.

Ben-Shalom, N., Ardi, R., Pinto, R., Aki, C., & Fallik, E. (2003). Controlling grey mould caused by *Botrytis cinerea* in cucumber plants by means of chitosan. *Crop Protection*, *22*, 285–290.

Berger, L. R. R., Stamford, T. C. M., Stamford-Arnaud, T. M., Alcântara, S. R. C., Silva, A. C., Silva, A. M., Nascimento, A. E., & Campos-Takaki, G. M. (2014). Green conversion of agro-industrial wastes into chitin and chitosan by *Rhizopus arrhizus* and *Cunninghamella elegans* strains. *International Journal of Molecular Sciences*, *15*(5), 9082–9102. https://doi.org/10.3390/ijms15059082 (accessed on 6th November 2025).

Brugnerotto, J., Lizardi, J., Goycoolea, F. M., Arguëlles-Monal, W., Desbrières, J., & Rinaudo, M. (2001). An infrared investigation in relation with chitin and chitosan characterization. *Polymer*, *42*, 3569–3580.

Cardoso, A., Lins, C. I. M., Santos, E. R., Silva, M. C. F., & Campos-Takaki, G. M. (2012). Microbial enhance of chitosan production by *Rhizopus arrhizus* using agro-industrial substrates. *Molecules*, *17*, 4904–4914.

Chakraborty, M., Hasanuzzaman, M., Rahman, M., Khan, M. A. R., Bhowmik, P., Mahmud, N. U., Tanveer, M., & Islam, T. (2020). Mechanism of plant growth promotion and disease suppression by chitosan biopolymer. *Agriculture*, *10*(12), 624. https://doi.org/10.3390/agriculture10120624 (accessed on 6th November 2025).

Chandra, S., Chakraborty, N., Dasgupta, A., Sarkar, J., Panda, K., & Acharya, K. (2015). Chitosan nanoparticles: A positive modulator of innate immune responses in plants.

Scientific Reports, *5*, 15195. https://doi.org/10.1038/srep15195 (accessed on 6th November 2025).

Chatterjee, S., Chatterjee, S., Chatterjee, B. P., & Guha, A. K. (2008). Enhancement of growth and chitosan production by *Rhizopus oryzae* in whey medium by plant growth hormones. *International Journal of Biological Macromolecules*, *42*(2), 120–126. https://doi.org/10.1016/j.ijbiomac.2007.10.006 (accessed on 6th November 2025).

Chen, X., Yang, H., Zhong, Z., & Yan, N. (2017). Base-catalyzed, one-step mechanochemical conversion of chitin and shrimp shells into low molecular weight chitosan. *Green Chemistry*, *19*, 2783–2792.

Chien, P. J., & Chou, C. C. (2006). Antifungal activity of chitosan and its application to control post-harvest quality and fungal rotting of Tankan citrus fruit (*Citrus tankan* Hayata). *Journal of the Science of Food and Agriculture*, *86*, 1964–1969.

Choudhary, R. C., Kumaraswamy, R. V., Kumari, S., Sharma, S. S., Pal, A., Raliya, R., & Saharan, V. (2017). Cu-chitosan nanoparticle boost defense responses and plant growth in maize (*Zea mays* L.). *Scientific Reports*, *7*, 9754. https://doi.org/10.1038/s41598-017-08571-0 (accessed on 6th November 2025).

Cooper, A. M., Song, H., Yu, Z., Biondi, M., Bai, J., Shi, X., Ren, Z., Weerasekara, S. M., Hua, D. H., Silver, K., & Zhang, J. (2021). Comparison of strategies for enhancing RNA interference efficiency in *Ostrinia nubilalis*. *Pest Management Science*, *77*, 635–645.

Crognale, S., Russo, C., Petruccioli, M., & D'Annibale, A. (2022). Chitosan production by fungi: Current state of knowledge, future opportunities and constraints. *Fermentation*, *8*(2), 76. https://doi.org/10.3390/fermentation8020076 (accessed on 6th November 2025).

Dapkekar, A., Deshpande, P., Oak, M. D., Paknikar, K. M., & Rajwade, J. M. (2018). Zinc use efficiency is enhanced in wheat through nano fertilization. *Scientific Reports*, *8*, 6832. https://doi.org/10.1038/s41598-018-25247-5 (accessed on 6th November 2025).

Dar, T. A., Uddin, M., Khan, M. M. A., Ali, A., Mir, S. R., & Varshney, L. (2015). Effect of Co-60 gamma irradiated chitosan and phosphorus fertilizer on growth, yield and trigonelline content of *Trigonella foenum-graecum* L. *Journal of Radiation Research and Applied Sciences*, *8*, 446–458.

Deshpande, P., Dapkekar, A., Oak, M. D., Paknikar, K. M., & Rajwade, J. M. (2017). Zinc complexed chitosan/TPP nanoparticles: A promising micronutrient nanocarrier suited for foliar application. *Carbohydrate Polymers*, *165*, 394–401.

Dhillon, G. S., Kaur, S., Brar, S. K., & Verma, M. (2012). Green synthesis approach: Extraction of chitosan from fungus mycelia. *Critical Reviews in Biotechnology*, *33*(4), 379–403. https://doi.org/10.3109/07388551.2012.717217 (accessed on 6th November 2025).

Dhlamini, B., Paumo, H. K., Katata-Seru, L., & Kutu, F. R. (2020). Sulphate-supplemented NPK nanofertilizer and its effect on maize growth. *Materials Research Express*, *7*(9), 095011. https://doi.org/10.1088/2053-1591/abb69d (accessed on 6th November 2025).

Doiphode, N., Rajamohanan, P. R., Ghormade, V., & Deshpande, M. V. (2009). Chitosan production using a dimorphic fungus *Benjaminiella poitrasii*: Role of chitin deacetylase. *Asian Chitin Journal*, *5*, 19–26.

Du, W. L., Niu, S. S., Xu, Y. L., Xu, Z. R., & Fan, C. L. (2009). Antibacterial activity of chitosan tripolyphosphate nanoparticles loaded with various metal ions. *Carbohydrate Polymers*, *75*, 385–389.

El Ghaouth, A., Arul, J., Asselin, A., & Benhamou, N. (1992). Antifungal activity of chitosan on post-harvest pathogens: Induction of morphological and cytological alterations in *Rhizopus stolonifer*. *Mycological Research*, *96*, 769–779.

El Hadrami, A., Adam, L. R., El Hadrami, I., & Daayf, F. (2010). Chitosan in plant protection. *Marine Drugs*, *8*, 968–987.

El-Mohamedy, R. S. R., Shafeek, M. R., Abd El-Samad, E. E. D. H., Salama, D. M., & Rizk, F. A. (2017). Field application of plant resistance inducers (PRIs) to control important root rot diseases and improvement growth and yield of green bean (*Phaseolus vulgaris* L.). *Australian Journal of Crop Science*, *11*(5), 496–505. https://doi.org/10.21475/AJCS.17.11.05.P260 (accessed on 6th November 2025).

El-Sawy, N. M., Abd El-Rehim, H. A., Elbarbary, A. M., & Hegazy, E. A. (2010). Radiation-induced degradation of chitosan for possible use as a growth promoter in agricultural purposes. *Carbohydrate Polymers*, *79*, 555–562.

Elsherbiny, A. S., Galal, A., Ghoneem, K. M., & Salahuddin, N. A. (2022). Novel chitosan-based nanocomposites as ecofriendly pesticide carriers: Synthesis, root rot inhibition and growth management of tomato plants. *Carbohydrate Polymers*, *282*, 119111. https://doi.org/10.1016/j.carbpol.2022.119111 (accessed on 6th November 2025).

Escudero, N., Lopez-Moya, F., Ghahremani, Z., Zavala-Gonzalez, E. A., Alaguero-Cordovilla, A., Ros-Ibañez, C., Lacasa, A., Sorribas, F. J., & Lopez-Llorca, L. V. (2017). Chitosan increases tomato root colonization by *Pochonia chlamydosporia* and their combination reduces root-knot nematode damage. *Frontiers in Plant Science*, *8*, 1415. https://doi.org/10.3389/fpls.2017.01415 (accessed on 6th November 2025).

Esyanti, R. R., Dwivany, F. M., Mahani, S., Nugrahapraja, H., & Meitha, K. (2019). Foliar application of chitosan enhances growth and modulates expression of defense genes in chili pepper (*Capsicum annuum* L.). *Australian Journal of Crop Science*, *13*(1), 55–60. https://doi.org/10.21475/ajcs.19.13.01.p1169 (accessed on 6th November 2025).

Fan, Z., Wang, L., Qin, Y., & Li, P. (2023). Activity of chitin/chitosan/chitosan oligosaccharide against plant pathogenic nematodes and potential modes of application in agriculture: A review. *Carbohydrate Polymers*, 120592. https://doi.org/10.1016/j.carbpol.2023.120592 (accessed on 6th November 2025).

Fellet, G., Pilotto, L., Marchiol, L., & Braidot, E. (2021). Tools for nano-enabled agriculture: Fertilizers based on calcium phosphate, silicon, and chitosan nanostructures. *Agronomy*, *11*(6), 1239. https://doi.org/10.3390/agronomy11061239 (accessed on 6th November 2025).

Gong, C., Hasnain, A., Wang, Q., Liu, D., Xu, Z., Zhan, X., Liu, X., Pu, J., Sun, M., & Wang, X. (2023). Eco-friendly deacetylated chitosan base siRNA biological-nano pesticide loading cyromazine for efficiently controlling *Spodoptera frugiperda*. *International Journal of Biological Macromolecules*, *241*, 124575. https://doi.org/10.1016/j.ijbiomac.2023.124575 (accessed on 6th November 2025).

Gonzalez-Gomez, H., Ramirez-Godina, F., Ortega-Ortiz, H., Benavides-Mendoza, A., & Robledo-Torres, V., Cabrera-De la Fuente, M. (2017). Use of chitosan-PVA hydrogels with copper nanoparticles to improve the growth of grafted watermelon. *Molecules*, *22*(7), 1031. https://doi.org/10.3390/molecules22071031 (accessed on 6th November 2025).

Grifoll-Romero, L., Pascual, S., Aragunde, H., Biarnés, X., & Planas, A. (2018). Chitin deacetylases: Structures, specificities, and biotech applications. *Polymers (Basel)*, *10*(4), 352. https://doi.org/10.3390/polym10040352 (accessed on 6th November 2025).

Hai, N. T. T., Thu, L. H., Nga, N. T. T., Hoa, T. T., Tuan, L. N. A., Phu, D. V., & Hien, N. Q. (2019). Preparation of Chito oligosaccharide by hydrogen peroxide degradation of chitosan and its effect on soybean seed germination. *Journal of Polymers and the Environment*, *27*, 2098–2104.

Hameed, A., Sheikh, M. A., Hameed, A., Farooq, T., Basra, S. M. A., & Jamil, A. (2013). Chitosan priming enhances the seed germination, antioxidants, hydrolytic enzymes, soluble proteins and sugars in wheat seeds. *Agrochemical*, *LVII*(1), 32–46.

Hassnain, M., Alam, I., Ahmad, A., Basit, I., Ullah, N., Alam, I., Ullah, I., Khalid, M. A., Shair, M., & ul Ain, N. (2020). Efficacy of chitosan on performance of tomato (*Lycopersicon esculentum* L.) plant under water stress condition. *Pakistan Journal of Agricultural Research*, *33*(1), 27–41. http://doi.org/10.17582/journal.pjar/2020/33.1.27.41 (accessed on 6th November 2025).

Hawrylak-Nowak, B., Dresler, S., Rubinowska, K., & Matraszek-Gawrona, R. (2021). Eliciting effect of foliar application of chitosan lactate on the phytochemical properties of *Ocimum basilicum* L. and *Melissa officinalis* L. *Food Chemistry*, *342*, 128358. https://doi.org/10.1016/j.foodchem.2020.128358 (accessed on 6th November 2025).

Huq, T., Khana, A., Brown, D., Dhayagude, N., He, Z., & Ni, Y. (2022). Sources, production and commercial applications of fungal chitosan: A review. *Journal of Bioresources and Bioproducts*, *7*, 85–98.

Ibrahim, E. A., & Ramadan, W. A. (2015). Effect of zinc foliar spray alone and combined with humic acid or/and chitosan on growth, nutrient elements content and yield of dry bean (*Phaseolus vulgaris* L.) plants sown at different dates. *Scientia Horticulture*, *184*, 101–105. https://doi.org/10.1016/j.scienta.2014.11.010 (accessed on 6th November 2025).

Ingle, P. U., Shende, S. S., Shingote, P. R., Mishra, S. S., Sarda, V., Wasule, D. L., Rajput, V. D., Minkina, T., Rai, M., Sushkova, S., Mandzhieva, S., & Gade, A. (2022). Chitosan nanoparticles (ChNPs): A versatile growth promoter in modern agricultural production. *Helion*, *8*(11), e11893. https://doi.org/10.1016/j.heliyon.2022.e11893 (accessed on 6th November 2025).

Kahromi, S., & Khara, J. (2021). Chitosan stimulates secondary metabolite production and nutrient uptake in medicinal plant *Dracocephalum kotschyi*. *Journal of the Science of Food and Agriculture*, *101*, 3898–3907.

Kannan, M., Nesakumari, M., Rajarathinam, R., & Ranjit Singh, A. J. A. (2010). Production and characterization of mushroom chitosan under solid-state fermentation conditions. *Advances in Biological Research*, *4*(1), 10–13. https://www.idosi.org/abr/4(1)/2.pdf (accessed on 6th November 2025).

Kassai, M. (2008). A review of several reported procedures to determine the degree of N-acetylation for chitin and chitosan using infrared spectroscopy. *Carbohydrate Polymers*, *71*, 497–508.

Katas, H., & Alpar, H. O. (2006). Development and characterization of chitosan nanoparticles for siRNA delivery. *Journal of Controlled Release*, *115*, 216–225.

Kiirika, L. M., Stahl, F., & Wydra, K. (2013). Phenotypic and molecular characterization of resistance induction by single and combined application of chitosan and silicon in tomato against *Ralstonia solanacearum*. *Physiological and Molecular Plant Pathology*, *81*, 1–2.

Kim, Y. J., Zhao, Y., Oh, K. T., Nguyen, V. N., & Park, R. D. (2008). Enzymatic deacetylation of chitin by extracellular chitin deacetylase from a newly screened *Mortierella* sp. DY-52. *Journal of Microbiology and Biotechnology*, *18*, 759–766.

Kolge, H., Kadam, K., Galande, S., Lanjekar, V., & Ghormade, V. (2021). New frontiers in pest control: Chitosan nanoparticles-shielded dsRNA as an effective topical RNAi spray for gram pod borer biocontrol. *ACS Applied Bio Materials*, *4*, 5145–5157.

Kolodziejska, I., Wojtasz-Pajak, A., Ogonowska, G., & Sikorski, Z. E. (2000). Deacetylation of chitin in a two-stage chemical and enzymatic process. *Bulletin of the Sea Fisheries Institute*, *2*, 15–24.

Kondal, R., Kalia, A., Krejcar, O., Kuca, K., Sharma, S. P., Luthra, K., & Gomes, C. L. (2021). Chitosan-urea nanocomposite for improved fertilizer applications: The effect on the soil enzymatic activities and microflora dynamics in N cycle of potatoes (*Solanum tuberosum* L.). *Polymers*, *13*(17), 2887. https://doi.org/10.3390/polym13172887 (accessed on 6th November 2025).

Kong, M., Chen, X. G., Xing, K., & Park, H. J. (2010). Antimicrobial properties of chitosan and mode of action: A state-of-the-art review. *International Journal of Food Microbiology*, *144*, 51–63.

Kou, S. G., Peters, L. M., & Mucalo, M. R. (2021). Chitosan: A review of sources and preparation methods. *International Journal of Biological Macromolecules*, *169*, 85–94.

Kumar, D. R., Kumar, P. S., Gandhi, M. R., Al-Dhabi, N. A., Paulraj, M. G., & Ignacimuthu, S. (2016). Delivery of chitosan/dsRNA nanoparticles for silencing of wing development vestigial (vg) gene in *Aedes aegypti* mosquitoes. *International Journal of Biological Macromolecules*, *86*, 89–95.

Lichtenberg, S. S., Tsyusko, O. V., Palli, S. R., & Unrine, J. M. (2019). Uptake and bioactivity of chitosan/double-stranded RNA polyplex nanoparticles in *Caenorhabditis elegans*. *Environmental Science & Technology*, *53*, 3832–3840.

Lin, W., Hu, X., Zhang, W., Rogers, W. J., & Cai, W. (2005). Hydrogen peroxide mediates defense responses induced by chitosan's of different molecular weights in rice. *Journal of Plant Physiology*, *162*, 937–944.

Liu, H., Zheng, Z., Han, X., Zhang, C., Li, H., & Wu, M. (2022). Chitosan soaking improves seed germination of *Platycodon grandiflorus* and enhances its growth, photosynthesis, resistance, yield, and quality. *Horticulture*, *8*(10), 943. https://doi.org/10.3390/horticulturae8100943 (accessed on 6th November 2025).

Liu, J., Tian, S., Meng, X., & Xu, Y. (2007). Effects of chitosan on control of postharvest diseases and physiological responses of tomato fruit. *Postharvest Biology and Technology*, *44*, 300–306.

Liu, S., Yang, S., & Ho, P. C. (2018). Intranasal administration of carbamazepine-loaded carboxymethyl chitosan nanoparticles for drug delivery to the brain. *Asian Journal of Pharmaceutics*, *13*, 72–81.

Liu, Y., Jiang, F., Zhang, Y., Xu, C., Fu, L., & Lin, B. (2023). Phosphate-modified cellulose/chitosan with high drug loading for effective prevention of rice leaf folder (*Cnaphalocrocis medinalis*) outbreaks in fields. *International Journal of Biological Macromolecules*, *243*, 125145. https://doi.org/10.1016/j.ijbiomac.2023.125145 (accessed on 6th November 2025).

Lv, M., Hu, Y., Gänzle, M. G., Lin, J., Wang, C., & Cai, J. (2016). Preparation of Chito oligosaccharides from fungal waste mycelium by recombinant chitinase. *Carbohydrate Research*, *430*, 1–7. https://doi.org/10.1016/j.carres.2016.04.019 (accessed on 6th November 2025).

Lyu, Z., Xiong, M., Mao, J., Li, W., Jiang, G., & Zhang, W. (2023). A dsRNA delivery system based on the rosin-modified polyethylene glycol and chitosan induces gene silencing and mortality in *Nilaparvata lugens*. *Pest Management Science*, *79*, 1518–1527.

Ma, Z., Yang, L., Yan, H., Kennedy, J. F., & Meng, X. (2013). Chitosan and oligochaetes enhance the resistance of peach fruit to brown rot. *Carbohydrate Polymers*, *94*, 272–277.

Mahmood, N., Abbasi, N. A., Hafiz, I. A., Ali, I., & Zakia, S. (2017). Effect of bio stimulants on growth, yield and quality of bell pepper cv. Yolo Wonder. *Pakistan Journal of Agricultural Sciences*, *54*(2), 311–317.

Malmo, J., Sørgård, H., Vårum, K. M., & Strand, S. P. (2012). siRNA delivery with chitosan nanoparticles: Molecular properties favoring efficient gene silencing. *Journal of Controlled Release*, *158*, 261–268.

Mane, S., Pathan, E., Tupe, S., Deshmukh, S., Kale, D., Ghormade, V., Chaudhari, B., & Deshpande, M. (2022). Isolation and characterization of chitosan's from different fungi with special emphasis on Zygomycetes dimorphic fungus *Benjaminiella poitrasii*: Evaluation of its chitosan nanoparticles for the inhibition of human pathogenic fungi. *Biomacromolecules*, *23*, 808–815.

Mao, S., Sun, W., & Kissel, T. (2010). Chitosan-based formulations for delivery of DNA and siRNA. *Advanced Drug Delivery Reviews*, *62*, 12–27.

Maruyama, C. R., Guilger, M., Pascoli, M., Bileshy-José, N., Abhilash, P. C., Fraceto, L. F., & De Lima, R. (2016). Nanoparticles based on chitosan as carriers for the combined herbicides imazapic and imazapyr. *Scientific Reports*, *6*, 19768.

Meitha, K., Pramesti, Y., Signorelli, S., & Kriswantoro, J. A. (2022). Postharvest chitosan application maintains the quality of spinach through suppression of bacterial growth and elicitation. *Journal of Horticultural Environment and Biotechnology*, *63*, 217–227.

Mejdoub-Trabelsi, B., Touihri, S., Ammar, N., Riahi, A., & Daami-Remadi, M. (2020). Effect of chitosan for the control of potato diseases caused by *Fusarium* species. *Journal of Phytopathology*, *168*(1), 18–27. https://doi.org/10.1111/jph.12847 (accessed on 6th November 2025).

Merzendorfer, H. (2011). The sugar code: Fundamentals of glycosciences. In H. J. Gabius (Ed.), *The Sugar Code: Fundamentals of Glycosciences* (p. 597). John Wiley & Sons.

Mohammadi, N., Shariatmadari, H., Khademi, H., & Bazarganipour, M. (2020). Coating of sepiolite-chitosan nanocomposites onto urea increases nitrogen availability and its use efficiency in maize. *Archives of Agronomy and Soil Science*, *66*(7), 884–896.

Mondal, M., Puteh, A. B., & Dafader, N. C. (2016). Foliar application of chitosan improved morphophysiological attributes and yield in summer tomato (*Solanum lycopersicum*). *Pakistan Journal of Agricultural Sciences*, *53*(2), 339–344. https://doi.org/10.21162/PAKJAS/16.2011 (accessed on 6th November 2025).

Mujtaba, M., Khawar, K. M., Camara, M. C., Carvalho, L. B., Fraceto, L. F., Morsi, R. E., & Wang, D. (2020). Chitosan-based delivery systems for plants: A brief overview of recent advances and future directions. *International Journal of Biological Macromolecules*, *154*, 683–697.

Mukarram, M., Khan, M. M. A., Uddin, M., & Corpas, F. J. (2022). Irradiated chitosan (ICH): An alternative tool to increase essential oil content in lemongrass (*Cymbopogon flexuosus*). *Acta Physiologiae Plantarum*, *44*, 2. https://doi.org/10.1007/s11738-021-03335-w (accessed on 6th November 2025).

Nair, M. S., Tomar, M., Punia, S., Kukula-Koch, W., & Kumar, M. (2020). Enhancing the functionality of chitosan- and alginate-based active edible coatings/films for the preservation of fruits and vegetables: A review. *International Journal of Biological Macromolecules*, *164*, 304–320.

Namasivayam, S. K. R., Bharani, R. A., & Karunamoorthy, K. (2018). Insecticidal fungal metabolites fabricated chitosan nanocomposite (IM-CNC) preparation for the enhanced larvicidal activity—An effective strategy for green pesticide against economic important insect pests. *International Journal of Biological Macromolecules*, *120*, 921–944.

Naqvi, S., Cord-Landwehr, S., Singh, R., Bernard, F., Kolkenbrock, S., El Gueddari, N. E., & Moerschbacher, B. M. (2016). A recombinant fungal chitin deacetylase produces fully defined chitosan oligomers with novel patterns of acetylation. *Applied and Environmental Microbiology*, *82*, 6645–6655.

New, N., Furuike, T., Osaka, I., Fujimori, H., Kawasaki, H., Arakawa, R., Tokura, S., Stevens, W. F., Kurozumi, S., Takamori, Y., Fukuda, M., & Tamura, H. (2011). Laboratory scale production of 13C labeled chitosan by fungi *Absidia coerulea* and *Gongronella butleri* grown in solid substrate and submerged fermentation. *Carbohydrate Polymers*, *84*, 743–750.

Niederhofer, A., & Muller, B. W. (2004). A method for direct preparation of chitosan with low molecular weight from fungi. *European Journal of Pharmaceutics and Biopharmaceutics*, *57*, 101–105.

Obianom, C., Romanazzi, G., & Sivakumar, D. (2019). Effects of chitosan treatment on avocado postharvest diseases and expression of phenylalanine ammonia-lyase, chitinase and lipoxygenase genes. *Postharvest Biology and Technology*, *147*, 214–221.

Ochoa-Meza, L. C., Quintana-Obregón, E. A., Vargas-Arispuro, I., Falcón-Rodríguez, A. B., Aispuro-Hernández, E., Virgen-Ortiz, J. J., & Martínez-Téllez, M. A. (2021). Oligosaccharins as elicitors of defense responses in wheat. *Polymers*, *13*(18), 3105. https://doi.org/10.3390/polym13183105 (accessed on 6th November 2025).

Olawuyi, I. F., Park, J. J., Lee, J. J., & Lee, W. Y. (2019). Combined effect of chitosan coating and modified atmosphere packaging on fresh-cut cucumber. *Food Science & Nutrition*, *7*(3), 1043–1052. https://doi.org/10.1002/fsn3.937 (accessed on 6th November 2025).

Palacio-Márquez, A., Ramirez-Estrada, C. A., Sanchez, E., Ojeda-Barrios, D. L., Chavezmendoza, C., Sida-Arreola, J. P., & Preciado-Rangel, P. (2022). Use of bio stimulant compounds in agriculture: Chitosan as a sustainable option for plant development. *Notulae Scientia Biological*, *14*(1), 11124. https://www.notulaebiologicae.ro/index.php/nsb/article/view/11124/9453 (accessed on 6th November 2025).

Pareek, N., Vivekanand, V., Agarwal, P., Saroj, S., & Singh, R. P. (2013). Bioconversion to chitosan: A two-stage process employing chitin deacetylase from *Penicillium oxalicum* SAEM-51. *Carbohydrate Polymers*, *96*, 417–425.

Peian, Z., Haifeng, J., Peijie, G., Sadeghnezhad, E., Qianqian, P., Tianyu, D., & Jinggui, F. (2021). Chitosan induces jasmonic acid production leading to resistance of ripened fruit against *Botrytis cinerea* infection. *Journal of Food Chemistry*, *337*, 127772. https://doi.org/10.1016/j.foodchem.2020.127772 (accessed on 6th November 2025).

Pellis, A., Guebitz, G. M., & Nyanhongo, G. S. (2022). Chitosan: Sources, processing and modification techniques. *Gels*, *8*(7), 393. https://doi.org/10.3390/gels8070393 (accessed on 6th November 2025).

Pérez-de-Luque, A., Cifuentes, Z., Beckstead, J. A., Sillero, J. C., Ávila, C., Rubio, J., & Ryan, R. O. (2012). Effect of amphotericin B nanodisks on plant fungal diseases. *Pest Management Science*, *68*, 67–74.

Photchanachai, S., Singkaew, J., & Thamthong, J. (2006). Effects of chitosan seed treatment on *Colletotrichum* sp. and seedling growth of chili cv. jinda. *IV International Conference on Managing Quality in Chains—The Integrated View on Fruits and Vegetables Quality*, *712*, 585–590.

Pichyangkura, R., & Chadchawan, S. (2015). Bio stimulant activity of chitosan in horticulture. *Scientia Horticulture*, *196*, 49–65. https://doi.org/10.1016/j.scienta.2015.09.031 (accessed on 6th November 2025).

Pimentel, D. (2009). Pesticides and pest control. *Journal of Integrated Pest Management: Innovation-Development Process*, 83–87.

Pochanavanich, P., & Suntornsuk, W. (2002). Fungal chitosan production and its characterization. *Letters in Applied Microbiology*, *35*, 17–21.

Qiao, H., Zhao, J., Wang, X., Xiao, L., Zhu-Salzman, K., Lei, J., Xu, D., Xu, G., Tan, Y., & Hao, D. (2023). An oral dsRNA delivery system based on chitosan induces G protein-coupled receptor kinase 2 gene silencing for *Apolygus lucorum* control. *Pesticide Biochemistry and Physiology*, *194*, 105481. https://doi.org/10.1016/j.pestbp.2023.105481 (accessed on 6th November 2025).

Qu, B., & Luo, Y. (2020). Chitosan-based hydrogel beads: Preparations, modifications and applications in food and agriculture sectors–A review. *International Journal of Biological Macromolecules*, *152*, 437–448.

Rahman, M., Mukta, J. A., Sabir, A. A., Gupta, D. R., Mohi-Ud-Din, M., Hasanuzzaman, M., Miah, M. G., Rahman, M., & Islam, M. D. (2018). Chitosan biopolymer promotes yield and stimulates accumulation of antioxidants in strawberry fruit. *PLoS ONE*, *13*, e0203769. https://doi.org/10.1371/journal.pone.0203769 (accessed on 6th November 2025).

Rendina, N., Nuzzaci, M., Scopa, A., Cuypers, A., & Sofo, A. (2019). Chitosan-elicited defense responses in *Cucumber mosaic virus* (CMV)-infected tomato plants. *Journal of Plant Physiology*, *234–235*.

Riseh, R. S., Vatankhah, M., Hassanisaadi, M., & Kennedy, J. F. (2023). Chitosan/silica: A hybrid formulation to mitigate phytopathogens. *International Journal of Biological Macromolecules*, *239*, 124192. https://doi.org/10.1016/j.ijbiomac.2023.124192 (accessed on 6th November 2025).

Roberts, G. A. F., & Domszy, J. G. (1982). Determination of the viscometry constants for chitosan. *International Journal of Biological Macromolecules*, *4*, 374–377.

Rodríguez, A. T., Ramírez, M. A., Cárdenas, R. M., Hernández, A. N., Velázquez, M. G., & Bautista, S. (2007). Induction of defense response of *Oryza sativa* L. against *Pyricularia grisea* (Cooke) Sacc. by treating seeds with chitosan and hydrolyzed chitosan. *Pesticide Biochemistry and Physiology*, *89*, 206–215.

Romanazzi, G., & Moumni, M. (2022). Chitosan and other edible coatings to extend shelf life, manage postharvest decay, and reduce loss and waste of fresh fruits and vegetables. *Current Opinion in Biotechnology*, *78*, 102834. https://doi.org/10.1016/j.copbio.2022.102834 (accessed on 6th November 2025).

Román-Doval, R., Torres-Arellanes, S. P., Tenorio-Barajas, A. Y., Gómez-Sánchez, A., & Valencia-Lazcano, A. A. (2023). Chitosan: Properties and its application in agriculture in context of molecular weight. *Polymers*, *15*(13), 2867. https://doi.org/10.3390/polym15132867 (accessed on 6th November 2025).

Salachna, P., & Łopusiewicz, L. (2022). Chitosan oligosaccharide lactate increases productivity and quality of baby leaf red perilla. *Agronomy*, *12*(5), 1182. https://doi.org/10.3390/agronomy12051182 (accessed on 6th November 2025).

Sandal, S., Singh, S., Bansal, G., Kaur, R., Mogilicherla, K., Pandher, S., Roy, A., Kaur, G., Rathore, P., & Kalia, A. (2023). Nanoparticle-shielded dsRNA delivery for enhancing RNAi efficiency in cotton spotted bollworm *Earias vittella* (Lepidoptera: Nolidae). *International Journal of Molecular Sciences*, *24*(11), 9161. https://doi.org/10.3390/ijms24119161 (accessed on 6th November 2025).

Sanzari, I., Leone, A., & Ambrosone, A. (2019). Nanotechnology in plant science: To make a long story short. *Frontiers in Bioengineering and Biotechnology*, *7*, 120. https://doi.org/10.3389/fbioe.2019.00120 (accessed on 6th November 2025).

Sathiyabama, M., & Parthasarathy, R. (2016). Biological preparation of chitosan nanoparticles and it's *in vitro* antifungal efficacy against some phytopathogenic fungi. *Carbohydrate Polymers*, *151*, 321–325.

Sebastian, J., Rouissi, T., Brar, S. K., Hegde, K., & Verma, M. (2019). Microwave-assisted extraction of chitosan from *Rhizopus oryzae* NRRL 1526 biomass. *Carbohydrate Polymers*, *219*, 431–404. https://doi.org/10.1016/j.carbpol.2019.05.047 (accessed on 6th November 2025).

Shehata, S. A., Fawzy, Z. F., & El-Ramady, H. R. (2012). Response of cucumber plants to foliar application of chitosan and yeast under greenhouse conditions. *Australian Journal of Basic and Applied Sciences*, *6*(4), 63–71.

Singh, R., & Lillard Jr, J. W. (2009). Nanoparticle-based targeted drug delivery. *Experimental and Molecular Pathology*, *86*, 215–223.

Sivamani, E., DeLong, R. K., & Qu, R. (2009). Protamine-mediated DNA coating remarkably improves bombardment transformation efficiency in plant cells. *Plant Cell Reports*, *28*, 213–221.

Stasinska-Jakubas, M., & Hawrylak-Nowak, B. (2022). Protective, bio stimulating, and eliciting effects of chitosan and its derivatives on crop plants. *Molecules*, *27*(9), 2801. https://doi.org/10.3390/molecules27092801 (accessed on 6th November 2025).

Stasinska-Jakubas, M., Hawrylak-Nowak, B., Dresler, S., Wojciak, M., & Rubinowska, K. (2023). Application of chitosan lactate, selenite, and salicylic acid as an approach to induce biological responses and enhance secondary metabolism in *Melissa officinalis* L. *Industrial Crops and Products*, *205*, 117571. https://doi.org/10.1016/j.indcrop.2023.117571 (accessed on 6th November 2025).

Sudarshan, N. R., Hoover, D. G., & Knorr, D. (1992). Antibacterial action of chitosan. *Food Biotechnology*, *6*, 257–272.

Suntornsuk, W., Pochanavanich, P., & Suntornsuk, L. (2002). Fungal chitosan production on food processing by-products. *Process Biochemistry*, *37*, 727–729.

Suriyakala, G., Sathiyaraj, S., Balasundaram, M., Murugan, K., Babujanarthanam, R., & Gandhi, A. D. (2023). *Plumeria alba* flower extract-mediated synthesis of recyclable chitosan-coated cadmium nanoparticles for pest control and dye degradation. *Bioprocess and Biosystems Engineering*, *46*, 1483–1498.

Van, S. N., Minh, H. D., & Anh, D. N. (2013). Study on chitosan nanoparticles on biophysical characteristics and growth of *Robusta* coffee in green house. *Biocatalysis and Agricultural Biotechnology*, *2*, 289–294.

Wang, W., Bo, S. Q., Li, S. Q., & Qin, W. (1991). Determination of the Mark–Hou wink equation for chitosan's with different degrees of deacetylation. *International Journal of Biological Macromolecules*, *13*, 281–285.

Wu, T., Zivanovic, S., Draughon, F. A., Conway, W. S., & Sams, C. E. (2005). Physicochemical properties and bioactivity of fungal chitin and chitosan. *Journal of Agricultural and Food Chemistry*, *53*, 3888–3894.

Yang, L., Lia, X., Laia, C., Fan, Y., Ouyang, J., & Yong, Q. (2017). Fungal chitosan production using xylose rich of corn stover prehydrolysate by *Rhizopus oryzae*. *Biotechnology & Biotechnological Equipment*, *31*, 1160–1166.

Yin, H., Zhao, X., Bai, X., & Du, Y. (2010). Molecular cloning and characterization of a *Brassica napus* L. MAP kinase involved in oligochitosan-induced defense signaling. *Plant Molecular Biology Reporter*, *28*, 292–301.

Yu, J., Wang, D., Geetha, N., Khawar, K. M., Jogaiah, S., & Mujtaba, M. (2021). Current trends and challenges in the synthesis and applications of chitosan-based nanocomposites for plants: A review. *Carbohydrate Polymers*, *261*, 117904. https://doi.org/10.1016/j.carbpol.2021.117904 (accessed on 6th November 2025).

Zhang, X., Zhang, J., & Zhu, K. Y. (2010). Chitosan/double-stranded RNA nanoparticle-mediated RNA interference to silence chitin synthase genes through larval feeding in the African malaria mosquito (*Anopheles gambiae*). *Insect Molecular Biology*, *19*, 683–693.

Zhao, X. M., She, X. P., Yu, W., Liang, X. M., & Du, Y. G. (2007). Effects of oligochitosans on tobacco cells and role of endogenous nitric oxide burst in the resistance of tobacco-to-tobacco mosaic virus. *Plant Pathology Journal*, *1*, 55–65.

Zhao, Y. G., Shen, H. Y., Shi, J. W., Chen, X. H., & Jin, M. C. (2011). Preparation and characterization of amino functionalized nano-composite material and its application for multi-residue analysis of pesticides in cabbage by gas chromatography–triple quadrupole mass spectrometry. *Journal of Chromatography A*, *1218*(33), 5568–5580. https://doi.org/10.1016/j.chroma.2011.06.090 (accessed on 6th November 2025).

CHAPTER 7

Fungal-Based Biostimulants and Growth Regulators for Plant Growth and Disease Resistance

DEEPAK KUMAR,[1,2,*] SANJAY K. SINGH,[3] VISHNU D. RAJPUT,[4] DEEPTI SRIVASTAVA,[5] MD. SHAMIM,[6] RAJA HUSAIN,[7] and SHACHI AGRAWAL[8*]

[1]R&D Division, Nextnode Bioscience Pvt. Ltd., Gujarat, India

[2]Shri Ram Solvent Extraction Pvt. Ltd., Jaspur, Uttarakhand, India

[3]Biodiversity and Paleobiology Group, MACS' Agharkar Research Institute, Pune, Maharashtra, India

[4]Academy of Biology and Biotechnology, Southern Federal University, Rostov-on-Don, Russia

[5]Department of Agriculture, Integral Institute of Agricultural Science and Technology, Integral University, Lucknow, Uttar Pradesh, India

[6]Department of Molecular Biology and Genetic Engineering, Dr. Kalam Agricultural College, Agricultural University, Sabour, Bhagalpur, Bihar, India

[7]Department of Agriculture, Himalayan University, Itanagar, Arunachal Pradesh, India

[8]Department of Botany, Dr. Manmohan Singh Bengaluru City University, Bangaluru, Karnataka, India

*Corresponding authors

Plant Biostimulants and Growth Regulators in Sustainable Agriculture.
Deepak Kumar, Sanjay K. Singh, Nutan Kaushik, and Mahaveer P. Sharma (Eds.)

DOI: 10.1201/9781779641458-9

ABSTRACT

In recent years, there has been a growing interest in harnessing the potential of fungal-based biostimulants (BS) and growth regulators to enhance plant growth and bolster disease resistance. Rising demand for sustainable agriculture and a shift away from hazardous chemical fertilizers are driving the biostimulant market. Fungi play multifaceted roles in the soil microbiome, influencing nutrient cycling, soil structure, and plant health through symbiotic relationships with plants. BS produced by fungi can successfully stimulate nutrient uptake, improve stress tolerance, and enhance overall plant vigor by utilizing the symbiotic interactions between fungi and plants. Furthermore, fungal-based growth regulators offer promising avenues for sustainable agriculture by modulating plant physiology and defense mechanisms. Fungi produce an array of bioactive compounds, including secondary metabolites and volatile organic compounds (VOC), which can elicit systemic resistance against pathogens. These regulators induce the expression of defense-related genes, triggering plant immune responses and priming systemic acquired resistance. Additionally, fungal-based growth regulators can mitigate the impact of plant pathogens by antagonizing their growth or producing antimicrobial compounds.

7.1 INTRODUCTION

Fungal-based biostimulants (BS) and growth regulators are products derived from various types of fungi that are used to enhance plant growth, productivity, and overall health (Anand et al., 2022; du Jardin, 2015; Fadiji et al., 2022). These products utilize the beneficial effects of fungi on plant growth and development, often through mechanisms such as improved nutrient uptake, hormone regulation, disease resistance, and stress tolerance (Begum et al., 2019; Zin and Badaluddin, 2020). BS derived from fungi contain a rich array of bioactive compounds, including hormones, enzymes, vitamins, and secondary metabolites, which exert multifaceted effects on plant physiology. These compounds can stimulate root development, enhance nutrient assimilation, improve water utilization efficiency, and activate defense mechanisms against pests and pathogens. By bolstering plant health from within, fungal-based BS enable crops to thrive even under challenging environmental conditions, such as drought, salinity, or nutrient deficiency (Anand et al., 2023; Melini et al., 2023). Moreover, fungal growth regulators offer a targeted approach to modulating plant growth and development, thereby optimizing resource allocation and maximizing yield potential. These regulators act on specific metabolic pathways within plants, regulating processes such as cell division, elongation, differentiation, and flowering (Ma et al., 2022; Verma et al., 2022). By fine-tuning growth patterns and synchronizing developmental stages, fungal growth regulators facilitate optimal plant architecture, fruit set, and ripening, resulting in higher-quality crops with enhanced market value (Goyal and Kalia, 2020; Vedenicheva and Kosakivska; 2023).

The adoption of fungal-based BS and growth regulators represents a paradigm shift towards sustainable and eco-friendly agricultural practices (Gupta et al., 2023; Magnabosco et al., 2023). Unlike conventional synthetic inputs, fungal formulations are biodegradable, non-toxic, and pose minimal risks to human health and the environment (Singh et al., 2024; Vaksmaa et al., 2023). Furthermore, by harnessing the natural symbiotic relationships between fungi and plants, these products promote soil biodiversity and resilience, contributing to the long-term health and fertility of agricultural ecosystems (Frąc et al., 2018; Zhang et al., 2021). In this era of climate change and resource scarcity, the integration of fungal-based BS and growth regulators into agricultural systems offers a holistic approach to enhancing crop productivity while minimizing environmental impacts (Bhupenchandra et al., 2022). By unlocking the untapped potential of fungi, we can cultivate resilient and sustainable food systems that nourish both people and the planet (Herath et al., 2024; Kržišnik and Gonçalves, 2023; Ulian et al., 2020). In this chapter, all aspects of fungal-based BS and growth regulators are briefly discussed in the context of types, formulations, regulatory frameworks, and environmental sustainability.

7.2 TYPES OF FUNGAL BASED BIOSTIMULANTS AND GROWTH REGULATORS

Fungal-based BS and growth regulators are derived from various species of fungi and are utilized to enhance plant growth, yield, and overall health (Anand et al., 2022; du Jardin, 2015; Fadiji et al., 2022). These biostimulants (BS) can be categorized based on the type of fungi used and their mode of action. There are some common types of fungal-based BS:

7.2.1 MYCORRHIZAL FUNGI

Mycorrhizal fungi are a group of fungi that form mutualistic associations with the roots of most plants. This symbiotic relationship benefits both the fungi and the plants involved. The form and function of these relationships vary greatly, but the arbuscular mycorrhizal (AM) association is the most prevalent type of interaction. This kind of linkage is thought to exist in over 80% of all terrestrial plants. The word "*mycorrhiza*" is derived from the Greek words "*mycos*," meaning fungus, and "*rhiza*," meaning root (Goltapeh et al., 2008). The two main functional forms of mycorrhizal fungi that develop symbioses with the roots of most plants on Earth are ectomycorrhizal (EM) and AM (Tedersoo, 2017). In ecosystem ecology, the concept that AM and EM linkages reflect different approaches to acquiring nutrients has been around for several years (Averilla et al., 2019; Phillips et al., 2013; Read, 1991). Moreover, mycorrhizae, which help plants absorb nutrients and water, are present in all soil types where plants can grow. Essential components of the plant microbiome, mycorrhizal fungi improve plants' uptake of water and nutrients from the soil (Van der Heijden et al., 2015). Fungi receive sugar from plants through their leaves. In addition, mycorrhizae have the ability to expand root surface area, which enables plants to absorb nutrients and water from a wide volume of soil more effectively (Nadeem et al., 2014). Furthermore, research has demonstrated that the reactions displayed by various mycorrhizal species vary according to the plant species that they are linked with (Ortas and Ustuner 2014). This is due to the large variety of mycorrhizal fungal species, which can alter the degree of plant-plant interactions and induce variations in plant growth. It is well known that mycorrhizal inoculation causes plants to develop more quickly than control plants due to increased photosynthetic activity (Huey et al., 2020).

Based on the hyphae's structure, two types of mycorrhizae are defined. According to Szabo et al. (2014), endomycorrhizal fungi are fungi that infiltrate the cell membrane and break through the cell wall, whereas ectomycorrhizal fungi are fungi that do not penetrate the individual cells within the roots. van der Heijden et al. (2015) further state that four primary categories of mycorrhizal fungi have been identified: ectomycorrhizae, arbuscular mycorrhizae, ericoid mycorrhizae, and orchid mycorrhizae. Arbuscular, orchid, and ericoid mycorrhizae are examples of endomycorrhizae; arbutoid mycorrhizae are categorized as ectomycorrhizae; and monotropoid mycorrhizae belong to a special category. Vogelsang et al. (2004) proposed two alternative names for arbuscular mycorrhizal fungi (AMF): vesicular-arbuscular mycorrhizal fungi (VAM) and soil fungi. These fungi are thought to have an asexual reproduction system and are members of the Glomeromycota class. These fungi are essential for plants to flourish to their full potential. Over 80% of vascular plant roots have AM symbiosis, which is the most prevalent non-pathogenic symbiosis in the soil (Brundrett 2002). Furthermore, the growth of arbuscular mycorrhizae varies depending on the host plant and can only occur in combination with certain host plants. Arbuscules are fungal entities that proliferate within the individual cells of plants. Angiosperm phyla comprise nearly all species of AM fungus (Duhoux et al., 2001). According to Hart and Forsythe (2012), AM fungi increase the uptake of nitrogen, zinc, iron, and copper, in addition to improving the nutrition of plants with phosphorus. They endure drought and are resistant to a few root diseases.

Although ectomycorrhizae are a big and widely distributed group of fungi (Szabo et al., 2014), only 3–4% of the families of vascular plants are known to be connected with them (Brundrett, 2004). Primarily, EM belongs to the phyla Ascomycota and Basidiomycota, and it is believed that EM mutualism originated independently from saprophytic lineages on many occasions (Merckx, 2012). It has been demonstrated that certain plant species that create EM mutualisms possess antibacterial elements that shield the plants from root infections. The growth of these fungi is unique in that it occurs on the outside surfaces of roots (Schnepf et al., 2008). Consequently, fungus covers roots with tissue that is referred to as a hyphal mantle. Fungal filaments, or strands of mycelium, emerge from the hyphal mantle into the soil, where they take up nutrients and minerals to function as the plant's roots (Huey et al., 2020).

When AMF generate mycorrhizae, they go through a number of developmental phases. Because there are no host plants present during the symbiotic stage, spores germinate, and AMF have been seen to produce little hyphal growth. In contrast, in the presence of root exudates, the spores undergo considerable hyphal branching during germination and arrive at the presymbiotic stage (Kuo et al., 2014). Additionally, before the hyphae pierce the root epidermis, appressoria form when the fungus comes into contact with a root surface. Subsequently, a distinct extraradical mycelium is produced along with the symbiotic colonization of the root cortical tissue. This entails the formation of intracellular arbuscules, which are densely branched structures resembling trees, or hyphal coils. It is conceivable that analogous alterations take place when the cortical cells are colonized, given that host plants are crucial to the AM infection process. Such developmental processes between the plant and arbuscular mycorrhizae generally depend on the exchange of signals and molecular communication between the symbiotic partners (Wang et al., 2022). Prior to the development of mycorrhizal symbiosis, symbiotic partners engaged in molecular communication. The two symbionts subsequently release signaling chemicals and tiny secretory proteins. Myc factors, indole 3-acetic acid, hypaphorine, and other signaling molecules are secreted by the fungi. They also secrete proteins with conserved LysM and CFEM domains as well as several carbohydrate-active enzymes, including lipases and proteases. The effectiveness with which mycorrhizal symbiosis is established is enhanced by these released signaling chemicals and proteins, which operate outside of the host cell wall (Wu et al., 2021).

It is well known that mycorrhizal symbioses are crucial for ionic transport and plant nutrition, especially for phosphorus uptake (Etesami et al., 2021). Modern agricultural systems rely heavily on continuous supplies of fertilizers based on phosphates to sustain crop yields. Phosphate rock is a naturally occurring resource that is not renewable and is processed into these fertilizers. As a result, a crisis of resource scarcity that could compromise global food security may soon threaten the planet. Arbuscular mycorrhizae coexist harmoniously with the roots of almost all vascular plants and may be a vital component in resolving the phosphate deficiency issue. Furthermore, mycorrhizal symbiosis can promote plant development and lessen the demand for phosphate-based fertilizers by enhancing plant tolerance to pathogens and abiotic stressors and increasing the efficiency of nutrient uptake (Huey et al., 2020; Khrieba, 2019). According to recent studies, the mycorrhizal fungus *Glomus intraradices* solubilizes rock phosphates by producing organic acid anions that may function as chelating agents and/or localized pH changes. While the majority of studies on mycorrhizae's function in P supply have focused on inorganic P, few have also examined how AM-colonized plants use organic P molecules (Khrieba, 2019; Yadav et al., 2021). Moreover, the contribution of mycorrhizal colonization to the enhancement of plant nutrition remains incompletely understood. Mycorrhizal colonization affects how other nutrients, such as Zn, Fe, Mn, Mg, Cu, Ca, K, B, and Na, are absorbed by developing plants (Etesami et al., 2021; Sardans et al., 2023) (Figure 7.1).

By producing glomalin and organic acids that chelate heavy metals, enhance carbon sequestration, prevent soil erosion, and stabilize soil macro-aggregation, AMF benefits the soil. Moreover, AMF attracts microorganisms that generate alkaline phosphatase, an enzyme linked

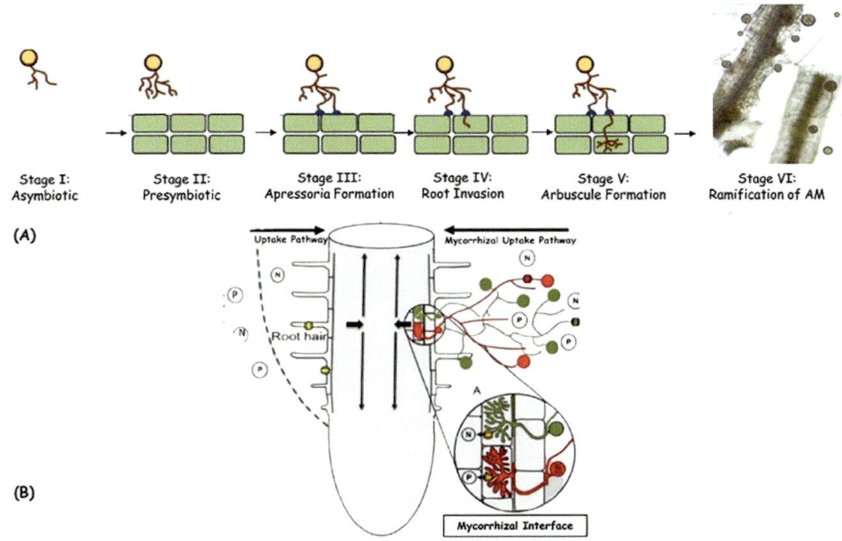

FIGURE 7.1 Mycorrhizal plant interactions and nutrient uptake: (a) developmental stages of arbuscular mycorrhizae; and (b) nutrient uptake pathways in the plant's root and mycorrhizae.

Source: Adapted from Huey et al. (2020).

to soil mineralization and organic phosphorus availability. Furthermore, through mechanisms of cooperation or antagonistic interactions, AMFs affect the diversity, activity, and composition of microbial communities in the soil (Fall et al., 2022). Furthermore, a common belief is that AMF inoculation confers tolerance on host plants against a range of stressful conditions, including salinity, heat, drought, metals, and extremely high or low temperatures. AMF may prevent important metabolic pathways from being down-regulated and help host plants up-regulate tolerance mechanisms (Begum et al., 2019; Beslemes et al., 2023). Kim et al. (2017) examined the impact of different species of AMF on soils from diverse sources, and they identified AMF species that are appropriate for environmentally friendly farming in household settings. The type of AMF used has different effects on plants. *Scutellospora heterogama*, *Funneliformis mosseae*, and *Acaulospora longula* all exhibited favorable effects on host growth in carrots, but AMF only marginally affected the growth of leek and red pepper.

7.2.2 TRICHODERMA SP.

The filamentous fungi *Trichoderma* is frequently found in soil, and several species of this genus are well-known for their abilities to promote plant development and act as a biocontrol agent. The potential of *Trichoderma* sp. to control a range of plant diseases via distinct mechanisms has been thoroughly investigated (Yao et al., 2023; Zin and Badaluddin, 2020). Additionally, *Trichoderma* spp. exhibits eco-friendly, cost-efficient, safe, and effective biocontrol properties for a variety of crop species (Guzmán-Guzmán et al., 2023).

The genus *Trichoderma* is responsible for many of the various strains' capacities as multipurpose fungi that are present in a wide range of environments. They are usually found in soils from forests or farms. Common morphological traits of *Trichoderma* strains include fast growth, a vivid green conidial pigment, and repetitive branching, as illustrated in Figure 7.2 (Zin and Badaluddin, 2020). According to Kubicek et al. (2019), *Trichoderma* belongs to Eumycota, Deuteromycotina, Hyphomycetes, Hyphomycetales, and Moniliaceae, commonly one of the rhizosphere inhabitants. Since the physical variations are uncommon and difficult to see, the challenges faced in the process of identifying *Trichoderma* isolates at the species level have become increasingly substantial. Morphological characteristics were typically employed to categorize *Trichoderma* species. According to Waghunde et al. (2016), there are around 10,000 species in the *Trichoderma* genus, the majority of which are expanding quickly. Some common types of *Trichoderma* species include:

1. ***Trichoderma harzianum:*** It is known for its biocontrol abilities and is used in agriculture to suppress plant diseases caused by various pathogens. It has also been studied for its ability to promote plant growth (Pani et al., 2021; Saravanakumar et al., 2017).

2. ***Trichoderma viride:*** It is another biocontrol agent that is used to manage plant diseases caused by different phytopathogenic fungi. It is widely used as a biological fungicide in agriculture (Surekha et al., 2013).

3. ***Trichoderma atroviride:*** Specifically, against soil-borne plant diseases, this species is employed in biocontrol. Its efficacy in

preventing diseases like root rot and damping-off has been established (Longa et al., 2009).

4. ***Trichoderma hamatum:*** This species is used as a biocontrol agent. *T. hamatum* has demonstrated efficacy against soil-borne pathogens. It has been researched for its ability to enhance plant growth and suppress diseases (Li et al., 2022; Lodi et al., 2023).
5. ***Trichoderma longibrachiatum:*** This species' potential for biological control has been investigated, and it has been demonstrated that it possesses antifungal qualities against a variety of plant diseases (Degani et al., 2021; Zhang et al., 2015).
6. ***Trichoderma reesei:*** This species is well-known for its cellulase production and is used in the production of enzymes for industrial applications, particularly in the biofuel and textile industries (Druzhinina and Kubicek, 2017; Kumar et al., 2014).
7. ***Trichoderma koningii:*** This species is known for its mycoparasitic activity. *T. koningii* is employed as a biofungicide to control the harmful diseases caused by pathogenic fungi. It can parasitize and inhibit the growth of other fungi (Samuels et al., 2006; Sumayku et al., 2015).
8. ***Trichoderma virens:*** It is known for its biocontrol and plant growth-promoting abilities. *T. virens* is well-known for its ability

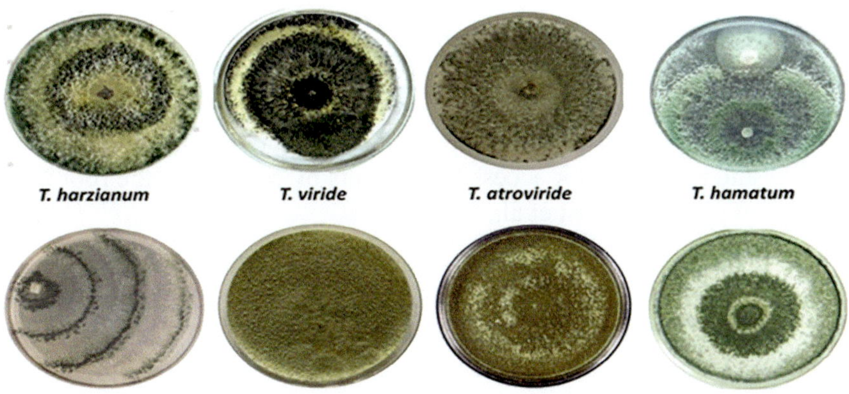

FIGURE 7.2 Morphological variations in different strains of *Trichoderma* species.

to biocontrol a variety of plant diseases. It has the ability to prevent the formation of pathogenic fungi, such as those that cause root rot and other diseases spread by the soil. *T. virens* helps shield plants from diseases by producing antifungal compounds, developing systemic resistance in plants, and competing with other plants for nutrition and space (Contreras-Cornejo et al., 2009; Inayati et al., 2021).

It's been documented that certain species of *Trichoderma* can attack other fungi. Certain studies have found that *Trichoderma* sp. can both eliminate plant diseases and promote plant development (Garnica-Vergara et al., 2016; Yao et al., 2023). Also, it has been demonstrated that *Trichoderma* sp. can detoxify harmful substances and hasten the breakdown of organic materials (Sood et al., 2020; Vázquez et al., 2015; Zafra et al., 2015). *Trichoderma* species are effective in the soil ecosystem because of their capacity to accelerate development, absorb nutrients, and alter the rhizosphere (Guzmán-Guzmán et al., 2023). *Trichoderma* sp. plays an important role as a natural decomposer in the soil. It possesses strong destructive skills against plant pathogenic bacteria and can withstand unfavorable environmental conditions (Abdul Rahman et al., 2021; Alfiky and Weisskopf, 2021; Tyśkiewicz et al., 2022). *Trichoderma* sp. have been extensively studied for their ability to suppress various plant pathogens through different mechanisms (Manzar et al., 2022; Tyśkiewicz et al., 2022; Zin and Badaluddin, 2020) (Table 7.1).

There are some key aspects of *Trichoderma*'s mode of action:

1. **Competition for Resources:** The fungus *Trichoderma* sp. develops quickly and has the ability to outcompete other soil-borne diseases for resources and available space. *Trichoderma* creates a barrier that keeps pathogenic organisms from growing and restricts their access to nutrients by colonizing the root zones of plants (Oszust et al., 2020).

2. **Antibiosis:** *Trichoderma* species produce a variety of secondary metabolites, such as antibiotics, that have antifungal and antibacterial properties. These metabolites can inhibit the growth and development of plant pathogens. One well-known example is the production of the antifungal compound gliotoxin (Leylaie and Zafari, 2018; Stracquadanio et al., 2020).

TABLE 7.1 Role of *Trichoderma* Species in Biological Control of Plant Diseases

***Trichoderma* sp.**	**Target Phytopathogen**	**Plant Disease**	**Significant Findings**	**References**
T. harzianum	*F. oxysporum* f. sp. tuberose	bulb rot of tuberose	-Better quality and quantity of flowers due to increased plant health and growth	Nosir (2016)
T. harzianum (Strain MH027645.1)	*Fusarium oxysporum*, *Pythium* sp., *Macrophomina phaseolina*, *Sclerotium rolfsii*, *Phytophthora infestans*	Soil borne diseases	-Mycoparasitism via hyphae penetration, hydrolytic enzyme production, coiling and appressoria development, and pathogen lysis -By restricting pathogens access to sufficient space and nutrients, it also prevents plant diseases from spreading.	Jadav et al. (2022)
T. virens *T. harzianum*	*Pythium* sp.	Damping off	-A metabolite known as gliotoxin imitates the plant growth hormone gibberellic acid, which speeds up seed germination and increases the vigor of seedlings. It also reduces the chance of a pathogen attacking.	Uddin et al. (2011)
T. asperellum	*Rhizoctonia solani*	Sheath blight of rice	-A reduced rate of sheath blight progression leads to increased yield and grain weight.	de França et al. (2015)
T. harzianum (Strain M10)	*Sclerotium rolfsii*	Soil borne diseases	-A novel siderophore that promotes seed germination and enhances seedling growth is harzianic acid.	Manganiello et al. (2018)
T. hamatum (Strain Th23)	Tobacco mosaic virus (TMV)	Viral disease of tomato	-Systemic resistance and innate immunity were increased, and plant growth was accelerated.	Abdelkhalek et al. (2022)
T. citrinoviride *T. harzianum* T22	*Meloidogyne incognita*	Root-knot nematodes of tomato and soybean	-High virulence and significant suppression of egg hatching against second-stage juveniles (J2) with elevated plant biomass.	Izuogu and Abiri (2015)
T. atroviride (Strain HB20111)	*Rhizoctonia cerealis* and *Fusarium pseudograminearum*	Sharp eyespot of wheat and *Fusarium* crown rot	-Seed dressing can decrease the amount of plant pathogens in the rhizosphere of wheat plants, which may improve the micro-ecological conditions for growth.	Sui et al. (2022)
T. virens NBAII Tvs12 *T. spirale* T76-1	*Sclerotium rolfsii*, *Curvularia aeria* or *Corynespora cassiicola*	Stem rot in groundnut and leaf spot on lettuce	-Pathogen growth was prevented by the increased activity of chitinase and β-1,3 glucanase.	Wonglom et al. (2019)

3. **Mycoparasitism:** As mycoparasites, *Trichoderma* species are able to directly parasitize and attack other fungi. They release enzymes that cause the target organism to perish after penetrating the cell walls of pathogenic fungi. *Trichoderma* mycoparasitism entails wrapping around the pathogen's hyphae and then breaking through the cell walls (Mukherjee et al., 2022; Omann and Zeilinger, 2010).
4. **Induction of Plant Defense Mechanisms:** *Trichoderma* species have the ability to stimulate a plant's defensive mechanisms, increasing its resistance to pathogenic invasions. This involves the creation of defense-related proteins and secondary metabolites in the host plant, as well as the activation of many signaling pathways (Pacheco-Trejo et al., 2022; Yao et al., 2023).
5. **Promotion of Plant Growth:** *Trichoderma* species not only perform biocontrol functions but also augment nutrient uptake, generate auxins and plant growth hormones, and enhance overall plant health. A higher ability to withstand abiotic stresses and higher crop yields may result from this (Sood et al., 2020; Zin and Badaluddin, 2020).
6. **Siderophore Production:** *Trichoderma* sp. can produce siderophores, which are molecules that chelate iron and make it less available to competing pathogens. This iron competition further inhibits the growth of other fungi that rely on iron for their metabolic processes (Tyśkiewicz et al., 2022).

Trichoderma sp. can indirectly influence mineral solubilization by enhancing the availability of nutrients to plants. This is often achieved through the production of organic acids, siderophores, and enzymes (Tyśkiewicz et al., 2022). These substances can facilitate the solubilization of minerals in the soil, making them more accessible to plants (Li et al., 2015). Some *Trichoderma* strains have been reported to possess phosphate-solubilizing abilities (Sood et al., 2020). They can produce organic acids, such as citric and oxalic acids, which can help in the solubilization of insoluble phosphates, making phosphorus more available to plants (Li et al., 2015). Moreover, *Trichoderma* sp. can form symbiotic relationships with plants, increasing the absorption of essential nutrients (like nitrogen, phosphorus, and potassium) and producing plant growth hormones like auxins through various mechanisms. They can induce systemic resistance, making plants more resilient to stressful conditions (Andrzejak and Janowska, 2022; Tyśkiewicz et al., 2022). In agriculture, *Trichoderma* species-based biocontrol agents are widely used to manage soil-borne diseases and promote plant health. As eco-friendly substitutes for chemical pesticides, they support integrated and sustainable approaches to pest management (Yao et al., 2023; Zin and Badaluddin, 2020).

7.2.3 *ASPERGILLUS SP.*

Aspergillus sp. refers to a group of fungi that includes various species commonly found in soil, decaying organic matter, and other environments. Some species of *Aspergillus* have been studied for their potential applications in agriculture as BS and growth regulators (Nji et al., 2023; Perrone et al., 2007). *Aspergillus* sp. can produce various metabolites, such as organic acids, enzymes, and secondary metabolites, that can positively influence plant growth. These substances may enhance nutrient availability, stimulate root development, and improve overall plant performance. Khuna et al. (2023) investigated the ability of three phosphate-solubilizing fungal strains of *Aspergillus* sp. (viz., *Aspergillus chiangmaiensis* SDBR-CMUI4, *Aspergillus pseudoliparis* SDBR-CMUI1, and *Aspergillus pseudotubin-gensis* SDBR-CMUO2 or pseudotumoral aspergillosis) to promote plant growth. The results showed that indole-3-acetic acid synthesis and positive siderophore expression were present in all fungi. Moreover, all fungal strains produced phosphatase enzymes and organic acids that allowed them to solubilize the insoluble phosphate minerals [$Ca_3(PO_4)_2$ and $FePO_4$]. Inoculating plants with microorganisms that produce indole-3-acetic acid (IAA) has been proven in many studies to help in plant growth and development. Da-Prado et al. (2019) assessed 16 microbial strains for their ability to produce IAA and its analogues. High IAA production in the strain *Aspergillus flavipes* (ATCC® 16814™) was reported for the first time, and LC-MS/MS revealed that *A. flavipes* produces IAA via a tryptophan-dependent biosynthetic pathway. Moreover, Attia et al. (2022) investigated the use of eco-friendly inducers (*Aspergillus alabamensis*, *Aspergillus tubingensis*, and *Aspergillus oryzae*) to stimulate physiological immune responses in pepper plants resistant to *Fusarium* wilt. The findings show that testing fungal endophytes via two different modes—foliar and soil—produced substantial protection and decreased the percent disease severity (PDI). *A. tubingensis* (via the soil) had the greatest influence on protection, accounting for 83.33% of the protection. Additionally, the findings revealed that all fungal isolates, *A. alabamensis*, *A. tubingensis*, and *A. oryzae*, can be employed as ecofriendly BS on a commercial scale to protect pepper seedlings from *Fusarium* wilt disease. Phytopathogen-induced crop loss is a serious problem that has an international impact. For years, synthetic pesticides have been employed to stop phytopathogen-caused economic losses, which has raised serious concerns about the environment and public health. In this context, Daigham et al. (2023) isolated *Aspergillus chevalieri* AGH-1 OP593083 and *Aspergillus egyptiacus* AGH-4 OP593080 from the wheat rhizosphere and used them as plant growth-promoting fungi (PGPF) to fight *Alternaria solani*-caused early blight disease and to induce growth promotion strategies and the defense system in *Vicia faba*. It's important to consider that the effectiveness of *Aspergillus* sp. as BS and growth regulators can vary depending on the specific species/strain, environmental conditions, and the target crops. Research in this field is ongoing, and scientists are continually exploring the potential benefits of fungal-based BS for sustainable agriculture.

7.2.4 CHAETOMIUM SP.

The widely distributed genus *Chaetomium* is well-known for its qualities that can be used in agriculture (Kedves et al., 2021). Because they produce bioactive compounds, *Chaetomium* species are reported to be effective in controlling plant diseases by inhibiting many plant pathogens (Moya et al., 2020). Ascomycota is a phylum that includes the filamentous species *Chaetomium globosum* Kunze. Known to be a saprotroph and a plant endophyte with a broad host plant range, *Chaetomium globosum* exhibits a worldwide distribution (Linkies et al., 2021; Větrovský et al., 2020). The species *C. globosum* has been extensively researched, and several studies have shown that it produces a wide range of bioactive metabolites, including phenols, hydrocarbons, terpenoids, and sulfur compounds. These include major bioactive compounds such as 4-methyl-(1,5-dimethyl-4-hexenyl)-benzene, tetradecane, dodecane, hexadecane, β-bisabolene, and dimethyl-propyl-disulfide, and derivatives of chlorinated azaphilones, such as chaetoglobosins and chaetomugilins, which have antifungal properties (Kumar et al., 2021, 2020). Moreover, Spinelli et al. (2022) examined the growth performance and metabolomic characteristics of chicory (*Cichorium intybus*) plants using the biostimulant effect of fungal culture filtrates derived from *Chaetomium globosum* and *Minimededusa polyspora*. The findings demonstrated that *M. polyspora* and *Chaetomium globosum* culture filtrates directly promote plant growth by increasing biomass in the form of shoots, roots, and leaf area. Globally, *Chaetomium* sp. was first identified as a new broad-spectrum biological fungicide and patented (No. 6266, International Code: AO1N25/12) and registered as Ketomium® mycofungicide for the management of plant diseases in Thailand, Cambodia, Laos, Vietnam, and China (Kumar et al., 2023; Soytong et al., 2021). Integrated plant disease control is achieved through the use of *Chaetomium* biofungicides and BS. In preventing plant disease and encouraging plant growth, it demonstrated both therapeutic and protective properties. It has proven effective when applied to infected soils in conjunction with integrated cultural control to provide long-term protection against a variety of pathogenic diseases, including tomato wilt (*Fusarium oxysporum* f. sp. *lycopersici*), citrus rot (*Phytophthora parasitica*), black pepper rot (*Phytophthora palmivora*), and rice blast (*Magnaporte oryzae*) (Soytong et al., 2021).

7.2.5 PENICILLIUM SP.

With more than 200 species, the *Penicillium* genus is among the biggest and most interesting classifications of fungi. It is especially well-known for being a reliable source of antibiotics. Endophytic *Penicillium* is known to colonize its ecological niches and shield its host plant from various pressures through a variety of biological capabilities that can be used for a wide range of purposes, such as biotechnological, pharmacological, and agricultural (Toghueo and Boyom, 2020). Some of the *Penicillium* sp. have been explored for their potential applications in agriculture as BS and growth regulators. *Penicillium* sp. can produce various metabolites and secondary compounds that may act as plant growth promoters. These substances include organic acids, enzymes, and phytohormones that can enhance nutrient uptake, root development, and overall plant growth (Adedayo and Babalola, 2023; Hossain et al., 2014; Radhakrishnan et al., 2014).

While the research in this area is ongoing, there is evidence to suggest that certain strains of *Penicillium* can have positive effects on plant growth and development (Babu et al., 2015; Landa, 2022). In order to promote growth and disease resistance in the cucumber plant, Hossain et al. (2014) evaluated a plant growth-promoting fungus (PGPF) isolate GP15-1 of *Penicillium* sp., which was obtained from zoysiagrass rhizospheres. *Penicillium* sp. GP15-1 significantly decreased damping-off disease and the causative pathogen, *Rhizoctonia solani*. *Penicillium* sp. may help plants tolerate various environmental stresses, such as drought, salinity, or heavy metal toxicity. The presence of these fungi can induce stress resistance mechanisms in plants, leading to better survival and growth under adverse conditions (Leitão and Enguita, 2016). From the halophyte plant *Aeluropus littoralis*' rhizosphere, Tarroum et al. (2022) identified the halotolerant *Penicillium olsonii* fungal strain (A3), which increases the salt tolerance of tobacco plants, promotes their growth, and lowers the amount of chemical fertilizer inputs needed in hydroponic farming systems by half, according to the study's findings.

7.2.6 ENDOPHYTIC FUNGI

Endophytic fungi are microorganisms that live inside plant tissues without causing any apparent harm to the host plant (Alam et al., 2021; Wilson, 1995). These microorganisms, which typically cohabit with diseases, comprise both bacteria and fungi (Gouda et al., 2016). Classifying endophytic fungi into facultative and obligatory groups is based on how they colonies an environment. Plants are colonized by facultative endophytes at specific stages of their life cycles, but they can also live outside the plant at other times and establish a relationship with the host plant's immediate rhizosphere (Abreu-Tarazi et al., 2010; Alam et al., 2021). Endophytic fungi (EF) communities can be divided into two groups based on host occurrence and reproductive pattern: the non-Clavicipitaceous/non-Balansiaceous group (NC-group) and the Clavicipitaceous/Balansiaceous group (C-group). Ascomycota and Basidiomycota, two non-grass host-related groups of endophytic fungi (NC-group EFs), are widely distributed and have a high biodiversity, ranging from tropical to polar regions. Woody, vascular, and nonvascular plant communities serve as their hosts. By generating spores or conidia, they spread asexually or sexually and aid in the induction of symbiosis, or horizontal proliferation. Because these EFs are capable of existing in a quiescent state until they detect chemical changes from host plants suffering from wounds, injuries, or other environmental stresses, they are not directly linked with the host plants (Carroll, 1998; Mishra et al., 2021). Whereas, C-group endophytic fungi (EFs) infect host plant ovules and spread vertically through host seeds, from parents to progenies. C-group species are classic examples of obligatory endophytes (Alam et al., 2021). By secreting supporting or defensive bioactive compounds, respectively, these organisms help their hosts endure dry conditions or defend against herbivore attacks (Carroll, 1988; Poveda, 2021; Roberts and Lindow, 2014). The NC-group typically colonizes aerial organs in a localized, confined, limited, and mostly intercellular manner; however, it colonizes roots or the rhizosphere in a widespread, organized, systematic, intercellular, and intracellular manner. The NC group includes *Piriformospora indica*, *Fusarium* sp., and dark septate mycobiota as examples of illustrative endophytic mycobiomes (Alam et al., 2021; Schulz et al., 2002).

Endophytic fungi (EFs) have garnered significant attention in recent years of research because of their ability to yield BS for the production of essential oils, as well as novel sources of cytotoxic, antibacterial, and anticarcinogenic compounds (El-Enshasy et al., 2019; Uzma et al., 2018). EFs can serve as biological control agents (BCAs), stimulate plant growth, improve nutrient solubilization in the rhizosphere, or trigger systemic plant resistances to biotic or abiotic stresses (Alam et al., 2021; Cui et al., 2021; Poveda and Baptista, 2021; Poveda et al., 2020a; 2021). The use of endophytic fungi as BS and growth regulators holds great promise in sustainable agriculture (Baron and Rigobelo, 2021) (Table 7.2). Within this perspective, fungal endophytes (viz., *Metarhizium anisopliae* and *Beauveria bassiana*) are frequently used as BCAs against pests in different crops (Baron et al., 2019; Moraga et al., 2020). On the other hand, a number of studies have demonstrated that these fungi, along with other recognized entomopathogens, have the ability to directly protect plants (Baron and Rigobelo, 2021; El-Enshasy et al., 2019). Endophytic fungi can work as antagonists against plant pathogens by a variety of strategies, such as the synthesis of metabolites (volatile compounds, antibiotics, and enzymes), competing with the pathogens for resources (carbon, nitrogen, minerals, and space), parasitism, inducing systemic resistance in the plant, and increasing plant growth, which lowers the pathogens' activity (Lr, 2018;

TABLE 7.2 Biostimulant and Biocontrol Properties of Different Fungal Endophytes

Fungal Endophyte(s)	Host Plant	Function	References
Alternaria sp., *Aspergillus* sp., *Arthrinium* sp., *Ascorhizoctonia* sp., *Chaetomium* sp., *Penicillium* sp.	*Vitis vinifera* cv. Summerblack	-Endophytes have antifungal activities against some fungal diseases and growth-promoting activities for plant growth and development.	Dissanayake et al. (2018)
Epichloë bromicola	*Hordeum brevisubulatum*	-Plant polyamine metabolism is closely linked to microbial infection and tolerance to various abiotic stimuli; therefore, *E. bromicola* improves *H. brevisubulatum*'s resistance to salt stress.	Chen et al. (2019)
Gilmaniella sp.	*Atractylodes lancea*	-In *Atractylodes lancea* host plantlets, inoculation with the endophytic fungus *Gilmaniella* sp. caused a variety of reactions, such as an increase in volatile oil, a burst of nitric oxide (NO), salicylic acid (SA), and hydrogen peroxide (H_2O_2).	Wang et al. (2011)
Epichloe gansuensis	*Achnatherum inebrians*	-Improved seed longevity and seed vigor -Increased proline, soluble sugar content, and antioxidant enzymes under abiotic stresses	Li et al. (2020)
Epichloe gansuensis	*Achnatherum inebrians*	-Accelerate the rate of seed germination in response to temperature stress, pH, salt, and different light conditions.	Ahmad et al. (2020)
Epicoccum nigrum	*Dysphania ambrosioides*	-Increased production of auxin, gibberellin, and jasmonate by endophyte upregulation of the decreased glutathione concentration -Accelerate the rate of seed germination	Parmar et al. (2022)
Fusarium oxysporum, Fusarium sp., *Fusarium solani*	*Senna alata*	-Increased seed germination rate	Pradhan et al. (2023)
Penicillium sp.	*Triticum turgidum*	-During the heat and drought stress stages of seed development and germination, the endophyte *Penicillium* sp. increased the number and weight of seeds.	Hubbard et al. (2014)
Acremonium coenophialum, Acremonium loliae	*Festuca arundinacea, Lolium perenne*	-Fungal endophytes synthesized Ergot alkaloïd and, as a result, showed deterrent effects against flour beetles (*Tribolium castaneum*) and fall armyworms (*Spodoptera frugiperda*).	Cheplick and Clay (1988)

Moraga, 2020; Vega, 2018). Interaction with endophytic fungi has direct benefits for the plant, such as increased nutrient uptake and phytohormone levels, which are directly linked to increased biomass production, expanded plant height, weight reproduction, root system development, and yield. These advantages allow them to be regarded as biofertilizers (Bamisile et al., 2018; Baron and Rigobelo, 2021). Moreover, endophytic fungi have the ability to enhance the plant host's intake of macronutrients like nitrogen, phosphorus, potassium, and magnesium, as well as micronutrients like iron, zinc, and copper from the soil and organic matter (Chaudhary et al., 2022; Rana et al., 2020; Sarkar et al., 2021).

7.2.7 FUNGAL-BASED ELICITORS

Fungal-based elicitors are substances derived from fungi that can induce a defense response in plants (Guo and Cheng, 2022). These elicitors play a crucial role in plant-microbe interactions and are used in agriculture to enhance plant resistance to diseases. Plants identify certain molecules, referred to as elicitors, linked to fungi when they come into contact with fungal infections (Thakur and Sohal, 2013). The plant responds to the invasive pathogen by going through a sequence of defensive mechanisms as a result of this recognition (Guan et al., 2022). There are different types of fungal-based elicitors, and they can be broadly classified into two categories: pathogen-associated molecular patterns (PAMPs) and effector molecules (Guo and Cheng, 2022; Jamiołkowska, 2020).

7.2.7.1 PATHOGEN-ASSOCIATED MOLECULAR PATTERNS (PAMP)

PAMPs are conserved molecular structures found on the surface of pathogens, including fungi. Plants have evolved to recognize these structures as foreign and activate defense responses upon detection (Patel et al., 2020; Sák et al., 2021). The existence of these patterns usually results in an immunological response in the host organism and is necessary for the pathogens to survive. Multiple PAMPs have been found in the case of fungi (Lee et al., 2021). There are a few examples of fungal-based PAMPs:

1. **β-Glucans:** β-glucans are glucose polymers that form the structural component of the fungal cell wall. β-1,3-glucan, which binds to β-1,6-glucan in the shape of a branch, has the largest percentage of all the β-glucans in the fungal cell wall, ranging

from 65 to 90% (Ruiz-Herrera and Ortiz-Castellanos, 2019). They are potent stimulators of the immune system and are recognized by receptors like Dectin-1 on the surface of immune cells (Fesel and Zuccaro, 2016; Yamaguchi et al., 2000). Dectin-1 activation leads to various immune responses, including the production of pro-inflammatory cytokines and the initiation of phagocytosis (Elder et al., 2017). According to Ayers et al. (1976), β-glucan can strengthen plant resistance as a PAMP by encouraging the buildup of phytoalexins and enhancing their ability to fend off pathogen invasion. Moreover, it was observed in many studies that a particular receptor in the membrane of many legumes, such as pea, soybean, lupin, alfalfa, and bean, was shown to bind with β-1,3/-1,6-glucan heptaglucoside (Cosio et al., 1996; Côté et al., 2000; Guo and Cheng, 2022).

2. **Chitin:** Chitin is another major component of the fungal cell wall (Lee et al., 2021). It is a polymer of N-acetylglucosamine (GlcNAc) and is recognized by receptors such as Chitin Receptor 1 (CHIT1) on immune cells (Geoghegan and Gurr, 2017). Chitin recognition triggers immune responses, contributing to the defense against fungal infections (Pusztahelyi, 2018).

3. **Mannans:** Mannans are polysaccharides found in the fungal cell wall. They can be recognized by pattern recognition receptors (PRRs), like Dectin-2. Recognition of mannans by Dectin-2 can lead to the activation of immune responses (Vendele et al., 2020).

4. **Fungal DNA and RNA:** Nucleic acids from fungi, including fungal DNA and RNA, can act as PAMPs. These molecules are detected by various receptors, such as Toll-like receptors (TLRs) and retinoic acid-inducible gene I (RIG-I)-like receptors, which trigger signaling pathways that result in the production of interferons and other immune responses (Duan et al., 2022).

5. **Fungal Proteins:** Some fungal proteins can also act as PAMPs. For example, certain cell surface proteins or secreted proteins from fungi may be recognized by the immune system and trigger an immune response (Guo and Cheng, 2022).

Recognition of these fungal PAMPs by the host's immune system is crucial for the initiation of an effective immune response against fungal infections. It allows the innate immune system to detect and respond to the presence of fungal pathogens, activating various defense mechanisms to eliminate or control the infection (Guan et al., 2022; Guo and Cheng, 2022).

7.2.7.2 FUNGAL EFFECTOR MOLECULES

Fungal effector molecules play a crucial role in the interactions between fungi and their hosts, whether the hosts are plants, animals, or other organisms (Sun and Zhang, 2020). Fungi synthesize chemical molecules called effectors that are used to control the host's biological functions and initiate a successful infection. These molecules can aid in the fungus's survival within the host by facilitating the absorption of nutrients and suppressing the host's immunological response (Lee et al., 2021). There are some common types of fungal effector molecules and their functions:

1. **Proteases and Glycosidases:** In many different ecosystems, fungi are essential to the breakdown and cycling of nutrients. Proteases and glycosidases are two of the many enzymes that fungi make in order to collect nutrients from their surroundings (Bezerra et al., 2021; Guo and Cheng, 2022; Wang et al., 2022). These enzymes break down complex chemical compounds into simpler forms that the fungus can absorb. Proteases are enzymes that hydrolyze peptide bonds in proteins, dividing them into smaller peptides or individual amino acids. They are often referred to as peptidases or proteinases. To break down exterior proteinaceous materials like plant cell walls, animal tissues, or other organic substances, fungi release proteases. Proteins are broken down into amino acids, which gives the fungus a source of nitrogen—a nutrient that is necessary for fungal growth and development (Bezerra et al., 2021; de Souza et al., 2015). On the other hand, enzymes called glycosidases catalyze the breakdown of complex carbohydrates into simpler sugars or oligosaccharides by hydrolyzing the glycosidic linkages in those carbohydrates. To break down complex carbohydrates in their surroundings, fungi create glycosidases. Breaking down fungal, plant, and other substrates rich in polysaccharides are part of this process. The fungus then uses the produced simple sugars or oligosaccharides as a source of carbon (Wang et al., 2022).

2. **Cellulases:** Cellulases are a type of glycosidase specifically involved in the breakdown of cellulose, a complex polysaccharide found in the cell walls of plants. Fungi use cellulases to degrade plant cell walls and access the glucose molecules within cellulose. This process is crucial for the recycling of plant biomass in ecosystems (Ma et al., 2015; Saravanakuma et al., 2016).
3. **Chitinases:** Chitinases are enzymes that hydrolyze chitin, a polysaccharide found in the cell walls of fungi and the exoskeletons of arthropods. Fungi release chitinases to break down their own cell walls during growth and development. They can also use chitinases to exploit the nutrients present in the cell walls of other fungi or the exoskeletons of arthropods (Fiorin et al., 2018; Pusztahelyi et al., 2018).
4. **Amidases and peptidases:** Fungi may produce amidases and peptidases to hydrolyze amide and peptide bonds in various nitrogen-containing compounds, recycling nitrogen for fungal growth (Li et al., 2022; Zhai et al., 2017).
5. **Mycotoxins:** Fungal toxins, also known as mycotoxins, can play a role in suppressing the defense mechanisms of hosts (Brown et al., 2021). Mycotoxins are secondary metabolites produced by fungi, and some of them have been found to interfere with the host's defense responses. Toxins can facilitate the spread of the fungus within the host (Bennett and Klich, 2003). Fungal toxins may interfere with the activity of enzymes that play a crucial role in plant defense responses (Jones et al., 2023). For example, some toxins inhibit enzymes involved in the synthesis of cell wall components, making the plant more susceptible to fungal penetration (Ismaiel and Papenbrock, 2015; Tsuge et al., 2013). Additionally, mycotoxins have the ability to obstruct plant signal transduction pathways. The ability of the plant to develop a successful defense response may be compromised by the toxins' disruption of these pathways, which may restrict the activation of genes relevant to defense and impede the production of defense proteins (Brown et al., 2021; Ismaiel and Papenbrock, 2015). Reactive oxygen species (ROS) are a common defense mechanism used by plants against infections. Mycotoxins can obstruct the production of ROS, which are necessary to initiate defense

reactions (Liu et al., 2010). The plant's capacity to produce an oxidative burst—a sudden rise in ROS levels linked to defense responses—may be hampered by this interference. Fungal toxins also have the potential to disrupt the host's metabolic functions by rerouting resources from defense-related pathways. Toxins have the ability to modify the metabolism of the host, which can facilitate the growth of fungus colonization (Ali et al., 2018).

6. **Small RNAs:** Some fungi release small RNA molecules into the host cells, which can regulate gene expression in the host. This manipulation helps the fungus evade host defenses and establish infection (Derbyshire et al., 2019; Mapuranga et al., 2023).
7. **Phytotoxins:** In plant-fungal interactions, certain fungi produce compounds known as phytotoxins that can cause damage to plant tissues (Agudo-Jurado et al., 2023). These phytotoxins are harmful to other living organisms, such as animals and other plants. These substances are essential to plants' defense mechanisms because they aid in the deterrence of diseases, herbivores, and rival plant species (War et al., 2012). The chemical structures and modes of action of phytotoxins can vary widely. Moreover, these toxins contribute to the virulence of the fungus and help in nutrient release (Agudo-Jurado et al., 2023).

Understanding the molecular mechanisms behind these effector molecules is crucial for developing strategies to control fungal infections in agriculture, medicine, and other fields. Researchers study these interactions to identify potential targets for antifungal treatments and to enhance our understanding of host-pathogen dynamics (Lee et al., 2021; Sun and Zhang, 2020; Vandeputte et al., 2012).

7.3 FUNGAL BIOSTIMULANTS FORMULATIONS AND REGULATORY FRAMEWORK

Fungal BS, or fungal-based growth enhancers, are formulations containing beneficial fungi that promote plant growth by forming symbiotic relationships with plant roots (Adedayo and Babalola, 2023; Sharma et al., 2023). These fungi can improve nutrient uptake, enhance resistance to diseases, and increase tolerance to environmental stresses (Sood et al., 2020; Zin and Badaluddin, 2020). In a fungal-based commercial biostimulant formulation, appropriate fungal species such as *Glomus* spp., *Rhizopogon* spp., *Trichoderma* spp., or *Mycorrhizae* spp. that are compatible with a wide range of plant species and suitable for the target soil conditions are used along with suitable carriers (Kumar et al., 2023a). A carrier material is essential for providing a medium for the fungal spores to adhere to and protecting them during storage and application. Common carrier materials include vermiculite, perlite, peat moss, or a mixture of these materials. Moreover, nutrient sources (such as molasses, humic acids, or seaweed extracts) are also added to formulations to enhance the viability of selected fungi for growth and establishment during application (Kumar et al., 2023b). Commercial fungal biostimulant formulations can be applied to seeds, seedlings, or directly to the soil around plant roots. The types of fungal biostimulant formulations can vary widely depending on their intended use, target crops, and specific fungal strains utilized (Kumar et al., 2023a, b). There are some common types of commercial fungal biostimulant formulations:

1. **Liquid Formulations**: These are liquid-based formulations containing fungal spores, mycelium, or fungal metabolites. They may also contain other ingredients like nutrients, humic acids, or plant growth regulators (PGRs). Liquid formulations of fungal BS are becoming increasingly popular in agriculture and horticulture due to their convenience, efficacy, low contamination, and longer shelf life during storage. These formulations aim to improve soil health, nutrient cycling, and plant growth through microbial interactions. They are typically applied through irrigation systems or as foliar sprays. Additionally, proper storage conditions should be maintained to ensure product viability and efficacy (Kumar et al., 2023b; Vassilev et al., 2017).

2. **Powder Formulations**: These are dry formulations, usually containing fungal spores or mycelium in powdered or granular form. They can be easily mixed with water for application as soil drenches, seed treatments, or foliar sprays. Powder formulations of fungal BS are commonly used for agricultural and horticultural purposes. Powder formulations can offer advantages such as improved stability, precise dosing, and ease of handling. However, challenges may include ensuring uniform distribution of active ingredients, controlling particle size and shape, and addressing

issues related to dustiness and handling safety (Kumar et al., 2023a, b; Oancea et al., 2016).

3. **Granular Formulations**: Granular formulations refer to solid formulations of chemicals or substances that are broken down into small, granular particles. These particles can vary in size, from very fine granules to larger pellets. These formulations consist of granules or pellets containing fungal spores, mycelium, or fungal metabolites. They are typically applied to the soil or incorporated into growing media. Granular formulations are often easier to handle than powders or liquids. They can be poured, spread, or dispensed more accurately and with less risk of spillage or waste (Kumar et al., 2023a, b).
4. **Soil Amendments**: Some fungal BS are formulated as soil amendments, either alone or in combination with other organic materials like compost or biochar. These formulations aim to improve soil health and promote beneficial interactions between fungi and plants (Visconti et al., 2023).
5. **Custom Blends**: Many commercial formulations may combine various fungal strains with other beneficial microbes, nutrients, and growth enhancers to create custom blends tailored to specific crops, growing conditions, or desired outcomes (Kumar et al., 2023a, b).

The market for fungal BS has been steadily growing, driven by the increasing demand for sustainable agricultural practices and organic products (Critchley et al., 2021). Fungal BS, such as mycorrhizal fungi and *Trichoderma* spp., have gained attention for their ability to enhance nutrient uptake, improve soil structure, and promote plant growth. Farmer perceptions of fungal BS vary depending on factors such as effectiveness, ease of use, and cost-effectiveness. Several companies have developed commercial formulations of fungal BS for agricultural use, which are very popular among farmers and in the global market and are shown in Table 7.3.

The BS regulation in India was notified by the government through the official Gazette under the Essential Commodities Act, 1955, by the Ministry of Agriculture and Farmers Welfare, Department of Agriculture, Co-operation, and Farmers Welfare, Government of India. This order is called the Fertilizer (Inorganic, Organic, or Mixed) (Control) Amendment

TABLE 7.3 Some of the Selected Fungal-Based biostimulant Formulations Available in the Global Market

Fungal Inoculant	Species Involved	Product Name	Producing Company	Country of Origin	Mode of Action	References
Trichoderma sp.	*T. harzanium* strain T-22	RootShield®	Arbico Organics Inc	United States	Combat the most prevalent root diseases	https://www.arbico-organics.com/
	T. harzanium strain T-22 and *T. virens* strain G-41 (Consortium)	RootShield *PLUS*+	Arbico Organics Inc	United States	At a wider range of temperatures, Root Shield PLUS+ prevents a broader spectrum of root diseases.	https://www.arbico-organics.com/
	T. harzanium	Bioharz™	IPL Biologicals Ltd.	India	It increases crop yields by effectively managing soil-borne phytopathogenic pathogen populations in crops.	https://www.iplbiologicals.com/
	T. harzanium	Neemoderma-H®	Shri Ram Solvent Extractions Pvt Ltd.	India	It acts efficiently against many soil- and seed-borne phytopathogenic diseases. Moreover, it has a positive effect on plant growth and development.	https://neemplus.com/
	T. harzanium/ T. hamatum	Bioorganic Plus	Nova Science Co. Ltd	Thailand	It acts as a mycofertilizer, decomposer, and mycofungicide.	Kumar et al. (2023a)
	T. harzanium strain T-22	Trianum-G	Koppert Biological Systems	United Kingdom	Trianum-G uses multiple mechanisms of action to shield plants from a variety of soil-borne root diseases. Additionally, it enhances nutrient uptake, root system development, and plant growth.	https://www.koppert.co.uk/
Mycorrhiza sp.	*Vesicular Arbuscular Mycorrhiza*	Premium Mycorrhizal Inoculant	Wallace Organic Wonder	United States	It mobilizes macro- and micronutrients, encouraging growth, and lowering stress	https://wallacewow.com/

TABLE 7.3 *(Continued)*

Fungal Inoculant	Species Involved	Product Name	Producing Company	Country of Origin	Mode of Action	References
	Vesicular Arbuscular Mycorrhiza (*Glomus intraradices*)	Josh Super	CADAgro	India	It sticks to plant roots and encourages the growth of white roots, which greatly expands the rhizosphere. It makes inaccessible types of phosphorus and other nutrients easier to dissolve.	https://cadagro.co.in/
	Combination of *Mycorrhiza* sp. + Rhizobacteria and other beneficial fungi	Mikro-Myco	Arbico Organics Inc	United States	Mycorrhizae and other beneficial microbes combine to foster the ideal exchange between roots, soil, and plants. By generating an organic acid that solubilizes this vital mineral, the helpful fungi in this combination are able to release fixed phosphorus.	https://www.arbico-organics.com/
	Vesicular Arbuscular Mycorrhiza (*Glomas intraradices)*	Mycozone	Agriland Biotech Ltd.	India	A combination of three distinct strains of *Glomas intraradices*, Mycozone is able to function in a variety of environments, including waterlogged soils, adverse soils, and drought.	https://www.agrilandbiotech.com/
Aspergillus sp.	*Aspergillus Niger* (strain AN-27)	Kalisena FS Kalisena FA	CADAgro	India	It functions as a phosphorus solubilizer, prevents the growth of common soil-borne fungal diseases, and promotes better germination and growth by generating two growth-promoting compounds.	https://cadagro.co.in/
Chaetomium sp.	*Chaetomium globosum*	Alglow	Amruth Organic Fertilizers	India	It is a biological fungicide used as a preventative measure against pathogens that cause diseases.	https://amruthgroups.com/
Penicillium sp.	*Penicillium citrinum*	Mn Sol-B®	Agrilife	India	It works as a manganese (Mn) solubilizer and nitrate assimilatior.	http://agrilife.in/

Order, 2021. This FCO Amendment Order, 2021, sets down the regulatory framework for bio-stimulant registration in India. India is now among the few nations in the world that have separate regulations governing BS (Press Trust of India, 2023). The commercialization of BS falls under Schedule VI of the Fertilizer (Inorganic, Organic, or Mixed) (Control) Order (FCO), 1985. The criteria state that before BS can be sold, they must be registered and proven to be effective. As per the FCO amendment Order, 2021, the biostimulant committee is accountable for overseeing the quality and specifications of all BS and making sure that safe molecules are utilized during the production process (The Fertilizer Association of India, 2021, https://www.faidelhi.org/). Moreover, those fungal inoculants that do not fit under the biostimulant definition of the FCO Amendment Order 2021 but have biofertilizer abilities can be included under Schedule III of the FCO, 1985. Schedule III defines the specifications of biofertilizers, and some fungal inoculants, such as mycorrhizal fungi and phosphate-solubilizing fungi (PSF), are covered under this category (Kumar et al., 2023b).

Regulation of BS in the European Union (EU) is governed by the European Union Regulation (EU) 2019/1009 on fertilizing products. This regulation came into effect on July 16, 2019, and it lays down rules for placing fertilizing products, including BS, on the EU market (European Biostimulants Industry Council (EBIC), 2024). The regulation (EU) 2019/1009 provides a definition of BS as products containing substances and/or microorganisms whose function, when applied to plants or the rhizosphere, is to stimulate natural processes to enhance nutrient uptake, nutrient efficiency, tolerance to abiotic stress, or crop quality characteristics. BS need to be registered and labelled according to the provisions of the regulation before they can be placed on the EU market (Fertilizers Europe, 2023). This includes providing information about the product's composition, intended use, and any claims made about its effects. Moreover, BS must meet certain quality standards and demonstrate efficacy in order to be authorized for sale in the EU. Once a biostimulant has been registered and authorized in accordance with EU regulations, it can be placed on the EU market and sold to farmers and other users (European Biostimulants Industry Council (EBIC), 2024; Fertilizers Europe, 2023).

Biostimulant regulation in the United States is still evolving. In the United States, BS were not subject to a specific regulatory framework like pesticides or fertilizers. Instead, they fell under various regulatory authorities depending on their ingredients and intended use. However, the regulation of BS has been a topic of increasing interest and concern as their usage grows in agricultural practices. The U.S. Environmental Protection Agency (EPA) and the U.S. Department of Agriculture (USDA) are the primary federal agencies involved in regulating agricultural inputs, including BS (https://www.usda.gov/). The EPA regulates products that make pesticidal claims, while the USDA oversees products containing plant nutrients. BS often blur the lines between these categories, making regulation complex (https://www.epa.gov/). In March 2019, the U.S. Congress passed the Agriculture Improvement Act of 2018, commonly known as the 2018 Farm Bill, which included provisions related to BS (USDA (2019)). The bill directed the USDA to study the regulation of BS and report back to Congress on potential regulatory pathways. This study was intended to inform future regulatory decisions regarding BS. Since then, there have been ongoing efforts to develop a clearer regulatory framework for BS (USDA, 2019).

7.4 ENVIRONMENTAL SUSTAINABILITY

Fungal-based BS and growth regulators offer promising avenues for enhancing environmental sustainability in agriculture. These BS can decrease the reliance on synthetic chemical fertilizers and pesticides (Asif et al., 2023; Singh et al., 2024). Moreover, these products harness the symbiotic relationship between fungi and plants, promoting nutrient uptake, stress resistance, and overall plant health without the need for chemical additives (Wahab et al., 2023). By reducing chemical inputs, fungal-based BS contribute to minimizing environmental pollution and mitigating the negative impacts of agrochemicals on soil, water, and air quality (Lamichhane et al., 2016). Certain fungi possess antagonistic properties against plant pathogens (Bhupenchandra et al., 2022). By utilizing fungal-based BS with biocontrol capabilities, farmers can manage plant diseases in an environmentally friendly manner, reducing the need for chemical fungicides. This approach contributes to the preservation of beneficial microbial communities in the soil and minimizes the development of pesticide-resistant pathogens, thereby supporting long-term agricultural sustainability (Bertola et al., 2021; Sangiorgio et al., 2020).

Mycorrhizal fungi play a significant role in carbon sequestration by forming extensive networks of hyphae in the soil, which facilitate the transfer of carbon compounds between plants and soil. Increased fungal activity can enhance carbon sequestration in agricultural soils, helping to mitigate climate change by reducing atmospheric carbon dioxide levels (Sun and Shahrajabian, 2023). Fungal-based BS also promote biodiversity by fostering the growth of diverse microbial communities in the soil. This microbial diversity is essential for maintaining ecosystem resilience, as it enhances nutrient cycling, improves soil structure, and contributes to overall ecosystem stability (Frąc et al., 2018; Sardans et al., 2023). By preserving and enhancing soil biodiversity, fungal-based BS support sustainable agricultural systems that are more resilient to environmental stresses (Asif et al., 2023; Fadiji et al., 2022; Tarroum et al., 2022). However, it is essential to continue researching and developing these products to ensure their efficacy and safety while maximizing their positive impacts on the environment and agricultural sustainability (Kumar et al., 2023a, b).

7.5 FUTURE PROSPECTS

Fungal-based BS and growth regulators offer a myriad of benefits to agricultural systems (Sharma et al., 2023). Firstly, they enhance soil fertility and structure by promoting the growth of beneficial microorganisms and improving nutrient availability (Sardans et al., 2023). Fungi such as mycorrhizae form symbiotic associations with plant roots, facilitating the uptake of water and nutrients, thereby enhancing plant growth and resilience to environmental stressors (Sun and Shahrajabian, 2023). Additionally, fungal metabolites can stimulate plant growth, induce systemic resistance against pathogens, and alleviate abiotic stresses such as drought and salinity (Baron and Rigobelo, 2021; Fadiji et al., 2022). These multifaceted benefits contribute to increased crop yields, improved quality of produce, and reduced reliance on synthetic inputs (Lamichhane et al., 2016).

Despite their potential, fungal-based BS and growth regulators face several challenges that hinder their widespread adoption (Castiglione et al., 2021). One of the primary concerns is the variability in efficacy observed under different environmental conditions and crop species. Factors such as soil type, climate, and agronomic practices can influence the performance of fungal inoculants, making it challenging to achieve consistent results across diverse agricultural systems. Moreover, the production and formulation of fungal-based products pose technical and economic challenges, including the scalability of production methods, stability of formulations, and cost-effectiveness compared to conventional inputs. Addressing these challenges requires interdisciplinary research efforts spanning microbiology, agronomy, biotechnology, and economics to optimize product performance, streamline production processes, and enhance market competitiveness (Kumar et al., 2023a, b). Despite the challenges, the future prospects of fungal-based BS and growth regulators are promising, driven by advances in research, technology, and market demand for sustainable agriculture solutions. Key areas for future exploration include:

1. **Strain Selection and Optimization:** Identifying and selecting fungal strains with superior growth-promoting and stress-tolerance traits through screening and genetic engineering approaches (Kumar et al., 2023a, b).

2. **Formulation and Delivery:** Developing innovative formulations and delivery systems to improve the shelf-life, efficacy, and convenience of fungal-based products, including granules, emulsions, and coated seeds (Kumar et al., 2023a, b).

3. **Integration with Precision Agriculture:** Harnessing digital technologies and precision agriculture techniques to tailor fungal applications based on site-specific conditions, crop requirements, and agronomic practices (Traversari et al., 2021).

4. **Regulatory Framework and Market Access:** Establishing regulatory frameworks and standards to ensure the safety, efficacy, and quality of fungal-based products, facilitating their approval and commercialization in global markets (Kumar et al., 2023a, b).

5. **Knowledge Transfer and Capacity Building:** Promoting awareness, knowledge dissemination, and capacity building among farmers, agronomists, and policymakers regarding the benefits and best practices of using fungal-based BS and growth regulators (Kumar et al., 2023a).

Overall, fungal-based BS and growth regulators offer significant potential to address key challenges facing modern agriculture while promoting sustainability, resilience, and productivity in food production systems. Continued investment in research, development, and market adoption is likely to further enhance their role in shaping the future of agriculture.

7.6 CONCLUSION

Environmentally friendly and sustainable solutions have significantly changed farming practices in recent years. Among these, fungal-based BS and growth regulators have shown promise as viable substitutes for conventional chemical inputs. In a nutshell, fungal-based growth regulators and BS seem to have bright futures and have the potential to revolutionize farming practices. Fungi's unique features minimize their negative effects on the environment while providing a range of advantages, including better nutrient uptake, increased agricultural productivity, and enhanced stress tolerance. We should expect novel formulations and application methods that are suited to certain crops and environmental circumstances as research into the complexities of fungal interactions with plants and soil ecosystems advances. Moreover, the use of fungal BS in sustainable farming systems responds to the growing need for resource- and environmentally-conscious agricultural solutions. Fungal-based BS and growth regulators have the potential to be crucial tools for tackling global food security issues and building resilient agroecosystems in the coming years due to continuous developments in agronomy and biotechnology.

KEYWORDS

- **arbuscular mycorrhizal**
- **biostimulants**
- **plant growth-promoting fungi**
- **vesicular-arbuscular mycorrhizal fungi**
- **volatile organic compounds**

REFERENCES

Abdelkhalek, A., Al-Askar, A. A., Arishi, A. A., & Behiry, S. I. (2022). *Trichoderma hamatum* strain Th23 promotes tomato growth and induces systemic resistance against tobacco mosaic virus. *Journal of Fungi (Basel)*, *8*(3), 228. https://doi.org/10.3390/jof8030228 (accessed on 6th November 2025).

Abdul Rahman, N. S. N., Abdul Hamid, N. W., & Nadarajah, K. (2021). Effects of abiotic stress on soil microbiome. *International Journal of Molecular Sciences*, *22*(16), 9036. https://doi.org/10.3390/ijms22169036 (accessed on 6th November 2025).

Abreu-Tarazi, M. F., Navarrete, A. A., Andreote, F. D., Almeida, C. V., Tsai, S. M., & Almeida, M. (2010). Endophytic bacteria in long-term *in vitro* cultivated "axenic" pineapple micro plants revealed by PCR–DGGE. *World Journal of Microbiology and Biotechnology*, *26*, 555–560.

Adedayo, A. A., & Babalola, O. O. (2023). Fungi that promote plant growth in the rhizosphere boost crop growth. *Journal of Fungi*, *9*(2), 239. https://doi.org/10.3390/jof9020239 (accessed on 6th November 2025).

Agudo-Jurado, F. J., Reveglia, P., Rubiales, D., Evidente, A., & Barilli, E. (2023). Status of phytotoxins isolated from necrotrophic fungi causing diseases on grain legumes. *International Journal of Molecular Sciences*, *24*(6), 5116. https://doi.org/10.3390/ijms24065116 (accessed on 6th November 2025).

Ahmad, R. Z., Khalid, R., Aqeel, M., Ameen, F., & Li, C. J. (2020). Fungal endophytes trigger *Achnatherum inebrians* germination ability against environmental stresses. *South African Journal of Botany*, *134*, 230–236.

Alam, B., Lǐ, J., Gě, Q., Khan, M. A., Gōng, J., Mehmood, S., Yuán, Y., & Gǒng, W. (2021). Endophytic fungi: From symbiosis to secondary metabolite communications or vice versa? *Frontiers in Plant Science*, *12*, 791033. https://doi.org/10.3389/fpls.2021.791033 (accessed on 6th November 2025).

Alfiky, A., & Weisskopf, L. (2021). Deciphering *Trichoderma*-plant-pathogen interactions for better development of biocontrol applications. *Journal of Fungi (Basel)*, *7*(1), 61. https://doi.org/10.3390/jof7010061 (accessed on 6th November 2025).

Ali, M., Cheng, Z., Ahmad, H., & Hayat, S. (2018). Reactive oxygen species (ROS) as defenses against a broad range of plant fungal infections and case study on ROS employed by crops against *Verticillium dahliae* wilts. *Journal of Plant Interactions*, *13*(1), 353–363. https://doi.org/10.1080/17429145.2018.1484188 (accessed on 6th November 2025).

Anand, G., Gupta, R., Marash, I., Leibman-Markus, M., & Bar, M. (2022). Cytokinin production and sensing in fungi. *Microbiological Research*, *262*, 127103. https://doi.org/10.1016/j.micres.2022.127103 (accessed on 6th November 2025).

Anand, U., Pal, T., Yadav, N., Singh, V. K., Tripathi, V., Choudhary, K. K., Shukla, A. K., Sunita, K., Kumar, A., Bontempi, E., Ma, Y., Kolten, M., & Singh, A. K. (2023). Current scenario and future prospects of endophytic microbes: Promising candidates for abiotic and biotic stress management for agricultural and environmental sustainability. *Microbial Ecology*, *86*, 1455–1486. https://doi.org/10.1007/s00248-023-02190-1 (accessed on 6th November 2025).

Andrzejak, R., & Janowska, B. (2022). *Trichoderma* spp. improves flowering, quality, and nutritional status of ornamental plants. *International Journal of Molecular Sciences*,

23(24), 15662. https://doi.org/10.3390/ijms232415662 (accessed on 6th November 2025).

Asif, A. E., Babalola, O. O., Santoyo, G., & Perazzolli, M. (2022). The potential role of microbial bio stimulants in the amelioration of climate change-associated abiotic stresses on crops. *Frontiers in Microbiology*, *12*, 829099. https://doi.org/10.3389/fmicb.2021.829099 (accessed on 6th November 2025).

Asif, A., Ali, M., Qadir, M., Karthikeyan, R., Singh, Z., Khangura, R., Di Gioia, F., & Ahmed, Z. F. R. (2023). Enhancing crop resilience by harnessing the synergistic effects of bio stimulants against abiotic stress. *Frontiers in Plant Science*, *14*, 1276117. https://doi.org/10.3389/fpls.2023.1276117 (accessed on 6th November 2025).

Attia, M. S., Salem, M. S., & Abdelaziz, A. M. (2022). Endophytic fungi *Aspergillus* spp. reduces fusaria wilt disease severity, enhance growth, metabolism and stimulate the plant defense system in pepper plants. *Biomass Conversion and Biorefinery*. https://doi.org/10.1007/s13399-022-03607-6 (accessed on 6th November 2025).

Averilla, C., Bhatnagara, J. M., Dietze, M. C., Pearse, W. D., & Kivlin, S. N. (2019). Global imprint of mycorrhizal fungi on whole-plant nutrient economics. *Proceedings of the National Academy of Sciences (PNAS)*, *116*(46), 23163–23168. https://www.pnas.org/cgi/doi/10.1073/pnas.1906655116 (accessed on 6th November 2025).

Ayers, A. R., Ebel, J., Valent, B., & Albersheim, P. (1976). Host-pathogen interactions: X. Fractionation and biological activity of an elicitor isolated from the mycelial walls of *Phytophthora megasperma* var. *sojae*. *Plant Physiology*, *57*, 760–765.

Babu, A. G., Kim, S. W., Yadav, D. R., Hyum, U., Adhikari, M., & Lee, Y. S. (2015). *Penicillium menonorum*: A novel fungus to promote growth and nutrient management in cucumber plants. *Mycobiology*, *43*(1), 49–56. https://doi.org/10.5941/MYCO.2015.43.1.49 (accessed on 6th November 2025).

Bamisile, B. S., Dash, C. K., Akutse, K. S., Keppanan, R., & Wang, L. D. (2018). Fungal endophytes: Beyond herbivore management. *Frontiers in Microbiology*, *9*, 544. https://doi.org/10.3389/fmicb.2018.00544 (accessed on 6th November 2025).

Baron, N. C., & Rigobelo, E. C. (2021). Endophytic fungi: A tool for plant growth promotion and sustainable agriculture. *Mycology*, *13*(1), 39–55.

Baron, N. C., Rigobelo, E. C., & Zied, D. C. (2019). Filamentous fungi in biological control: Current status and future perspectives. *Chilean Journal of Agricultural Research*, *79*(2), 307–315.

Begum, N., Qin, C., Ahanger, M. A., Raza, S., Khan, M. I., Ashraf, M., Ahmed, N., & Zhang, L. (2019). Role of arbuscular mycorrhizal fungi in plant growth regulation: Implications in abiotic stress tolerance. *Frontiers in Plant Science*, *10*, 1068. https://doi.org/10.3389/fpls.2019.01068 (accessed on 6th November 2025).

Bennett, J. W., & Klich, M. (2003). Mycotoxins. *Clinical Microbiology Reviews*, *16*(3), 497–516.

Bertola, M., Ferrarini, A., & Visioli, G. (2021). Improvement of soil microbial diversity through sustainable agricultural practices and its evaluation by -omics approaches: A perspective for the environment, food quality and human safety. *Microorganisms*, *9*(7), 1400. https://doi.org/10.3390/microorganisms9071400 (accessed on 6th November 2025).

Beslemes, D., Tigka, E., Roussis, I., Kakabouki, I., Mavroeidis, A., & Vlachostergios, D. (2023). Effect of arbuscular mycorrhizal fungi on nitrogen and phosphorus uptake efficiency and crop productivity of two-rowed barley under different crop production systems. *Plants (Basel)*, *12*(9), 1908. https://doi.org/10.3390/plants12091908 (accessed on 6th November 2025).

Bezerra, V. H. S., Cardoso, S. L., Fonseca-Bazzo, Y., Silveira, D., Magalhães, P. O., & Souza, P. M. (2021). Protease produced by endophytic fungi: A systematic review. *Molecules*, *26*(22), 7062. https://doi.org/10.3390/molecules26227062 (accessed on 6th November 2025).

Bhupenchandra, I., Chongtham, S. K., Devi, E. L., R. R., Choudhary, A. K., Salam, M. D., Sahoo, M. R., et al. (2022). Role of bio stimulants in mitigating the effects of climate change on crop performance. *Frontiers in Plant Science*, *13*, 967665. https://doi.org/10.3389/fpls.2022.967665 (accessed on 6th November 2025).

Brown, R., Priest, E., Naglik, J. R., & Richardson, J. P. (2021). Fungal toxins and host immune responses. *Frontiers in Microbiology*, *12*, 643639. https://doi.org/10.3389/fmicb.2021.643639 (accessed on 6th November 2025).

Brundrett, M. (2004). Diversity and classification of mycorrhizal associations. *Biological Reviews of the Cambridge Philosophical Society*, *79*, 473–495. https://doi.org/10.1017/S1464793103006316 (accessed on 6th November 2025).

Carroll, G. (1988). Fungal endophytes in stems and leaves: From latent pathogen to mutualistic symbiont. *Ecology*, *69*, 2–9.

Castiglione, A. M., Mannino, G., Contartese, V., Bertea, C. M., & Ertani, A. (2021). Microbial bio stimulants as response to modern agriculture needs: Composition, role and application of these innovative products. *Plants (Basel)*, *10*(8), 1533. https://doi.org/10.3390/plants10081533 (accessed on 6th November 2025).

Chaudhary, P., Agri, U., Chaudhary, A., Kumar, A., & Kumar, G. (2022). Endophytes and their potential in biotic stress management and crop production. *Frontiers in Microbiology*, *13*, 933017. https://doi.org/10.3389/fmicb.2022.933017 (accessed on 6th November 2025).

Chen, T., Li, C., White, J. F., & Nan, Z. (2019). Effect of the fungal endophyte *Epichloë bromicola* on polyamines in wild barley (*Hordeum brevisubulatum*) under salt stress. *Plant and Soil*, *436*, 29–48. https://doi.org/10.1007/s11104-018-03913-x (accessed on 6th November 2025).

Cheplick, G. P., & Clay, K. (1988). Acquired chemical defenses in grasses: The role of fungal endophytes. *Oikos*, *52*, 309–318.

Contreras-Cornejo, H. A., Macías-Rodríguez, L., Cortés-Penagos, C., & López-Bucio, J. (2009). *Trichoderma virens*, a plant beneficial fungus, enhances biomass production and promotes lateral root growth through an auxin-dependent mechanism in *Arabidopsis*. *Plant Physiology*, *149*(3), 1579–1592. https://doi.org/10.1104/pp.108.130369 (accessed on 6th November 2025).

Cosio, E. G., Feger, M., Miller, C. J., Antelo, L., & Ebel, J. (1996). High-affinity binding of fungal β-glucan elicitors to cell membranes of species of the plant family *Fabaceae*. *Planta*, *200*, 92–99.

Côté, F., Roberts, K. A., & Hahn, M. G. (2000). Identification of high-affinity binding sites for the hepta-beta-glucoside elicitor in membranes of the model legumes *Medicago truncatula* and *Lotus japonicus*. *Planta*, *211*, 596–605.

Critchley, A. T., Critchley, J. S. C., Norrie, J., Gupta, S., & van Staden, J. (2021). Perspectives on the global bio stimulant market: Applications, volumes, and values,

2016 data and projections to 2022. In S. Gupta & J. van Staden (Eds.), *Bio Stimulants for Crops from Seed Germination to Plant Development: A Practical Approach* (pp. 89–296). Academic Press. https://doi.org/10.1016/B978-0-12-823048-0.00012-5 (accessed on 6th November 2025).

Cui, R., Lu, X., Chen, X., Malik, W. A., Wang, D., Wang, J., et al. (2021). A novel raffinose biological pathway is observed by symbionts of cotton = *Verticillium dahliae* to improve salt tolerance genetically on cotton. *Journal of Agronomy and Crop Science*, *207*, 956–969.

Daigham, G. E., Mahfouz, A. Y., Abdelaziz, A. M., Nofel, M. M., & Atta, M. S. (2023). Protective role of plant growth-promoting fungi *Aspergillus chevalieri* OP593083 and *Aspergillus egyptiacus* OP593080 as biocontrol approach against *Alternaria* leaf spot disease of *Vicia faba* plant. *Biomass Conversion and Biorefinery*. https://doi.org/10.1007/s13399-023-04510-4 (accessed on 6th November 2025).

da-Prado, D. Z., Oliveira, S. L., Okino-Delgado, C. H., Auer, S., Ludwig-Müller, J., da Silva, M. R., Fernandes, C. J. C., Carbonari, C. A., Zambuzzi, W. F., & Fleuri, L. F. (2019). *Aspergillus flavipes* as a novel bio stimulant for rooting-enhancement of *Eucalyptus*. *Journal of Cleaner Production*, *234*, 681–689. https://doi.org/10.1016/j.jclepro.2019.06.211 (accessed on 6th November 2025).

de França, S. K. S., Cardoso, A. F., Lustosa, D. C., Ramos, E. M. L. S., de Filippi, M. C. C., & da Silva, G. B. (2015). Biocontrol of sheath blight by *Trichoderma asperellum* in tropical lowland rice. *Agronomy for Sustainable Development*, *35*, 317–324.

de Souza, P. M., Bittencourt, M. L., Caprara, C. C., de Freitas, M., de Almeida, R. P., Silveira, D., Fonseca, Y. M., Ferreira Filho, E. X., Pessoa Junior, A., & Magalhães, P. O. (2015). A biotechnology perspective of fungal proteases. *Brazilian Journal of Microbiology*, *46*(2), 337–346.

Degani, O., Rabinovitz, O., Becher, P., Gordani, A., & Chen, A. (2021). *Trichoderma longibrachiatum* and *Trichoderma asperellum* confer growth promotion and protection against late wilt disease in the field. *Journal of Fungi (Basel)*, *7*(6), 444. https://doi.org/10.3390/jof7060444 (accessed on 6th November 2025).

Derbyshire, M., Mbengue, M., Barascud, M., Navaud, O., & Raffaele, S. (2019). Small RNAs from the plant pathogenic fungus *Sclerotinia sclerotiorum* highlight host candidate genes associated with quantitative disease resistance. *Molecular Plant Pathology*, *20*(9), 1279–1297. https://doi.org/10.1111/mpp.12841 (accessed on 6th November 2025).

Dissanayake, A. J., Purahong, W., Wubet, T., Hyde, K. D., Zhang, W., Xu, H., Zhang, G., Fu, C., Liu, M., Xing, Q., Li, X., & Yan, J. (2018). Direct comparison of culture-dependent and culture-independent molecular approaches reveal the diversity of fungal endophytic communities in stems of grapevine (*Vitis vinifera*). *Fungal Diversity*, *90*, 85–107. https://doi.org/10.1007/s13225-018-0399-3 (accessed on 6th November 2025).

Druzhinina, I. S., & Kubicek, C. P. (2017). Genetic engineering of *Trichoderma reesei* cellulases and their production. *Microbial Biotechnology*, *10*(6), 1485–1499. https://doi.org/10.1111/1751-7915.12726 (accessed on 6th November 2025).

du Jardin, P. (2015). Plant bio stimulants: Definition, concept, main categories and regulation. *Scientia Horticulture*, *196*, 3–14. https://doi.org/10.1016/j.scienta.2015.09.021 (accessed on 6th November 2025).

Duan, T., Du, Y., Xing, C., Wang, H. Y., & Wang, R. F. (2022). Toll-like receptor signaling and its role in cell-mediated immunity. *Frontiers in Immunology*, *13*, 812774. https://doi.org/10.3389/fimmu.2022.812774 (accessed on 6th November 2025).

Duhoux, E., Rinaudo, G., Diem, H. G., Auguy, F., Fernandez, D., Bogusz, D., Franche, C., Dommergues, Y., & Huguenin, B. (2001). Angiosperm *Gymnostoma* trees produce root nodules colonized by arbuscular mycorrhizal fungi related to *Glomus*. *New Phytologist*, *149*, 115–125.

Elder, M. J., Webster, S. J., Chee, R., Williams, D. L., Hill Gaston, J. S., & Goodall, J. C. (2017). β-glucan size controls dectin-1-mediated immune responses in human dendritic cells by regulating IL-1β production. *Frontiers in Immunology*, *8*, 791. https://doi.org/10.3389/fimmu.2017.00791 (accessed on 6th November 2025).

El-Enshasy, H. A., Hanapi, S. Z., Malek, R. A., Abdelgalil, S. A., & Leng, O. M. (2019). Endophytic fungi: The desired bio stimulants for essential oil production. In B. P. Singh (Ed.), *Advances in Endophytic Fungal Research: Present Status and Future Challenges* (pp. 211–232). Springer International Publishing.

Etesami, H., Jeong, B. R., & Glick, B. R. (2021). Contribution of arbuscular mycorrhizal fungi, phosphate–solubilizing bacteria, and silicon to P uptake by plant. *Frontiers in Plant Science*, *12*, 699618. https://doi.org/10.3389/fpls.2021.699618 (accessed on 6th November 2025).

Fall, A. F., Nakabonge, G., Ssekandi, J., Founoune-Mboup, H., Apori, S. O., Ndiaye, A., Badji, A., & Ngom, K. (2022). Roles of arbuscular mycorrhizal fungi on soil fertility: Contribution in the improvement of physical, chemical, and biological properties of the soil. *Frontiers in Fungal Biology*, *3*, 723892. https://doi.org/10.3389/ffunb.2022.723892 (accessed on 6th November 2025).

Fesel, P. H., & Zuccaro, A. (2016). Beta-glucan: Crucial component of the fungal cell wall and elusive MAMP in plants. *Fungal Genetics and Biology*, *90*, 53–60.

Fiorin, G. L., Sanchéz-Vallet, A., Thomazella, D. P. T., do Prado, P. F. V., do Nascimento, L. C., Figueira, A. V. O., Thomma, P. H. J., Pereira, G. A. G., & Teixeira, P. J. L. (2018). Suppression of plant immunity by fungal chitinase-like effectors. *Current Biology*, *28*(18), 3023–3030.e5. https://doi.org/10.1016/j.cub.2018.07.055 (accessed on 6th November 2025).

Frąc, M., Hannula, S. E., Bełka, M., & Jędryczka, M. (2018). Fungal biodiversity and their role in soil health. *Frontiers in Microbiology*, *9*, 707. https://doi.org/10.3389/fmicb.2018.00707 (accessed on 6th November 2025).

Garnica-Vergara, A., Barrera-Ortiz, S., Muñoz-Parra, E., Raya-González, J., Méndez-Bravo, A., Macías-Rodríguez, L., & López-Bucio, J. (2016). The volatile 6-pentyl-2H-pyran-2-one from *Trichoderma atroviride* regulates *Arabidopsis thaliana* root morphogenesis via auxin signaling and ethylene insensitive 2 functioning. *New Phytologist*, *209*, 1496–1512.

Geoghegan, I. A., & Gurr, S. J. (2017). Investigating chitin deacetylation and chitosan hydrolysis during vegetative growth in *Magnaporthe oryzae*. *Cellular Microbiology*, *19*(9), e12743. https://doi.org/10.1111/cmi.12743 (accessed on 6th November 2025).

Goltapeh, E. M., Danesh, Y. R., Prasad, R., & Varma, A. (2008). Mycorrhizal fungi: What we know and what should we know? In A. Varma (Ed.), *Mycorrhiza* (pp. 3–23). Springer, Berlin, Heidelberg. https://doi.org/10.1007/978-3-540-78826-3_1 (accessed on 6th November 2025).

Gouda, S., Das, G., Sen, S. K., Shin, H. S., & Patra, J. K. (2016). Endophytes: A treasure house of bioactive compounds of medicinal importance. *Frontiers in Microbiology*, *7*, 1538. https://doi.org/10.3389/fmicb.2016.01538 (accessed on 6th November 2025).

Goyal, A., & Kalia, A. (2020). Fungal phytohormones: Plant growth-regulating substances and their applications in crop productivity. In A. Yadav, S. Mishra, D. Kour, N. Yadav, & A. Kumar (Eds.), *Agriculturally Important Fungi for Sustainable Agriculture* (pp. 143–169). Springer, Cham. https://doi.org/10.1007/978-3-030-45971-0_7 (accessed on 6th November 2025).

Guan, P., Schmidt, F., Fischer, J., Riemann, M., Thines, E., & Nick, P. (2022). The fungal elicitor eutypine from *Eutypa lata* activates basal immunity through its phenolic side chains. *Horticulture Research*, *9*, uhac120. https://doi.org/10.1093/hr/uhac120 (accessed on 6th November 2025).

Guo, J., & Cheng, Y. (2022). Advances in fungal elicitor-triggered plant immunity. *International Journal of Molecular Sciences*, *23*(19), 12003. https://doi.org/10.3390/ijms231912003 (accessed on 6th November 2025).

Gupta, S., Bhattacharyya, P., Kulkarni, M. G., & Doležal, K. (2023). Editorial: Growth regulators and bio stimulants: Upcoming opportunities. *Frontiers in Plant Science*, *14*, 1209499. https://doi.org/10.3389/fpls.2023.1209499 (accessed on 6th November 2025).

Guzmán-Guzmán, P., Kumar, A., de Los Santos-Villalobos, S., Parra-Cota, F. I., Orozco-Mosqueda, M. D. C., Fadiji, A. E., Hyder, S., Babalola, O. O., & Santoyo, G. (2023). *Trichoderma* species: Our best fungal allies in the biocontrol of plant diseases—A review. *Plants (Basel)*, *12*(3), 432. https://doi.org/10.3390/plants12030432 (accessed on 6th November 2025).

Hart, M. M., & Forsythe, J. A. (2012). Using arbuscular mycorrhizal fungi to improve the nutrient quality of crops; nutritional benefits in addition to phosphorus. *Scientia Horticulture*, *148*, 206–214. https://doi.org/10.1016/j.scienta.2012.09.018 (accessed on 6th November 2025).

Herath, B. M., Kalamulla, K. W. Y. R., Mayadunna, T. A. N., Perera, M. A. E. G., Jayamanna, I. K., & Yapa, P. N. (2024). Applications of arbuscular mycorrhizal fungi for sustainable agricultural systems. In G. J. Ahammed & R. Hajiboland (Eds.), *Arbuscular mycorrhizal Fungi and Higher Plants* (pp. 319–340). Springer, Singapore. https://doi.org/10.1007/978-981-99-8220-2_14 (accessed on 6th November 2025).

Hossain, M. M., Sultana, F., Miyazawa, M., & Hyakumachi, M. (2014). Plant growth-promoting fungus *Penicillium* spp. GP15-1 enhances growth and confers protection against damping-off and anthracnose in the cucumber. *Journal of Oleo Science*, *63*(4), 391–400. https://doi.org/10.5650/jos.ess13143 (accessed on 6th November 2025).

Hubbard, M., Germida, J., & Vujanovic, V. (2012). Fungal endophytes improve wheat seed germination under heat and drought stress. *Botany*, *90*, 137–149.

Huey, C. J., Gopinath, S. C. B., Uda, M. N. A., Zulhaimi, H. J., Jaafar, M. N., Kasim, F. H., & Yaakub, A. R. W. (2020). Mycorrhiza: A natural resource assists plant growth under varied soil conditions. *3 Biotech*, *10*, 204. https://doi.org/10.1007/s13205-020-02188-3 (accessed on 6th November 2025).

Inayati, A., Setyowati, L., Aini, L. Q., & Yusnawan, E. (2021). Plant growth promoter produced by *Trichoderma virens* and its effect on moonbeam (*Vigna radiata* (L.) Wilczek) seedling. *IOP Conference Series: Earth and Environmental Science*, *803*,

012013. https://iopscience.iop.org/article/10.1088/1755-1315/803/1/012013 (accessed on 6th November 2025).

Ismaiel, A. A., & Papenbrock, J. (2015). Mycotoxins: Producing fungi and mechanisms of phytotoxicity. *Agriculture*, *5*(3), 492–537. https://doi.org/10.3390/agriculture5030492 (accessed on 6th November 2025).

Izuogu, N. B., & Abiri, T. O. (2015). Efficacy of *Trichoderma harzianum* T22 as a biocontrol agent against root-knot nematode (*Meloidogyne incognita*) on some soybean varieties. *Croatian Journal of Food Science and Technology*, *7*, 47–51.

Jadav, P. C., Hegde, G. M., Jahagirdar, S., & Navi, V. (2022). Evaluation of microbial consortia against *Sclerotium rolfsii* Sacc. causing foot rot of wheat. *Journal of Farm Sciences*, *35*(2), 219–222.

Jamiołkowska, A. (2020). Natural compounds as elicitors of plant resistance against diseases and new biocontrol strategies. *Agronomy*, *10*(2), 173. https://doi.org/10.3390/agronomy10020173 (accessed on 6th November 2025).

Jones, R. W., & Perez, F. (2023). Differential plant response to toxins and elicitor proteins released by the potato and tomato pathogens *Alternaria solani* and *Alternaria alternata*. *Journal of Plant Pathology*, *105*, 21–28. https://doi.org/10.1007/s42161-022-01286-w (accessed on 6th November 2025).

Kedves, O., Kocsubé, S., Bata, T., Andersson, M. A., Salo, J. M., Mikkola, R., Salonen, H., Szűcs, A., Kedves, A., Kónya, Z., Vágvölgyi, C., Magyar, D., & Kredics, L. (2021). *Chaetomium* and *Chaetomium*-like species from European indoor environments include *Dichotomopilus finlandicus* sp. nov. *Pathogens*, *10*(9), 1133. https://doi.org/10.3390/pathogens10091133 (accessed on 6th November 2025).

Khrieba, M. I. (2019). Mycorrhizae's role in plant nutrition and protection from pathogens. *Current Investigations in Agriculture and Current Research*, *8*(1), 1037–1045. https://lupinepublishers.com/agriculture-journal/pdf/CIACR.MS.ID.000277.pdf (accessed on 6th November 2025).

Khuna, S., Kumla, J., Srinuanpan, S., Lumyong, S., & Suwannarach, N. (2023). Multifarious characterization and efficacy of three phosphate-solubilizing *Aspergillus* species as bio stimulants in improving root induction of cassava and sugarcane stem cuttings. *Plants (Basel)*, *12*(20), 3630. https://doi.org/10.3390/plants12203630 (accessed on 6th November 2025).

Kim, S. J., Eo, J. K., Lee, E. H., Park, H., & Eom, A. H. (2017). Effects of arbuscular mycorrhizal fungi and soil conditions on crop plant growth. *Mycobiology*, *45*(1), 20–24. https://doi.org/10.5941/MYCO.2017.45.1.20 (accessed on 6th November 2025).

Kržišnik, D., & Gonçalves, J. (2023). Environmentally conscious technologies using fungi in a climate-changing world. *Earth*, *4*(1), 69–77. https://doi.org/10.3390/earth4010005 (accessed on 6th November 2025).

Kubicek, C. P., Steindorff, A. S., Chenthamara, K., Manganiello, G., Henrissat, B., Zhang, J., Cai, F., Kopchinskiy, A. G., Kubicek, E. M., Kuo, A., Baroncelli, R., Sarrocco, S., Noronha, E. F., Vannacci, G., Shen, Q., Grigoriev, I. V., & Druzhinina, I. S. (2019). Evolution and comparative genomics of the most common *Trichoderma* species. *BMC Genomics*, *20*(1), 485. https://doi.org/10.1186/s12864-019-5680-7 (accessed on 6th November 2025).

Kumar, D., Singh, P. N., Willer, H., Sharma, S. K., Singh, U. B., Lagashetti, A. C., & Husain, R. (2023a). Fungal biofertilizers and biopesticides and their roles in sustainable agriculture. In K. Singh, D. Kumar, M. Shamim, & R. Sharma (Eds.), *Applied mycology for agriculture and foods: Industrial applications* (pp. 165–216). Apple Academic Press, Inc. https://doi.org/10.1201/9781003369868 (accessed on 6th November 2025).

Kumar, D., Singh, S. K., Arya, S. K., Srivastava, D., Rajput, V. D., & Husain, R. (2023b). Multifunctional growth-promoting microbial consortium-based biofertilizers and their techno-commercial feasibility for sustainable agriculture. In J. A. Parray, N. Shameem, D. Egamberdieva, & R. Z. Sayyed (Eds.), *Rhizobiome: Ecology, Management and Application, Microbiome Research in Plants and Soil* (pp. 167–208). https://doi.org/10.1016/B978-0-443-16030-1.00010-9 (accessed on 6th November 2025).

Kumar, M. R., Kumaran, M. D. B., Balashanmugam, A., Rebecca, A. I. N., Kumar, D. J. M., & Kalaichelvan, P. T. (2014). Production of cellulase enzyme by *Trichoderma reesei* Cef19 and its application in the production of bio-ethanol. *Pakistan Journal of Biological Sciences*, *17*, 735–739. https://scialert.net/abstract/?doi=pjbs.2014.735.739 (accessed on 6th November 2025).

Kumar, R., Kundu, A., Dutta, A., Saha, S., & Das, A. (2020). Profiling of volatile secondary metabolites of *Chaetomium globosum* for potential antifungal activity against soil borne fungi. *Journal of Pharmacognosy and Phytochemistry*, *9*, 922–927.

Kumar, R., Kundu, A., Dutta, A., Saha, S., Das, A., & Bhowmik, A. (2021). Chemo-profiling of bioactive metabolites from *Chaetomium globosum* for biocontrol of *Sclerotinia* rot and plant growth promotion. *Fungal Biology*, *125*, 167–176.

Kuo, A., Kohler, A., Martin, F. M., & Grigoriev, I. V. (2014). Expanding genomics of mycorrhizal symbiosis. *Frontiers in Microbiology*, *5*, 1–7. https://doi.org/10.3389/fmicb.2014.00582 (accessed on 6th November 2025).

Lamichhane, J. R., Dachbrodt-Saaydeh, S., Kudsk, P., & Messéan, A. (2016). Toward a reduced reliance on conventional pesticides in European agriculture. *Plant Disease*, *100*(1), 10–24. https://doi.org/10.1094/PDIS-05-15-0574-FE (accessed on 6th November 2025).

Landa, F. A. (2022). Agricultural applications of *Penicillium* genera. *Environmental Analysis & Ecology Studies*, *9*(5), 1060–1063. https://crimsonpublishers.com/eaes/fulltext/EAES.000724.php (accessed on 6th November 2025).

Lee, D. H., Lee, H. S., & Belkhadir, Y. (2021). Coding of plant immune signals by surface receptors. *Current Opinion in Plant Biology*, *62*, 102044. https://doi.org/10.1016/j.pbi.2021.102044 (accessed on 6th November 2025).

Leitão, A. L., & Enguita, F. J. (2016). Gibberellins in *Penicillium* strains: Challenges for endophyte-plant host interactions under salinity stress. *Microbiological Research*, *183*, 8–18. https://doi.org/10.1016/j.micres.2015.11.004 (accessed on 6th November 2025).

Leylaie, S., & Zafari, D. (2018). Antiproliferative and antimicrobial activities of secondary metabolites and phylogenetic study of endophytic *Trichoderma* species from *Vinca* plants. *Frontiers in Microbiology*, *9*, 1484. https://doi.org/10.3389/fmicb.2018.01484 (accessed on 6th November 2025).

Li, R. X., Cai, F., Pang, G., Shen, Q. R., Li, R., & Chen, W. (2015). Solubilization of phosphate and micronutrients by *Trichoderma harzianum* and its relationship with the promotion of tomato plant growth. *PLoS ONE*, *10*(6), e0130081. https://doi.org/10.1371/journal.pone.0130081 (accessed on 6th November 2025).

Li, W., Lu, J., Yang, C., Arildsen, K., Li, X., & Xia, S. (2022). An amidase contributes to full virulence of *Sclerotinia sclerotiorum*. *International Journal of Molecular Sciences*,

23(19), 11207. https://doi.org/10.3390/ijms231911207 (accessed on 6th November 2025).

Li, X. Z., Simpson, W. R., Song, M. L., Bao, G. S., Niu, X. L., Zhang, Z. H., Xu, H. F., Liu, X., Li, Y. L., & Li, C. J. (2020). Effects of seed moisture content and *Epichloe* endophyte on germination and physiology of *Achnatherum inebrians*. *South African Journal of Botany*, *134*, 407–414. https://doi.org/10.1016/j.sajb.2020.03.022 (accessed on 6th November 2025).

Li, X., Leng, J., Yu, L., Bai, H., Li, X., Wisniewski, M., Liu, J., & Sui, Y. (2022). Efficacy of the biocontrol agent *Trichoderma hamatum* against *Lasiodiplodia theobromae* on macadamia. *Frontiers in Microbiology*, *13*, 994422. https://doi.org/10.3389/fmicb.2022.994422 (accessed on 6th November 2025).

Linkies, A., Jacob, S., Zink, P., Maschemer, M., Maier, W., & Koch, E. (2021). Characterization of cultural traits and fungicidal activity of strains belonging to the fungal genus *Chaetomium*. *Journal of Applied Microbiology*, *131*, 375–391.

Liu, X., Williams, C. E., Nemacheck, J. A., Wang, H., Subramanyam, S., Zheng, C., & Chen, M. S. (2010). Reactive oxygen species are involved in plant defense against a gall midge. *Plant Physiology*, *152*(2), 985–999.

Lodi, R. S., Peng, C., Dong, X., Deng, P., & Peng, L. (2023). *Trichoderma hamatum* and its benefits. *Journal of Fungi (Basel)*, *9*(10), 994. https://doi.org/10.3390/jof9100994 (accessed on 6th November 2025).

Longa, C. M. O., Savazzini, F., Tosi, S., Elad, Y., & Pertot, I. (2009). Evaluating the survival and environmental fate of the biocontrol agent *Trichoderma atroviride* SC1 in vineyards in northern Italy. *Journal of Applied Mycology*, *106*(5), 1549–1557. https://doi.org/10.1111/j.1365-2672.2008.04117.x (accessed on 6th November 2025).

Lr, J. (2018). Seed inoculation with endophytic fungal entomopathogens promotes plant growth and reduces crown and root rot (CRR) caused by *Fusarium culmorum* in wheat. *Planta*, *248*(6), 1525–1535.

Ma, Y. A., Han, C., Chen, J. Y., Li, H. Y., He, K., Liu, A. X., & Li, D. C. (2015). Fungal cellulase is an elicitor but its enzymatic activity is not required for its elicitor activity. *Molecular Plant Pathology*, *16*, 14–26.

Ma, Y., Freitas, H., & Dias, M. C. (2022). Strategies and prospects for bio stimulants to alleviate abiotic stress in plants. *Frontiers in Plant Science*, *13*, 1024243. https://doi.org/10.3389/fpls.2022.1024243 (accessed on 6th November 2025).

Magnabosco, P., Masi, A., Shukla, R., Bansal, V., & Carletti, P. (2023). Advancing the impact of plant bio stimulants to sustainable agriculture through nanotechnologies. *Chemical and Biological Technologies in Agriculture*, *10*, 117. https://doi.org/10.1186/s40538-023-00491-8 (accessed on 6th November 2025).

Manganiello, G., Sacco, A., Ercolano, M. R., Vinale, F., Lanzuise, S., Pascale, A., Napolitano, M., Lombardi, N., Lorito, M., & Woo, S. L. (2018). Modulation of tomato response to *Rhizoctonia solani* by *Trichoderma harzianum* and its secondary metabolite harzianic acid. *Frontiers in Microbiology*, *9*, 1966. https://doi.org/10.3389/fmicb.2018.01966 (accessed on 6th November 2025).

Manzar, N., Kashyap, A. S., Goutam, R. S., Rajawat, M. V. S., Sharma, P. K., Sharma, S. K., & Singh, H. V. (2022). *Trichoderma*: Advent of versatile biocontrol agent, its secrets and insights into mechanism of biocontrol potential. *Sustainability*, *14*(19), 12786. https://doi.org/10.3390/su141912786 (accessed on 6th November 2025).

Mapuranga, J., Chang, J., Zhang, L., Zhang, N., & Yang, W. (2022). Fungal secondary metabolites and small RNAs enhance pathogenicity during plant-fungal pathogen interactions. *Journal of Fungi (Basel)*, *9*(1), 4. https://doi.org/10.3390/jof9010004 (accessed on 6th November 2025).

Melini, F., Melini, V., Luziatelli, F., Abou Jaoudé, R., Ficca, A. G., & Ruzzi, M. (2023). Effect of microbial plant bio stimulants on fruit and vegetable quality: current research lines and future perspectives. *Frontiers in Plant Science*, *14*, 1251544. https://doi.org/10.3389/fpls.2023.1251544 (accessed on 6th November 2025).

Merckx, V. S. F. T. (2012). *Mycoheterotrophy: The Biology of Plants Living on Fungi*. Springer, New York, NY. https://doi.org/10.1007/978-1-4614-5209-6 (accessed on 6th November 2025).

Mishra, Y., Sharma, L., Dhiman, M., & Sharma, M. M. (2021). Endophytic fungal diversity of selected medicinal plants and their bio-potential applications. In V. K. Sharma, M. P. Shah, S. Parmar, & A. Kumar (Eds.), *Fungi Bio-Prospects in Sustainable Agriculture, Environment and Nano-Technology* (pp. 227–283). Academic Press.

Moraga, E. Q. (2020). Entomopathogenic fungi as endophytes: their broader contribution to IPM and crop production. *Biocontrol Science and Technology*, *30*(9), 864–877.

Moya, P., Cipollone, J., & Sisterna, M. (2020). The fungal genus *Chaetomium* and its agricultural applications. In J. M. Mérillon & K. G. Ramawat (Eds.), *Plant Defense: Biological Control* (*Progress in Biological Control*, Vol. 22). Springer, Cham. https://doi.org/10.1007/978-3-030-51034-3_12 (accessed on 6th November 2025).

Mukherjee, P. K., Mendoza-Mendoza, A., Zeilinger, S., & Horwitz, B. A. (2022). Mycoparasitism as a mechanism of *Trichoderma*-mediated suppression of plant diseases. *Fungal Biology Reviews*, *39*, 15–33. https://doi.org/10.1016/j.fbr.2021.11.004 (accessed on 6th November 2025).

N/A. (2019). Report to the President of the United States Congress on plant bio stimulants submitted by the United States Department of Agriculture (USDA) in consortium with the Environmental Protection Agency (EPA) on December XX, 2019. https://agriculture.house.gov/uploadedfiles/usda_report_on_plant_bio stimulants_12.20.2019.pdf (accessed on 6th November 2025).

N/A. (2023). Agriculture Ministry issues draft guidelines for registration of bio-stimulants. *Business Standards*. https://www.business-standard.com/industry/agriculture/agri-min-issues-draft-guidelines-for-registration-of-bio-stimulants-123061400625_1.html (accessed on 6th November 2025).

N/A. (2023). EU fertilizing product regulation. https://www.fertilizerseurope.com/agriculture-environment/fertilizing-products-regulation/ (accessed on 6th November 2025).

N/A. (2024). All plant bio stimulants deserve access to the single market. https://bio stimulants.eu/issue/all-plant-bio stimulants-deserve-access-to-the-single-market/ (accessed on 6th November 2025).

Nadeem, S. M., Ahmad, M., Zahir, Z. A., Javaid, A., & Ashraf, M. (2014). The role of mycorrhizae and plant growth-promoting rhizobacteria (PGPR) in improving crop productivity under stressful environments. *Biotechnology Advances*, *32*, 429–448. https://doi.org/10.1016/j.biotechadv.2013.12.005 (accessed on 6th November 2025).

Nji, Q. N., Babalola, O. O., & Mwanza, M. (2023). Soil *Aspergillus* species, pathogenicity and control perspectives. *Journal of Fungi (Basel)*, *9*(7), 766. https://doi.org/10.3390/jof9070766 (accessed on 6th November 2025).

Nosir, W. S. (2016). *Trichoderma harzianum* as a growth promoter and bio-control agent against *Fusarium oxysporum* f. sp. *tuberosi*. *Advances in Crop Science and Technology*, *4*, 217.

Oancea, F., Raut, I., Sesan, T. E., & Cornea, P. C. (2016). Dry flowable formulation of bio stimulants *Trichoderma* strains. *Agriculture and Agricultural Science Procedia*, *10*, 494–502.

Omann, M., & Zeilinger, S. (2010). How a mycoparasite employs G-protein signaling: using the example of *Trichoderma*. *Journal of Signal Transduction*, *2010*, 123126. https://doi.org/10.1155/2010/123126 (accessed on 6th November 2025).

Ortas, I., & Ustuner, O. (2014). Determination of different growth media and various mycorrhizae species on citrus growth and nutrient uptake. *Scientia Horticulture (Amsterdam)*, *166*, 84–90. https://doi.org/10.1016/j.scienta.2013.12.014 (accessed on 6th November 2025).

Oszust, K., Cybulska, J., & Frąc, M. (2020). How do *Trichoderma* genus fungi win a nutritional competition battle against soft fruit pathogens? A report on niche overlap nutritional potentiates. *International Journal of Molecular Sciences*, *21*(12), 4235. https://doi.org/10.3390/ijms21124235 (accessed on 6th November 2025).

Pacheco-Trejo, J., Aquino-Torres, E., Reyes-Santamaría, M. I., Islas-Pelcastre, M., Pérez-Ríos, S. R., Madariaga-Navarrete, A., & Saucedo-García, M. (2022). Plant defensive responses triggered by *Trichoderma* spp. as tools to face stressful conditions. *Horticulture*, *8*(12), 1181. https://doi.org/10.3390/horticulturae8121181 (accessed on 6th November 2025).

Pani, S., Kumar, A., & Sharma, A. (2021). *Trichoderma harzianum*: An overview. *Bulletin of Environment, Pharmacology and Life Sciences*, *10*(6), 32–39. https://bepls.com/beplsmay2021/5.pdf (accessed on 6th November 2025).

Parmar, S., Sharma, V. K., Li, T., Tang, W., & Li, H. (2022). Fungal seed endophyte FZT214 improves *Dysphania ambrosioides* Cd tolerance throughout different developmental stages. *Frontiers in Microbiology*, *12*, 783475. https://doi.org/10.3389/fmicb.2021.783475 (accessed on 6th November 2025).

Patel, Z. M., Mahapatra, R., & Jampala, S. S. M. (2020). Role of fungal elicitors in plant defense mechanism. In *Molecular Aspects of Plant Beneficial Microbes in Agriculture* (pp. 143–158).

Perrone, G., Susca, A., Cozzi, G., Ehrlich, K., Varga, J., Frisvad, J. C., Meijer, M., Noonim, P., Mahakarnchanakul, W., & Samson, R. A. (2007). Biodiversity of *Aspergillus* species in some important agricultural products. *Studies in Mycology*, *59*, 53–66.

Phillips, R. P., Brzostek, E., & Midgley, M. G. (2013). The mycorrhizal-associated nutrient economy: A new framework for predicting carbon–nutrient couplings in temperate forests. *New Phytologist*, *199*, 41–51.

Poveda, J. (2021). Beneficial effects of microbial volatile organic compounds (MVOCs) in plants. *Applied Soil Ecology*, *168*, 104118. https://doi.org/10.1016/j.apsoil.2021.104118 (accessed on 6th November 2025).

Poveda, J., & Baptista, P. (2021). Filamentous fungi as biocontrol agents in olive (*Olea europaea* L.) diseases: Mycorrhizal and endophytic fungi. *Crop Protection*, *146*, 105672. https://doi.org/10.1016/j.cropro.2021.105672 (accessed on 6th November 2025).

Poveda, J., Abril-Urias, P., & Escobar, C. (2020a). Biological control of plant-parasitic nematodes by filamentous fungi inducers of resistance: *Trichoderma*, mycorrhizal and endophytic fungi. *Frontiers in Microbiology*, *11*, 992. https://doi.org/10.3389/fmicb.2020.00992 (accessed on 6th November 2025).

Poveda, J., Eugui, D., Abril-Urías, P., & Velasco, P. (2021). Endophytic fungi as direct plant growth promoters for sustainable agricultural production. *Symbiosis*, *85*, 1–19.

Pusztahelyi, T. (2018). Chitin and chitin-related compounds in plant–fungal interactions. *Mycology*, *9*, 189–201.

Radhakrishnan, R., Kang, S. M., Baek, I. Y., & Lee, I. J. (2014). Characterization of plant growth-promoting traits of *Penicillium* species against the effects of high soil salinity and root disease. *Journal of Plant Interactions*, *9*(1), 754–762. https://doi.org/10.1080/17429145.2014.930524 (accessed on 6th November 2025).

Rana, K. L., Kour, D., Kaur, T., Devi, R., Yadav, A. N., Yadav, N., Dhaliwal, H. S., & Saxena, A. K. (2020). Endophytic microbes: biodiversity, plant growth-promoting mechanisms and potential applications for agricultural sustainability. *Antonie van Leeuwenhoek International Journal of General and Molecular Microbiology*, *113*(8), 1075–1107.

Read, D. J. (1991). Mycorrhizas in ecosystems. *Experiential, 47*, 376–391.

Roberts, E., & Lindow, S. (2014). Loline alkaloid production by fungal endophytes of Fescue species select for particular epiphytic bacterial microflora. *ISME Journal, 8*, 359–368.

Ruiz-Herrera, J., & Ortiz-Castellanos, L. (2019). Cell wall glucans of fungi. A review. *Cell Surface, 5*, 100022. https://doi.org/10.1016/j.tcsw.2019.100022 (accessed on 6th November 2025).

Sák, M., Dokupilová, I., Kaňuková, A., Mrkvová, M., & Kraic, J. (2021). Biotic and abiotic elicitors of stilbenes production in *Vitis vinifera* L. cell culture. *Plants, 10*(3), 490. https://doi.org/10.3390/plants10030490 (accessed on 6th November 2025).

Samuels, G. J., Dodd, S. L., Lu, B. S., Petrini, O., Schroers, H. J., & Druzhinina, I. S. (2006). The *Trichoderma koningii* aggregate species. *Studies in Mycology, 56*, 67–133. https://doi.org/10.3114/sim.2006.56.03 (accessed on 6th November 2025).

Sangiorgio, D., Cellini, A., Donati, I., Pastore, C., Onofrietti, C., & Spinelli, F. (2020). Facing climate change: Application of microbial bio stimulants to mitigate stress in horticultural crops. *Agronomy, 10*(6), 794. https://doi.org/10.3390/agronomy10060794 (accessed on 6th November 2025).

Saravanakumar, K., Fan, L., Fu, K., Yu, C., Wang, M., Xia, H., Sun, J., Li, Y., & Chen, J. (2016). Cellulase from *Trichoderma harzianum* interacts with roots and triggers induced systemic resistance to foliar disease in maize. *Scientific Reports, 6*, 3554. https://doi.org/10.1038/srep35543 (accessed on 6th November 2025).

Saravanakumar, K., Li, Y., Yu, C., Wang, Q. Q., Wang, M., Sun, J., Gao, J. X., & Chen, J. (2017). Effect of *Trichoderma harzianum* on maize rhizosphere microbiome and biocontrol of *Fusarium* stalk rot. *Scientific Reports, 7*, 1771. https://doi.org/10.1038/s41598-017-01680-w (accessed on 6th November 2025).

Sardans, J., Lambers, H., Preece, C., Alrefaei, A. F., & Peñuelas, J. (2023). Role of mycorrhizas and root exudates in plant uptake of soil nutrients (calcium, iron, magnesium, and potassium): Has the puzzle been completely solved? *The Plant Journal, 114*(6), 1227–1242. https://doi.org/10.1111/tpj.16184 (accessed on 6th November 2025).

Sarkar, S., Dey, A., Kumar, V., Batiha, G. E. S., El-Esawi, M. A., Tomczyk, M., & Ray, P. (2021). Fungal endophyte: An interactive endosymbiont with the capability of modulating host physiology in myriad ways. *Frontiers in Plant Science, 12*, 701800. https://doi.org/10.3389/fpls.2021.701800 (accessed on 6th November 2025).

Schnepf, A., Roose, T., & Schweiger, P. (2008). Growth model for arbuscular mycorrhizal fungi. *Journal of the Royal Society Interface, 5*, 773–784.

Schulz, B., Boyle, C., Draeger, S., Römmert, A. K., & Krohn, K. (2002). Endophytic fungi: A source of novel biologically active secondary metabolites. *Mycological Research, 106*, 996–1004.

Sharma, I., Raina, A., Choudhary, M., Apra, Kaul, S., & Dhar, M. K. (2023). Fungal endophyte bioinoculants as a green alternative towards sustainable agriculture. *Helion, 9*(9), e19487. https://doi.org/10.1016/j.heliyon.2023.e19487 (accessed on 6th November 2025).

Singh, R., Kaur, S., Bhullar, S. S., Singh, H., & Sharma, L. K. (2024). Bacterial bio stimulants for climate smart agriculture practices: Mode of action, effect on plant growth and roadmap for commercial products. *Journal of Sustainable Agriculture and Environment, 3*(1), e12085. https://doi.org/10.1002/sae2.12085 (accessed on 6th November 2025).

Sood, M., Kapoor, D., Kumar, V., Sheteiwy, M. S., Ramakrishnan, M., Landi, M., Araniti, F., & Sharma, A. (2020). *Trichoderma*: The "secrets" of a multitalented biocontrol agent. *Plants (Basel), 9*(6), 762. https://doi.org/10.3390/plants9060762 (accessed on 6th November 2025).

Soytong, K., Kahonokmedhakul, S., Song, J., & Tongon, R. (2021). *Chaetomium* application in agriculture. *IntechOpen*. https://www.intechopen.com/chapters/78015 (accessed on 6th November 2025).

Spinelli, V., Brasili, E., Sciubba, F., Ceci, A., Giampaoli, O., Miccheli, A., Pasqua, G., & Persiani, A. M. (2022). Bio stimulant effects of *Chaetomium globosum* and *Minimedusa polyspora* culture filtrates on *Cichorium intybus* plant: Growth performance and metabolomic traits. *Frontiers in Plant Science, 13*, 879076. https://doi.org/10.3389/fpls.2022.879076 (accessed on 6th November 2025).

Stracquadanio, C., Quiles, J. M., Meca, G., & Cacciola, S. O. (2020). Antifungal activity of bioactive metabolites produced by *Trichoderma asperellum* and *Trichoderma atroviride* in liquid medium. *Journal of Fungi (Basel), 6*(4), 263. https://doi.org/10.3390/jof6040263 (accessed on 6th November 2025).

Sui, L., Li, J., Philp, J., Yang, K., Wei, Y., Li, H., Li, J., Li, L., Ryder, M., Toh, R., Zhou, Y., Denton, M. D., Hu, J., & Wang, Y. (2022). *Trichoderma atroviride* seed dressing influenced the fungal community and pathogenic fungi in the wheat rhizosphere. *Scientific Reports, 12*(1), 9677. https://doi.org/10.1038/s41598-022-13669-1 (accessed on 6th November 2025).

Sumayku, B. R. A., Waridiyati, T., Afandhi, A., & Polii-Mandang, J. (2015). The effectiveness of *Trichoderma koningii* in improving the quality of strawberry fruits in

Rurukan Plantation, Tomohon. *International Journal of Advanced Research in Biological Sciences, 2*(3), 25–29.

Sun, L., & Zhang, J. (2020). Regulatory role of receptor-like cytoplasmic kinases in early immune signaling events in plants. *FEMS Microbiology Reviews, 44*, 845–856.

Sun, W., & Shahrajabian, M. H. (2023). The application of arbuscular mycorrhizal fungi as microbial bio stimulant, sustainable approaches in modern agriculture. *Plants (Basel), 12*(17), 3101. https://doi.org/10.3390/plants12173101 (accessed on 6th November 2025).

Surekha, C., Neelapu, N. R. R., Kamala, G., Prasad, B. S., & Ganesh, P. S. (2013). Efficacy of *Trichoderma viride* to induce disease resistance and antioxidant responses in legume *Vigna mungo* infested by *Fusarium oxysporum* and *Alternaria alternata*. *International Journal of Agricultural Science and Research, 3*(2), 285–294.

Szabo, K., Böll, S., & Erős-Honti, Z. S. (2014). Applying artificial mycorrhizae in planting urban trees. *Applied Ecology, 12*, 835–853. https://doi.org/10.15666/aeer/1204 (accessed on 6th November 2025).

Tarroum, M., Romdhane, W. B., Al-Qurainy, F., Ali, A. A. M., Al-Doss, A., Fki, L., & Hassairi, A. (2022). A novel PGPF *Penicillium olsonii* isolated from the rhizosphere of *Aeluropus littoralis* promotes plant growth, enhances salt stress tolerance, and reduces chemical fertilizers inputs in hydroponic system. *Frontiers in Microbiology, 13*, 996054. https://doi.org/10.3389/fmicb.2022.996054 (accessed on 6th November 2025).

Tedersoo, L. (2017). Global biogeography and invasions of ectomycorrhizal plants: Past, present and future. In L. Tedersoo (Ed.), *Biogeography of Mycorrhizal Symbiosis* (pp. 469–531). Springer International Publishing.

Thakur, M., & Sohal, B. S. (2013). Role of elicitors in inducing resistance in plants against pathogen infection: A review. *ISRN Biochemistry*, *2013*, 762412. https://doi.org/10.1155/2013/762412 (accessed on 6th November 2025).

Toghueo, R. M. K., & Boyom, F. F. (2020). Endophytic *Penicillium* species and their agricultural, biotechnological, and pharmaceutical applications. *3 Biotech, 10*(3), 107. https://doi.org/10.1007/s13205-020-2081-1 (accessed on 6th November 2025).

Traversari, S., Cacini, S., Galieni, A., Nesi, B., Nicastro, N., & Pane, C. (2021). Precision agriculture digital technologies for sustainable fungal disease management of ornamental plants. *Sustainability, 13*(7), 3707. https://doi.org/10.3390/su13073707 (accessed on 6th November 2025).

Tsuge, T., Harimoto, Y., Akimitsu, K., Ohtani, K., Kodama, M., Akagi, Y., Egusa, M., Yamamoto, M., & Otani, H. (2013). Host-selective toxins produced by the plant pathogenic fungus *Alternaria alternata*. *FEMS Microbiology Reviews, 37*(1), 44–66. https://doi.org/10.1111/j.1574-6976.2012.00350.x (accessed on 6th November 2025).

Tyśkiewicz, R., Nowak, A., Ozimek, E., & Jaroszuk-Ściseł, J. (2022). *Trichoderma*: The current status of its application in agriculture for the biocontrol of fungal phytopathogens and stimulation of plant growth. *International Journal of Molecular Sciences, 23*(4), 2329. https://doi.org/10.3390/ijms23042329 (accessed on 6th November 2025).

Uddin, M. M., Akhtar, N., Islamand, M. T., & Faruq, A. N. (2011). Effect of *Trichoderma harzianum* and some selected soil amendment against damping off disease complex of potato and chili. *The Agriculturists, 9*(1&2), 106–116.

Ulian, T., Diazgranados, M., Pironon, S., Padulos, S., Liu, U., Davies, L., Howes, M. R. R., Borrell, J. S., Ondo, I., Pérez-Escobar, O. A., Sharrock, S., Ryan, P., Hunter, D., Lee,

M. A., Barstow, C., Łuczaj, Ł., Pieroni, A., Cámara-Leret, R., Noorani, A., Mba, C., Womdim, R. N., Muminjanov, H., Antonelli, A., Pritchard, H. W., & Mattana, E. (2020). Unlocking plant resources to support food security and promote sustainable agriculture. *Plants, People, Planet, 2*(5), 421–445. https://doi.org/10.1002/ppp3.10145 (accessed on 6th November 2025).

Uzma, F., Mohan, C. D., Hashem, A., Konappa, N. M., Rangappa, S., Kamath, P. V., Singh, B. P., Mudili, V., Gupta, V. K., Siddaiah, C. N., Chowdappa, S., Alqarawi, A. A., & Abd-Allah, E. F. (2018). Endophytic fungi–Alternative sources of cytotoxic compounds: A review. *Frontiers in Pharmacology, 9*, 309. https://doi.org/10.3389/fphar.2018.00309 (accessed on 6th November 2025).

Vaksmaa, A., Guerrero-Cruz, S., Ghosh, P., Zeghal, E., Hernando-Morales, V., & Niemann, H. (2023). Role of fungi in bioremediation of emerging pollutants. *Frontiers in Marine Science, 10*, 1070905. https://doi.org/10.3389/fmars.2023.1070905 (accessed on 6th November 2025).

van der Heijden, M. G. A., Martin, F. M., Selosse, M. A., & Sanders, I. R. (2015). Mycorrhizal ecology and evolution: The past, the present, and the future. *New Phytologist, 205*, 1406–1423. https://doi.org/10.1111/nph.13288 (accessed on 6th November 2025).

Vandeputte, P., Ferrari, S., & Coste, A. T. (2012). Antifungal resistance and new strategies to control fungal infections. *International Journal of Microbiology, 2012*, 713687. https://doi.org/10.1155/2012/713687 (accessed on 6th November 2025).

Vassilev, N., Eichler-Löbermann, B., Flor-Peregrin, E., Martos, V., Reyes, A., & Vassileva, M. (2017). Production of a potential liquid plant bio-stimulant by immobilized *Piriformospora indica* in repeated-batch fermentation process. *AMB Express, 7*(1), 106. https://doi.org/10.1186/s13568-017-0408-z (accessed on 6th November 2025).

Vázquez, M. B., Barrera, V., & Bianchinotti, V. (2015). Molecular identification of three isolates of *Trichoderma harzianum* isolated from agricultural soils in Argentina, and their abilities to detoxify *in vitro* metsulfuron methyl. *Botany, 93*, 793–800.

Vedenicheva, N., & Kosakivska, I. (2023). In search of the phytohormone functions in fungi: Cytokinin's. *Fungal Biology Reviews, 45*, 100309. https://doi.org/10.1016/j.fbr.2023.100309 (accessed on 6th November 2025).

Vega, V. (2018). The use of fungal entomopathogens as endophytes in biological control: A review. *Mycologic, 110*(1), 4–30.

Vendele, I., Willment, J. A., Silva, L. M., Palma, A. S., Chai, W., Liu, Y., Feizi, T., Spyrou, M., Stappers, M. H. T., Brown, G. D., & Gow, N. A. R. (2020). Mannan detecting C-type lectin receptor probes recognize immune epitopes with diverse chemical, spatial and phylogenetic heterogeneity in fungal cell walls. *PLoS Pathogens, 16*(1), e1007927. https://doi.org/10.1371/journal.ppat.1007927 (accessed on 6th November 2025).

Verma, A., Shameem, N., Jatav, H. S., Sathyanarayana, E., Parray, J. A., Poczai, P., & Sayyed, R. Z. (2022). Fungal endophytes to combat biotic and abiotic stresses for climate-smart and sustainable agriculture. *Frontiers in Plant Science, 13*, 953836. https://doi.org/10.3389/fpls.2022.953836 (accessed on 6th November 2025).

Větrovský, T., Morais, D., Kohout, P., Lepinay, C., Algora, C., Awokunle Hollá, S., et al. (2020). GlobalFungi, a global database of fungal occurrences from high-throughput-sequencing metabarcoding studies. *Scientific Data, 7*(1), 228. https://doi.org/10.1038/s41597-020-0567-7 (accessed on 6th November 2025).

Visconti, D., Ventorino, V., Fagnano, M., Woo, S. L., Pepe, O., Adamo, P., Caporale, A. G., Carrino, L., & Fiorentino, N. (2023). Compost and microbial bio stimulant applications improve plant growth and soil biological fertility of a grass-based Phyto stabilization system. *Environmental Geochemistry and Health, 45*, 787–807. https://doi.org/10.1007/s10653-022-01235-7 (accessed on 6th November 2025).

Vogelsang, K. M., Bever, J. D., Griswold, M., & Schultz, P. A. (2004). The use of mycorrhizal fungi in erosion control applications. *Contract*, 1–150.

Waghunde, R. R., Shelake, R. M., & Sabalpara, A. N. (2016). *Trichoderma*: A significant fungus for agriculture and environment. *African Journal of Agricultural Research, 11*, 1952–1965.

Wahab, A., Muhammad, M., Munir, A., Abdi, G., Zaman, W., Ayaz, A., Khizar, C., & Reddy, S. P. P. (2023). Role of arbuscular mycorrhizal fungi in regulating growth, enhancing productivity, and potentially influencing ecosystems under abiotic and biotic stresses. *Plants (Basel), 12*(17), 3102. https://doi.org/10.3390/plants12173102 (accessed on 6th November 2025).

Wang, X. Q., Wang, Y. H., Song, Y. B., & Dong, M. (2022). Formation and functions of arbuscular mycorrhizae in coastal wetland ecosystems: A review. *Ecosystem Health and Sustainability, 8*(1), 2144465. https://doi.org/10.1080/20964129.2022.2144465 (accessed on 6th November 2025).

Wang, X., Li, J., Shang, J., Bai, J., Wu, K., Liu, J., Yang, Z., Ou, H., & Shao, L. (2022). Metabolites extracted from microorganisms as potential inhibitors of glycosidases (α-glucosidase and α-amylase): A review. *Frontiers in Microbiology, 13*, 1050869. https://doi.org/10.3389/fmicb.2022.1050869 (accessed on 6th November 2025).

Wang, Y., Dai, C. C., Zhao, Y. W., & Peng, Y. (2011). Fungal endophyte-induced volatile oil accumulation in *Atractylodes lancea* plantlets is mediated by nitric oxide, salicylic acid and hydrogen peroxide. *Process Biochemistry, 46*(3), 730–735. https://doi.org/10.1016/j.procbio.2010.11.020 (accessed on 6th November 2025).

War, A. R., Paulraj, M. G., Ahmad, T., Buhroo, A. A., Hussain, B., Ignacimuthu, S., & Sharma, H. C. (2012). Mechanisms of plant defense against insect herbivores. *Plant Signaling & Behavior, 7*(10), 1306–1320.

Wilson, D. (1995). Endophyte: The evolution of a term, and clarification of its use and definition. *Oikos, 73*(2), 274–276.

Wonglom, P., Daengsuwan, W., Ito, S. I., & Sunpapao, A. (2019). Biological control of *Sclerotium* fruit rot of snake fruit and stem rot of lettuce by *Trichoderma* sp. T76-12/2 and the mechanisms involved. *Physiological and Molecular Plant Pathology, 107*, 1–7. https://doi.org/10.1016/j.pmpp.2019.04.007 (accessed on 6th November 2025).

Wu, C., Qu, J., Liu, L., Kang, H., Sun, H., Zhang, Y., Ghorbani, A., & Pehlivan, N. (2021). *Quo vadis*: Signaling molecules and small secreted proteins from mycorrhizal fungi at the early stage of mycorrhiza formation. *Symbiosis, 85*, 123–143. https://doi.org/10.1007/s13199-021-00793-1 (accessed on 6th November 2025).

Yadav, A., Saini, I., Kaushik, P., Ansari, M. A., Khan, M. R., & Haq, N. (2021). Effects of arbuscular mycorrhizal fungi and P-solubilizing *Pseudomonas fluorescens* (ATCC-17400) on morphological traits and mineral content of sesame. *Saudi Journal of Biological Sciences, 28*(5), 2649–2654. https://doi.org/10.1016/j.sjbs.2021.03.024 (accessed on 6th November 2025).

Yamaguchi, T., Yamada, A., Hong, N., Ogawa, T., & Shibuya, I. N. (2000). Differences in the recognition of glucan elicitor signals between rice and soybean: Beta-glucan fragments from the rice blast disease fungus *Pyricularia oryzae* that elicit phytoalexin biosynthesis in suspension-cultured rice cells. *Plant Cell, 12*, 817–826.

Yao, X., Guo, H., Zhang, K., Zhao, M., Ruan, J., & Chen, J. (2023). *Trichoderma* and its role in biological control of plant fungal and nematode disease. *Frontiers in Microbiology, 14*, 1160551. https://doi.org/10.3389/fmicb.2023.1160551 (accessed on 6th November 2025).

Zafra, G., Moreno-Montaño, A., Absalón, Á. E., & Cortés-Espinosa, D. V. (2015). Degradation of polycyclic aromatic hydrocarbons in soil by a tolerant strain of *Trichoderma asperellum*. *Environmental Science and Pollution Research, 22*, 1034–1042.

Zhai, X., Jia, M., Chen, L., Zheng, C. J., Rahman, K., Han, T., & Qin, L. P. (2017). The regulatory mechanism of fungal elicitor-induced secondary metabolite biosynthesis in medical plants. *Critical Reviews in Microbiology, 43*(2), 238–261. https://doi.org/10.1080/1040841X.2016.1201041 (accessed on 6th November 2025).

Zhang, J., Cook, J., Nearing, J. T., Zhang, J., Raudonis, R., Glick, B. R., Langille, M. G. I., & Cheng, Z. (2021). Harnessing the plant microbiome to promote the growth of agricultural crops. *Microbiological Research, 245*, 126690. https://doi.org/10.1016/j.micres.2020.126690 (accessed on 6th November 2025).

Zhang, S., Gan, Y., & Xu, B. (2015). Biocontrol potential of a native species of *Trichoderma longibrachiatum* against *Meloidogyne incognita*. *Applied Soil Ecology, 94*, 21–29. https://doi.org/10.1016/j.apsoil.2015.04.010 (accessed on 6th November 2025).

Zin, N. A., & Badaluddin, N. A. (2020). Biological functions of *Trichoderma* spp. for agriculture applications. *Annals of Agricultural Sciences, 65*(2), 168–178. https://doi.org/10.1016/j.aoas.2020.09.003 (accessed on 6th November 2025).

PART III

Plant Biostimulants in Bioactive Compound Production

CHAPTER 8

How Biostimulants Enhance Artemisinin Production: A Review of Current Knowledge and Future Perspectives

DEXTER ACHU MOSOH[1,2,3,*]

[1]*Center for Biodiversity Exploration and Conservation (CBEC), Jabalpur, Madhya Pradesh, India*

[2]*Indian Institute of Technology Gandhinagar, Palaj Campus, Gandhinagar, Gujarat, India*

[3]*School of Science, SAGE University, Bhopal, Madhya Pradesh, India*

*Corresponding author

ABSTRACT

This chapter provides a comprehensive and in-depth exploration of the utilization of biostimulants (BS) to enhance artemisinin production in *Artemisia annua* plants. Artemisinin, a potent plant antimalarial compound, is crucial against malaria. The chapter systematically examines the current state of knowledge, elucidating the mechanisms through which BS enhance artemisinin yield and quality. It discusses the diverse array of BS, their origins, compositions, and modes of action, offering insights into their potential applications for sustainable and efficient secondary metabolite production. Furthermore, the chapter delves into the translational bottlenecks in biostimulant research and application, including standardization, regulation, and elucidation of modes of action. Additionally, it identifies critical knowledge gaps and research needs,

Plant Biostimulants and Growth Regulators in Sustainable Agriculture.
Deepak Kumar, Sanjay K. Singh, Nutan Kaushik, and Mahaveer P. Sharma (Eds.)

DOI: 10.1201/9781779641458-11

advocating for further investigations in screening novel BS, understanding molecular targets, and evaluating cost-effectiveness and sustainability. By addressing these challenges, the chapter underscores the potential benefits of biostimulant use, including enhancing plant resilience to climate change, improving human health through increased artemisinin availability, and contributing to global development goals. Ultimately, the chapter emphasizes the urgent need for continued research and innovation in biostimulant technologies, issuing a call to action for researchers, policymakers, and industry stakeholders to drive the translation of research findings into practical field-level interventions. As we stand at the nexus of agricultural innovation, health advancement, and sustainable development, utilizing BS in artemisinin production emerges as a promising avenue to fortify our agrarian systems, combat malaria, and contribute to the broader global development agenda.

8.1 INTRODUCTION

8.1.1 BACKGROUND AND CONTEXT

Artemisinin, a sesquiterpene lactone derived from the plant *Artemisia annua* L., holds profound significance in global health as a pivotal weapon against malaria. This disease continues to exact a heavy toll on human lives, particularly in tropical and subtropical regions. Malaria, caused by Plasmodium parasites and transmitted through infected Anopheles mosquitoes, remains a major global health challenge. In 2020 alone, an estimated 245 million cases and 625,000 deaths were attributed to malaria (WHO, 2020). The discovery of artemisinin's antimalarial properties represented a critical turning point in infectious disease control. Artemisinin and its derivatives have demonstrated exceptional efficacy in rapidly reducing parasite biomass in infected individuals, alleviating symptoms, and preventing severe complications (White, 1997). This unique mode of action, involving the generation of reactive oxygen species (ROS) following activation by iron, contributes to the rapid and potent antimalarial effects of artemisinin (Meshnick, 2002). Artemisinin-based combination therapies (ACTs), which combine artemisinin derivatives with other antimalarial drugs, have become the cornerstone of contemporary malaria treatment and control strategies endorsed by the World Health Organization (WHO, 2015). ACTs ensure swift parasite clearance and minimize the risk of drug resistance—a historical challenge associated with monotherapies. Introducing artemisinin and ACTs has substantially reduced malaria-related morbidity and mortality in various endemic regions (Dondorp et al., 2009). Furthermore, artemisinin has catalyzed a paradigm shift in malaria prevention and management infrastructures. Its exceptional potency against Plasmodium parasites, combined with its relatively short half-life in the human body, has led to the development of novel therapeutic regimens such as artemisinin-based intermittent preventive treatment (IPT) and seasonal malaria chemoprevention (SMC). These strategies aim to reduce malaria in vulnerable populations, particularly pregnant women and children (Gosling et al., 2009; WHO, 2012).

8.1.2 THE ROLE OF PLANT GROWTH ENHANCERS IN OPTIMIZING ARTEMISININ PRODUCTION

The role of plant growth enhancers in optimizing artemisinin production presents an intriguing avenue in medicinal plant cultivation. Artemisinin, derived from *Artemisia annua* L., is globally recognized for its potent antimalarial properties (WHO, 2015). As the demand for this crucial compound continues to escalate due to its pivotal role in malaria treatment, there is a growing impetus to enhance artemisinin yield through innovative agricultural practices. Plant growth enhancers, encompassing a spectrum of substances such as biostimulants (BS), hormones, and elicitors, offer promising strategies to amplify artemisinin biosynthesis. Research has revealed that these enhancers can exert multifaceted influences on plant physiology, affecting various stages of growth and development (Khan et al., 2009a). Through their modulation of gene expression, hormonal signaling pathways, and stress responses, plant growth enhancers can orchestrate a series of intricate biochemical and metabolic shifts that culminate in augmented secondary metabolite production (Ghazal et al., 2018; Jamwal et al., 2018). Specifically, BS, a subset of growth enhancers, have demonstrated remarkable potential for optimizing artemisinin accumulation. BS encompass various substances derived from natural sources, such as seaweed extracts, humic substances, and microbial consortia. Their application has been associated with increased photosynthetic efficiency, nutrient uptake, and overall plant health, thereby establishing conducive conditions for enhanced secondary metabolite biosynthesis (du Jardin, 2015). Moreover, these growth enhancers often operate synergistically with environmental cues, optimizing the plant's response to external factors and stressors. For instance, light intensity, temperature, and nutrient availability interact intricately with biostimulant-mediated effects, shaping artemisinin biosynthesis and accumulation (Bulgari et al., 2019; Chen et al., 2022). The interplay between BS and these factors adds complexity to optimizing artemisinin production and underscores the importance of a comprehensive approach that integrates intrinsic and extrinsic determinants.

8.1.3 BIOSTIMULANTS

BS represent a diverse group of substances or microorganisms that benefit plant growth, development, yield, quality, and stress tolerance without being considered traditional nutrients or pesticides (Calvo et al., 2014; du Jardin, 2015). These multifunctional agents harness various mechanisms to enhance plant performance, such as promoting nutrient uptake and assimilation, modulating hormonal signaling pathways, stimulating root and shoot growth, and fostering microbial interactions within the rhizosphere (Cansu and Erdal, 2018; Rouphael and Colla, 2020a). BS encompass many compounds, including humic and fulvic acids, seaweed extracts, microbial inoculants, and other organic and inorganic substances (Schiavon et al., 2010).

In artemisinin production, the application of BS holds immense promise as a strategy to optimize both the quantity and quality of artemisinin-rich plants. Through their intricate influence on plant physiological processes, BS can, directly and indirectly, impact artemisinin biosynthesis and accumulation. For instance, BS can enhance photosynthetic efficiency by increasing carbon assimilation and energy availability for secondary metabolite production (Savvas and Ntatsi, 2015). Additionally, BS' stimulation of root development and nutrient uptake can improve nutrient availability for artemisinin biosynthesis (Bulgari et al., 2019). Moreover, BS' modulation of hormonal balances, particularly the jasmonic acid and auxin pathways, can potentially trigger the activation of artemisinin biosynthetic genes (Liu et al., 2017). Furthermore, ameliorating abiotic stress responses through biostimulant application can bolster plant resilience, safeguarding artemisinin-producing plants against environmental adversities and further optimizing secondary metabolite production (Nardi et al., 2016). This book chapter aims to comprehensively analyze the role of BS in enhancing artemisinin production in *Artemisia annua* L. and explore their potential applications in sustainable malaria treatment. The review thoroughly examines the diverse classes of BS, their underlying mechanisms of action, and their effects on various aspects of plant physiology, including growth, yield, and secondary metabolite production. Special emphasis is placed on elucidating the intricate interactions between BS and artemisinin biosynthesis pathways and integrating BS into holistic approaches for optimizing artemisinin-rich plant cultivation. By synthesizing current research and emerging trends, this review offers novel insights into BS' transformative potential in advancing artemisinin production and malaria treatment strategies.

8.2 ARTEMISININ BIOSYNTHESIS AND IMPORTANCE

Artemisinin is a medicinal natural compound derived from the plant *Artemisia annua*, commonly known as sweet wormwood (Krishna et al., 2008). Significant progress has been made in the fight against malaria due to its discovery and subsequent therapeutic applications, which cause a substantial burden on global health. Artemisinin's remarkable potency against Plasmodium parasites, the causative agent of malaria, has elevated it to a frontline antimalarial agent crucial in treating and preventing the disease (WHO, 2015). The compound's therapeutic significance lies in its distinctive mode of action, which involves generating ROS upon activation by iron, leading to parasite eradication (Meshnick, 2002). Artemisinin's rapid and potent action is particularly effective against drug-resistant strains of malaria, addressing a critical need in regions where resistance to conventional antimalarials has emerged (White, 1997). Furthermore, artemisinin and its derivatives have exhibited potential beyond malaria treatment. Emerging research has explored their efficacy in treating other diseases, such as certain types of cancer, owing to their ability to induce apoptosis and inhibit tumor growth (Efferth et al., 2008). Artemisinin's versatility extends to its immunomodulatory properties, with studies suggesting its potential role in boosting immune responses and reducing inflammation (Dhama et al., 2018). These multifaceted therapeutic applications highlight the compound's immense promise in addressing global health challenges.

8.2.1 ARTEMISININ BIOSYNTHESIS PATHWAY IN ARTEMISIA ANNUA L.

The artemisinin biosynthesis pathway in *Artemisia annua* L., thc plant source of this valuable compound, is a complex and intricately regulated series of enzymatic reactions that culminate in the production of artemisinin, a sesquiterpene lactone with potent antimalarial properties. Central to this pathway is the precursor molecule, amorpha-4,11-diene, which undergoes a series of enzymatic transformations, many of which are catalyzed by cytochrome P450 monooxygenases (CYPs), to yield artemisinin (Teoh et al., 2009) (Figure 8.1).

The pathway to amorpha-4,11-diene, a terpene precursor, begins with farnesyl diphosphate converted by amorpha-4,11-diene synthase (ADS) (Ro et al., 2006). This initial step is a crucial branch point that diverts the metabolic flux towards artemisinin biosynthesis. Multiple enzymatic steps, including the actions of CYP71AV1 and CYP71AV8 enzymes, lead to the formation of intermediate dihydroartemisinic acid through hydroxylation, oxidation, and reduction reactions (Zhang et al., 2008). Dihydroartemisinic acid then converts to artemisinic alcohol through a series of reductions. A critical enzymatic conversion from artemisinic alcohol to artemisinin involves the oxidation of the alcohol group to a lactone, a reaction catalyzed by artemisinic aldehyde Δ11(13) reductase (DBR2) (Bouwmeester et al., 1999). This step represents a key regulatory point in the pathway, influencing the flux towards artemisinin production. Subsequently, artemisinin is produced from artemisinic aldehyde via a non-enzymatic photochemical reaction involving the endoperoxide bridge's cleavage (Ho et al., 2014). This endoperoxide bridge is a crucial structural feature responsible for artemisinin's potent antimalarial activity, as it is implicated in generating ROS upon activation by iron, leading to parasite death (Meshnick, 2002). Various factors, including developmental stage, environmental cues, and genetic variations, influence the regulation of the artemisinin biosynthesis pathway. Transcription factors AaERF1 and AaORA regulate artemisinin biosynthesis by controlling gene expression in response to external stimuli (Chen et al., 2016; Lv et al., 2017). Additionally, post-transcriptional regulation mediated by microRNAs has been implicated in modulating artemisinin biosynthesis (Wang, 2019). Understanding these regulatory mechanisms holds promise for the targeted enhancement of artemisinin production through genetic and metabolic engineering approaches.

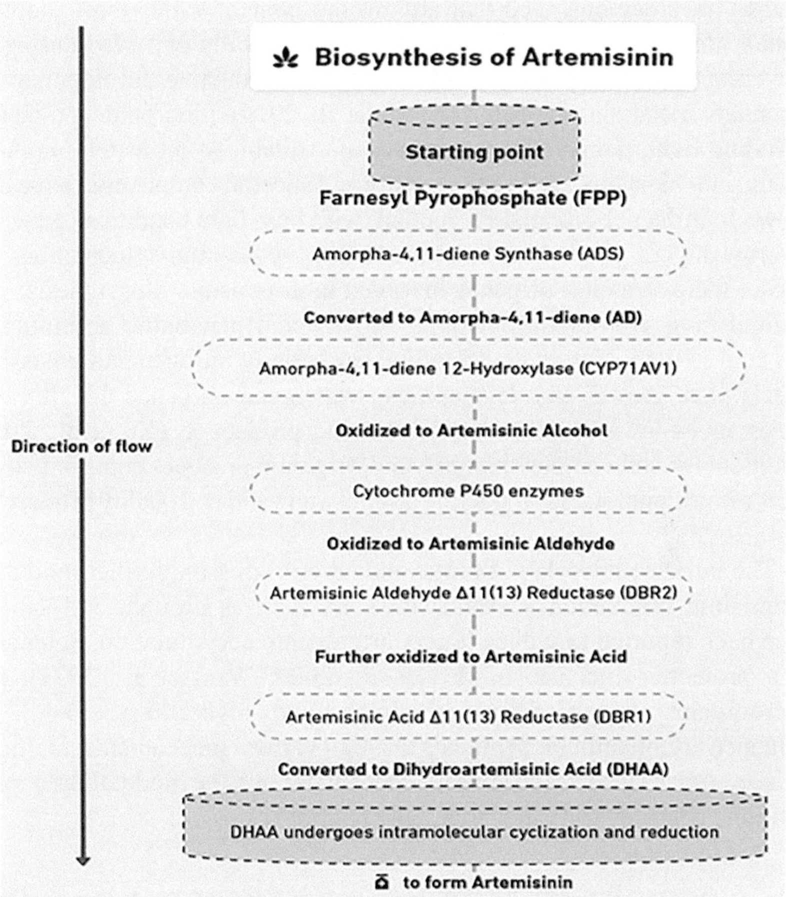

FIGURE 8.1 The biosynthetic pathway of artemisinin in *Artemisia annua* L.

8.2.2 FACTORS INFLUENCING ARTEMISININ YIELD AND QUALITY

The yield and quality of artemisinin, a pivotal antimalarial compound derived from *Artemisia annua* L., are subject to many intricate factors that collectively shape its production. The plant's genetic makeup plays a fundamental role, as different cultivars exhibit varying capacities for artemisinin biosynthesis. Genetic variability arises from geographical origin, breeding efforts, and selection pressures, influencing the accumulation of critical intermediates and enzymes within the biosynthetic pathway (Olsson et al., 2009). Furthermore, the developmental stage of the plant significantly impacts artemisinin yield. Studies have demonstrated that artemisinin content varies throughout the plant's lifecycle, with higher concentrations typically observed during the flowering stage, suggesting a link between reproductive development and secondary metabolite synthesis (Rydén et al., 2010). Environmental factors, including light, temperature, and nutrient availability, intricately modulate artemisinin biosynthesis. Light intensity and spectral composition have been shown to influence artemisinin content, with high light conditions generally favoring higher yields (Lopes et al., 2020). Temperature fluctuations can impact the expression of genes involved in artemisinin biosynthesis, with optimal ranges promoting enzyme activity and metabolite accumulation (Lu et al., 2018). Nutrient availability, particularly nitrogen and potassium, affects plant growth and development, which can influence the allocation of resources towards secondary metabolite production (Xu et al., 2020). Additionally, soil composition, pH, and irrigation practices contribute to the plant's nutritional status, impacting artemisinin yield and quality (Todeschini et al., 2022).

The intricate interplay of biotic and abiotic stressors further modulates artemisinin production. Certain stressors, such as drought and salinity, have been reported to induce higher artemisinin accumulation, potentially as a protective mechanism (Fu et al., 2020; Wani et al., 2021). Soil microorganisms, including endophytes and mycorrhizal fungi, can influence artemisinin biosynthesis through various mechanisms, including the secretion of growth-promoting compounds and the modulation of plant hormone (PH) levels (Chen et al., 2021).

8.3 BIOSTIMULANTS AND THEIR MODES OF ACTION ON PLANTS

BS are substances or microorganisms applied to plants to enhance their growth, development, and productivity. Unlike fertilizers, which provide essential nutrients directly to plants, BS work through various mechanisms to improve nutrient uptake, stress tolerance, and overall plant health (Castiglione et al., 2021; Rouphael and Colla, 2020).

8.3.1 CATEGORIES, COMPOSITION, AND FUNCTIONS OF DIFFERENT BIOSTIMULANTS

BS, a diverse group of substances or microorganisms that enhance plant growth, yield, quality, and stress tolerance, can be broadly classified into distinct categories based on their origin, composition, and function. Humic substances derived from decayed organic matter constitute one such category. These complex mixtures of organic compounds encompass humic and fulvic acids, which play a crucial role in improving soil structure, nutrient retention, and water-holding capacity, enhancing plant growth and stress resilience (du Jardin, 2015). Seaweed extracts, another prominent category, are derived from marine algae and are rich in bioactive compounds like polysaccharides, amino acids, and plant growth regulators (PGRs). Seaweed extracts have been reported to stimulate plant growth, nutrient uptake, and secondary metabolite synthesis, making them valuable tools for enhancing crop productivity and quality (Khan et al., 2009a).

Amino acids, whether in free or peptide-bound forms, represent another class of BS (Nephali et al., 2020). These organic compounds are building blocks for protein synthesis and participate in various metabolic processes. Amino acid-based BS have been shown to influence plant growth, nutrient assimilation, and stress responses, thereby contributing to improved crop performance and stress tolerance (Cansu and Erdal, 2018). Microbial inoculants, including beneficial bacteria, fungi, and mycorrhizal associations, constitute yet another category. These microorganisms form symbiotic relationships with plants, facilitating nutrient uptake, hormone signaling, and pathogen resistance. Microbial BS have demonstrated the potential to promote plant growth, enhance nutrient availability, and induce systemic resistance, thus enhancing overall plant health and productivity (Castiglione et al., 2021; Fadiji et al., 2022). Moreover, BS can include plant-derived extracts from plant growth-promoting rhizobacteria (PGPR), which produce bioactive compounds that stimulate plant growth and modulate root architecture (Bhattacharyya and Jha, 2012). Chitin and chitosan derivatives, derived from chitin-rich sources like crustacean shells, exhibit biostimulant properties through their involvement in plant defense responses and growth regulation (Cansu and Erdal, 2018). These diverse categories of BS collectively offer multifaceted mechanisms to optimize plant growth, development, and stress resilience, making them integral components of sustainable agriculture and crop management strategies.

8.3.2 MODE OF ACTION OF BIOSTIMULANTS

The mode of action of BS is multifaceted and often involves complex interactions between the plant, soil, and associated microorganisms, resulting in improved growth, yield, and stress tolerance (Drobek et al., 2019; Kumari et al., 2021). They act through various mechanisms, which are discussed in subsections.

8.3.2.1 NUTRIENT UPTAKE AND ASSIMILATION

BS affect plant growth and development through multifaceted mechanisms spanning plant physiology, biochemistry, and molecular biology. These mechanisms contribute to enhanced nutrient uptake and assimilation, improving plant performance. One primary mode of action involves the stimulation of root development and architecture. BS, such as humic substances and microbial inoculants, promote root elongation and branching, expanding the root surface area and facilitating greater nutrient absorption (Canellas et al., 2002). This enhanced root system also enables plants to explore a larger volume of soil, accessing nutrients and water that might otherwise remain untapped (Cansu and Erdal, 2018). Furthermore, BS play a pivotal role in enhancing nutrient uptake efficiency. Seaweed extracts, for instance, contain natural growth-promoting compounds that can chelate essential nutrients and micronutrients, rendering them more available for plant uptake (Khan et al., 2009a). BS also influence ion exchange processes at the root-soil interface, improving the plant's ability to acquire nutrients from the soil solution. Amino acid-based BS have been shown to enhance the uptake of both macronutrients and micronutrients through their interactions with transport proteins and membrane channels (Bulgari et al., 2019).

Beyond nutrient uptake, BS impact plant physiology by modulating hormonal signaling pathways. Seaweed extracts and microbial inoculants can induce the synthesis and release of PGRs, such as auxins, cytokinins, and gibberellins, which regulate various developmental processes (Cansu and Erdal, 2018). This hormone-mediated response influences cell division, elongation, and differentiation, enhancing overall plant growth. Additionally, BS can trigger the production of secondary metabolites, including those involved in plant defense against stressors, through the activation of jasmonic acid and ethylene signaling pathways (Divekar et al., 2022; Pérez-Llorca et al., 2023; Vargas-Hernandez et al., 2017). At the molecular level, BS can influence gene expression patterns that underlie plant growth and stress responses. According to Fadiji et al. (2022), through symbiotic associations, microbial BS can enhance nutrient uptake, stress tolerance, and hormone signaling in plants. Humic substances have been reported to affect gene expression related to nutrient transporters and enzymes involved in nutrient assimilation (Trevisan et al., 2010). These changes in gene expression ultimately lead to alterations in metabolic pathways that contribute to improved plant growth and resilience.

8.3.2.2 MODULATING HORMONAL BALANCE AND SIGNALING PATHWAYS

BS affect plant growth through a range of mechanisms encompassing plant physiology, biochemistry, and molecular biology. One pivotal mode of action involves the modulation of hormonal balance and signaling pathways. BS, such as seaweed extracts and humic substances, have been reported to induce the synthesis and release of phytohormones like auxins, cytokinins, and gibberellins, which regulate critical developmental processes (Ali et al., 2021a; Shahrajabian et al., 2021). BS can amplify cell division, elongation, and differentiation by interacting with hormone receptors and signal transduction pathways, thereby contributing to enhanced plant growth and overall vigor (Baltazar et al., 2021). Furthermore, BS play a pivotal role in ameliorating plant stress responses. These stress-alleviating effects are often mediated by the activation of abscisic acid (ABA), an essential stress hormone that modulates various physiological and biochemical reactions. BS have been shown to enhance ABA synthesis and signaling, resulting in improved water use efficiency and drought tolerance (Bhupenchandra et al., 2022; Rouphael and Colla, 2020). The ability of BS to fine-tune hormonal balance equips plants with enhanced resilience against abiotic stressors, ultimately contributing to sustained growth under adverse conditions.

At the molecular level, BS influence gene expression patterns associated with defense and stress responses. Seaweed extracts, for instance, can upregulate genes associated with nutrient uptake, hormonal regulation, and stress tolerance (Deolu-Ajayi et al., 2022). This molecular reprogramming leads to heightened metabolic activities, increased synthesis of antioxidants, and enhanced defense mechanisms. Additionally, BS can mediate epigenetic modifications, altering the accessibility of chromatin and influencing gene expression profiles that contribute to growth promotion and stress adaptation (Yakhin et al., 2017). Moreover, BS interact with plant-microbe symbioses, particularly mycorrhizal associations and PGPR, further augmenting their impact on plant physiology. BS can enhance the colonization of roots by beneficial microorganisms, thereby enhancing nutrient acquisition, stress tolerance, and growth-promoting activities (Drobek et al., 2019). This interaction involves intricate crosstalk between plant hormonal signaling and microbial-derived compounds, culminating in synergistic effects on plant growth and health.

8.3.2.3 ACTIVATING DEFENSE RESPONSES AND ANTIOXIDANT SYSTEMS

BS influence plant growth and development through intricate mechanisms that span physiological, biochemical, and molecular processes. A pivotal mode of action involves their capacity to activate defense responses and antioxidant systems within plants. BS can trigger the production of secondary metabolites like phenolics, flavonoids, and phytoalexins, which help plants defend against environmental stressors and pathogens (Divekar et al., 2022; Ertani et al., 2018; Pereira et al., 2021). These compounds act as potent antioxidants, scavenging ROS and minimizing oxidative damage caused by stress. BS help maintain cellular integrity and promote plant health by enhancing antioxidant enzyme activity, including superoxide dismutase (SOD), catalase, and peroxidase (Hasanuzzaman et al., 2021).

Furthermore, BS modulate plant defense through the priming of systemic acquired resistance (SAR) and induced systemic resistance (ISR) mechanisms (Choudhary et al., 2007; Yu et al., 2022). SAR involves the activation of immune responses in systemic tissues following localized pathogen attack, while ISR is mediated by beneficial microorganisms that elicit defense pathways against diverse stressors (Conrath et al., 2006). BS, such as microbial inoculants and humic substances, have been shown to induce these resistance mechanisms, leading to enhanced resistance against pathogens and pests. These biostimulant-induced defenses involve the jasmonic acid (JA) and salicylic acid (SA) signaling pathways, culminating in the transcriptional activation of pathogenesis-related (PR) genes (Tamaoki et al., 2013; Zhang et al., 2010). At the molecular level, BS influence gene expression patterns related to defense and stress responses. They can modulate the expression of genes encoding PR proteins, chitinases, glucanases, and other defense-related molecules (Yakhin et al., 2017). This molecular reprogramming primes plants to mount a faster and more robust defense response upon subsequent stress exposure. Moreover, BS have been reported to induce epigenetic modifications that contribute to establishing a primed state, enhancing the capacity of plants to respond effectively to stressors (Lephatsi et al., 2021).

8.3.3 POTENTIAL OF BIOSTIMULANTS TO INFLUENCE SECONDARY METABOLITE PRODUCTION

The potential of BS to influence secondary metabolite production in plants presents an intriguing avenue for enhancing the synthesis of valuable compounds with various pharmaceutical, nutritional, and ecological applications. Their impact on terpenoid pathways is fascinating, which play a central role in producing diverse secondary metabolites, including artemisinin in *Artemisia annua* L. BS have been shown to exert specific effects on these pathways, thereby influencing the biosynthesis of terpenoids and related compounds. For instance, humic substances have been demonstrated to enhance gene expression in terpenoid biosynthesis. In one study, the humic acid treatment led to increased transcript levels of key enzymes in the terpenoid pathway, including those associated with the mevalonate and non-mevalonate pathways, which are critical for artemisinin production (Trevisan et al., 2010). This upregulation of gene expression suggests that BS can promote the availability of precursors required for terpenoid synthesis, potentially leading to elevated levels of artemisinin and related terpenes. Furthermore, BS have been reported to influence the activity of enzymes directly involved in terpenoid biosynthesis (Cotas et al., 2020). This enzymatic stimulation may produce higher flux through the terpenoid pathway, promoting increased artemisinin production (Cotas et al., 2020).

Moreover, BS have been linked to the modulation of hormonal signaling pathways that intersect with terpenoid biosynthesis. Phytohormones, including jasmonic acid (JA) and gibberellins (GAs), regulate secondary metabolite production. BS can influence the balance of these hormones, potentially leading to altered terpenoid profiles. For instance, microbial BS have been shown to induce JA synthesis and signaling, activating terpenoid biosynthesis in response to stressors (Bukhat et al., 2020; Kaushal et al., 2023). This hormone-mediated regulatory network offers a potential mechanism for BS to impact artemisinin production.

8.4 UTILIZATION OF BIOSTIMULANTS FOR ENHANCING ARTEMISININ PRODUCTION

Using BS to enhance artemisinin production in *Artemisia annua* L. has garnered significant attention across diverse cultivation environments *in vitro*, greenhouse, and field conditions. *In vitro,* applications of BS have exhibited promising potential for elevating artemisinin accumulation, as exemplified by studies highlighting the efficacy of seaweed extracts in stimulating artemisinin content through the activation of secondary metabolism pathways (Ghatas et al., 2021; Pu et al., 2009). The beneficial impact of BS on artemisinin synthesis has been further substantiated through greenhouse studies, such as the augmentation of artemisinin yield and growth parameters of *A. annua* by humic substances (Gupta et al., 2016; Jha et al., 2011). Similarly, field trials have revealed the promise of BS in enhancing artemisinin production under realistic agricultural conditions, wherein microbial BS contributed to increased artemisinin levels, potentially due to improved soil health and nutrient uptake (Aftab et al., 2016; Kapoor et al., 2007; Rouphael and Colla, 2020; Zhai et al., 2019). Moreover, these BS have been demonstrated to influence plant physiology, nutrient utilization, and stress responses, collectively contributing to enhanced artemisinin content (Rouphael and Colla, 2018). The interactions between BS and environmental factors have been a subject of investigation, particularly under varying climatic conditions. The adaptability of biostimulant applications to influence artemisinin production in response to different environmental stressors has been observed, indicating the potential to enhance water use efficiency and mitigate drought stress, ultimately enhancing artemisinin synthesis in regions with limited water availability (Dangash et al., 2014; Naeem et al., 2017; Soni and Abdin, 2017; Vasconcelos, 2020). Notably, the intricate interplay between BS and their effectiveness in distinct environmental contexts highlights the need for further research to understand their mechanisms and impacts comprehensively.

The intricacy of artemisinin production in *A. annua* arises from factors such as plant genotype, environmental stress, and nutrient availability. In particular, the influence of BS on artemisinin biosynthesis, dependent on diverse interactions, underscores the complexity of this process. Different genotypes of *A. annua* exhibit specific responses to BS due to their distinct capacities and mechanisms for artemisinin biosynthesis. Selecting genotypes responsive to BS is paramount for optimizing artemisinin yield (Padalia et al., 2011; Sajwan et al., 2020; Shen et al., 2018). Environmental stress conditions, which can induce oxidative stress and trigger artemisinin synthesis, further accentuate the role of BS in enhancing stress tolerance and promoting artemisinin production (Ghatas et al., 2021; Hasanuzzaman et al., 2021; Ma et al., 2022). Additionally, elicitors, compounds that stimulate plant defense responses, synergize with BS to trigger enhanced secondary metabolite synthesis, exemplified by jasmonic acid's capacity to amplify artemisinin content in conjunction with humic acids or seaweed extracts (Caretto et al., 2011; Shen et al., 2016; Xiang et al., 2015). BS also mediate nutrient dynamics and uptake, optimizing nutrient availability, assimilation, and translocation, thereby impacting artemisinin production (Boavida and Wetzel, 1998; Curie and Briat, 2003; Davies et al., 2009). In conclusion, BS exhibit multifaceted effects on artemisinin biosynthesis in *A. annua* through intricate interactions with various factors. Understanding these interactions is pivotal for tailored experiments and mechanistic insights to maximize artemisinin production, with the potential to address global demand through innovative biotechnological strategies (Ikram and Simonsen, 2017). The dynamic interplay between BS, plant physiology, soil conditions, and growth-promoting agents underscores the intricate nature of these interactions, necessitating in-depth exploration for optimal outcomes.

8.5 MOLECULAR AND PHYSIOLOGICAL INSIGHTS

The application of BS holds significant promise in enhancing artemisinin production in *Artemisia annua* L., with molecular and physiological insights shedding light on the mechanisms underlying this enhancement. BS exert notable effects on gene expression and metabolic pathways, providing deeper insight into artemisinin synthesis. Studies indicate that BS can influence the expression of critical genes tied to secondary metabolite biosynthesis, as exemplified by the upregulation of genes associated with the mevalonate and non-mevalonate pathways upon seaweed extract application (Dubey et al., 2003; Khan et al., 2009a). This transcriptional modulation signifies the potential for BS to impact artemisinin precursor production directly. Furthermore, BS extend their effects to broader metabolic pathways encompassing phytohormone regulation. Investigations reveal their capacity to influence hormonal signaling pathways, potentially contributing to augmented artemisinin biosynthesis (Rouphael and Colla, 2018). This may enhance artemisinin production through hormonal signalling (Ghanati and Bakhtiarian, 2013; Shahzad et al., 2015; Wang et al., 2010).

BS improve the ability of plants to tolerate stress, which in turn affects the synthesis of artemisinin. Microbial inoculants, for example, promote stress-responsive gene expression in *A. annua*, often linked to secondary metabolite biosynthesis, indicating a potential diversion of resources toward secondary metabolism (Awasthi et al., 2011; Chaudhary et al., 2008). Moreover, nutrient uptake and assimilation are pivotal in secondary metabolite synthesis. BS influence nutrient acquisition, potentially impacting artemisinin production. Humic substances, for instance, improve nutrient uptake efficiency, highlighting the multifaceted nature of biostimulant effects (Jha et al., 2011). The use of BS regulates hormonal signaling networks, such as jasmonic acid (JA), salicylic acid (SA), and ABA, leading to a significant increase in artemisinin production. The involvement of JA signaling is well documented, wherein the application of seaweed extracts increases endogenous JA levels, which is known to play a pivotal role in secondary metabolism regulation (Rouphael and Colla, 2018). The implications of SA and ABA in biostimulant-mediated artemisinin production have also been observed. The interaction of these hormonal pathways contributes to a complex regulatory framework that orchestrates artemisinin biosynthesis in response to biostimulant application. This intricate hormonal interplay underscores the potential of BS to enhance the production of valuable secondary metabolites such as artemisinin.

BS induce physiological changes and stress responses in plants, culminating in enhanced artemisinin production in *A. annua*. These alterations, encompassing improvements in photosynthesis, antioxidant defense mechanisms, and osmotic regulation, improve the plant's ability to cope with stressors and optimize resource allocation towards secondary metabolite biosynthesis (Khan and Shri, 2020; Takshak and Agrawal, 2015). Enhanced photosynthesis, marked by elevated chlorophyll content and net photosynthetic rates, has been observed in biostimulant-treated plants, mitigating stress-induced reductions in photosynthesis and potentially bolstering carbon availability for artemisinin biosynthesis. BS also stimulate antioxidant defense systems, alleviating oxidative stress and contributing to improved artemisinin production. Osmotic regulation, crucial for plant adaptation to changing environmental conditions, is influenced by biostimulant application. By enhancing osmoprotectant synthesis and maintaining turgor pressure, BS likely support artemisinin biosynthesis under stress conditions (Jiménez-Arias et al., 2021; Singh et al., 2015). In conclusion, BS exert diverse effects on gene expression, metabolic pathways, and physiological responses in *A. annua*, underscoring their potential to enhance artemisinin production through intricate mechanisms. These insights offer valuable perspectives for utilizing BS as strategic tools to optimize artemisinin yield and quality, contributing to the sustainable production of this essential antimalarial compound.

8.6 SYNERGISTIC APPROACHES: BIOSTIMULANTS AND ENVIRONMENTAL FACTORS

Synergistic approaches involving BS and environmental factors are gaining attention in various fields, particularly in agriculture and environmental science. BS are substances or microorganisms applied to plants or soils to enhance nutrient uptake, improve stress tolerance, and promote growth and development. Environmental factors encompass a wide range of variables such as temperature, humidity, light, soil quality, and water availability, among others, that influence the growth and performance of organisms within ecosystems (Asif et al., 2023; Rouphael and Colla, 2018). Combining BS with considerations of environmental factors can lead to enhanced efficacy and sustainability in agricultural practices, ecosystem management, and environmental restoration efforts (Rouphael and Colla, 2018).

8.6.1 INTERACTIONS BETWEEN BIOSTIMULANTS AND ABIOTIC STRESSORS

The pursuit of synergistic approaches that harness the interactions between BS and environmental factors holds immense potential for advancing artemisinin production in *Artemisia annua* L. BS play a pivotal role as facilitators in enabling plant adaptation to abiotic stressors, including light, temperature, and nutrient availability. By modulating physiological responses, BS effectively empower plants to manage stress-induced challenges, thus optimizing the synthesis of secondary metabolites, including artemisinin. The role of light as a crucial driver of photosynthesis and plant growth is particularly noteworthy when considering its synergy with biostimulant applications. Demonstrated by their ability to enhance light capture and utilization, BS promote leaf area expansion and increased chlorophyll content. These effects are exemplified by studies showcasing that seaweed extracts increase leaf area and chlorophyll levels in *A. annua*, potentially bolstering the plant's photosynthetic capacity (Blunden et al., 1996; Ghatas et al., 2021). This synergy could amplify the energy available for artemisinin biosynthesis, contributing to enhanced yield.

Temperature, another critical abiotic factor, can engage in synergistic interactions with BS to influence artemisinin production. BS have been reported to enhance temperature stress tolerance by bolstering antioxidant defenses and osmotic regulation. Notable studies have observed that humic acid application ameliorates oxidative stress induced by temperature fluctuations (Abdellatif et al., 2017; Chakraborty and Pradhan, 2011). Combining the stress-mitigating effects of BS with temperature responses can alleviate constraints on artemisinin biosynthesis imposed by suboptimal temperatures. Nutrient availability, a pivotal determinant of plant growth and metabolite accumulation, can synergistically interact with BS to influence artemisinin production. BS can help plants absorb nutrients better, reducing nutrient limitations. Further studies showcase that lignin-based BS have increased nutrient uptake efficiency, hinting at improved nutrient availability for artemisinin biosynthesis (Cesarino et al., 2012; Fatriasari et al., 2020). This intricate interplay between BS and nutrient dynamics underscores their potential to optimize artemisinin yield collaboratively.

8.6.2 IMPACT OF BIOSTIMULANTS ON PLANT-MICROBE INTERACTIONS AND NUTRIENT UPTAKE

The symbiotic interplay between BS and environmental factors extends its scope to encompass plant-microbe interactions and nutrient uptake, establishing a holistic approach to boosting artemisinin production in *Artemisia annua* L. BS are acknowledged for fostering favorable associations with microorganisms, fostering a balanced rhizosphere microbiome that not only supports nutrient acquisition but also creates an environment conducive to artemisinin biosynthesis. The role of BS in enhancing the abundance and diversity of beneficial microorganisms within the rhizosphere is of significance. Studies underline that microbial inoculants can amplify rhizosphere microbial diversity, contributing to enhanced nutrient cycling and availability (Shi et al., 2022; Trabelsi and Mhamdi, 2013). This elevated microbial activity can lead to superior nutrient mineralization, making vital elements more accessible for incorporation into artemisinin biosynthesis pathways.

BS facilitate symbiotic relationships in plants with mycorrhizal fungi, increasing nutrient absorption. Notable studies indicate that seaweed extracts positively influence mycorrhizal colonization, potentially augmenting nutrient uptake efficiency (Rouphael and Colla, 2018). This symbiotic partnership aligns with our intention to enhance nutrient availability for artemisinin synthesis. Additionally, BS can indirectly influence nutrient uptake and allocation by modulating plant responses to stressors. BS have been shown to enhance root architecture and exudation, creating a fertile rhizosphere environment for nutrient uptake. Studies highlight how humic acid application increases root biomass and plant exudation, improving nutrient mobilization (Canellas et al., 2019; Nikbakht et al., 2008). This alteration in root morphology and exudation patterns may contribute to heightened nutrient availability for artemisinin biosynthesis. The intricate synergy binding BS, plant-microbe interactions, and nutrient uptake unveil a multidimensional strategy for bolstering artemisinin production. By nurturing a diverse and beneficial rhizosphere microbiome, optimizing nutrient cycling, and influencing plant responses, BS create an environment amenable to artemisinin biosynthesis. These holistic approaches align with sustainable and efficient secondary metabolite production objectives, accentuating the potential of BS as tools to harness the intricate interplay among plants, microbes, and nutrients.

8.6.3 STRATEGIES FOR OPTIMIZING BIOSTIMULANT APPLICATION UNDER VARYING ENVIRONMENTAL CONDITIONS

Optimal utilization of BS under variable environmental conditions emerges as a strategic imperative for amplifying artemisinin production in *Artemisia annua* L. To fully utilize BS, the precise application of factors such as light, temperature, and nutrients is crucial. Several strategies have emerged to achieve this synergy, offering insights into the judicious deployment of BS for maximizing artemisinin yield. One approach involves synchronizing biostimulant applications with favorable environmental conditions. Environmental cues like photoperiod and temperature can influence plant physiology and metabolism. Plants can optimally leverage their advantages by administering BS during heightened metabolic activity. A study illustrating this concept showcased that the exogenous application of BS during suboptimal temperature periods enhanced stress tolerance in tomato plants (Zhang et al., 2013). This strategic timing may be applied to *A. annua* cultivation, ensuring BS are used when plants are most receptive to their effects.

An alternative strategy involves tailoring biostimulant dosages in response to nutrient availability. Since BS rely on a foundational supply of essential elements for optimal function, supplementing biostimulant application with targeted nutrient management offers a synergistic approach. Other studies provided evidence of this concept by demonstrating that lignin-based BS improved nutrient uptake efficiency in plants (Cesarino et al., 2012; Fatriasari et al., 2020). Implementing this approach in *A. annua* cultivation could ensure that biostimulant-enhanced nutrient uptake supports artemisinin biosynthesis. Moreover, integrating BS with precision agriculture techniques presents a forward-looking strategy for optimizing applications under varying conditions. Remote sensing technologies and data-driven methodologies enable real-time plant health and environmental parameter monitoring. By harnessing this information, the application of BS can be dynamically adjusted to align with plant-specific needs and current environmental conditions. This approach offers the potential to fine-tune biostimulant applications for achieving optimal artemisinin production while minimizing waste and environmental impact.

8.7 FUTURE PERSPECTIVES AND CHALLENGES

BS have shown promising potential for enhancing the production of artemisinin, a crucial compound used in the treatment of malaria. Artemisinin is primarily derived from *Artemisia annua*, commonly known as sweet wormwood. The compound's production is influenced by various factors, including environmental conditions, soil quality, and plant health. BS offer a sustainable and eco-friendly approach to optimizing these factors, thereby enhancing artemisinin yield. Here are some future prospects for the use of BS in artemisinin production:

8.7.1 HURDLES OF BIOSTIMULANT USE FOR ARTEMISININ ENHANCEMENT

While BS hold substantial promise for enhancing artemisinin production, their research and application in this context encounter several challenges and limitations that warrant careful consideration. A primary challenge is the lack of standardized biostimulant definitions, classifications, and evaluation methods. This lack of a universally accepted framework limits clear comparisons among studies and complicates efforts to identify optimal biostimulant formulations and application strategies (Yakhin et al., 2017). As noted by a study, this need for more standardization limits the ability to draw robust conclusions and hampers the translation of research findings into practical applications (Rouphael and Colla, 2018). Regulation and quality control emerge as crucial concerns in biostimulant research and application. BS encompass various substances, including plant extracts, microbial products, and organic materials. The inconsistent regulatory standards pose challenges in ensuring product quality, efficacy, and safety (Caradonia et al., 2019; La Torre et al., 2016). The absence of rigorous quality control measures can result in variations in biostimulant effectiveness, leading to inconsistent outcomes in artemisinin production. Addressing this challenge necessitates collaborative efforts involving researchers, regulatory bodies, and industry stakeholders to establish comprehensive quality assurance protocols.

Another limitation pertains to the need for a deeper comprehension of BS' precise modes of action. While the broad benefits of BS are acknowledged, the intricate molecular and physiological mechanisms through which they exert their effects remain largely enigmatic. Deciphering these mechanisms would facilitate the design of more targeted and effective biostimulant formulations. Prior studies underscored the significance of elucidating the signaling pathways and molecular interactions that mediate biostimulant-induced enhancements in artemisinin production (Khan and Shri, 2020; Khan et al., 2009b). Furthermore, the application of BS in artemisinin production can be influenced by the intricate interplay between BS and other factors, such as plant genotype and environmental conditions. These interactions introduce an additional layer of variability that must be accounted for in experimental designs and practical applications. The challenge of unravelling and harnessing these interactions to consistently achieve desired outcomes underscores the necessity for multidisciplinary research approaches that integrate plant physiology, molecular biology, and environmental science.

8.7.2 BRIDGING THE GAPS AND BOOSTING THE POTENTIAL OF BIOSTIMULANTS FOR ARTEMISININ PRODUCTION

The endeavor to advance biostimulant utilization for optimizing artemisinin production reveals critical knowledge gaps and research needs that warrant focused attention. A primary knowledge gap pertains to the identification and screening of novel BS. Despite the promise exhibited by several biostimulant formulations, the diverse spectrum of potential sources, including various plant-derived extracts and microbial products, suggests that untapped biostimulant candidates may exist. Comprehensive screening efforts are required to unearth new, effective BS tailored to enhance artemisinin production. A systematic approach to biostimulant discovery, as highlighted by a study, has the potential to yield innovative solutions (Rouphael and Colla, 2018). Another significant research need is to understand the molecular targets and mechanisms through which BS enhance artemisinin production. Investigating signaling pathways and molecular interactions underpinning biostimulant-induced responses is pivotal for designing targeted and effective strategies. Delving deeper into the molecular underpinnings of biostimulant-mediated improvements can provide insights into key regulatory nodes, enabling the rational design of biostimulant formulations tailored for artemisinin biosynthesis (Baltazar et al., 2021). A previous study underscores the necessity of elucidating these mechanisms to guide the development of precise biostimulant interventions (Wang et al., 2003).

Evaluating the cost-effectiveness and sustainability of biostimulant-assisted artemisinin production represents a practical concern that requires comprehensive investigation. While BS hold the potential to enhance yield, their economic feasibility and environmental impact warrant meticulous assessment. Rigorous studies are essential to quantifying the cost-benefit ratio of biostimulant application, considering factors such as production expenses, improved yield, and environmental benefits (Calvo et al., 2014; Drobek et al., 2019). These studies advocate for comprehensive evaluations of biostimulant efficacy and sustainability to guide practical implementation. Challenges and opportunities related to scaling up biostimulant-assisted artemisinin production also require exploration. Translating laboratory findings to field applications involves complexities associated with agronomic practices, large-scale production, and commercial viability. Strategies for optimizing biostimulant application under real-world conditions must be developed while simultaneously addressing challenges such as product consistency, distribution, and compatibility with existing agricultural practices (Jindo et al., 2022; Le Mire et al., 2016; Malik et al., 2021; Moreno-Hernández et al., 2020). This necessitates interdisciplinary research that integrates agronomy, economics, and technology transfer. These studies emphasize the importance of bridging the gap between research and practical application.

8.7.3 BIOSTIMULANTS FOR ARTEMISININ PRODUCTION: FUTURE PROSPECTS AND PRACTICAL IMPLICATIONS

Future research on BS in artemisinin production should prioritize interdisciplinary approaches to address these bioactive substances' complex challenges and opportunities. Firstly, researchers should intensify efforts to screen and identify novel BS from diverse sources, including unexplored plant extracts and microbial communities, through systematic screening and characterization processes (Kocira et al., 2021; Povero et al., 2016; Xu and Geelen, 2018). This could expand the biostimulant toolbox and provide tailored solutions for enhancing artemisinin yield. Elucidating the molecular mechanisms underpinning biostimulant-mediated improvements in artemisinin production is crucial. Advanced molecular and omics techniques, such as transcriptomics and metabolomics, can uncover intricate signaling networks and regulatory pathways, shedding light on critical targets for intervention (Huang and Fraenkel, 2009; Li et al., 2018; Patti et al., 2012). Combining this knowledge with synthetic biology approaches could enable the rational design of custom biostimulant formulations optimized for artemisinin biosynthesis.

The practical application of BS in artemisinin-rich plant cultivation warrants exploration under varying environmental conditions. Research should focus on developing site-specific biostimulant application strategies, considering local climate, soil conditions, and agricultural practices. Collaborations between agronomists, biologists, and farmers are essential to successfully integrating BS into existing cultivation practices. The potential impact of BS extends beyond artemisinin production, with significant implications for both the pharmaceutical and agricultural sectors. The use of BS in artemisinin-rich plant cultivation aligns with the global push for sustainable and environmentally friendly farming practices (Ali et al., 2021b; Rajabi et al., 2020; Rouphael and Colla, 2018; Zulfiqar et al., 2020). BS can contribute to reduced chemical input usage, minimized environmental impact, and enhanced crop resilience, reflecting the broader goals of sustainable agriculture. Furthermore, the pharmaceutical sector could benefit from biostimulant-enhanced artemisinin production. The increased availability of artemisinin, a crucial antimalarial compound, can bolster global efforts to combat malaria. As the demand for artemisinin-derived pharmaceuticals grows, BS offer a valuable tool to ensure a reliable supply chain for this life-saving medication.

8.8 CONCLUSION

In conclusion, this review has examined the multifaceted role of biostimulant use in artemisinin production, shedding light on its immense potential to revolutionize agricultural practices and global health efforts. Through a thorough exploration of current knowledge and future perspectives, we have underscored the remarkable ability of BS to enhance artemisinin yield and quality by optimizing plant growth, nutrient uptake, and secondary metabolite biosynthesis. The intricate modes of action, encompassing hormonal modulation, defense activation, and microbial interactions, offer insights into the underlying mechanisms that drive this enhancement. The significance and implications of biostimulant use in artemisinin production extend far beyond the field's confines. In the face of climate change and its detrimental impacts on agricultural productivity, BS emerge as powerful tools to enhance plant resilience and adaptation. By promoting stress tolerance and fostering efficient resource utilization, BS equip artemisinin-rich plants to withstand environmental challenges, ultimately securing a more stable supply of this vital antimalarial compound. Furthermore, using BS to produce artemisinin has far-reaching benefits for human health and well-being. The increased availability of artemisinin contributes to the global fight against malaria, a disease that continues to exact a devastating toll on human lives. BS amplify this impact by ensuring a steady and sustainable source of artemisinin, thereby bolstering the world's arsenal against malaria and improving the health and well-being of vulnerable populations.

As we stand at the intersection of agricultural innovation, global health, and sustainable development, the significance of biostimulant use in artemisinin production is substantial. This review emphasizes the critical role of ongoing research and innovation in biostimulant technologies. Although significant progress has been achieved, there is still a pressing need to unravel the complexities of biostimulant mechanisms, standardize their application, and explore unexplored sources for new BS. Collaborative efforts among researchers, policymakers, and industry stakeholders are essential to driving the translation of research findings into practical applications that can positively impact agriculture, health, and development on a global scale. In closing, this review serves as a call for the scientific community to seize the untapped potential of biostimulant-assisted artemisinin production. By harnessing the synergistic power of biostimulants, we can fortify our agricultural systems, safeguard human health, and contribute to achieving global development goals. As we navigate the complex challenges of our time, the path forward is illuminated by the promise of BS, beckoning us to continue our journey of exploration, discovery, and innovation.

KEYWORDS

- **artemisinin-based combination therapies**
- **biostimulants**
- **induced systemic resistance**
- **intermittent preventive treatment**
- **seasonal malaria chemoprevention**
- **systemic acquired resistance**

REFERENCES

Abdellatif, I. M. Y., Abdel-Ati, Y. Y., Abdel-Mageed, Y. T., & Hassan, M. A. M. M. (2017). Effect of humic acid on growth and productivity of tomato plants under heat stress. *Journal of Horticultural Research, 25*(2), 59–66. https://doi.org/10.1515/johr-2017-0022 (accessed on 6th November 2025).

Aftab, T., Naeem, M., Idrees, M., Khan, M. M. A., Moinuddin, & Varshney, L. (2016). Simultaneous use of irradiated sodium alginate and nitrogen and phosphorus fertilizers enhance growth, biomass and artemisinin biosynthesis in *Artemisia annua* L. *Journal of Applied Research on Medicinal and Aromatic Plants, 3*(4), 186–194. https://doi.org/10.1016/j.jarmap.2016.05.001 (accessed on 6th November 2025).

Ali, O., Ramsubhag, A., & Jayaraman, J. (2021). Bio stimulant properties of seaweed extracts in plants: Implications towards sustainable crop production. *Plants, 10*(3), 3. https://doi.org/10.3390/plants10030531 (accessed on 6th November 2025).

Asif, A., Ali, M., Qadir, M., Karthikeyan, R., Singh, Z., Khangura, R., Di Gioia, F., & Ahmed, Z. F. R. (2023). Enhancing crop resilience by harnessing the synergistic effects of bio stimulants against abiotic stress. *Frontiers in Plant Science, 14*, 1276117. https://doi.org/10.3389/fpls.2023.1276117 (accessed on 6th November 2025).

Awasthi, A., Bharti, N., Nair, P., Singh, R., Shukla, A. K., Gupta, M. M., Darokar, M. P., & Kalra, A. (2011). Synergistic effect of *Glomus mosseae* and nitrogen-fixing *Bacillus subtilis* strain Daz26 on artemisinin content in *Artemisia annua* L. *Applied Soil Ecology, 49*, 125–130. https://doi.org/10.1016/j.apsoil.2011.06.005 (accessed on 6th November 2025).

Baltazar, M., Correia, S., Guinan, K. J., Sujeeth, N., Bragança, R., & Gonçalves, B. (2021). Recent advances in the molecular effects of bio stimulants in plants: An overview. *Biomolecules, 11*(8), 8. https://doi.org/10.3390/biom11081096 (accessed on 6th November 2025).

Bhattacharyya, P. N., & Jha, D. K. (2012). Plant growth-promoting rhizobacteria (PGPR): Emergence in agriculture. *World Journal of Microbiology and Biotechnology, 28*(4), 1327–1350. https://doi.org/10.1007/s11274-011-0979-9 (accessed on 6th November 2025).

Bhupenchandra, I., Chongtham, S. K., Devi, E. L., R. R., Choudhary, A. K., Salam, M. D., Sahoo, M. R., et al. (2022). Role of bio stimulants in mitigating the effects of climate change on crop performance. *Frontiers in Plant Science, 13*, 967665. https://doi.org/10.3389/fpls.2022.967665 (accessed on 6th November 2025).

Blunden, G., Jenkins, T., & Liu, Y. W. (1996). Enhanced leaf chlorophyll levels in plants treated with seaweed extract. *Journal of Applied Phycology, 8*(6), 535–543. https://doi.org/10.1007/BF02186333 (accessed on 6th November 2025).

Boavida, M. J., & Wetzel, R. G. (1998). Inhibition of phosphatase activity by dissolved humic substances and hydrolytic reactivation by natural ultraviolet light. *Freshwater Biology, 40*(2), 285–293. https://doi.org/10.1046/j.1365-2427.1998.00349.x (accessed on 6th November 2025).

Bouwmeester, H. J., Wallaart, T. E., Janssen, M. H., van Loo, B., Jansen, B. J., Posthumus, M. A., Schmidt, C. O., De Kraker, J. W., König, W. A., & Franssen, M. C. (1999).

Amorpha-4,11-diene synthase catalysis the first probable step in artemisinin biosynthesis. *Phytochemistry, 52*(5), 843–854.

Bukhat, S., Imran, A., Javaid, S., Shahid, M., Majeed, A., & Naqqash, T. (2020). Communication of plants with microbial world: Exploring the regulatory networks for PGPR mediated defense signaling. *Microbiological Research, 238*, 126486. https://doi.org/10.1016/j.micres.2020.126486 (accessed on 6th November 2025).

Bulgari, R., Franzoni, G., & Ferrante, A. (2019). Bio stimulants application in horticultural crops under abiotic stress conditions. *Agronomy, 9*(6), 6. https://doi.org/10.3390/agronomy9060306 (accessed on 6th November 2025).

Calvo, P., Nelson, L., & Kloepper, J. W. (2014). Agricultural uses of plant bio stimulants. *Plant and Soil, 383*(1), 3–41. https://doi.org/10.1007/s11104-014-2131-8 (accessed on 6th November 2025).

Canellas, L. P., Olivares, F. L., Canellas, N. O. A., Mazzei, P., & Piccolo, A. (2019). Humic acids increase the maize seedlings exudation yield. *Chemical and Biological Technologies in Agriculture, 6*(1), 3. https://doi.org/10.1186/s40538-018-0139-7 (accessed on 6th November 2025).

Canellas, L. P., Olivares, F. L., Okorokova-Façanha, A. L., & Façanha, A. R. (2002). Humic acids isolated from earthworm compost enhance root elongation, lateral root emergence, and plasma membrane H+-ATPase activity in maize roots. *Plant Physiology, 130*(4), 1951–1957.

Cansu, M., & Erdal, İ. (2018). Effect of humic substance applications on mineral nutrition and yield of Granny Smith and Jersey Mac apple variety. *Journal of Agricultural Sciences, 24*(2), 162–169.

Caradonia, F., Battaglia, V., Righi, L., Pascali, G., & La Torre, A. (2019). Plant bio stimulant regulatory framework: Prospects in Europe and current situation at international level. *Journal of Plant Growth Regulation, 38*(2), 438–448. https://doi.org/10.1007/s00344-018-9853-4 (accessed on 6th November 2025).

Caretto, S., Quarta, A., Durante, M., Nisi, R., De Paolis, A., Blando, F., & Mita, G. (2011). Methyl jasmonate and miconazole differently affect artemisinin production and gene expression in *Artemisia annua* suspension cultures. *Plant Biology, 13*(1), 51–58. https://doi.org/10.1111/j.1438-8677.2009.00306.x (accessed on 6th November 2025).

Castiglione, A. M., Mannino, G., Contartese, V., Bertea, C. M., & Ertani, A. (2021). Microbial bio stimulants as response to modern agriculture needs: Composition, role and application of these innovative products. *Plants (Basel), 10*(8), 1533. https://doi.org/10.3390/plants10081533 (accessed on 6th November 2025).

Cesarino, I., Araújo, P., Domingues Júnior, A. P., & Mazzafera, P. (2012). An overview of lignin metabolism and its effect on biomass recalcitrance. *Brazilian Journal of Botany, 35*, 303–311. https://doi.org/10.1590/S0100-84042012000400003 (accessed on 6th November 2025).

Chakraborty, U., & Pradhan, D. (2011). High temperature-induced oxidative stress in *Lens culinaris*, role of antioxidants and amelioration of stress by chemical pre-treatments. *Journal of Plant Interactions, 6*(1), 43–52. https://doi.org/10.1080/17429145.2010.513484 (accessed on 6th November 2025).

Chaudhary, V., Kapoor, R., & Bhatnagar, A. K. (2008). Effectiveness of two arbuscular mycorrhizal fungi on concentrations of essential oil and artemisinin in three accessions of *Artemisia annua* L. *Applied Soil Ecology, 40*(1), 174–181. https://doi.org/10.1016/j.apsoil.2008.04.003 (accessed on 6th November 2025).

Chen, T. T., Yao, X. H., Liu, H., Li, Y. P., Qin, W., Yan, X., Wang, X. Y., Peng, B. W., Zhang, Y. J., Shao, J., Hu, X. Y., Miao, Q., Fu, X. Q., Wang, Y. L., Li, L., & Tang, K. X. (2022). MADS-box gene AaSEP4 promotes artemisinin biosynthesis in *Artemisia annua*. *Frontiers in Plant Science, 13*, 982317. https://doi.org/10.3389/fpls.2022.982317 (accessed on 6th November 2025).

Chen, T., Li, Y., Xie, L., Hao, X., Liu, H., Qin, W., Wang, C., Yan, X., Wu-Zhang, K., Yao, X., Peng, B., Zhang, Y., Fu, X., Li, L., & Tang, K. (2021). AaWRKY17, a positive regulator of artemisinin biosynthesis, is involved in resistance to *Pseudomonas syringae* in *Artemisia annua*. *Horticulture Research, 8*, 217. https://doi.org/10.1038/s41438-021-00652-6 (accessed on 6th November 2025).

Chen, Y., Chen, C., Tan, Z., Liu, J., Zhuang, L., Yang, Z., & Huang, B. (2016). Functional identification and characterization of genes cloned from halophyte seashore paspalum conferring salinity and cadmium tolerance. *Frontiers in Plant Science, 7*, 102. https://doi.org/10.3389/fpls.2016.00102 (accessed on 6th November 2025).

Choudhary, D. K., Prakash, A., & Johri, B. N. (2007). Induced systemic resistance (ISR) in plants: Mechanism of action. *Indian Journal of Microbiology, 47*(4), 289–297. https://doi.org/10.1007/s12088-007-0054-2 (accessed on 6th November 2025).

Conrath, U., Beckers, G. J., Flors, V., García-Agustín, P., Jakab, G., Mauch, F., Newman, M. A., Pieterse, C. M., Poinssot, B., & Pozo, M. J. (2006). Priming: Getting ready for battle. *Molecular Plant-Microbe Interactions, 19*(10), 1062–1071.

Cotas, J., Leandro, A., Monteiro, P., Pacheco, D., Figueirinha, A., Gonçalves, A. M. M., da Silva, G. J., & Pereira, L. (2020). Seaweed phenolics: From extraction to applications. *Marine Drugs, 18*(8), 384. https://doi.org/10.3390/md18080384 (accessed on 6th November 2025).

Curie, C., & Briat, J. F. (2003). Iron transport and signaling in plants. *Annual Review of Plant Biology, 54*(1), 183–206. https://doi.org/10.1146/annurev.arplant.54.031902.135018 (accessed on 6th November 2025).

Dangash, A. J., Pandya, N., Bharillya, A., Jhala, A., & Jain, D. (2014). Impact of exogenous elicitors on artemisinin production and trichome density in *Artemisia annua* L. under subtropical conditions. *Notulae Scientia Biologicae, 6*(3), 3. https://doi.org/10.15835/nsb639109 (accessed on 6th November 2025).

Davies, M. J., Atkinson, C. J., Burns, C., Woolley, J. G., Hipps, N. A., Arroo, R. R. J., Dungey, N., Robinson, T., Brown, P., Flockart, I., Hill, C., Smith, L., & Bentley, S. (2009). Enhancement of artemisinin concentration and yield in response to optimization of nitrogen and potassium supply to *Artemisia annua*. *Annals of Botany, 104*(2), 315–323. https://doi.org/10.1093/aob/mcp126 (accessed on 6th November 2025).

Deolu-Ajayi, A. O., van der Meer, I. M., van der Werf, A., & Karlova, R. (2022). The power of seaweeds as plant bio stimulants to boost crop production under abiotic stress. *Plant, Cell & Environment, 45*(9), 2537–2553.

Dhama, K., Karthik, K., Khandia, R., Munjal, A., Tiwari, R., Rana, R., Khurana, S. K., Ullah, S., Khan, R. U., & Alagawany, M. (2018). Medicinal and therapeutic potential of herbs and plant metabolites/extracts countering viral pathogens—Current knowledge and future prospects. *Current Drug Metabolism, 19*(3), 236–263.

Divekar, P. A., Narayana, S., Divekar, B. A., Kumar, R., Gadratagi, B. G., Ray, A., Singh, A. K., Rani, V., Singh, V., Singh, A. K., Kumar, A., Singh, R. P., & Meena, R. S., & Behera, T. K. (2022). Plant secondary metabolites as defense tools against herbivores for sustainable crop protection. *International Journal of Molecular Sciences, 23*(5), 2690. https://doi.org/10.3390/ijms23052690 (accessed on 6th November 2025).

Dondorp, A. M., Nosten, F., Yi, P., Das, D., Phyo, A. P., Tarning, J., Lwin, K. M., Ariey, F., Hanpithakpong, W., & Lee, S. J. (2009). Artemisinin resistance in *Plasmodium falciparum* malaria. *New England Journal of Medicine, 361*(5), 455–467.

Drobek, M., Frąc, M., & Cybulska, J. (2019). Plant bio stimulants: Importance of the quality and yield of horticultural crops and the improvement of plant tolerance to abiotic stress—A review. *Agronomy, 9*(6), 335. https://doi.org/10.3390/agronomy9060335 (accessed on 6th November 2025).

du Jardin, P. (2015). Plant bio stimulants: Definition, concept, main categories and regulation. *Scientia Horticulture, 196*, 3–14. https://doi.org/10.1016/j.scienta.2015.09.021 (accessed on 6th November 2025).

Dubey, V. S., Bhalla, R., & Luthra, R. (2003). An overview of the non-mevalonate pathway for terpenoid biosynthesis in plants. *Journal of Biosciences, 28*(5), 637–646. https://doi.org/10.1007/BF02703339 (accessed on 6th November 2025).

Efferth, T., Romero, M. R., Wolf, D. G., Stamminger, T., Marin, J. J., & Marschall, M. (2008). The antiviral activities of artemisinin and artesunate. *Clinical Infectious Diseases, 47*(6), 804–811.

Ertani, A., Francioso, O., Tinti, A., Schiavon, M., Pizzeghello, D., & Nardi, S. (2018). Evaluation of seaweed extracts from *Laminaria* and *Ascophyllum nodosum* spp. as bio stimulants in *Zea mays* L. using a combination of chemical, biochemical and morphological approaches. *Frontiers in Plant Science, 9*, 428. https://doi.org/10.3389/fpls.2018.00428 (accessed on 6th November 2025).

Fadiji, A. E., Babalola, O. O., Santoyo, G., & Perazzolli, M. (2022). The potential role of microbial bio stimulants in the amelioration of climate change-associated abiotic stresses on crops. *Frontiers in Microbiology, 12*, 829099. https://doi.org/10.3389/fmicb.2021.829099 (accessed on 6th November 2025).

Fatriasari, W., Nurhamzah, F., Raniya, R., Laksana, R. P. B., Anita, S. H., Iswanto, A. H., & Hermiati, E. (2020). Enzymatic hydrolysis performance of biomass by the addition of a lignin-based biosurfactant. *Journal of the Korean Wood Science and Technology, 48*(5), 651–665. https://doi.org/10.5658/WOOD.2020.48.5.651 (accessed on 6th November 2025).

Fu, X., Liu, H., Hassani, D., Peng, B., Yan, X., Wang, Y., Wang, C., Li, L., Liu, P., Pan, Q., Zhao, J., Qian, H., Sun, X., & Tang, K. (2020). AaABCG40 enhances artemisinin content and modulates drought tolerance in *Artemisia annua. Frontiers in Plant Science, 11*, 950. https://doi.org/10.3389/fpls.2020.00950 (accessed on 6th November 2025).

Ghanati, F., & Bakhtiarian, S. (2013). Changes of natural compounds of *Artemisia annua* L. by methyl jasmonate and silver nanoparticles. *Advances in Environmental Biology*, 2251–2259.

Ghatas, Y., Ali, M., Elsadek, M., & Mohamed, Y. (2021). Enhancing growth, productivity and artemisinin content of *Artemisia annua* L. plant using seaweed extract and micronutrients. *Industrial Crops and Products, 161*, 113202. https://doi.org/10.1016/j.indcrop.2020.113202 (accessed on 6th November 2025).

Ghazal, B., Saif, S., Farid, K., Khan, A., Rehman, S., Reshma, A., Fazal, H., Ali, M., Ahmad, A., & Rahman, L. (2018). Stimulation of secondary metabolites by copper and gold nanoparticles in submerged adventitious root cultures of *Stevia rebaudiana* (Bert.). *IET Nanobiotechnology, 12*(5), 569–573.

Gosling, R. D., Gesase, S., Mosha, J. F., Carneiro, I., Hashim, R., Lemnge, M., Mosha, F. W., Greenwood, B., & Chandramohan, D. (2009). Protective efficacy and safety of three antimalarial regimens for intermittent preventive treatment for malaria in infants: A randomised, double-blind, placebo-controlled trial. *The Lancet, 374*(9700), 1521–1532.

Gupta, R., Singh, A., Gupta, M. M., & Pandey, R. (2016). Cumulative role of bioinoculants on growth, antioxidant potential and artemisinin content in *Artemisia annua* L. under organic field conditions. *World Journal of Microbiology and Biotechnology, 32*(10), 167. https://doi.org/10.1007/s11274-016-2130-4 (accessed on 6th November 2025).

Hasanuzzaman, M., Parvin, K., Bardhan, K., Nahar, K., Anee, T. I., Masud, A. A. C., & Fotopoulos, V. (2021). Bio stimulants for the regulation of reactive oxygen species metabolism in plants under abiotic stress. *Cells, 10*(10), 2537. https://doi.org/10.3390/cells10102537 (accessed on 6th November 2025).

Ho, W. E., Peh, H. Y., Chan, T. K., & Wong, W. S. F. (2014). Artemisinin's: Pharmacological actions beyond antimalarial. *Pharmacology & Therapeutics, 142*(1), 126–139. https://doi.org/10.1016/j.pharmthera.2013.12.001 (accessed on 6th November 2025).

Huang, S. C., & Fraenkel, E. (2009). Integrating proteomic, transcriptional, and interactome data reveals hidden components of signaling and regulatory networks. *Science Signaling, 2*(81), ra40–ra40. https://doi.org/10.1126/scisignal.2000350 (accessed on 6th November 2025).

Ikram, N. K. B. K., & Simonsen, H. T. (2017). A review of biotechnological artemisinin production in plants. *Frontiers in Plant Science, 8*, 1966. https://doi.org/10.3389/fpls.2017.01966 (accessed on 6th November 2025).

Jamwal, K., Bhattacharya, S., & Puri, S. (2018). Plant growth regulator mediated consequences of secondary metabolites in medicinal plants. *Journal of Applied Research on Medicinal and Aromatic Plants*, *9*, 26–38.

Jha, P., Ram, M., Khan, M. A., Kiran, U., Mahmooduzzafar, & Abdin, M. Z. (2011). Impact of organic manure and chemical fertilizers on artemisinin content and yield in *Artemisia annua* L. *Industrial Crops and Products*, *33*(2), 296–301. https://doi.org/10.1016/j.indcrop.2010.12.011 (accessed on 6th November 2025).

Jiménez-Arias, D., García-Machado, F. J., Morales-Sierra, S., García-García, A. L., Herrera, A. J., Valdés, F., Luis, J. C., & Borges, A. A. (2021). A beginner's guide to Osmo protection by bio stimulants. *Plants*, *10*(2), 363. https://doi.org/10.3390/plants10020363 (accessed on 6th November 2025).

Jindo, K., Goron, T. L., Pizarro-Tobías, P., Sánchez-Monedero, M. Á., Audette, Y., Deolu-Ajayi, A. O., et al. (2022). Application of bio stimulant products and biological control agents in sustainable viticulture: A review. *Frontiers in Plant Science*, *13*, 932311. https://doi.org/10.3389/fpls.2022.932311 (accessed on 6th November 2025).

Kapoor, R., Chaudhary, V., & Bhatnagar, A. K. (2007). Effects of arbuscular mycorrhiza and phosphorus application on artemisinin concentration in *Artemisia annua* L. *Mycorrhiza, 17*(7), 581–587. https://doi.org/10.1007/s00572-007-0135-4 (accessed on 6th November 2025).

Kaushal, P., Ali, N., Saini, S., Pati, P. K., & Pati, A. M. (2023). Physiological and molecular insight of microbial bio stimulants for sustainable agriculture. *Frontiers in Plant Science*, *14*, 1041413. https://doi.org/10.3389/fpls.2023.1041413 (accessed on 6th November 2025).

Khan, H., & Shri, R. (2020). Application of abiotic stresses for enhancing secondary metabolite production: A review. *American Journal of Phytomedicine and Clinical Therapeutics*, *8*(2), p7. https://www.imedpub.com/articles-pdfs/application-of-abiotic-stresses-for-enhancing-secondary-metabolite-production-a-review.pdf (accessed on 6th November 2025).

Khan, W., Rayirath, U. P., Subramanian, S., Jithesh, M. N., Rayorath, P., Hodges, D. M., Critchley, A. T., Craigie, J. S., Norrie, J., & Prithiviraj, B. (2009b). Seaweed extracts as bio stimulants of plant growth and development. *Journal of Plant Growth Regulation*, *28*(4), 386–399. https://doi.org/10.1007/s00344-009-9103-x (accessed on 6th November 2025).

Kocira, S., Szparaga, A., Krawczuk, A., Bartoš, P., Zaguła, G., Plawgo, M., & Černý, P. (2021). Plant material as a novel tool in designing and formulating modern bio stimulants—Analysis of botanical extract from *Linum usitatissimum* L. *Materials*, *14*(21), 6661. https://doi.org/10.3390/ma14216661 (accessed on 6th November 2025).

Krishna, S., Bustamante, L., Haynes, R. K., & Staines, H. M. (2008). Artemisinin's: Their growing importance in medicine. *Trends in Pharmacological Sciences*, *29*(10), 520–527.

Kumari, M., Swarupa, P., Kesari, K. K., & Kumar, A. (2022). Microbial inoculants as plant bio stimulants: A review on risk status. *Life (Basel)*, *13*(1), 12. https://doi.org/10.3390/life13010012 (accessed on 6th November 2025).

La Torre, A., Battaglia, V., & Caradonia, F. (2016). An overview of the current plant bio stimulant legislations in different European Member States. *Journal of the Science of Food and Agriculture*, *96*(3), 727–734. https://doi.org/10.1002/jsfa.7358 (accessed on 6th November 2025).

Le Mire, G., Luan Nguyen, M., Fassotte, B., du Jardin, P., Verheggen, F., Delaplace, P., & Haissam Jijakli, M. (2016). Review: Implementing plant bio stimulants and biocontrol strategies in the agroecological management of cultivated ecosystems. *Biotechnologies, Agronomy, Société et Environnement*, *20*(S1), 299–313. https://popups.uliege.be/1780-4507/index.php?id=16764&file=1&pid=12717 (accessed on 6th November 2025).

Lephatsi, M. M., Meyer, V., Piater, L. A., Dubery, I. A., & Tugizimana, F. (2021). Plant responses to abiotic stresses and rhizobacterial bio stimulants: Metabolomics and epigenetics perspectives. *Metabolites*, *11*(7), 457. https://doi.org/10.3390/metabo11070457 (accessed on 6th November 2025).

Li, H., Wang, X., Rukina, D., Huang, Q., Lin, T., Sorrentino, V., et al. (2018). An integrated systems genetics and omics toolkit to probe gene function. *Cell Systems*, *6*(1), 90–102. e4. https://doi.org/10.1016/j.cels.2017.10.016 (accessed on 6th November 2025).

Liu, W., Wang, H., Chen, Y., Zhu, S., Chen, M., Lan, X., Chen, G., & Liao, Z. (2017). Cold stress improves the production of artemisinin depending on the increase in endogenous jasmonate. *Biotechnology and Applied Biochemistry*, *64*(3), 305–314.

Lopes, E. M., Guimarães-Dias, F., Gama, T. do S., Macedo, A. L., Valverde, A. L., de Moraes, M. C., de Aguiar-Dias, A. C. A., Bizzo, H. R., Alves-Ferreira, M., & Tavares, E. S. (2020). *Artemisia annua* L. and photo response: From artemisinin accumulation, volatile profile and anatomical modifications to gene expression. *Plant Cell Reports*, *39*, 101–117.

Lu, J. N., Zhang, D., Ding, D. D., Gao, H., Han, Z. X., Liu, X., & Xiang, L. (2018). Mechanism of high temperature promoting artemisinin biosynthesis in *Artemisia annua*. *Zhongguo Zhong Yao Za Zhi = Zhongguo Zhongyao Zazhi = China Journal of Chinese Materia Medica, 43*(20), 4169–4176. https://doi.org/10.19540/j.cnki.cjcmm.20180726.011 (accessed on 6th November 2025).

Lv, Z., Zhang, L., & Tang, K. (2017). New insights into artemisinin regulation. *Plant Signaling & Behavior, 12*(10), e1366398. https://doi.org/10.1080/15592324.2017.1366398 (accessed on 6th November 2025).

Ma, Y., Freitas, H., & Dias, M. C. (2022). Strategies and prospects for bio stimulants to alleviate abiotic stress in plants. *Frontiers in Plant Science, 13*, 1024243. https://doi.org/10.3389/fpls.2022.1024243 (accessed on 6th November 2025).

Malik, A., Mor, V. S., Tokas, J., Punia, H., Malik, S., Malik, K., et al. (2021). Bio stimulant-treated seedlings under sustainable agriculture: A global perspective facing climate change. *Agronomy, 11*(1), 14. https://doi.org/10.3390/agronomy11010014 (accessed on 6th November 2025).

Meshnick, S. R. (2002). Artemisinin: Mechanisms of action, resistance and toxicity. *International Journal for Parasitology, 32*(13), 1655–1660.

Moreno-Hernández, J. M., Benítez-García, I., Mazorra-Manzano, M. A., Ramírez-Suárez, J. C., & Sánchez, E. (2020). Strategies for production, characterization and application of protein-based bio stimulants in agriculture: A review. *Chilean Journal of Agricultural Research, 80*(2), 274–289. https://doi.org/10.4067/S0718-58392020000200274 (accessed on 6th November 2025).

N/A. (2012). *Seasonal Malaria Chemoprevention with Sulfoxide–Pyrimethamine Plus Amodiaquine in Children: A Field Guide*. World Health Organization.

N/A. (2020). *World malaria report 2020: 20 years of global progress and challenges. World Malaria Report*, 1–151.

Naeem, M., Aftab, T., & Khan, M. M. A. (2017). Strategies for enhancing artemisinin production in *Artemisia annua* under changing environment. In Ghorbanpour, M., & Varma, A. (Eds.), *Medicinal Plants and Environmental Challenges* (pp. 227–246). Springer International Publishing. https://doi.org/10.1007/978-3-319-68717-9_13 (accessed on 6th November 2025).

Nardi, S., Pizzeghello, D., Schiavon, M., & Ertani, A. (2016). Plant bio stimulants: Physiological responses induced by protein hydrolyzed-based products and humic substances in plant metabolism. *Scientia Agricola, 73*, 18–23.

Nephali, L., Piater, L. A., Dubery, I. A., Patterson, V., Huyser, J., Burgess, K., & Tugizimana, F. (2020). Bio stimulants for plant growth and mitigation of abiotic stresses: A metabolomics perspective. *Metabolites, 10*(12), 505. https://doi.org/10.3390/metabo10120505 (accessed on 6th November 2025).

Nikbakht, A., Kafi, M., Babalar, M., Xia, Y. P., Luo, A., & Etemadi, N. (2008). Effect of humic acid on plant growth, nutrient uptake, and postharvest life of gerbera. *Journal of Plant Nutrition, 31*(12), 2155–2167. https://doi.org/10.1080/01904160802462819 (accessed on 6th November 2025).

Olsson, M. E., Olofsson, L. M., Lindahl, A. L., Lundgren, A., Brodelius, M., & Brodelius, P. E. (2009). Localization of enzymes of artemisinin biosynthesis to the apical cells of glandular secretory trichomes of *Artemisia annua* L. *Phytochemistry, 70*(9), 1123–1128.

Organization, W. H. (2015). *Guidelines for the Treatment of Malaria. Third Edition.* World Health Organization. https://www.afro.who.int/publications/guidelines-treatment-malaria-third-edition (accessed on 6th November 2025).

Padalia, R. C., Verma, R. S., Chauhan, A., Chanotiya, C. S., & Yadav, A. (2011). Variation in the volatile constituents of *Artemisia annua* var. CIM-Arogya during plant ontogeny. *Natural Product Communications*, *6*(2), 239–242. https://doi.org/10.1177/1934578X1100600221 (accessed on 6th November 2025).

Patti, G. J., Yanes, O., & Siuzdak, G. (2012). Metabolomics: The apogee of the omics trilogy. *Nature Reviews Molecular Cell Biology*, *13*(4), 263–269. https://doi.org/10.1038/nrm3314 (accessed on 6th November 2025).

Pereira, R. V., Filgueiras, C. C., Dória, J., Peñaflor, M. F. G., & Willett, D. S. (2021). The effects of bio stimulants on induced plant defense. *Frontiers in Agronomy*, *3*, 630596. https://doi.org/10.3389/fagro.2021.630596 (accessed on 6th November 2025).

Pérez-Llorca, M., Pollmann, S., & Müller, M. (2023). Ethylene and jasmonates signaling network mediating secondary metabolites under abiotic stress. *International Journal of Molecular Sciences*, *24*(6), 5990. https://doi.org/10.3390/ijms24065990 (accessed on 6th November 2025).

Povero, G., Mejia, J. F., Di Tommaso, D., Piaggesi, A., & Warrior, P. (2016). A systematic approach to discover and characterize natural plant bio stimulants. *Frontiers in Plant Science*, *7*, 435. https://doi.org/10.3389/fpls.2016.00435 (accessed on 6th November 2025).

Pu, G. B., Ma, D. M., Chen, J. L., Ma, L. Q., Wang, H., Li, G. F., Ye, H. C., & Liu, B. Y. (2009). Salicylic acid activates artemisinin biosynthesis in *Artemisia annua* L. *Plant Cell Reports*, *28*(7), 1127–1135. https://doi.org/10.1007/s00299-009-0713-3 (accessed on 6th November 2025).

Rajabi Hamedani, S., Rouphael, Y., Colla, G., Colantoni, A., & Cardarelli, M. (2020). Bio stimulants as a tool for improving environmental sustainability of greenhouse vegetable crops. *Sustainability*, *12*(12), 5101. https://doi.org/10.3390/su12125101 (accessed on 6th November 2025).

Ro, D. K., Paradise, E. M., Ouellet, M., Fisher, K. J., Newman, K. L., Ndungu, J. M., Ho, K. A., Eachus, R. A., Ham, T. S., & Kirby, J. (2006). Production of the antimalarial drug precursor artemisinic acid in engineered yeast. *Nature*, *440*(7086), 940–943.

Rouphael, Y., & Colla, G. (2018). Synergistic bio stimulatory action: Designing the next generation of plant bio stimulants for sustainable agriculture. *Frontiers in Plant Science, 9*, 1655. https://doi.org/10.3389/fpls.2018.01655 (accessed on 6th November 2025).

Rouphael, Y., & Colla, G. (2020). Editorial: Bio stimulants in agriculture. *Frontiers in Plant Science*, *11*, 40. https://doi.org/10.3389/fpls.2020.00040 (accessed on 6th November 2025).

Rydén, A. M., Ruyter-Spira, C., Quax, W. J., Osada, H., Muranaka, T., Kayser, O., & Bouwmeester, H. (2010). The molecular cloning of dihydroartemisinic aldehyde reductase and its implication in artemisinin biosynthesis in *Artemisia annua*. *Planta Medica*, *76*(15), 1778–1783. https://doi.org/10.1055/s-0030-1249930 (accessed on 6th November 2025).

Sajwan, M. Y., Mishra, D., Negi, M., & Bisht, P. (2020). Effect of organic manures on antioxidant activity and essential oil composition of *Artemisia annua* cv. CIM Arogya. *International Journal of Current Microbiology and Applied Sciences*, *9*(10), 3922–3930.

Savvas, D., & Ntatsi, G. (2015). Bio stimulant activity of silicon in horticulture. *Scientia Horticulture*, *196*, 66–81.

Schiavon, M., Pizzeghello, D., Muscolo, A., Vaccaro, S., Francioso, O., & Nardi, S. (2010). High molecular size humic substances enhance phenylpropanoid metabolism in maize (*Zea mays* L.). *Journal of Chemical Ecology*, *36*, 662–669.

Shahrajabian, M. H., Chaski, C., Polyzos, N., & Petropoulos, S. A. (2021). Bio stimulants application: A low input cropping management tool for sustainable farming of vegetables. *Biomolecules*, *11*(5), 698. https://doi.org/10.3390/biom11050698 (accessed on 6th November 2025).

Shahzad, R., Waqas, M., Khan, A. L., Hamayun, M., Kang, S. M., & Lee, I. J. (2015). Foliar application of methyl jasmonate induced physio-hormonal changes in *Pisum sativum* under diverse temperature regimes. *Plant Physiology and Biochemistry*, *96*, 406–416. https://doi.org/10.1016/j.plaphy.2015.08.020 (accessed on 6th November 2025).

Shen, Q., Lu, X., Yan, T., Fu, X., Lv, Z., Zhang, F., Pan, Q., Wang, G., Sun, X., & Tang, K. (2016). The jasmonate-responsive AaMYC2 transcription factor positively regulates artemisinin biosynthesis in *Artemisia annua*. *New Phytologist*, *210*(4), 1269–1281. https://doi.org/10.1111/nph.13874 (accessed on 6th November 2025).

Shen, Q., Zhang, L., Liao, Z., Wang, S., Yan, T., Shi, P., et al. (2018). The genome of *Artemisia annua* provides insight into the evolution of Asteraceae family and artemisinin biosynthesis. *Molecular Plant*, *11*(6), 776–788. https://doi.org/10.1016/j.molp.2018.03.015 (accessed on 6th November 2025).

Shi, Y., Pan, Y., Xiang, L., Zhu, Z., Fu, W., Hao, G., Geng, Z., Chen, S., Li, Y., & Han, D. (2022). Assembly of rhizosphere microbial communities in *Artemisia annua*: Recruitment of plant growth-promoting microorganisms and inter-kingdom interactions between bacteria and fungi. *Plant and Soil*, *470*(1), 127–139. https://doi.org/10.1007/s11104-021-04829-9 (accessed on 6th November 2025).

Singh, M., Kumar, J., Singh, S., Singh, V. P., & Prasad, S. M. (2015). Roles of Osmo protectants in improving salinity and drought tolerance in plants: A review. *Reviews in Environmental Science and Bio/Technology*, *14*(3), 407–426. https://doi.org/10.1007/s11157-015-9372-8 (accessed on 6th November 2025).

Soni, P., & Abdin, M. Z. (2017). Water deficit-induced oxidative stress affects artemisinin content and expression of proline metabolic genes in *Artemisia annua* L. *FEBS Open Bio*, *7*(3), 367–381. https://doi.org/10.1002/2211-5463.12184 (accessed on 6th November 2025).

Takshak, S., & Agrawal, S. B. (2015). Defense strategies adopted by the medicinal plant *Coleus forskohlii* against supplemental ultraviolet-B radiation: Augmentation of secondary metabolites and antioxidants. *Plant Physiology and Biochemistry*, *97*, 124–138. https://doi.org/10.1016/j.plaphy.2015.09.018 (accessed on 6th November 2025).

Tamaoki, D., Seo, S., Yamada, S., Kano, A., Miyamoto, A., Shishido, H., Miyoshi, S., Taniguchi, S., Akimitsu, K., & Gomi, K. (2013). Jasmonic acid and salicylic acid activate a common defense system in rice. *Plant Signaling & Behavior*, *8*(6), e24260. https://doi.org/10.4161/psb.24260 (accessed on 6th November 2025).

Teoh, K. H., Polichuk, D. R., Reed, D. W., & Covello, P. S. (2009). Molecular cloning of an aldehyde dehydrogenase implicated in artemisinin biosynthesis in *Artemisia annua*.

Botany, *87*(6), 635–642. https://doi.org/10.1139/B09-032 (accessed on 6th November 2025).

Todeschini, V., Anastasia, F., Massa, N., Marsano, F., Cesaro, P., Bona, E., Gamalero, E., Oddi, L., & Lingua, G. (2022). Impact of phosphatic nutrition on growth parameters and artemisinin production in *Artemisia annua* plants inoculated or not with *Funneliformis mosseae*. *Life*, *12*(4), 497. https://doi.org/10.3390/life12040497 (accessed on 6th November 2025).

Trabelsi, D., & Mhamdi, R. (2013). Microbial inoculants and their impact on soil microbial communities: A review. *BioMed Research International*, *2013*, e863240. https://doi.org/10.1155/2013/863240 (accessed on 6th November 2025).

Trevisan, S., Pizzeghello, D., Ruperti, B., Francioso, O., Sassi, A., Palme, K., Quaggiotti, S., & Nardi, S. (2010). Humic substances induce lateral root formation and expression of the early auxin-responsive IAA19 gene and DR5 synthetic element in *Arabidopsis*. *Plant Biology (Stuttgart, Germany)*, *12*(4), 604–614. https://doi.org/10.1111/j.1438-8677.2009.00248.x (accessed on 6th November 2025).

Vargas-Hernandez, M., Macias-Bobadilla, I., Guevara-Gonzalez, R. G., de Romero-Gomez, S. J., Rico-Garcia, E., Ocampo-Velazquez, R. V., de Alvarez-Arquieta, L. L., & Torres-Pacheco, I. (2017). Plant hormesis management with bio stimulants of biotic origin in agriculture. *Frontiers in Plant Science*, *8*, 1762. https://doi.org/10.3389/fpls.2017.01762 (accessed on 6th November 2025).

Vasconcelos, A. C. F. de. (2020). Amelioration of drought stress on plants under bio stimulant sources. In *Plant Stress Physiology*. Intech Open. https://doi.org/10.5772/intechopen.91975 (accessed on 6th November 2025).

Wang, H., Ma, C., Li, Z., Ma, L., Wang, H., Ye, H., Xu, G., & Liu, B. (2010). Effects of exogenous methyl jasmonate on artemisinin biosynthesis and secondary metabolites in *Artemisia annua* L. *Industrial Crops and Products*, *31*(2), 214–218. https://doi.org/10.1016/j.indcrop.2009.10.008 (accessed on 6th November 2025).

Wang, H., Ye, H. C., Liu, B. Y., Li, Z. Q., & Li, G. F. (2003). [Advances in molecular regulation of artemisinin biosynthesis]. *Sheng Wu Gong Cheng Xue Bao*, *19*(6), 646–650. Chinese.

Wang, M. (2019). Novel strategy for disease control of rice and examination of the associated microbiome. *North Carolina State University*. https://portal.nifa.usda.gov/web/crisprojectpages/0231543-novel-strategies-for-managing-blast-diseases-on-rice-and-wheat.html (accessed on 6th November 2025).

Wani, K. I., Choudhary, S., Zehra, A., Naeem, M., Weathers, P., & Aftab, T. (2021). Enhancing artemisinin content in and delivery from *Artemisia annua*: A review of alternative, classical, and transgenic approaches. *Planta*, *254*(2), 29. https://doi.org/10.1007/s00425-021-03676-3 (accessed on 6th November 2025).

White, N. (1997). Assessment of the pharmacodynamic properties of antimalarial drugs *in vivo*. *Antimicrobial Agents and Chemotherapy*, *41*(7), 1413–1422.

Xiang, L., Zhu, S., Zhao, T., Zhang, M., Liu, W., Chen, M., Lan, X., & Liao, Z. (2015). Enhancement of artemisinin content and relative expression of genes of artemisinin biosynthesis in *Artemisia annua* by exogenous MeJA treatment. *Plant Growth Regulation*, *75*(2), 435–441. https://doi.org/10.1007/s10725-014-0004-z (accessed on 6th November 2025).

Xu, L., & Geelen, D. (2018). Developing bio stimulants from agro-food and industrial by-products. *Frontiers in Plant Science*, *9*, 416258. https://doi.org/10.3389/fpls.2018.01567 (accessed on 6th November 2025).

Xu, X., Du, X., Wang, F., Sha, J., Chen, Q., Tian, G., Zhu, Z., Ge, S., & Jiang, Y. (2020). Effects of potassium levels on plant growth, accumulation and distribution of carbon, and nitrate metabolism in apple dwarf rootstock seedlings. *Frontiers in Plant Science*, *11*, 904.

Yakhin, O. I., Lubyanov, A. A., Yakhin, I. A., & Brown, P. H. (2017). Bio stimulants in plant science: A global perspective. *Frontiers in Plant Science*, *7*, 2049. https://doi.org/10.3389/fpls.2016.02049 (accessed on 6th November 2025).

Yu, Y., Gui, Y., Li, Z., Jiang, C., Guo, J., & Niu, D. (2022). Induced systemic resistance for improving plant immunity by beneficial microbes. *Plants*, *11*(3), 386. https://doi.org/10.3390/plants11030386 (accessed on 6th November 2025).

Zhai, T., Wang, Y., Liu, C., Liu, Z., Zhao, M., Chang, Y., & Zhang, R. (2019). *Trichoderma asperellum* ACCC30536 inoculation improves soil nutrition and leaf artemisinin production in *Artemisia annua*. *Acta Physiologiae Plantarum*, *41*(4), 46. https://doi.org/10.1007/s11738-019-2836-7 (accessed on 6th November 2025).

Zhang, J., Du, X., Wang, Q., Chen, X., Lv, D., Xu, K., Qu, S., & Zhang, Z. (2010). Expression of pathogenesis related genes in response to salicylic acid, methyl jasmonate and 1-aminocyclopropane-1-carboxylic acid in *Malus hupehensis* (Pamp.) Rehd. *BMC Research Notes*, *3*(1), 208. https://doi.org/10.1186/1756-0500-3-208 (accessed on 6th November 2025).

Zhang, Y., Jiang, J., & Yang, Y. L. (2013). Acetyl salicylic acid induces stress tolerance in tomato plants grown at a low night-time temperature. *The Journal of Horticultural Science and Biotechnology*, *88*(4), 490–496. https://doi.org/10.1080/14620316.2013.11512996 (accessed on 6th November 2025).

Zhang, Y., Teoh, K. H., Reed, D. W., Maes, L., Goossens, A., Olson, D. J. H., Ross, A. R. S., & Covello, P. S. (2008). The molecular cloning of artemisinin aldehyde Δ11(13) reductase and its role in glandular trichome-dependent biosynthesis of artemisinin in *Artemisia annua*. *Journal of Biological Chemistry*, *283*(31), 21501–21508. https://doi.org/10.1074/jbc.M803090200 (accessed on 6th November 2025).

Zulfiqar, F., Casadesús, A., Brockman, H., & Munné-Bosch, S. (2020). An overview of plant-based natural bio stimulants for sustainable horticulture with a particular focus on moringa leaf extracts. *Plant Science*, *295*, 110194. https://doi.org/10.1016/j.plantsci.2019.110194 (accessed on 6th November 2025).

PART IV

Plant Growth Regulators

CHAPTER 9

A Review of the Diverse Roles of Brassinosteroids and Jasmonates as Plant Growth Regulators

RAJA HUSAIN,[1] SONIKA PANDEY,[2] NITIN VIKRAM,[3] SUBIR BOSE,[4] NARENDRA KUMAR,[5] LIPI RINA,[1] DEEPAK KUMAR,[6,7] MD. SHAMIM,[8] SONBEER CHACK,[1] N. A. KHAN,[9] and S. P. TIWARI[10]

[1]*Department of Agriculture, Himalayan University, Itanagar, Arunachal Pradesh, India*

[2]*Division of Plant Protection, ICAR–Indian Institute of Pulses Research (IIPR), Kanpur, Uttar Pradesh, India*

[3]*Zila Parishad Krishi Mahavidyalaya, Banda, Uttar Pradesh, India*

[4]*Department of Biotechnology and Microbiology, MIET College, Meerut, Uttar Pradesh, India*

[5]*Faculty of Agriculture, Oriental University, Indore, Madhya Pradesh, India*

[6]*R&D Division, Nextnode Bioscience Pvt. Ltd., Kadi, Gujarat, India*

[7]*Shri Ram Solvent Extraction Pvt. Ltd., Jaspur, Uttarakhand, India*

[8]*Department of Molecular Biology and Genetic Engineering, Dr. Kalam Agricultural College, Bihar Agricultural University, Sabour, Bihar, India*

[9]*Department of Plant Molecular Biology and Genetic Engineering, Acharya Narendra Deva University of Agriculture and Technology (ANDUAT), Ayodhya, Uttar Pradesh, India*

[10]*Faculty of Agriculture, Mangalayatan University, Jabalpur, Madhya Pradesh, India*

*Corresponding author

Plant Biostimulants and Growth Regulators in Sustainable Agriculture.
Deepak Kumar, Sanjay K. Singh, Nutan Kaushik, and Mahaveer P. Sharma (Eds.)

DOI: 10.1201/9781779641458-13

ABSTRACT

Brassinosteroids (BRs) and jasmonates (JAs) are both important plant growth regulators (PGRs). Jasmonates were isolated from the essential oil of *Jasminum grandiflorum,* commonly known as royal jasmine. JAs are one of the plant hormones (PHs) that are derived from lipids and regulate plant growth and development. JAs play an important role in biotic and abiotic stress management, such as salinity, drought, solar radiation, cold stress, and so forth. JAs content increases in *A. thaliana,* tomatoes, and potatoes during salt and drought stress. It is evident that JAs can reduce drought, salt, and heat stress by boosting endogenous hormones and the antioxidative system. The JA content of salt-, drought-, and heat-stressed plants is significantly higher than that of normal plants due to increased APX, peroxidase, and SOD enzyme activities. Application of MeJA not only increases total carbohydrates, polysaccharides, total soluble sugars, free amino acids, total proline, and protein content but also the activities of catalase, POD, and SOD in plants. MeJA mitigates drought and cold tolerance by increasing organic osmoprotectants and antioxidative enzyme activity. Exogenous MeJA application can reduce cadmium toxicity in mangrove plants, European black nightshade, Chilli peppers, soybeans, and *A. thaliana* by boosting APX, SOD, and CAT activities. JAs have been shown in a number of studies to shield plants from the negative effects of micronutrient poisoning. Treatment of exogenous MeJA reduces lead and boron toxicity in plants by activating the antioxidative enzymes. In tomato plants, low ozone levels trigger the JA pathway, which dramatically increases the synthesis of volatile chemicals for plant defense. While rice has a drop in JA and defense-related metabolites, tobacco has an increase in these compounds when CO_2 levels are lower. By enhancing secondary metabolite content and antioxidant enzyme activity while lowering ROS and MDA concentration, jasmonic acid lessens the damage caused by heavy metals. *In vitro,* the jasmonates enhance the synthesis of secondary metabolites, such as the ginsenoside *Panax ginseng* via cell cultures, *Glycyrrhiza glabra* soy saponin, *Celastrus paniculatus* suspension cultures phenolics, and *Artemisia absinthium* leaf total phenols and flavonoids. A steroid PH, brassinolide is generated from campesterol and is essential for several plant physiological processes, including root formation, fruit ripening, seed germination, and plant anti-ageing.

Abiotic stresses like salinity, drought, high temperatures, and low temperatures have also been reported to affect BRs. By raising proline content and reducing MDA and ROS accumulation, EBR therapy improves cold tolerance in Elymus mutants. It has also been investigated that BRs are crucial in biotic stress. In the production of ethylene, BRs are crucial. Together, BRs and H_2O_2 contribute to plant development and stress reduction. BRs and hydrogen peroxide control photosynthesis and plant sugar metabolism to increase agricultural output. By raising SPAD chlorophyll, net photosynthetic rate, and carbonic anhydrase (CA) activity, BR and H_2O_2 increase plant resilience to cold stress in *Lycopersicon esculentum*.

9.1 INTRODUCTION

Both phytomolecules, jasmonates (JAs) and brassinosteroids (BRs), control the growth, development, and defense responses of plants; yet, not much is known about how these phytohormones mediate the growth-defense orchestra. In response to environmental cues, JAs, one of the plant hormones (PHs) made from lipids, work in concert with endogenous hormones to control development, plant growth, and defense (Sohn et al., 2022). Jasmonates (JAs) and their derivatives are PHs that regulate several aspects of plant development and defense mechanisms, including root growth, stamen development, leaf senescence, and flowering (Goossens et al., 2016; Howe and Jander, 2008; Wasternack and Hause, 2013). They also ensure that plants survive abiotic stresses like high temperatures, UV radiation, and ozone.

According to the study, JAs's foliage treatment also prevents seedlings from growing their primary roots, expanding their leaves, or lengthening their hypocotyls (Kim et al., 2015; Wasternack and Hause, 2013). *Brassica napus* (rapeseed) and its pollen yielded the most potent growth-promoting extracts, which were dubbed "brassins." The highest production was achieved when immature seedlings of potatoes, radishes, and leafy vegetables received a brassin foliage treatment (Gudesblat and Russinova, 2011; Mandava, 1988). Through extensive column chromatography and large-scale solvent extractions, the genuine chemical of brassin was identified from 227 kg of bee-collected *B. napus* pollen. Ultimately, a small amount (4 mg) of a pure substance (active component) was obtained, which was further identified by single crystal X-ray analysis as a steroidal lactone, which was given the name brassinolide (Steffens, 1991). Brassinosteroid (BR) synthesis in plants is unknown; however, since it is extensively expressed in all plant organs, it is likely that this hormone is created throughout the plant. Three routes work together as a cascade for BR biosynthesis: the early C-6 oxidation, early C-22 oxidation, and late C-6 oxidation pathways. 22, 23-tetrahydroxy-24-methyl-B-homo-7-oxa-5α-cholestan-6-one and (22R, 23R, 24S)-2α, 3α are polyhydroxylated derivatives of 5α-cholestan that are known as brassinolide (BL). Therefore, a growth-promoting steroid found in plants shares structural similarities with animal steroid hormones formed from cholesterol, including estrogens, androgens, corticosteroids from vertebrates, and ecdysteroids from insects in the Crustacea family.

In addition to BL and roughly 68 other free BRs, the BR family also includes a number of conjugates. BRs are known to aid in cell division and cell wall regeneration, as well as to cooperate with auxin to promote cell elongation and expansion. Signal transduction that encourages vascular differentiation is induced by it. A crucial stage in the production of pollen tubes, pollen elongation, is influenced by it. It causes dying tissue-cultured cells to undergo senescence more quickly. During periods of cold and drought stress, it protects plants. In plants, it also controls biotic and abiotic stresses. According to a research study, applying BRs to cucumbers boosted their metabolism and eliminated pesticides, reducing the quantity of pesticides that humans can digest. When the hormone was applied to rice seeds, their fresh weights increased in comparison to plants that were not treated. According to an experiment, the defensin gene PLANT DEFENSIN1 is expressed more frequently in the *Arabidopsis thaliana* dwarf at early stages 1 (dwe1) mutant (Liao et al., 2020). The sixth family of PHs is called BRs. These are derivatives of polyhydroxylated sterols that are found in many different parts of the plant world. Since their discovery in the 1970s, around 70 types of naturally occurring BRs have been identified (Yokota, 1999). This hormone is crucial for the growth and development of plants and also contributes to the defensive mechanisms of plant cells. Both structurally and functionally, they are very similar to animal steroid hormones. For the first time, Mitchell et al. (1970) demonstrated that treatment with organic extracts of *Brassica napus* (rapeseed pollen) elongated stems and promoted cell division.

Brassinolide, a physiologically active molecule and the first BR, was isolated from *Brassica napus* pollen in 1979. Since then, the BR family has grown to contain a large number of chemically similar substances. Following scientific communities' discoveries, significant progress was made in identifying negative regulators of the BR signaling pathway (Li and Nam, 2002; Perez-Perez et al., 2002), characterizing two new transcription factors that promptly regulate the activity of BR-regulated genes (Wang et al., 2002; Yin et al., 2002, 2005; Zhao et al., 2002), and exploring the basis for comprehending the molecular mechanisms of BR action and identifying its receptor and its co-receptor (Kinoshita et al., 2005; Li and Chory, 1997; Li et al., 2002; Nam and Li, 2002; Wang et al., 2001). The primary roles of JAs and their derivatives in regulating plant defenses against pathogen infection and herbivore attack, plant tolerability against abiotic stresses like UV light, high temperatures, and ozone, and regulating different facets of plant development and defense mechanisms have been extensively studied by scientists (Huang et al., 2017). Transcripts encoding the essential triterpene biosynthesis enzyme β-amyrin synthase were 50 times more abundant in *Medicago truncatula* cell suspension after MJ treatment, which increased crop production and productivity (Suzuki et al., 2005). Hairy root cultures of *Panax ginseng*, commonly referred to as Asian or Korean ginseng, treated with methyl jasmonate showed elevated expression of genes like dammarenediol synthase-II, squalene synthase, and squalene epoxidase (Kim et al., 2009). According to Shabani et al. (2010), two-month-old liquorice (*Glycyrrhiza glabra*) plantlets treated with varying concentrations of MJ 0.1, 1.0, and 2.0 mM showed elevated expression of two essential biosynthetic enzymes for terpenoid biosynthesis, such as beta-amyrin synthase and squalene synthase. The expression of the hmgr (HMGCoA reductase) gene, which is the primary regulator of the solasodine biosynthesis pathway, was elevated in hairy roots of *Solanum trilobatum* treated with 4 μM MJ (Shilpha et al., 2015). The mechanisms of action of BRs and JAs on plant growth and development are updated in this chapter, and their function in eliciting secondary metabolites in significant medicinal plants and other crops is further discussed.

9.2 BIOSYNTHESIS OF JASMONIC ACID AND BRASSINOSTEROIDS

Jasmonic Acid (JAs), which includes derivatives of jasmonic acid, are well-recognized lipid-derived plant stress hormones that control plant adaptations to biotic stresses as well as abiotic stresses, where biotic stress includes herbivore attack and pathogen infection, and abiotic stress includes wounding, high temperatures, freezing ultraviolet radiation, and also ozone molecules (Huang et al., 2017; Wasternack and Hause, 2013). JAs also play roles in plant developmental events, such as reproductive development, primary root growth, and leaf senescence, as numerous studies have shown. Royal jasmine, or Jasminum grandiflorum, is the plant from which jasmonates were first extracted from its essential oil (Demole et al., 1962). According to Brose (2009), Ho et al. (2020), and Wasternack and Hause (2013). JAs are cyclopentanone molecules that are produced through the octadecanoid pathway from α-linolenic acid (α-LeA 18:3). The enzymes fatty acid desaturase (FAD) and phospholipase A1 (PLA1) work together to produce α-linolenic acid, which is then converted by 13-lipoxygenase (LOX), and allene oxide cyclase (AOC), allene oxide synthase (AOS) into 12,13(S)-epoxyoctadecatrienoic acid (12,13-EOT), (13S)-hydroperoxyoctadecatrienoic acid (13-HPOT), and (9S,13S)-12-oxo-phytodienoic acid (OPDA) in turn. After being carried to peroxisomes, OPDA reductase (OPR) converts it to 3-oxo-2-(cis-2′-pentenyl)-cyclopentane-1-octanoic acid (OPC-8:0). After OPC-8:0 is converted to OPC-8:0 CoA by OPC-8:0 CoA ligase (OPCL), it undergoes three rounds of β-oxidation, which are catalyzed by three distinct enzymes: 3-ketoacyl-CoA thiolase (KAT), acyl-CoA oxidase (ACX), and multifunctional protein (MFP). After being exported to the cytoplasm, jasmonic acid is conjugated with isoleucine to create bioactive (+)-7-iso-JA-Ile. This compound can subsequently be metabolized to different inactive forms by methylation, glucosylation, or sulfation, or it can be inactivated to 12-hydroxy-JA-Ile by CYP94B3, a cytochrome P450 (Figure 9.1).

JA is biosynthesized by JA biosynthetic enzymes, which are present in plastids, peroxisomes, and the cytoplasm. The above pathway shows how to trigger JA responses and JA-responsive early genes. 13-HPOT, 13-hydroperoxyoctadecatrienoic acid; 12-13-EOT, 12-13(S)-epoxyoctadecatrienoic acid; JA-Ile, jasmonoyl-L-isoleucine; OPC-8:0, 3-oxo-2(cis-2′-pentenyl)-cyclopentane-1-octanoic acid; OPDA, (9S, 13S)-12-oxo-phytodienoic acid; and JAR1, JASMONATE RESISTANT 1.

The most potent of the BRs, brassinolide, is made from campesterol. The substance is created by a sequence of oxidations, hydroxylations, reductions, and epimerizations. There are two branches of the pathway: the early C-6 oxidation pathway and the late C-6 oxidation pathway. The

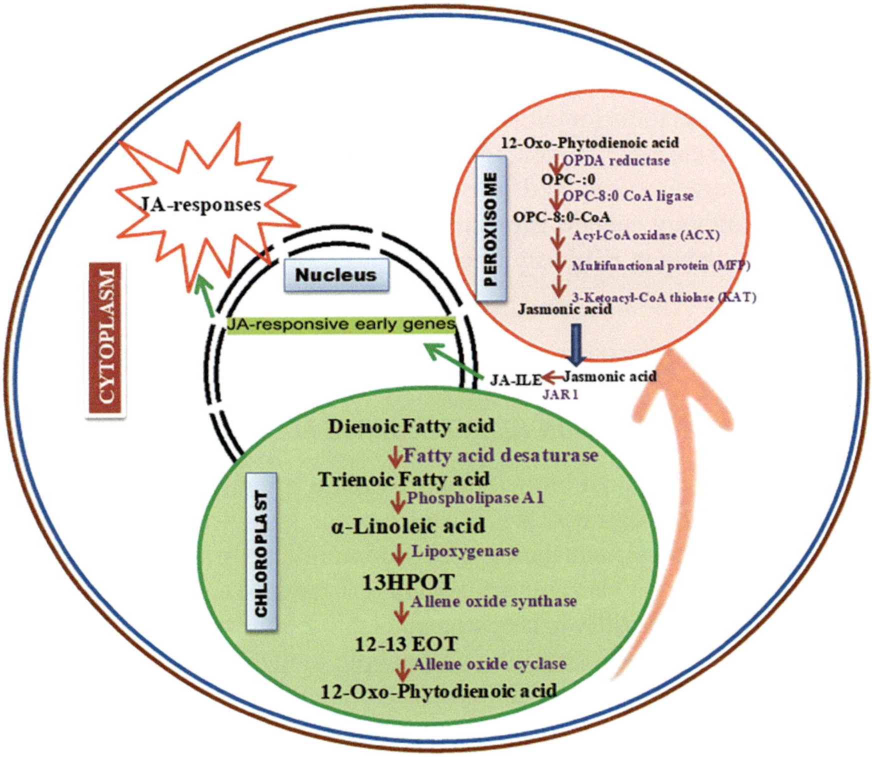

FIGURE 9.1 Jasmonic acid (JA) biosynthesis and signaling pathway.

addition of a ketone to the C6 carbon prior to side chain hydroxylation is known as early oxidation, whereas the addition of the ketone subsequent to hydroxylation is known as late oxidation. The biosynthetic pathways of BL were initially explained experimentally in *Catharanthus roseus* cell cultures. Recent characterization of BR-deficient mutants of the model plant *Arabidopsis thaliana*, *Pisum sativum*, and *Solanum lycopersicum* confirmed that the pathways operate in a wide variety of plant species. However, it was subsequently determined that it is not two distinct routes, but rather merely a metabolic grid. The cloned genes for the biosynthesis of BRs are found in *A. thaliana*. The biosynthesis of BRs in plants other than *Arabidopsis* and *C. roseus* has rarely been investigated. The pathways from teasterone to castasterone (CS) and from 6-deoxoCS to CS were demonstrated in rice and tobacco seedlings (Choe, 2001; Suzuki et al.,

1995). Recent studies of the BR biosynthesis pathway were explored using BR biosynthesis inhibitors (Asami and Yoshida, 1999; Fujioka and Yokota, 2003; Rademacher et al., 2000). Brassinolide mutants have been used as powerful tools for exploring the biosynthesis and functions of BRs. The chemical characterization, identification, and regulation of the various enzymes and genes involved in BR biosynthesis have been subjected to intensive study by genetic, molecular, and biochemical analyses of dwarf mutants defective in BR biosynthesis.

9.3 ROLE OF JASMONIC ACID

9.3.1 JASMONIC ACID IN PLANT DISEASE MANAGEMENT

In plant defense against biotic stress, JA plays a vital role, as supported by the provided references. It activates defense responses against various pathogens and pests, enhancing plant immunity. JA's involvement goes beyond activation; it also augments plant resistance to a range of stressors, showcasing its versatility in helping plants thrive amidst challenging biotic stress conditions. In essence, JA is a linchpin in the intricate web of plant defense mechanisms, emphasizing its paramount role in enhancing plant immunity and adapting to biotic stress complexities.

1. **Activation of Defense Responses:** JA is a key player in the activation of defense responses against biotrophic and necrotrophic pathogens (Guerreiro et al., 2016). It induces the expression of defense-related genes and the production of secondary metabolites, such as phytoalexins and terpenoids, which contribute to plant defense mechanisms (Aslam et al., 2022; Guerreiro et al., 2016). JA signaling is generally associated with the defense against necrotrophic pathogens and insect herbivores (Nguyen et al., 2019).
2. **Interaction with Other Hormones:** Jasmonic acid interacts with other PHs, such as salicylic acid (SA), to regulate plant defense responses. While SA is typically associated with defense against biotrophic pathogens, jasmonic acid-dependent signaling is involved in defense against necrotrophic pathogens. However, recent evidence suggests that jasmonic acid signaling can also play a role in resistance against certain biotrophic pathogens (Guerreiro et al., 2016).

3. **Variation in Pathogen-Induced Defense:** The ability of pathogens to trigger or avoid plant defenses mediated by jasmonate varies among species. The ability of the necrotrophic fungal pathogen Botrytis cinerea to suppress plant defenses mediated by jasmonate and to activate defense metabolites varied throughout its isolates (Rowe et al., 2010). This highlights the importance of considering pathogen variation in the context of jasmonate-mediated defense responses.
4. **Enhancement of Plant Resistance:** Exogenous application of JA acid has been shown to enhance plant resistance against aphids and improve physiological and biochemical attributes related to defense responses. JA treatment increased the production of antioxidant enzymes and reduced aphid infestation. It also positively influenced plant growth and improved resistance against biotic stress (Aslam et al., 2022).
5. **The Function of Jasmonate in Alleviating Fungal Stress in Plants:** Extensive research has explored the role of jasmonate signaling in plant defense against fungal pathogens. However, the potential intraspecific variability in pathogen responses to these defenses has often been overlooked. To address this, scientists investigated jasmonate-deficient and jasmonate-insensitive *Arabidopsis thaliana* plants exposed to different *Botrytis cinerea* isolates. Notably, the pathogen's reactions to jasmonate-mediated defenses and subsequent defense metabolite induction showed significant variations. These findings emphasize the complex interplay between plants and necrotrophic pathogens and underscore the importance of considering pathogen variability in comprehending plant defense dynamics.
6. **Jasmonic Acid's Impact on Plant Resilience Against Bacterial Biostress:** Jasmonic acid, particularly its analogue coronatine, plays a role in bacterial and biotic stress in plants. It suppresses the accumulation of salicylic acid, manipulates plant defense signaling, interacts with other defense pathways, and contributes to basal disease resistance against bacterial pathogens. Understanding the role of jasmonic acid in bacterial biotic stress can provide insights into plant-pathogen interactions and aid in the development of strategies for disease management.

7. **Suppression of Salicylic Acid Accumulation:** JA and its analogue coronatine (COR) play a crucial role in plant-pathogen interactions. COR, produced by *Pseudomonas* syringe, mimics JA-isoleucine (JA-Ile) and influences plant defenses, promoting bacterial entry and growth. This manipulation of PH signaling aids nutrient acquisition and suppresses defense responses (Zheng et al., 2012). JA signaling contributes to basal disease resistance against bacterial pathogens, with JA-related gene silencing reducing resistance (Tagami et al., 2021). This interaction highlights the importance of pathogen-induced manipulation of plant defense pathways in evolutionary processes.
8. **Bacterial Pathogens Counteract the Effects of Jasmonic Acid:** Bacterial pathogens have evolved strategies to undermine plant defenses by manipulating the jasmonic acid (JA) signaling pathway. For instance, *Pseudomonas syringae* produces coronatine, mimicking JA, to promote bacterial entry and disease progression. Additionally, bacterial effectors target transcriptional repressors, altering JA and salicylic acid pathways and impacting plant growth defense balance (Carvalhais et al., 2013; Gimenez-Ibanez et al., 2016; Jiang et al., 2013; Zhang et al., 2017).
9. **Jasmonic Acid's Role in Plant Resilience Against Biotic Stress: Viruses and Viroids:** JA is a key regulator of plant defenses against biotic stress, including plant viruses (Aguilar et al., 2022; Pieterse et al., 2012). It activates defense genes and compounds, but its interaction with other hormones, like salicylic acid (SA), is intricate, displaying both antagonistic and synergistic effects on viral defense. Additionally, auxin, abscisic acid (ABA), cytokinins (CKs), and BRs can affect JA signaling, influencing defense responses. Understanding this hormone crosstalk is crucial for plant defense against viruses (Robert-Seilaniantz et al., 2011). JA also enhances plant resilience to biotic stress caused by plant viroids (Gupta, 2020). Given viroids' disease-inducing capacity, JA's role in plant immunity suggests its contribution to viroid defense in maintaining precise hormonal balance for plant defenses against various pathogens.

9.3.2 JASMONIC ACID IN ABIOTIC STRESS MANAGEMENT

Complex biological networks in conjunction with catabolic and anabolic pathways are essential for the growth, development, and survival of plants (Altaf et al., 2015). Abiotic stress affects these pathways, resulting in the disturbance of plant growth and physiology. For example, extreme high or low temperatures can disrupt enzyme activity, which affects the plant's metabolism. Phytohormones, which play a key role in plant development, play a central role in mitigating abiotic stress. The following phytohormones—auxins, gibberellins, cyokinins, ABA, ethylene, BRs, jasmonic acid, salicylic acid, and strigolactones—are the major phytoregulators that generate different responses in plants. Jasmonic acid and its derivatives like jasmonyl isoleucine, cis-jasmone, JA-glucosyl ester, methyl jasmonate, jasmonyl amino acid, 12 hydroxy jasmonic acid sulphate, 12-o-glucosyl-JA, and JA-ILE methyl ester, JA-IL glucosyl ester, 12 carboxy-JA-ILE, 12-O-Glucosyl-JA-ILE, and lactones of 12 hydroxy-JA-Ile. These derivatives are collectively known as jasmonates (JAs). Jasmonates are fatty acid derivatives derived from cyclopentanones. Members of the oxidized lipid family known as oxylipins include cyclopentanes. Lipoxygenases, alpha-dioxygenases, or the autoxidation of polyunsaturated fatty acids can all enzymatically create oxylipins, which are signaling molecules (Gobel et al., 2009; Wasternack et al., 2018) (Figure 9.2).

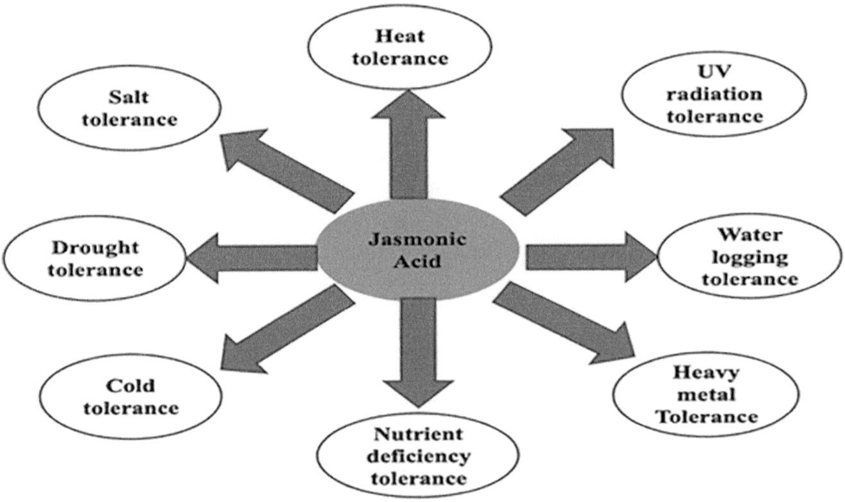

FIGURE 9.2 Role of jasmonic acid (JA) under different abiotic stresses.

Jasmonic acids are ubiquitous in nature; their level is high in reproductive tissues and flowers and low in mature leaves and roots (Dar et al., 2015; Wasternack et al., 2013). JA regulates much of the plant growth process. Fruit ripening, rubisco biosynthesis, anthocyanin biosynthesis, cell cycle regulation, glucose transport, N and P uptake, and stamen and trichome formation are all aided by JAs. They also modulate the expression of various genes during abiotic stress (cold, drought, low temperature, etc.) and elicit specific mechanisms. The biosynthesis of jasmonic acid has been well characterized in a variety of monocotyledons and dicotyledons. A series of enzymes found in the cytoplasm, peroxisome, and plastid carry out the manufacture of jasmonic acid. The phospholipases found in the plastid membrane are activated by abiotic stress, which encourages the production of linolenic acid. The action of beta oxidation (three cycles) and 12-oxo-phytodienoic acid reductase transforms this linolenic acid into 12-oxo-phytodienoic acid, which is then used to synthesize JA.

9.3.2.1 JASMONIC ACID IN MANAGEMENT OF SALINITY STRESS

Plant development and growth are impacted by salinity stress in both cytotoxic and osmotic ways. JA content increases in *A. thaliana,* tomatoes, and potatoes during salt stress. JA levels increase significantly to open up the effects of salinity. JA content increases significantly in salt-stressed plants compared to normal ones (De et al., 2019). Exogenous MeJA has been shown to enhance black locust trees' resistance to salt stress by boosting ascorbate peroxidase and superoxide dismutase (SOD) activity. These results are similar to those of Faghih et al. (2017), who found that MeJA enhances the activity of the enzymes SOD, peroxidase, and APX. This clearly shows that JAs can reduce salt stress by boosting the antioxidative system and endogenous hormones.

9.3.2.2 JASMONIC ACID IN MANAGEMENT OF DROUGHT AND COLD STRESS

Drought stress has various harmful effects on plants, like decreasing the turgor pressure, increasing ion toxicity, and inhibiting photosynthesis.

JA signaling is directly associated with the alleviation of drought stress. It has been observed that JA content increases in *A. thaliana* and citrus during drought stress (De et al., 2015) and decreases as soon as the drought stress elevates. It has also been observed that treatment with MeJA can improve drought resistance in peanut (Todaka et al., 2016), rice (Fu et al., 2017), soyabean (Mohamed et al., 2017), and broccoli (Wu et al., 2012). Application of MeJA not only increases total carbohydrates, polysaccharides, total soluble sugars, free amino acids, total proline, and protein content but also the activities of catalase, POD, and SOD in maize plants (Abdelgawad et al., 2014). In broad bean and barely drought, stress is regulated by the stomata closure (Evans et al., 2003; Horton, 1991). In cauliflower, MeJAs are also found to mitigate drought stress by activating various enzymes (SOD, POD, CAT, APX, and glutathione reductases), proline, and soluble sugar (Wu et al., 2012). MeJA mitigates drought tolerance by increasing organic osmoprotectants and antioxidative enzyme activity (Qiu et al., 2014).

Low temperatures cause the formation of extracellular ice crystals and cell dehydration. Plants' ability to adapt to cold stress is greatly influenced by signaling. It has been found that MeJA could mitigate the cold stress in tomato, pomegranate, mango, guava, cowpea, and peach (Fan et al., 2016; Gonzalez et al., 2000, 2004; Jin et al., 2009, 2014; Sayyari et al., 2011; Zhang et al., 2012) by increasing the synthesis of antioxidative enzymes and activation of defense compounds. These findings imply that JAs can lessen the effects of cold injury by producing antioxidants and defense chemicals.

9.3.2.3 ROLE OF JASMONIC ACID TO REDUCE THE EFFECTS OF HEAVY METAL AND MICRONUTRIENT TOXICITY

Heavy metal toxicity in plants generates free radicals in plants. Application of exogenous MeJA can alleviate the cadmium toxicity in soyabean (Poonam et al., 2013), *A. thaliana* (Maksymiec et al., 2007), European black nightshade (Yan et al., 2015), Chilli paper (Yan et al., 2013), and mangrove plants by increasing the activities of SOD, APX, and CAT. MeJA has been found to mitigate the boron toxicity in sweet wormwood by decreasing lipid peroxidation and increasing the activity of reducing enzymes (Aftab et al., 2011). In *B. napus,* oxidative stress is mitigated by the induction of genes encoding antioxidative and secondary metabolites (Farooq et al., 2016).

Several studies have suggested that JAs can protect plants from the effects of micronutrient toxicity. High boron concentrations are harmful for plant growth and development (apple, wheat, barley, and tomato) (Cervilla et al., 2007; Gunes et al., 2007; Karabal et al., 2003; Molassiotis et al., 2006; Papadakis et al., 2004). By promoting the antioxidative enzymes POD, CAT, and SOD and preventing lipid peroxidation, exogenous MeJA treatment can mitigate boron toxicity in plants (Aftab et al., 2011; Dar et al., 2015). JAs also play a pivotal role in toxicity against lead toxicity. According to Bali et al. (2019), applying JAs to tomato seeds has been proven to decrease Pb uptake and promote tomato plant growth.

9.3.2.4 ACTIVATION OF JASMONIC ACID BIOSYNTHESIS VIA OZONE AND THE SUNLIGHT EFFECT

ROS produced by ozone leads to plant lesions and programmed cell death in *Arabidopsis* of the wild type. Following ozone treatment, there was a notable rise in JA content (Rao et al., 2000). MeJA therapy may prevent the ozone-induced programmed cell death (Ismail et al., 2012; Kanna et al., 2003; Overmyer et al., 2000; Rao et al., 2000). After receiving exogenous MeJA treatment, hybrid poplar and tomato were shown to have decreased ozone sensitivity (Ismail et al., 2012; Kanna et al., 2003; Koch et al., 2000; Overmyer et al., 2000; Rao et al., 2000). Low levels of ozone activate the JA pathway in tomato plants, which significantly up-regulates the production of volatile compounds for plant protection (Cui et al., 2016).

Few reports exist that demonstrate how light and the JA signal affect the growth and development of plants. According to numerous findings, JA production initiates the JA signaling pathway in *Brassica* and *Nicotina* (Mewis et al., 2012; Svyatyna et al., 2012). According to several reports, JA signaling directly affects the red light-mediated photomorphogenesis in *A. thaliana* and rice (Mewis et al., 2012) and the blue light-mediated light morphogenesis in tomato and *A. thaliana* (Cerrudo et al., 2012; Gupta et al., 2014). There are not many publications that explicitly demonstrate how JA affects plants under CO_2 stress. Ballhorn et al. (2011) found that elevated CO_2 concentrations enhanced the concentration of MeJA and JA in lima beans. In tomato plants, decreased CO_2 levels have been found to increase the action against nematodes by activating the JA and SA-signaling pathways (Sun et al., 2011). In tobacco, decreased levels of CO_2 increase the JA and main defense-related metabolites, but in rice, there is a decrease in JA and defense-related metabolites (Lu et al., 2018). Plant defense and resilience to abiotic stressors are significantly influenced by JAs and their derivatives. By triggering plant defense systems, which include antioxidative enzymes and other protective substances, JAs promote tolerance in the face of abiotic stressors. Further studies are needed to investigate how different environmental signals help in the biosynthesis of JAs and also investigate the molecular mechanism of JA movement within plants.

9.3.2.5 ROLE OF JASMONIC ACID UNDER METAL/METALLOID TOXICITY

Jasmonates increase plant resistance to heavy metals by increasing the antioxidant defense system (Raza et al., 2020, 2021). Application of 0.1 μm MeJA on cadmium-treated tomato plants (40 mg dm^{-3} Cd) restores the R/S ratio and decreases Cd translocation and accumulation in both roots and shoots. MeJA treatment also increases the proline and glutathione content of POD and SOD as compared to the control plants (Yan et al., 2015). In an experiment conducted by Yan et al. (2013), seedling treatment with cadmium (50 mg/l) after seven days in *C. frutescens* resulted in decreased root dry weight and chlorophyll B concentration. Low MeJA application results in increased POD and CAT activities, while high MeJA application results in increased POD, APX, and SOD activities and inhibits CAT activities.

Exogenous MeJA application results in increased antioxidant enzyme activities like POD, CAT, and SOD, as well as non-enzymatic antioxidant enzymes GSH and AsA (Zhao et al., 2016). JA-mediated accumulation of osmolytes and carotenoids and antioxidant enzyme activities are the key mechanisms employed to mitigate the toxic effects of heavy metals. Defense against copper stress is provided by the accumulation of proline and antioxidant enzymes (SOD and POX) (Ponam et al., 2013). Under copper stress conditions, alfalfa plant root and leaf biomass significantly increased upon JA application. Chlorophyll conc., MDA, antioxidant enzyme activities, and H_2O_2 conc. also improved accordingly (Dai et al., 2020). In a study by Ali and Baek 2020, it was found that the application of MeJA mitigates the boron (B) toxicity in *Artemisia annua* by lowering lipid peroxidation and inducing enzymes. Under boron stress, hydrogen peroxide and oxygen activities are increased. POD and SOD activities are increased by the JA application, indicating that JA facilitates ROS elimination (Zhao et al., 2019). In soybean, application of JA increases tolerance towards nickel by reducing ROS levels, electrolyte leakage, and lipoxygenase activity. SOD and CAT activities are also increased by the JA application, which mitigates Ni stress damage (Mir et al., 2018). Application of 2 mM Ni and 1 nM MeJA results in a low accumulation of H_2O_2, NADPH oxidase, and MDA, which help in the stabilization of biomolecules. Application of JA results in increased osmolyte activity, antioxidant enzyme activity, and gene expression (Sirhindi et al., 2016).

Arsenic stress in plants causes oxidative damage by increasing root growth, ROS activity, and lipid peroxidation. MeJA treatment in plants results in decreased hydrogen peroxide and reactive oxygen content. Application of MeJA results in increased induction of secondary metabolites and PPO and PAL activity. MeJA alters the expression of AS3 transporters and increases arsenic detoxification by OsNRAMP1. Jasmonic acid mitigates the heavy metal damage by improving the antioxidant enzyme activity and secondary metabolite chl content and decreasing the ROS and MDA content (Yu et al., 2019). Under lead toxicity stress, JA treatment increases Asa, GSH, APOX, MDHAR, DHAR, and GR content (Bali et al., 2018). Application of JA enhances plant growth and reduces oxidative injury in heavy metal-stressed plants.

9.3.3 ROLE OF JASMONIC ACID IN IN VITRO PROPAGATION

Jasmonic acid is considered a new type of plant growth regulator in higher plants that influences different physiological processes in plant *in vitro* propagation (Yi et al., 2020). The physiological activities of jasmonic acid were evaluated in fruit ripening, plant growth, development, and senescence (Waternack and Hause, 2013), as well as in pathogen-induced defense responses (Anderson et al., 2004). Jasmonic acid is widely distributed in higher plants and is essential for plant growth and development, particularly under biotic and abiotic stress. The challenges that come from biotic and abiotic variables are salt stress, cold stress, temperature stress, heavy metal stress, ozone stress, UV stress, and so forth. (Anusha et al., 2016).

Regardless of several advantages, applying the *in vitro* plant using various strategies has been suggested to increase secondary metabolite production. In order to improve the yields of highly important bioactive molecules in different plants, a variety of strategies have been used, such as elicitation, genetic alteration, cell immobilization, and permeabilization (Isah et al., 2018). The germination of non-dormant seeds is restrained, whereas the germination of dormant seeds is stimulated. Additionally, it circulates within the plant in the liquid and vapor phases, impacting various areas of plant growth and development. Several developments regulated processes such as seed and seedling development, seed germination, flowering development, tuber formation, and senescence. The jasmonates increase the production of secondary metabolites *in vitro*, including the ginsenoside *Panax ginseng* using cell cultures, soya saponin from *Glycyrrhiza glabra*, phenolics from *Celastrus paniculatus* suspension cultures (Anusha et al., 2016), and total phenols and flavonoids in *Artemisia absinthium* leaf cultures (Ali et al., 2015).

Additionally, it severely restricts root growth and coleoptile elongation that is induced by indole acetic acid, which may be preventing the incorporation of glucose into cell wall polysaccharides (Ueda et al., 1995). The enzymes involved in biosynthesis, which are concentrated in chloroplasts, help in the formation of seeds, flowers, and fruit, which is due to the presence of high levels of this compound in developing plant reproductive tissues. Jasmonates regulate various aspects of reproductive processes; hormones generally promote defensive and plant development to inhibit the growth and photosynthesis of vegetative tissues (Devoto and Turner, 2005). It also helps in the development of storage organs in garlic bulbs (Ravnikar Maza et al., 1993). Additional developments that can be affected by jasmonate include fruit ripening, fruit carotenoid composition (Czapski and Saniewski, 1992), and the expression of genes encoding seed and vegetative storage proteins (VSPs), which have also been reported to inhibit stomata closure and cytokinin-induced callus growth.

The lipid derivatives known as jasmonic acid serve as essential signaling molecules in a variety of plant stress responses and developmental processes. The presence of jasmonates and similar derivatives of volatile fatty acids in the flower may suggest a role in insect attraction and compromised resistance to a wide range of arthropod herbivores, including leafhoppers (*Homoptera*), caterpillars (*Lepidoptera*), thrips (*Thysanoptera*), fungal gnats (*Diptera*), spider mites (*Acari*), and beetles (*Coleoptera*) (Howe and Jander, 2008). According to O'Brien and Benkova (2013), plants specifically produce jasmonates in response to damage to aid in defending against disease and insect attack, including chewing insects and necrotopic pathogens. Additionally, it has a significant impact on plant insect and pest resistance. For instance, potato treatment with jasmonate boosts resistance to the tomato mutant def1, which is vulnerable to injury caused by larvae of the tobacco hornworm and to the fungus *Phytophthora infestans* (Cohen et al., 1993). In response to threats like insect feeding, jasmonate promotes the transduction pathway and controls the expression of defense genes in plants during insect feeding (Walling, 2000). Triterpenes are produced when *Jatropha curcas* pale cell suspensions cultured in darkness are stimulated by jasmonic acid (Zaragoza-Martínez et al., 2016). When *Psoralea corylifolia* in *in vitro* culture is subjected to jasmonic acid elicitation, the amount of *psoralen* content in the leaves and roots increases when compared to non-elicited *in vitro*-grown plants (Siva et al., 2015).

Jasmonates control a variety of physiological and developmental processes, including male and female reproductive organ development (floral development), pollen development, embryo development, sex determination, root growth, tuberization, and fungal arbuscular mycorrhizal (AM) association (Schaller et al., 2005). Through cell division and proliferation in response to wound stress, jasmonates increasingly activate systems involved in the healing process (Leon et al., 2001). According to Lorenzo and Solano (2005), jasmonates are also strong inducers of gene expression in vegetative storage protein (VSP), and they are crucial for the development of vegetative storage organs (Jasik and de Klerk 2006).

9.4 CROSS-TALK BETWEEN BRASSINOSTEROIDS AND OTHER PHYTOHORMONES IN PLANT GROWTH, DEVELOPMENT, AND STRESS RESPONSES

Brassinosteroids (BRs) are steroid PHs that play an important role in various plant physiological responses. BRs are present in pollen, seeds, stems, and leaves. They play an irreplaceable role in plant physiology. BRs interact with other plant growth hormones and play an important role in plant development and stress resistance. The resilience of *Festuca arundinacea* Schreb was considerably enhanced by the co-treatment with melatonin and BRs. Reducing reactive oxygen species (ROS) and malondialdehyde levels can help alleviate heat stress (Alam et al., 2018). Small-molecule compounds (SMCs) like ethylene, nitric oxide, hydrogen peroxide, and so forth, interact with BRs and modulate plant growth and stress stimulus (Jiroutova et al., 2019; Tian et al., 2018). The present review summarizes the interrelationship of BRs and plant growth hormones.

In 1979, the most active BRs were discovered in *Brassica napus*. According to Li et al. (2020), BRs are currently regarded as the sixth most significant plant endogenous hormone, behind auxin, cytokinin, gibberellin, ethylene, and ABA. The biosynthesis pathway of BRs is stimulated by campesterol. BRs play an important role in seed germination, fruit development, fruit ripening, root development, and plant anti-ageing (Li et al., 2020; Wang et al., 2020, 2021). The co-interaction of BES1 and ABA insensitivity inhibits ABA signaling and promotes seed germination in *Arabidopsis* (Zhao et al., 2019). He et al. (2018) found that applying 10 micromolar EBR to *Diospyros kaki* L. stimulates fruit ripening by increasing ethylene production and activating enzymes that break down cell walls. EBR treatment in plants has been found to increase membrane integrity in flower buds by degrading chlorophyll concentration and MDA content (Yao et al., 2017). Through an increase in the total solid and soluble content and the stimulation of peroxidase, catalase, and SOD activity, EBR treatment postponed the kiwifruit's senescence (Lu et al., 2019).

9.5 ROLE OF BRASSINOSTEROIDS IN STRESS MANAGEMENT

BRs have also been found to play a role in abiotic stresses like drought, salinity, high temperature, and low temperature (Peres et al., 2019). Jan et al. (2020) showed that EBR treatment alleviates the chromium toxicity in *Solanum lycopersicum* L. In *Zea mays,* EBR content increases osmoprotectants and decreases MDA content (Rattan et al., 2020). Cold tolerance in Elymus mutants is enhanced by EBR treatment by increasing proline content and decreasing MDA and ROS accumulation (Fu et al., 2019). In *Solanum lycoperscium,* EBR and H_2O_2 ameliorate the chloroplast structure and stomatal behavior to improve photosynthetic efficiency (Nazir et al., 2020). In a study, it was found that the application of EBR, salicylic acid, and silicon significantly decreased the H_2O_2 content of MDA and EL and improved plant growth under arsenic stress. Choudhry et al. (2012) showed that BR signaling increases NO levels, triggers ABA biosynthesis, and promotes growth in *Raphanus sativus*.

These studies clearly indicate that BRs improve stress tolerance by interacting with other phytohormones and indicate a better relationship between BRs and other phytohormones. In *Allium sativum,* BRs affect lignin accumulation, which decreases salt stress (Kong et al., 2021). EBR and Si increase the high temperature tolerance of *Triticum aestivum* by decreasing the osmoprotectant and antioxidant systems (Hussain et al., 2019). BRs have also been found to play an important role in biotic stress. BRs application enhances nitric oxide (NO) content, which decreases the accumulation of the cucumber mosaic virus in *Arabidopsis thaliana* (Zou et al., 2017). BRs mediate the biotic stress response by managing photosynthetic mechanisms and secondary metabolism.

Nitric oxide (NO) is a gas molecule that plays an important role in plant development processes like seed germination, stomata development, adventitious root development, and various stresses like heat, drought, cold, salt, and so forth. Recent research has indicated that BRs interact with NO and play an important role in plant growth and stress management. Karpets et al.'s application of EBR and SNP in low amounts has been found to increase heat resistance in *Triticum aestivum* (Karpets et al., 2020). By stimulating *Cucumis sativus* L. to produce endogenous NO, BL encouraged the development of adventitious roots (Li et al., 2020). In *Atractylodes lancea,* co-application of NO and BL promotes fungal endophyte-induced volatile oil production, which improves its pharmacological value.

Ethylene is a gaseous plant growth-promoting hormone and plays a key role in physiological and stress responses in plants. According to Wei et al. (2015), BL caused *Cucumis sativus* L. seedlings to produce more ethylene, which helps in mitigating salt, drought, and chill stresses. BRs and ethylene interactions decrease oxidative damage in plant photosystems and increase tolerance under stress conditions. In *Nicotiana benthamiana,* treatment with BL and ACC increases resistance to *Pseudomonas syringae*. BRs play an important role in ethylene biosynthesis (Deng et al., 2015).

A tiny signaling molecule called hydrogen peroxide has an impact on a number of plant physiological and biochemical functions. BRs and H_2O_2 together play a role in plant growth and stress mitigation in plants. Tian et al. showed that hydrogen peroxide content was increased in BR-treated plants. BRs and hydrogen peroxide enhance crop yield by regulating photosynthesis and sugar metabolism in plants. Under cold stress in *Lycopersicon esculentum,* BR and H_2O_2 improve plant resistance by increasing SPAD chlorophyll, the net photosynthetic rate, and the activity of carbonic anhydrase (CA) (Khan et al., 2019). In *Solanum lycopersicum,* co-application of BRs and H_2O_2 increases resistance to copper toxicity (Nazir et al., 2020). These studies clearly indicate that BRs interact with plant growth hormones and improve plant quality.

9.6 FUTURE CHALLENGES

A class of polyhydroxylated plant steroid hormones known as BRs are essential to many facets of a plant's life. Although BRs were first identified for their role in cell elongation, it is now evident that they are important for plant development, growth, and responses to a variety of stresses, including drought and extremely high temperatures. In order for optimal plant growth in response to developmental phases and environmental changes, regulation of BR production, catabolism, or signaling at the tissue or organ level is essential. Although numerous internal variables have been found to mediate BR biosynthesis and signaling, jasmonic acids are lipid-derived phytohormones and play a role in development, abiotic stress responses, and interactions between plants and microorganisms in defense and symbiosis, such as gene expression, physiological variables, antioxidant metabolism, osmolyte production, and overall plant growth. To protect the plant and promote resistance against insects and pests, these hormones are only produced when usual conditions, such as heat, cold, or salinity, are not met.

Further research should be needed on how the JAs and BRs networks regulate other PHs under changing environmental conditions. Future crop improvements using conventional breeding or genetic engineering, such as CRISPR/Cas-based genome editing, would benefit from understanding the precise regulatory mechanisms of BRs and JAs biosynthesis and signaling. It is still unclear how these factors interact with specific developmental stages or a variety of environmental influences. Future research should focus on finding proof that BRs and JAs are carried outside of cells, as well as BR transporters and JAs. Finally, we need to discover more concerning the mechanisms by which BRs and JAs support plants' overall development programs in conditions like shadow, high temperatures, and drought. In order to help plants adapt to the stress brought on by higher variations in various environmental conditions, researchers should strive to modify the jasmonates and BRs pathway in crops and other plants.

9.7 CONCLUSION

JA serves as a pivotal PH, intricately involved in orchestrating defense responses against biotic as well as abiotic stresses. Its multifaceted role encompasses activating defense genes, regulating stress responses, and shaping plant growth and development. JA's diverse functions highlight its significance in enhancing plant immunity, resilience, and adaptability to a changing environment. Through its intricate signaling pathways and interactions with other hormones, JA contributes to the intricate balance between growth and defense, ultimately ensuring the survival and success of plants in challenging conditions. BRs have also been found to play a role in abiotic stresses like drought, salinity, high temperature, and low temperature. Cold tolerance in Elymus mutants is enhanced by EBR treatment by increasing proline content and decreasing MDA and ROS accumulation. BRs have also been found to play an important role in biotic stress.

KEYWORDS

- **brassinosteroids**
- **jasmonates**
- **multifunctional protein**
- **plant growth regulators**
- **plant hormones**

REFERENCES

Abdelgawad, Z. A., Khalafaallah, A. A., & Abdallah, M. M. (2014). Impact of methyl jasmonate on antioxidant activity and some biochemical aspects of maize plant grown under water stress condition. *Agricultural Sciences*, *5*(12), 1077–1088.

Aftab, T., Khan, M. M., Idrees, M., Naeem, M., Moinuddin, & Hashmi, N. (2011). Methyl jasmonate counteracts boron toxicity by preventing oxidative stress and regulating antioxidant enzyme activities and artemisinin biosynthesis in *Artemisia annua* L. *Protoplasma*, *248*(3), 601–612. https://doi.org/10.1007/s00709-010-0218-5 (accessed on 6th November 2025).

Aguilar, E., & Lozano-Duran, R. (2022). Plant viruses as probes to engineer tolerance to abiotic stress in crops. *Stress Biology*, *2*, 20. https://doi.org/10.1007/s44154-022-00043-4 (accessed on 6th November 2025).

Alam, M. N., Zhang, L., Yang, L., Islam, M. R., Liu, Y., Luo, H., Yang, P., Wang, Q., & Chan, Z. (2018). Transcriptomic profiling of tall fescue in response to heat stress improved thermotolerance by melatonin 24-epibrassinolide. *BMC Genomics*, *19*, 224. https://doi.org/10.1186/s12864-018-4588-y (accessed on 6th November 2025).

Ali, M., & Baek, K. (2020). Jasmonic acid signaling pathway in response to abiotic stresses in plants. *International Journal of Molecular Sciences*, *21*(2), 621. https://doi.org/10.3390/ijms21020621 (accessed on 6th November 2025).

Ali, M., Abbasi, B. H., & Ali, G. S. (2015). Elicitation of antioxidant secondary metabolites with jasmonates and gibberellic acid in cell suspension cultures of *Artemisia absinthium* L. *Plant Cell, Tissue and Organ Culture*, *120*, 1099–1106.

Altaf-Ul-Amin, M., Katsuragi, T., Sato, T., & Kanaya, S. (2015). A glimpse to background and characteristics of major molecular biological networks. *BioMed Research International*, *2015*, 540297. https://doi.org/10.1155/2015/540297 (accessed on 6th November 2025).

Anderson, J. P., Badruzsaufari, E., Peer, M. S., Manners, J. M., Desmond, O. J., Ehlet, C., Maclean, D. J., Ebert, P. R., & Kazan, K. (2004). Antagonistic interaction between abscisic acid and jasmonate-ethylene signaling pathways modulates defense gene expression and disease resistance in *Arabidopsis*. *Plant Cell*, *16*(12), 3460–3479.

Anusha, T. S., Joseph, M. V., & Elyas, K. K. (2016). Callus induction and elicitation of total phenolics in callus cell suspension culture of *Celastrus paniculatus*–wild, an endangered medicinal plant in India. *Pharmacognosy Journal*, *8*(5), 471–475. http://phcogj.com/article/193 (accessed on 6th November 2025).

Asami, T., & Yoshida, S. (1999). Brassinosteroid biosynthesis inhibitors. *Trends in Plant Science*, *4*, 348–353.

Aslam, H., Mushtaq, S., Maalik, S., Bano, N., Eed, E. M., Bibi, A., Tahir, A., Ijaz, I., Tanwir, S., & Khalifa, A. S. (2022). Exploring the effect of jasmonic acid for aphids' control for improving the yield of *Triticum aestivum* varieties. *PeerJ*, *10*, e14018. https://peerj.com/articles/14018/ (accessed on 6th November 2025).

Bali, S., Jamwal, V. L., Kaur, P., Kohli, S. K., Ohri, P., Gandhi, S. G., Bhardwaj, R., Al-Huqail, A. A., Siddiqui, M. H., & Ahmad, P. (2019). Role of P-type ATPase metal transporters and plant immunity induced by jasmonic acid against lead (Pb) toxicity in tomato. *Ecotoxicology and Environmental Safety*, *174*, 283–294.

Ballhorn, D. J., Reisdor, C., & Pfanz, H. (2011). Quantitative effects of enhanced CO_2 on jasmonic acid induced plant volatiles of lima bean (*Phaseolus lunatus* L.). *Journal of Applied Botany and Food Quality*, *84*, 65–71.

Carvalhais, L. C., Dennis, P. G., Badri, D. V., Tyson, G. W., Vivanco, J. M., & Schenk, P. M. (2013). Activation of the jasmonic acid plant defense pathway alters the composition of rhizosphere bacterial communities. *PLoS ONE*, *8*(2), e56457. https://doi.org/10.1371/journal.pone.0056457 (accessed on 6th November 2025).

Cerrudo, I., Keller, M. M., Cargnel, M. D., Demkura, P. V., de Wit, M., Patitucci, M. S., Pierik, R., Pieterse, C. M. J., & Ballaré, C. L. (2012). Low red/far-red ratios reduce *Arabidopsis* resistance to *Botrytis cinerea* and jasmonate responses via a COI1-JAZ10-dependent, salicylic acid-independent mechanism. *Plant Physiology*, *158*, 2042–2052.

Cervilla, L. M., Blasco, B., Ríos, J. J., Romero, L., & Ruiz, J. M. (2007). Oxidative stress and antioxidants in tomato (*Solanum lycopersicum*) plants subjected to boron toxicity. *Annals of Botany*, *100*, 747–756.

Choe, S., Fujioka, S., Noguchi, T., Takatsuto, S., Yoshida, S., & Feldmann, K. A. (2001). Overexpression of DWARF4 in the brassinosteroid biosynthetic pathway results in increased vegetative growth and seed yield in *Arabidopsis*. *Plant Journal*, *26*(6), 573–582. https://doi.org/10.1046/j.1365-313x.2001.01055.x (accessed on 6th November 2025).

Choudhary, S. P., Oral, H. V., Bhardwaj, R., Yu, J. Q., & Tran, L. S. P. (2012). Interaction of brassinosteroids polyamines enhances copper stress tolerance in *Raphanus sativus*. *Journal of Experimental Botany*, *63*(15), 5659–5675. https://doi.org/10.1093/jxb/ers219 (accessed on 6th November 2025).

Cohen, Y., Gisi, U., & Niderman, T. (1993). Local and systemic protection against *Phytophthora infestans* induced in potato and tomato plants by jasmonic acid and jasmonic methyl ester. *Phytopathology*, *83*(10), 1054–1062.

Cui, H., Wei, J., Su, J., Li, C., & Ge, F. (2016). Elevated O_3 increases volatile organic compounds via jasmonic acid pathway that promote the preference of parasitoid *Encarsia formosa* for tomato plants. *Plant Science*, *253*, 243–250.

Czapski, J., & Saniewski, M. (1992). Stimulation of ethylene production and ethylene-forming enzyme in fruits of the non-ripening nor and rin tomato mutants by methyl jasmonate. *Journal of Plant Physiology, 139*(3), 265–268. https://doi.org/10.1016/S0176-1617(11)80334-2 (accessed on 6th November 2025).

Dai, H., Wei, S., Pogrzeba, M., Rusinowski, S., Krzyżak, J., & Jia, G. (2020). Exogenous jasmonic acid decreased Cu accumulation by alfalfa and improved its photosynthetic pigments and antioxidant system. *Ecotoxicology and Environmental Safety, 190*, 110176. https://doi.org/10.1016/j.ecoenv.2020.110176 (accessed on 6th November 2025).

Dar, T. A., Uddin, M., Khan, M. M. A., Hakeem, K. R., & Jaleel, H. (2015). Jasmonates counter plant stress: A review. *Environmental and Experimental Botany, 115*, 49–57.

De Domenico, S., Taurino, M., Gallo, A., Poltronieri, P., Pastor, V., Flors, V., & Santino, A. (2019). Oxylipin dynamics in *Medicago truncatula* in response to salt and wounding stresses. *Physiologic Plantarum, 165*, 198–208.

De Ollas, C., Hernando, B., Arbona, V., & Gomez-Cadenas, A. (2015). Jasmonic acid transient accumulation is needed for abscisic acid increase in citrus roots under drought stress conditions. *Physiologic Plantarum, 147*, 296–306.

Demole, E., Lederer, E., & Mercier, D. (1962). Isolement et détermination de la structure du jasmonate de méthyle, constituant odorant caractéristique de l'essence de jasmin. *Helvetica Chimica Acta, 45*, 675–685.

Deng, X. G., Zhu, T., Zhang, D. W., & Lin, H. H. (2015). The alternative respiratory pathway is involved in brassinosteroid-induced environmental stress tolerance in *Nicotiana benthamiana*. *Journal of Experimental Botany, 66*, 6219–6232.

Devoto, A., & Turner, J. G. (2005). Jasmonate-regulated *Arabidopsis* stress signaling network. *Physiologic Plantarum, 123*, 161–172.

Evans, N. H. (2003). Modulation of guard cell plasma membrane potassium currents by methyl jasmonate. *Plant Physiology, 131*, 8–11.

Faghih, S., Ghobadi, C., & Zarei, A. (2017). Response of strawberry plant cv. 'Camarosa' to salicylic acid and methyl jasmonate application under salt stress condition. *Journal of Plant Growth Regulation, 36*, 651–659.

Fan, L., Wang, Q., Lv, J., Gao, L., Zuo, J., & Shi, J. (2016). Amelioration of postharvest chilling injury in cowpea (*Vigna sinensis*) by methyl jasmonate (MeJA) treatments. *Scientia Horticulture, 203*, 95–101.

Farooq, M. A., Gill, R. A., Islam, F., Ali, B., Liu, H., Xu, J., He, S., & Zhou, W. (2016). Methyl jasmonate regulates antioxidant defense and suppresses arsenic uptake in *Brassica napus* L. *Frontiers in Plant Science, 7*, 1–16.

Fu, J., Sun, P., Luo, Y., Zhou, H., Gao, J., Zhao, D., Pubu, Z., Liu, J., & Hu, T. (2019). Brassinosteroids enhance cold tolerance in *Elymus nutans* via mediating redox homeostasis and proline biosynthesis. *Environmental and Experimental Botany, 167*, 103831. https://doi.org/10.1016/j.envexpbot.2019.103831 (accessed on 6th November 2025).

Fu, J., Wu, H., Ma, S., Xiang, D., Liu, R., & Xiong, L. (2017). OSJAZ1 attenuates drought resistance by regulating JA and ABA signaling in rice. *Frontiers in Plant Science, 8*, 2108. https://doi.org/10.3389/fpls.2017.02108 (accessed on 6th November 2025).

Fujioka, S., & Yokota, T. (2003). Biosynthesis and metabolism of brassinosteroids. *Annual Review of Plant Biology, 54*, 137–164.

Gimenez-Ibanez, S., Chini, A., & Solano, R. (2016). How microbes twist jasmonate signaling around their little fingers. *Plants (Basel), 5*(1), 9. https://doi.org/10.3390/plants5010009 (accessed on 6th November 2025).

Gobel, C., & Feussner, I. (2009). Methods for the analysis of oxylipins in plants. *Phytochemistry, 70*, 1485–1503.

Gonzalez-Aguilar, G. A., Fortiz, J., Cruz, R., Baez, R., & Wang, C. Y. (2000). Methyl jasmonate reduces chilling injury and maintains postharvest quality of mango fruit. *Journal of Agricultural and Food Chemistry, 48*, 515–519.

Gonzalez-Aguilar, G. A., Tiznado-Hernández, M. E., Zavaleta-Gatica, R., & Martínez-Téllez, M. A. (2004). Methyl jasmonate treatments reduce chilling injury and activate the defense response of guava fruits. *Biochemical and Biophysical Research Communications, 313*, 694–701.

Goossens, J., Fernandez-Calvo, P., Schweizer, F., & Goossens, A. (2016). Jasmonates: Signal transduction components and their roles in environmental stress responses. *Plant Molecular Biology, 91*, 673–689.

Grove, M. D., Spencer, G. F., Rohwedder, W. K., Mandava, N. B., Worley, J. F., Warthen, J. D., Steffens, G. L., Flippen-Anderson, J. L., & Cook, J. C. (1979). Brassinolide, a plant growth-promoting steroid isolated from *Brassica napus* pollen. *Nature, 281*, 216–217.

Gudesblat, G. E., & Russinova, E. (2011). Plants grow on brassinosteroids. *Current Opinion in Plant Biology, 14*, 530–537.

Guerreiro, A., Figueiredo, J., Silva, M., & Figueiredo, A. (2016). Linking jasmonic acid to grapevine resistance against the biotrophic oomycete *Plasmopara viticola. Frontiers in Plant Science, 7*, 565. https://doi.org/10.3389/fpls.2016.00565 (accessed on 6th November 2025).

Gunes, A., Inal, A., Bagci, E. G., Coban, S., & Sahin, O. (2007). Silicon increases boron tolerance and reduces oxidative damage of wheat grown in soil with excess boron. *Biologic Plantarum, 51*, 571–574.

Gupta, A., Bhardwaj, M., & Tran, L. P. (2020). Jasmonic acid at the crossroads of plant immunity and *Pseudomonas syringae* virulence. *International Journal of Molecular Sciences, 21*(20), 7482. https://doi.org/10.3390/ijms21207482 (accessed on 6th November 2025).

Gupta, N., Prasad, V. B. R., & Chattopadhyay, S. (2014). LeMYC2 acts as a negative regulator of blue light mediated photomorphogenic growth, and promotes the growth of adult tomato plants. *BMC Plant Biology, 14*, 1–14.

He, J. X., Gendron, J. M., Sun, Y., Gampala, S. S., Gendron, N., Sun, C. Q., & Wang, Z. Y. (2005). BZR1 is a transcriptional repressor with dual roles in brassinosteroid homeostasis and growth responses. *Science, 307*, 1634–1638.

He, Y., Li, J., Ban, Q., Han, S., & Rao, J. (2018). Role of brassinosteroids in persimmon (*Diospyros kaki* L.) fruit ripening. *Journal of Agricultural and Food Chemistry, 66*, 2637–2644.

Ho, T. T., Murthy, H. N., & Park, S. Y. (2020). Methyl jasmonate induced oxidative stress and accumulation of secondary metabolites in plant cell and organ cultures. *International Journal of Molecular Sciences, 21*(3), 716. https://doi.org/10.3390/ijms21030716 (accessed on 6th November 2025).

Horton, R. F. (1991). Methyl jasmonate and transpiration in barley. *Plant Physiology, 96*, 1376–1378.

Howe, G. A., & Jander, G. (2008). Plant immunity to insect herbivores. *Annual Review of Plant Biology, 59*, 41–66.

Howe, G. A., Lightner, J., Browse, J., & Ryan, C. A. (1996). An octadecanoid pathway mutant (JL5) of tomato is compromised in signaling for defense against insect attack. *The Plant Cell, 8*, 2067–2077.

Huang, H., Liu, B., Liu, L., & Song, S. (2017). Jasmonate action in plant growth and development. *Journal of Experimental Botany, 68*(6), 1349–1359.

Hussain, M., Khan, T. A., Yusuf, M., & Fariduddin, Q. (2019). Silicon-mediated role of 24-epibrassinolide in wheat under high-temperature stress. *Environmental Science and Pollution Research, 26*, 17163–17172.

Isah, T., Umar, S., Mujib, A., Sharma, M. P., Rajasekharan, P. E., Zafar, N., & Frukh, A. (2018). Secondary metabolism of pharmaceuticals in the plant *in vitro* cultures: Strategies, approaches, and limitations to achieving higher yield. *Plant Cell, Tissue and Organ Culture, 132*, 239–265.

Ismail, A., Riemann, M., & Nick, P. (2012). The jasmonate pathway mediates salt tolerance in grapevines. *Journal of Experimental Botany, 63*, 2127–2139.

Jan, S., Noman, A., Kaya, C., Ashraf, M., Alyemeni, M. N., & Ahmad, P. (2020). 24-epibrassinolide alleviates the injurious effects of Cr (VI) toxicity in tomato plants: Insights into growth, physio-biochemical attributes, antioxidant activity regulation of ascorbate–glutathione glyoxalase cycles. *Journal of Plant Growth Regulation, 39*, 1587–1604.

Jasik, J., & de Klerk, G. J. (2006). Effect of methyl jasmonate on morphology and dormancy development in lily bulblets regenerated *in vitro*. *Journal of Plant Growth Regulation, 25*, 45–51. https://doi.org/10.1007/s00344-005-0048-4 (accessed on 6th November 2025).

Jiang, S., Yao, J., Ma, K. W., Zhou, H., Song, J., He, S. Y., & Ma, W. (2013). Bacterial effector activates jasmonate signaling by directly targeting JAZ transcriptional repressors. *PLoS Pathogens, 9*(10), e1003715. https://doi.org/10.1371/journal.ppat.1003715 (accessed on 6th November 2025).

Jin, P., Duan, Y., Wang, L., Wang, J., & Zheng, Y. (2014). Reducing chilling injury of loquat fruit by combined treatment with hot air and methyl jasmonate. *Food and Bioprocess Technology, 7*, 2259–2266.

Jin, P., Zheng, Y., Tang, S., Rui, H., & Wang, C. Y. (2009). A combination of hot air and methyl jasmonate vapor treatment alleviates chilling injury of peach fruit. *Postharvest Biology and Technology, 52*, 24–29.

Jiroutova, P., Mikulík, J., Novak, O., Strnad, M., & Oklestkova, J. (2019). Brassinosteroids induce strong, dose-dependent inhibition of etiolated pea seedling growth correlated with ethylene production. *Biomolecules, 9*(12), 849. https://doi.org/10.3390/biom9120849 (accessed on 6th November 2025).

Kanna, M., Tamaoki, M., Kubo, A., Nakajima, N., Rakwal, R., Agrawal, G. K., Tamogami, S., Ioki, M., Ogawa, D., Saji, H., & Aono, M. (2003). Isolation of an ozone-sensitive and jasmonate-semi-insensitive *Arabidopsis* mutant (oji1). *Plant and Cell Physiology, 44*(12), 1301–1310. https://doi.org/10.1093/pcp/pcg157 (accessed on 6th November 2025).

Karabal, E., Yücel, M., & Oktem, H. A. (2003). Antioxidant responses of tolerant and sensitive barley cultivars to boron toxicity. *Plant Science, 164*, 925–933.

Karpets, Y., Shkliarevskyi, M., Khripach, V. A., & Kolupaev, Y. E. (2020). State of enzymatic antioxidative system heat resistance of wheat plantlets treated by combination of 24-epibrassinolide NO donor. *Cereal Research Communications, 49*, 207–216.

Khan, T. A., Yusuf, M., Ahmad, A., Bashir, Z., Saeed, T., Fariduddin, Q., Shamsul, H., Mock, H. P., & Wu, T. Q. (2019). Proteomic physiological assessment of stress sensitive tolerant variety of tomato treated with brassinosteroids hydrogen peroxide under low-temperature stress. *Food Chemistry, 289*, 500–511.

Kim, H., Seomun, S., Yoon, Y., & Jang, G. (2021). Jasmonic acid in plant abiotic stress tolerance and interaction with abscisic acid. *Agronomy, 11*(9), 1886. https://doi.org/10.3390/agronomy11091886 (accessed on 6th November 2025).

Kim, J., Chang, C., & Tucker, M. L. (2015). To grow old: Regulatory role of ethylene and jasmonic acid in senescence. *Frontiers in Plant Science, 6*, 20. https://doi.org/10.3389/fpls.2015.00020 (accessed on 6th November 2025).

Kim, O. T., Bang, K. H., Kim, Y. C., Hyun, D. Y., Kim, M. Y., & Cha, S. W. (2009). Upregulation of ginsenoside and gene expression related to triterpene biosynthesis in ginseng hairy root cultures elicited by methyl jasmonate. *Plant Cell, Tissue and Organ Culture, 98*, 25–33.

Kinoshita, T., Cano-Delgado, A., Seto, H., Hiranuma, S., Fujioka, S., Yoshida, S., & Chory, J. (2005). Binding of brassinosteroids to the extracellular domain of plant receptor kinase BRI1. *Nature, 433*, 167–171.

Koch, J. R., Creelman, R. A., Eshita, S. M., Seskar, M., Mullet, J. E., & Davis, K. R. (2000). Ozone sensitivity in hybrid poplar correlates with insensitivity to both salicylic acid and jasmonic acid. The role of programmed cell death in lesion formation. *Plant Physiology, 123*, 487–496.

Kong, Q., Mostafa, H., Yang, W., Wang, J., Nuerawuti, M., Wang, Y., Song, J., Zhang, X., Ma, L., Wang, H., & Li, X. (2021). Comparative transcriptome profiling reveals that brassinosteroid-mediated lignification plays an important role in garlic adaption to salt stress. *Plant Physiology and Biochemistry, 158*, 34–42.

Leon, J., Rojo, E., & Sánchez-Serrano, J. J. (2001). Wound signaling in plants. *Journal of Experimental Botany, 52*, 1–9.

Li, J., & Chory, J. (1997). A putative leucine-rich repeat receptor kinase involved in brassinosteroid signal transduction. *Cell, 90*, 929–938.

Li, J., & Nam, K. H. (2002). Regulation of brassinosteroid signaling by a GSK3/SHAGGY-like kinase. *Science, 295*, 1299–1301.

Li, T., Lei, W., He, R., Tang, X., & Zhang, D. (2020). Brassinosteroids regulate root meristem development by mediating BIN2-UPB1 module in *Arabidopsis*. *PLoS Genetics, 16*, e1008883. https://doi.org/10.1371/journal.pgen.1008883 (accessed on 6th November 2025).

Li, Y. T., Wu, Y., Liao, W. B., Hu, L. L., & Yu, J. H. (2020). Nitric oxide is involved in the brassinolide-induced adventitious root development in cucumber. *BMC Plant Biology, 20*, 102. https://doi.org/10.1186/s12870-020-2320-y (accessed on 6th November 2025).

Liao, K., Peng, Y. J., Yuan, L. B., Dai, Y. S., Chen, Q. F., Yu, L. J., Bai, M. Y., Zhang, W. Q., Xie, L. J., & Xiao, S. (2020). Brassinosteroids antagonize jasmonate-activated plant defense responses through BRI1-EMS-SUPPRESSOR1 (BES1). *Plant Physiology, 182*(2), 1066–1082.

Lopez-Climent, M., & Gómez-Cadenas, A. (2015). Jasmonoyl isoleucine accumulation is needed for abscisic acid build-up in roots of *Arabidopsis* under water stress conditions. *Plant Cell and Environment, 38*(10), 2157–2170.

Lorenzo, O., & Solano, R. (2005). Molecular players regulating the jasmonate signaling network. *Current Opinion in Plant Biology, 8*, 532–540.

Lu, C., Qi, J., Hettenhausen, C., Lei, Y., Zhang, J., Zhang, M., Zhang, C., Song, J., Li, J., Cao, G., Malook, S. U., & Wu, J. (2018). Elevated CO_2 differentially effects tobacco and rice defense against lepidopteran larvae via the jasmonic acid signaling pathway. *Journal of Integrative Plant Biology, 60*(5), 412–431.

Lu, Z., Wang, X., Cao, M., Li, Y., Su, J., & Gao, H. (2019). Effect of 24-epibrassinolide on sugar metabolism delaying postharvest senescence of kiwifruit during ambient storage. *Scientia Horticulture, 253*, 1–7.

Macioszek, V., Jęcz, T., Ciereszko, I., & Kononowicz, A. K. (2023). Jasmonic acid as a mediator in plant response to necrotrophic fungi. *Cells, 12*(7), 1027. https://doi.org/10.3390/cells12071027 (accessed on 6th November 2025).

Maksymiec, W., & Krupa, Z. (2007). Effects of methyl jasmonate and excess copper on root and leaf growth. *Biologia Plantarum, 51*, 322–326.

Mandava, N. B. (1988). Plant growth-promoting brassinosteroids. *Annual Review of Plant Physiology and Plant Molecular Biology, 39*, 23–52.

Mewis, I., Schreiner, M., Nguyen, C. N., Krumbein, A., Ulrichs, C., Lohse, M., & Zrenner, R. (2012). UV-B irradiation changes specifically the secondary metabolite profile in broccoli sprouts: Induced signaling overlaps with defense response to biotic stressors. *Plant and Cell Physiology, 53*, 1546–1560.

Mir, M. A., Sirhindi, G., Alyemeni, M. N., Alam, P., & Ahmad, P. (2018). Jasmonic acid improves growth performance of soybean under nickel toxicity by regulating nickel uptake, redox balance, and oxidative stress metabolism. *Journal of Plant Growth Regulation, 37*, 1195–1209.

Mitchell, J. W., Mandava, N. B., Worley, J. F., Plimmer, J. R., & Smith, M. V. (1970). Brassins: A new family of plant hormones from rape pollen. *Nature (London), 225*, 1065–1066.

Mohamed, H. I., & Latif, H. H. (2017). Improvement of drought tolerance of soybean plants by using methyl jasmonate. *Physiology and Molecular Biology of Plants, 23*, 545–556.

Molassiotis, A., Sotiropoulos, T., Tanou, G., Diamantidis, G., & Therios, I. (2006). Boron-induced oxidative damage and antioxidant and nucleolytic responses in shoot tips culture of the apple rootstock EM 9 (*Malus domestica* Borkh). *Environmental and Experimental Botany, 56*, 54–62.

Nam, K. H., & Li, J. (2002). BRI1/BAK1, a receptor kinase pair mediating brassinosteroid signaling. *Cell, 110*, 203–212.

Nazir, F., Fariduddin, Q., Hussain, A., & Khan, T. A. (2020). Brassinosteroid and hydrogen peroxide improve photosynthetic machinery, stomatal movement, root morphology, cell viability and reduce Cu-triggered oxidative burst in tomato. *Ecotoxicology and Environmental Safety, 207*, 111081. https://doi.org/10.1016/j.ecoenv.2020.111081 (accessed on 6th November 2025).

Ncube, B., Finnie, J. F., & Van Staden, J. (2012). Quality from the field: The impact of environmental factors as quality determinants in medicinal plants. *South African Journal of Botany, 82*, 11–20.

Nguyen, H., To, H., Lebrun, M., Bellafiore, S., & Champion, A. (2019). Jasmonates—the master regulator of rice development, adaptation and defense. *Plants, 8*(9), 339. https://doi.org/10.3390/plants8090339 (accessed on 6th November 2025).

Ni, J., Zhao, Y., Tao, R., Yin, L., Gao, L., Strid, Å., & Bai, S. (2020). Ethylene mediates the branching of the jasmonate-induced flavonoid biosynthesis pathway by suppressing anthocyanin biosynthesis in red Chinese pear fruits. *Plant Biotechnology Journal, 18*(5), 1223–1240.

O'Brien, J. A., & Benková, E. (2013). Cytokinin cross-talking during biotic and abiotic stress responses. *Frontiers in Plant Science, 4*, 1–11.

Overmyer, K., Tuominen, H., Kettunen, R., Betz, C., Langebartels, C., Sandermann, H. J., & Kangasjärvi, J. (2000). Ozone-sensitive *Arabidopsis* rcd1 mutant reveals opposite roles for ethylene and jasmonate signaling pathways in regulating superoxide-dependent cell death. *The Plant Cell, 12*, 1849–1862.

Papadakis, I. E., Dimassi, K. N., Bosabalidis, A. M., Therios, I. N., Patakas, A., & Giannakoula, A. (2004). Effects of B excess on some physiological and anatomical parameters of "Navelina" orange plants grafted on two rootstocks. *Environmental and Experimental Botany, 51*, 247–257. https://doi.org/10.1016/j.envexpbot.2003.11.004 (accessed on 6th November 2025).

Peres, A. L., Soares, J. S., Tavares, R. G., Righetto, G., Zullo, M. A., Mandava, N. B., & Menossi, M. (2019). Brassinosteroids, the sixth class of phytohormones: A molecular view from the discovery to hormonal interactions in plant development and stress adaptation. *International Journal of Molecular Sciences, 20*(2), 331. https://doi.org/10.3390/ijms20020331 (accessed on 6th November 2025).

Perez-Perez, J. M., Ponce, M. R., & Micol, J. L. (2002). The UCU1 *Arabidopsis* gene encodes a SHAGGY/GSK3-like kinase required for cell expansion along the proximodistal axis. *Developmental Biology, 242*, 161–173.

Pieterse, C. M. J., Does, D. V. D., Zamioudis, C., Leon-Reyes, A., & Wees, S. C. M. V. (2012). Hormonal modulation of plant immunity. *Annual Review of Cell and Developmental Biology, 28*(1), 489–521.

Poonam, S., Kaur, H., & Geetika, S. (2013). Effect of jasmonic acid on photosynthetic pigments and stress markers in *Cajanus cajan* (L.) Millsp. seedlings under copper stress. *American Journal of Plant Sciences, 4*(4), 817–823. https://www.scirp.org/journal/paperinformation?paperid=29827 (accessed on 6th November 2025).

Qiu, Z., Guo, J., Zhu, A., Zhang, L., & Zhang, M. (2014). Exogenous jasmonic acid can enhance tolerance of wheat seedlings to salt stress. *Ecotoxicology and Environmental Safety, 104*, 202–208.

Rademacher, W. (2000). Growth retardants: Effects on gibberellin biosynthesis and other metabolic pathways. *Annual Review of Plant Physiology and Plant Molecular Biology, 51*, 501–531.

Rao, M. V., Lee, H., Creelman, R. A., Mullet, J. E., & Davis, K. R. (2000). Jasmonic acid signaling modulates ozone-induced hypersensitive cell death. *The Plant Cell, 12*(9), 1633–1647.

Rattan, A., Kapoor, D., Kapoor, N., Bhardwaj, R., & Sharma, A. (2020). Brassinosteroids regulate functional components of antioxidative defense system in salt-stressed maize seedlings. *Journal of Plant Growth Regulation, 39*, 1465–1475.

Ravnikar, M., Zel, J., Plaper, I., & Spacapan, A. (1993). Jasmonic acid stimulates shoot and bulb formation of garlic *in vitro*. *Journal of Plant Growth Regulation, 12*, 73–77. https://doi.org/10.1007/BF00193236 (accessed on 6th November 2025).

Raza, A. (2020). Eco-physiological and biochemical responses of rapeseed (*Brassica napus* L.) to abiotic stresses: Consequences and mitigation strategies. *Journal of Plant Growth Regulation, 40*, 1368–1388. https://doi.org/10.1007/s00344-020-10231-z (accessed on 6th November 2025).

Raza, A., Hussain, S., Javed, R., Hafeez, M. B., & Hasanuzzaman, M. (2021b). Antioxidant defense systems and remediation of metal toxicity in plants. In *Approaches to the Remediation of Inorganic Pollutants* (pp. 91–124). Springer, Singapore. https://doi.org/10.1007/978-981-15-6221-1 (accessed on 6th November 2025).

Robert-Seilaniantz, A., Grant, M., & Jones, J. D. G. (2011). Hormone crosstalk in plant disease and defense: More than just jasmonate-salicylate antagonism. *Annual Review of Phytopathology, 49*(1), 317–343.

Rogowska, A., Stpiczyńska, M., Pączkowski, C., & Szakiel, A. (2022). The influence of exogenous jasmonic acid on the biosynthesis of steroids and triterpenoids in *Calendula officinalis* plants and hairy root culture. *International Journal of Molecular Sciences, 23*(20), 12173. https://doi.org/10.3390/ijms232012173 (accessed on 6th November 2025).

Rowe, H., Walley, J. W., Corwin, J. A., Chan, E. K. F., Dehesh, K., & Kliebenstein, D. J. (2010). Deficiencies in jasmonate-mediated plant defense reveal quantitative variation in *Botrytis cinerea* pathogenesis. *PLoS Pathogens, 6*(4), e1000861. https://doi.org/10.1371/journal.ppat.1000861 (accessed on 6th November 2025).

Sayyari, M., Babalar, M., Kalantari, S., Martínez-Romero, D., Guillén, F., Serrano, M., & Valero, D. (2011). Vapor treatments with methyl salicylate or methyl jasmonate alleviated chilling injury and enhanced antioxidant potential during postharvest storage of pomegranates. *Food Chemistry, 124*, 964–970.

Schaller, F., Schaller, A., & Stintzi, A. (2005). Biosynthesis and metabolism of jasmonates. *Journal of Plant Growth Regulation, 23*, 179–199.

Schuman, M. C., Meldau, S., Gaquerel, E., Diezel, C., McGale, E., Greenfield, S., & Baldwin, I. T. (2018). The active jasmonate JA-Ile regulates a specific subset of plant jasmonate-mediated resistance to herbivores in nature. *Frontiers in Plant Science, 9*, 787. https://doi.org/10.3389/fpls.2018.00787 (accessed on 6th November 2025).

Shabani, L., Ehsanpour, A. A., & Esmaeili, A. (2010). Assessment of squalene synthase and beta-amyrin synthase gene expression in licorice roots treated with methyl jasmonate and salicylic acid using real-time qPCR. *Russian Journal of Plant Physiology, 57*, 480–484.

Shilpha, J., Satish, L., Kavikkuil, M., Joe Virgin Largia, M., & Ramesh, M. (2015). Methyl jasmonate elicits the solasodine production and antioxidant activity in hairy root cultures of *Solanum trilobatum* L. *Industrial Crops and Products, 71*, 54–64.

Sirhindi, G., Mir, M. A., Abd-Allah, E. F., Ahmad, P., & Gucel, S. (2016). Jasmonic acid modulates the physio biochemical attributes, antioxidant enzyme activity, and gene expression in *Glycine max* under nickel toxicity. *Frontiers in Plant Science, 7*, 591. https://doi.org/10.3389/fpls.2016.00591 (accessed on 6th November 2025).

Siva, G., Sivakumar, S., Premkumar, G., Senthil, K. T., & Jayabalan, N. (2015). Enhanced production of psoralen through elicitors treatment in adventitious root culture of *Psoralea corylifolia* (L). *International Journal of Pharmacy and Pharmaceutical Sciences, 7*, 146–149.

Sohn, S. I., Pandian, S., Rakkammal, K., Largia, M. J. V., Thamilarasan, S. K., Balaji, S., Zoclanclounon, Y. A. B., Shilpha, J., & Ramesh, M. (2022). Jasmonates in plant growth and development and elicitation of secondary metabolites: An updated overview. *Frontiers in Plant Science, 13*, 942789. https://doi.org/10.3389/fpls.2022.942789 (accessed on 6th November 2025).

Steffens, G. L. (1991). U.S. Department of Agriculture Brassins Project: 1970–1980. In Cutler, H. G., Yokota, T., & Adam, G. (Eds.), *Brassinosteroids: Chemistry, Bioactivity, & Applications* (pp. 2–17). American Chemical Society, Washington, D.C.

Sun, Y., Yin, J., Cao, H., Li, C., Kang, L., & Ge, F. (2011). Elevated CO_2 influences nematode-induced defense responses of tomato genotypes differing in the JA pathway. *PLoS ONE, 6*(5), e19751. https://doi.org/10.1371/journal.pone.0019751 (accessed on 6th November 2025).

Suzuki, H., Fujioka, S., Takatsuto, S., Yokota, T., Murofushi, N., & Sakurai, A. (1995). Biosynthesis of brassinosteroids in seedlings of *Catharanthus roseus*, *Nicotiana tabacum*, and *Oryza sativa*. *Bioscience, Biotechnology, and Biochemistry, 59*, 168–172. https://doi.org/10.1271/bbb.59.168 (accessed on 6th November 2025).

Suzuki, H., Reddy, M. S., Naoumkina, M., Aziz, N., May, G. D., Huhman, D. V., Sumner, L. W., Blount, J. W., Mendes, P., & Dixon, R. A. (2005). Methyl jasmonate and yeast elicitor induce differential transcriptional and metabolic re-programming in cell suspension cultures of the model legume *Medicago truncatula*. *Planta, 220*(5), 696–707. https://doi.org/10.1007/s00425-004-1387-2 (accessed on 6th November 2025).

Svyatyna, K., & Riemann, M. (2012). Light-dependent regulation of the jasmonate pathway. *Protoplasma, 249*, 137–145.

Tagami, S., Ohnishi, K., Hikichi, Y., & Kiba, A. (2021). Trigalactosyldiacylglycerol 3 protein orthologs are required for basal disease resistance in *Nicotiana benthamiana*. *Plant Biotechnology (Tokyo), 38*(3), 373–378.

Thaler, J. S., Owen, B., & Higgins, V. J. (2004). The role of the jasmonate response in plant susceptibility to diverse pathogens with a range of lifestyles. *Plant Physiology, 135*(1), 530–538.

Tian, Y., Min, F., Qin, Z., Lv, H., & Bai, M. Y. (2018). Hydrogen peroxide positively regulates brassinosteroid signaling through oxidation of the brassinazole-resistant1 transcription factor. *Nature Communications, 9*, 1063. https://doi.org/10.1038/s41467-018-03463-x (accessed on 6th November 2025).

Todaka, D., Shinozaki, K., & Yamaguchi-Shinozaki, K. (2016). Recent advances in the dissection of drought-stress regulatory networks and strategies for development of drought-tolerant transgenic rice plants. *Frontiers in Plant Science, 6*, 84. https://doi.org/10.3389/fpls.2015.00084 (accessed on 6th November 2025).

Ueda, J., Miyamoto, K., & Kamisaka, S. (1995). Inhibition of the synthesis of cell wall polysaccharides in oat coleoptile segments by jasmonic acid: Relevance to its growth inhibition. *Journal of Plant Growth Regulation, 14*, 69–76.

Walling, L. L. (2000). The myriad plant responses to herbivores. *Journal of Plant Growth Regulation, 19*, 195–216.

Wang, X., Lu, Z., Su, J., Li, Y., Cao, M., & Gao, H. (2020). 24-Epibrassinolide delays senescence in harvested kiwifruit through effects on mitochondrial membrane antioxidant activity. *LWT–Food Science and Technology, 118*, 108833. https://doi.org/10.1016/j.lwt.2019.108833 (accessed on 6th November 2025).

Wang, Y., Mostafa, S., & Jin, B. (2021). Function and mechanism of jasmonic acid in plant responses to abiotic and biotic stresses. *International Journal of Molecular Sciences, 22*(16), 8568. https://doi.org/10.3390/ijms22168568 (accessed on 6th November 2025).

Wang, Z. Y., Nakano, T., Gendron, J., He, J., Chen, M., Vafeados, D., Yang, Y., Fujioka, S., Yoshida, S., Asami, T., & Chory, J. (2002). Nuclear-localized BZR1 mediates brassinosteroid-induced growth and feedback suppression of brassinosteroid biosynthesis. *Developmental Cell, 2*, 505–513.

Wang, Z. Y., Seto, H., Fujioka, S., Yoshida, S., & Chory, J. (2001). BRI1 is a critical component of a plasma-membrane receptor for plant steroids. *Nature, 410*, 380–383.

Wasternack, C., & Feussner, I. (2018). The oxylipin pathways: Biochemistry and function. *Annual Review of Plant Biology, 69*, 363–386.

Wasternack, C., & Hause, B. (2013). Jasmonates: Biosynthesis, perception, signal transduction and action in plant stress response, growth and development. *Annals of Botany, 111*, 1021–1068.

Wei, L. J., Deng, X. G., Zhu, T., Zheng, T., Li, P. X., & Wu, J. Q. (2015). Ethylene is involved in brassinosteroids induced alternative respiratory pathway in cucumber (*Cucumis sativus* L.) seedlings response to abiotic stress. *Frontiers in Plant Science, 6*, 982. https://doi.org/10.3389/fpls.2015.00982 (accessed on 6th November 2025).

Wu, H., Wu, X., Li, Z., Duan, L., & Zhang, M. (2012). Physiological evaluation of drought stress tolerance and recovery in cauliflower (*Brassica oleracea* L.) seedlings treated with methyl jasmonate and coronatine. *Journal of Plant Growth Regulation, 31*, 113–123.

Yan, Z., Chen, J., & Li, X. (2013). Methyl jasmonate as modulator of Cd toxicity in *Capsicum frutescens* var. *fasciculatum* seedlings. *Ecotoxicology and Environmental Safety, 98*, 203–209.

Yan, Z., Li, X., Chen, J., & Tam, N. F. Y. (2015). Combined toxicity of cadmium and copper in *Avicennia marina* seedlings and the regulation of exogenous jasmonic acid. *Ecotoxicology and Environmental Safety, 113*, 124–132.

Yan, Z., Zhang, W., Chen, J., & Li, X. (2015). Methyl jasmonate alleviates cadmium toxicity in *Solanum nigrum* by regulating metal uptake and antioxidative capacity. *Biologic Plantarum, 59*, 373–381.

Yao, Y., Zhao, N., Xian, T., Tu, S., Pan, L., & Tu, K. (2017). Effect of 2,4-epibrassinolide treatment on the postharvest quality physiological metabolism of fresh daylily flower buds during storage. *Scientia Horticulture, 226*, 110–116.

Yi, M., Zhao, M., Wao, K., Liu, C., Deng, D., Khao, Z., Li, J., & Deng, A. (2020). Simultaneous detection of plant growth regulators jasmonic acid and methyl jasmonate in plant samples by a monoclonal antibody-based ELISA. *Analyst, 145*(11), 4004–4011.

Yin, Y., Vafeados, D., Tao, Y., Yoshida, S., Asami, T., & Chory, J. (2005). A new class of transcription factors mediates brassinosteroid-regulated gene expression in *Arabidopsis*. *Cell, 120*, 249–259.

Yin, Y., Wang, Z. Y., Mora-Garcia, S., Li, J., Yoshida, S., Asami, T., & Chory, J. (2002). BES1 accumulates in the nucleus in response to brassinosteroids to regulate gene expression and promote stem elongation. *Cell, 109*, 181–191.

Yokota, T. (1999). The history of brassinosteroids: Discovery to isolation of biosynthesis and signal transduction mutants. In Sakurai, A., Yokota, T., & Clouse, S. D. (Eds.), *Brassinosteroids: Steroidal Plant Hormones* (pp. 1–20). Springer, Tokyo.

Yu, X., Zhang, W., Zhang, Y., Zhang, X., Lang, D., & Zhang, X. (2019). The roles of methyl jasmonate to stress in plants. *Functional Plant Biology, 46*, 197–212.

Zaragoza-Martínez, F., Lucho-Constantino, G. G., Ponce-Noyola, T., Esparza-García, F., Poggi-Varaldo, H., Cerda-García-Rojas, C. M., Trejo-Tapia, G., & Ramos-Valdivia, A. C. (2016). Jasmonic acid stimulates the oxidative responses and triterpene production in *Jatropha curcas* cell suspension cultures through mevalonate as biosynthetic precursor. *Plant Cell, Tissue and Organ Culture, 127*, 47–56.

Zhang, L., Zhang, F., Melotto, M., Yao, J., & He, S. Y. (2017). Jasmonate signaling and manipulation by pathogens and insects. *Journal of Experimental Botany, 68*(6), 1371–1385.

Zhang, X., Sheng, J., Li, F., Meng, D., & Shen, L. (2012). Methyl jasmonate alters arginine catabolism and improves postharvest chilling tolerance in cherry tomato fruit. *Postharvest Biology and Technology, 64*, 160–167.

Zhao, J., Peng, P., Schmitz, R. J., Decker, A. D., Tax, F. E., & Li, J. (2002). Two putative BIN2 substrates are nuclear components of brassinosteroid signaling. *Plant Physiology, 130*, 1221–1229.

Zhao, Q., Sun, Q., Dong, P., Ma, C., Sun, H., & Liu, C. (2019). Jasmonic acid alleviates boron toxicity in *Puccinellia tenuiflora*, a promising species for boron phytoremediation. *Plant and Soil, 445*, 397–407.

Zhao, S., Ma, Q., Xu, X., Li, G., & Hao, L. (2016). Tomato jasmonic acid-deficient mutant spr2 seedling response to cadmium stress. *Journal of Plant Growth Regulation, 35*, 603–610.

Zheng, X. Y., Spivey, N. W., Zeng, W., Liu, P. P., Fu, Z. Q., Klessig, D. F., He, S. Y., & Dong, X. (2012). Coronatine promotes *Pseudomonas syringae* virulence in plants by activating a signaling cascade that inhibits salicylic acid accumulation. *Cell Host & Microbe, 11*(6), 587–596.

Zou, L. J., Deng, X. G., Zhang, L. E., Zhu, T., & Lin, H. H. (2017). Nitric oxide as a signaling molecule in brassinosteroid-mediated virus resistance to cucumber mosaic virus in *Arabidopsis thaliana*. *Physiologic Plantarum, 163*, 196–210.

CHAPTER 10

Unlocking the Potential of Triacontanol: A Natural Biostimulant for Sustainable Agriculture

PRINCE CHOYAL,* MAHAVEER P. SHARMA, RAKESH KUMAR VERMA, GIRIRAJ KUMAWAT, and GYANESH KUMAR SATPUTE

ICAR–National Soybean Research Institute, Indore, Madhya Pradesh, India

*Corresponding author

ABSTRACT

Triacontanol, a long-chain alcohol derived from plant cuticle waxes, has garnered attention as a potent biostimulant in agricultural practices. This chapter explores the multifaceted roles of triacontanol in enhancing plant growth, development, and stress tolerance, thereby promoting sustainable agriculture. Through elucidating its mechanisms of action, including modulation of gene expression, hormone levels, and physiological processes, this chapter underscores the versatile applications of triacontanol across various crops and environmental conditions. Furthermore, it highlights the eco-friendly nature of triacontanol, its compatibility with organic farming practices, and its potential to mitigate the adverse effects of chemical fertilizers and pesticides. By synthesizing recent advancements and emerging trends in triacontanol research, this review aims to provide insights into harnessing its full potential for optimizing crop productivity while minimizing environmental impacts, thus contributing to the paradigm shift towards sustainable agricultural practices.

Plant Biostimulants and Growth Regulators in Sustainable Agriculture.
Deepak Kumar, Sanjay K. Singh, Nutan Kaushik, and Mahaveer P. Sharma (Eds.)

DOI: 10.1201/9781779641458-14

10.1 INTRODUCTION

Triacontanol (TRIA) is a newly recognized plant growth stimulator that plays diverse physio-biochemical roles in plant growth and developmental processes. TRIA was first identified by Chibnall et al. (1933) in alfalfa (*Medicago sativa*) as a natural component of plants. Later, the work of Ries et al. (1977) established it as a natural plant growth regulator (PGR). TRIA is a saturated primary alcohol, primarily present in the epicuticular waxes of diverse plant species (Figure 10.1). The exogenous application of TRIA at low concentrations stimulates plant growth in most of the crop species (Table 10.1). The low concentration of TRIA improves gas exchange attributes, plant pigments, mineral nutrition, and enzyme activities like nitrate reductase, carbonic anhydrase (CA), and so forth. TRIA is well known for its role in abiotic stress mitigation, viz., salt stress, drought, temperature, heavy metal stress, oxidative stress, and so forth. A detailed mechanism of action and physio-biochemical effects on plant growth are discussed in the following sections.

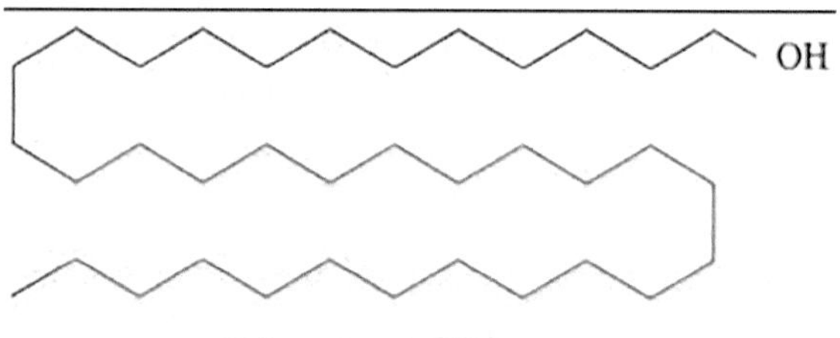

FIGURE 10.1 Structure of triacontanol (TRIA).

10.2 MODE OF ACTION

Application of TRIA can trigger various physiological processes like water uptake, nutrient uptake, photosynthesis, and metabolic processes such as the activation of enzymatic activities. TRIA elicits the secondary messenger TRIM and the TRIA-mediated formation of L (+)-adenosine, which is mainly supplied by AMP, which in turn is produced from adenosine diphosphate and triphosphate in plant tissues. The increased level of L (+)-adenosine in response to TRIA application simultaneously increases the Ca^{2+} concentration in the tonoplast. The binding of Ca^{2+} to calmodulin protein in response to TRIA-mediated increased Ca^{2+} levels in the tonoplast leads to modulation of the activation state of various transcription factors like GTL, MYB2, CAMTA3, and so forth, and thus enhances the activities of kinases and phosphatases. The activation of kinases and phosphatases results in the expression of various genes like *rbcS, pbsO, GDC*, and *SHMT* involved in photosynthesis and associated physiological processes. In short, TRIA mediates calcium signaling through the L (+)-adenosine secondary messenger and triggers the activity of various anti-stress genes through transcription factor modulation. Chen et al. (2002) identified in rice plants treated with triacontanol several genes that were associated with ABA synthesis and photosynthesis. Foliar spray of triacontanol stimulated the expression of the *rbcS* gene, while 2 down-regulated genes were found associated with ABA suppression and stress-related OsAsr1 in the shoot region (Llanes et al., 2015). Plasma membrane proteins (PIP) during stress regulate plant water status like aquaporin proteins and thus improve the osmotic potential (Alharbi et al., 2021). Rice plants, during osmotic stress, modulate aquaporin-associated genes, *viz.*, PIP1,1, PIP1,2, PIP2,4, and PIP2,5 after triacontanol treatment. In *Brassica juncea* (L.), triacontanol was found to regulate lignification over PEG-induced drought stress through elevated expression of *MYC46* and phenylalanine ammonia-lyase (PAL) genes (Ahmad et al., 2021). The lignification is mainly related to an upgraded plant defense system for mechanical strength. Transcriptome analysis in strawberry plants revealed that triacontanol affects physiological processes like photosynthesis, plant defense, hormonal interactions, and delayed senescence during fruit development. Up-regulation of *chlD, chlI*, and *PPOX* genes in triacontanol-treated plants encodes magnesium chelatase subunit and protoporphyrinogen enzymes, which are associated with chlorophyll metabolism. Furthermore, ACO (1-aminocyclopropane-1-carboxylic acid oxidase), JAZ (jasmonate ZIM domain-containing protein), and TGA (transcription factor) genes are expressed in plants treated with triacontanol (TRIA) that are known to influence ethylene, jasmonate, and salicylic acid synthesis and various stress signaling crosstalk (Pang et al., 2020).

10.3 PHYSIOLOGICAL AND BIOCHEMICAL EFFECTS OF TRIACONTANOL

Numerous studies have shown that application of TRIA, either through foliar spray or root medium, results in enhanced plant growth and yield in many field and horticultural crops. TRIA enhances the growth and yield of plants by modifying the membrane structure and function. TRIA stimulates calcium-dependent ATPase activity in the presence of calmodulin protein, thus modifying the cellular membrane structure and function (Lesniak et al., 1989). TRIA is an extremely lipophilic long-chain primary alcohol that modulates chloroplast membrane dynamics, plasma membrane fluidity, and permeability and thus activates various membrane-bound enzymes (Swamy et al., 2009). A detailed survey of the various physiological and biochemical effects of TRIA application is discussed in detail in the following section.

10.3.1 SEED GERMINATION AND EARLY SEEDLING VIGOR

Exogenous application of TRIA through seed priming results in an enhanced rate of seed germination. A lower concentration of TRIA stimulates seed germination; however, many reports claim that a higher concentration of TRIA results in a reduced seed germination percentage. Jin and Hong (1984) reported that TRIA application induced chlorophyll accumulation in cotyledons during the early seedling establishment stage and also minimized chlorophyll degradation during the later stage of cotyledon senescence by regulating the glycolate oxidase enzyme activity in radish. Morré et al. (1991) reported that TRIA application stimulated soybean hypocotyl elongation through enhancement of NADH oxidase activity in the plasma membrane. Xuan and Feng (2000) showed that seed dressing with TRIA resulted in a 10–12% higher seed germination rate in maize. The TRIA-treated seed had higher amylase and dehydrase activity. The seedlings showed increased chlorophyll content with good root system vigor. El-Shafey et al. (2018) reported that a low concentration of TRIA caused rapid radicle emergence with a significantly higher rate of seed germination. Further, the cytokinin and gibberellin content was also significantly higher in treated maize seedlings, whereas the ABA content was in relatively lower concentrations. The chlorophyll a and b and carotenoid content were also increased with concomitant enhancements in seedling fresh and dry weight. However, a relatively higher concentration of TRIA resulted in a lower seed germination percentage compared to the control.

10.3.2 GROWTH AND YIELD ATTRIBUTES

The TRIA application positively correlated with root growth. TRIA synergistically works with auxin and induces lateral root formation and increased root growth. Li et al. (2018) reported that TRIA induced lateral root formation through de novo formation of lateral root primordium. It also increased the lateral root density by 50–98% in *Arabidopsis*. TRIA treatment increased the biosynthesis of auxin (IAA). Thus, exogenous application of TRIA positively regulated lateral root formation by upregulating the auxin biosynthetic genes. Soundararajan et al. (2018) showed that TRIA promoted early root emergence by increasing the synthesis of growth-promoting compounds like IAA, serotonin, trigonelline, UDP-N-acetyl glucosamine, gallate, serotonin, and melatonin. It also enhances the synthesis of ATP and reduces power like NADPH. Similarly, TRIA application also promotes vegetative growth characteristics in a variety of plant species. Bhandari et al. (2021) showed that 1 ppm of TRIA foliar spray improved the growth and yield characteristics of *Brassica oleracea* var. *gongylodes*. It was shown that around 7–40% of yield was increased under TRIA foliar spray. The vegetative growth parameters like plant height, total leaf area, and number of leaves were increased. The TRIA application reduces the timing for flowering induction. The exogenous application of TRIA in micromolar concentration enhances the endogenous total cytokinin concentration in leaf and root tissues and thus induces early flowering (He and Loh, 2002). Yong (2000) reported that foliar sprays of TRIA caused flowering induction in duckweed (*Lemna aequinoctialis*), a short-day plant under long-day conditions. Sanches et al. (2023) showed that exogenous application of TRIA caused a significant increase in yield and quality of "Kent" mango fruits. The total amino acids, total proteins, and free proline, along with the antioxidant enzymes like CAT, POD, and SOD, were significantly increased under TRIA application. It also caused a 50–65% yield increase when applied through foliar spray.

10.3.3 PHOTOSYNTHETIC CHARACTERISTICS

The TRIA application causes a significant change in the photosynthesis processes. Ivanov and Angelov (1997) investigated the effect of TRIA application on photosynthesis and associated mesophyll membrane attributes. The application of TRIA to the isolated protoplast caused a very significant increment in the carboxylation reaction. The study showed that TRIA application caused a remarkable change in the chloroplast membrane fluidity to bring about the growth-promoting effects. It increases the concentration of chlorophyll, or, in other words, it minimizes chlorophyll degradation through a reduction in chlorophyllase activity (Kathuria et al., 2012).

TRIA increases the photosynthetic rate by upregulating photosynthesis-related genes. Chen et al. (2002) showed that TRIA upregulated the expression of *rbcS,* which encodes the rubisco small subunit, the *psbO* gene (encodes for the oxygen evolving complex of PSII), GDC (glycine decarboxylase), and *SHMT* genes. The higher transcript level of *rbcS* was associated with the improved photosynthetic rate, whereas the *GDC* and *SHMT* levels were correlated with the higher photorespiratory activity, thus the availability of more reducing power in the form of NADH. Borowski et al. (2000) studied the effect of TRIA on chlorophyll fluorescence and yield attributes in tomatoes. The TRIA application through the root drenching method caused a significant increase in the PSII efficiency, the quantum yield of the electron transport chain, and the photochemical quenching coefficient. Further, the increased photosynthetic efficiency resulted in an increased fruit yield. Maity et al. (2019) reported that exogenous application of TRIA resulted in increased green foliage yield in tea. It also increased the chlorophyll content of the leaves. Ramos-Zambrano et al. (2019) showed that foliar application of TRIA under both low and high irradiance levels caused a significant stimulation in photosynthetic attributes. The stomatal conductance was increased without affecting the stomatal density; however, the stomatal morphology and size were increased. The stomatal aperture was also increased under TRIA application. TRIA increased the electron transport rate and photochemical quenching.

10.3.4 SECONDARY METABOLITE SYNTHESIS

Apart from stimulating the normal growth and yield patterns of field and horticultural crops, TRIA application also causes a significant enhancement in the secondary metabolite content of medicinal and aromatic plant species. Khan et al. (2007) showed that foliar application of the TRIA enhanced the opium yield and morphine production in opium poppy. It also increased the capsule-bearing branches, thus increasing capsule yield. Naeem et al. (2011) reported that exogenous application of TRIA through foliar spray resulted in increased performance of mint plants in terms of herbage yield and active constituents of the plant. TRIA at 1 μM concentration showed a significant enhancement in herbage yield, essential oil, and the content of active constituents like methone, menthol, isomenthone, and menthyl acetate. However, higher concentrations of TRIA showed a negative impact on growth attributes. Khan et al. (2014) showed that foliar sprays of TRIA caused a significant increase in the growth and foliage yield of lemon grass. Further, the essential oil yield and the citral content in the oil were also increased. The TRIA increased the shoot biomass by stimulating photosynthesis with increased chlorophyll content (Misra and Srivastava, 1991). Naeem et al. (2019) reported that foliar application of TRIA resulted in a significant increase in the alkaloid yield and content of *Catharanthus roseus.* More than 70% of the increase in vincristine, vinblastine, and vindoline content was observed when TRIA was foliar sprayed in micromolar concentration.

10.4 TRIACONTANOL AS AN STRESS ALLEVIATING COMPANION

In recent times, abiotic stresses have been a major concern for agricultural production. Abiotic stresses refer to various abiotic or climatic factors that severely impact plant growth and development and, thus, productivity. These abiotic stresses include drought, salinity, heat, cold, heavy metals, UV stress, and so forth. In modern agriculture, PGRs are being utilized to combat various abiotic stresses. TRIA is a well-established plant growth stimulator and has the potential to minimize the deleterious effects of abiotic stresses by regulating the physiological and biochemical mechanisms of plants. The recent findings on the role of TRIA as an abiotic stress alleviator are listed in Table 10.1 and briefly discussed in the following sections.

10.4.1 SALINITY STRESS

Salinity stress is one of the most common abiotic stresses occurring worldwide. The area of productive land is decreasing significantly globally, and around 7.0% of the world's total area is covered under salinity, which

TABLE 10.1 Effect of TRIA Application on Physio-Biochemical Traits of Plants Under Abiotic Stresses

Plant Species	Stress	Application Rate	Effect	References
Zea mays	Drought	5 µM Foliar spray	Increased the oxidative stress tolerance potential through increased osmolyte accumulation and antioxidative enzyme synthesis.	Perveen et al. (2016)
Oryza sativa	Drought	35 ppm Seed priming	Seed priming with TRIA improved the water relation under drought stress through osmolyte accumulation and regulation of aquaporins genes.	Alharbi et al. (2021)
Helianthus annuus	Drought	25 µM Foliar spray	TRAI application counteracted the photosynthetic impairment under drought stress and thus improved the stress tolerance potential	Ismail and Younis (2021)
Spinacia oleracea	Salinity	1 µM foliar spray	TRIA application negated the salt stress and increased the leafy biomass of the plant.	Tompa and Fodorpataki (2021)
Zea mays	Salinity	3 µM Foliar spray	TRIA application improved salt tolerance potential through improving NDVI, SPAD and photosynthetic traits.	Lúcia et al. (2022)
Brassica juncea	Salinity	150 µM Foliar spray	The salinity stress tolerance potential was improved through enhancing antioxidant defense system along with phenolics, flavonoids, anthocyanin content.	Verma et al. (2023)
Coriandrum sativum	Salinity	10 µM Foliar spray	The vegetative growth was improved through minimizing the oxidative stress.	Asadi Karam and Keramat (2017)
Vigna radiata	Heat stress	11 µM Foliar spray	TRIA reversed the deleterious effect of heat stress through ABA and JA level modulation	Waqas et al. (2016)
Zea mays	Heavy metal (Ni) stress	25 µM Foliar spray	The deleterious effect of Ni toxicity was reversed through regulating the Ni uptake and enhancing antioxidant system.	Younis and Ismail (2019)
Triticum aestivum	Heavy metal (As) stress	1 µM Foliar spray	The seed yield, chlorophyll b, photosynthesis, flavonoids and anthocyanin content were significantly increased in response to TRIA application under arsenic (As) stress.	Ali and Perveen (2020)

TABLE 10.1 *(Continued)*

Plant Species	Stress	Application Rate	Effect	References
Mentha arvensis	Heavy metal (As) stress	1 μM Foliar spray	The toxicity of arsenic (As) was mitigated by TRIA foliar spray through combating the ROS scavenging ROS and improved the physiological attributes and essential oil content.	Nabi et al. (2022)
Coriandrum sativum	Heavy metal (Cd) stress	20 μM Seed priming	Increased yield attributes through minimizing the oxidative stress effects in seedlings of TRIA treated seeds.	Sardar et al. (2022)
Zea mays	Heavy metal (Pb) stress	10 μM Seed priming	Enhanced Pb stress tolerance through inducing the antioxidative machinery.	Iqbal et al. (2023)
Oryza sativa	Transplanting shock	10 μM Foliar spray	TRIA reduced the transplanting shock and resulted into yield enhancement of rice.	Li et al. (2016)

is more than 900 Mha of land (Sarwar et al., 2021). The excessive salt concentration in plant tissue causes osmotic and ionic stress and results in reactive oxygen species (ROS) generation. The amplification of ROS under saline conditions disrupts the plant's photosynthetic machinery, ultimately compromising the productivity of plants. Salinity adversely affects metabolic and physiological processes like photosynthesis, respiration, chlorophyll pigmentation, and so forth. Furthermore, the increase in salt concentration in the root zone causes soil moisture depletion, which affects root water uptake, leaf water potential, transpiration, and thus biomass accumulation.

The role of TRIA in imparting salinity tolerance is well documented, and its foliar application at low concentrations significantly improves plant growth characteristics under saline conditions in a variety of plant species. Sahbaz et al. (2013) observed that seed priming with TRIA caused yield enhancement in *Brassica napus* under both control and saline conditions. The plant growth-enhancing characteristics like number of seeds per plant, photosynthetic rate, transpiration rate, chlorophyll a/b ratio, electron transport rate, shoot and root K^+ concentration was increased. Further, the osmolytes like proline and glycine betaine were also increased to maintain the leaf water potential. Tompa and Fodorpataki (2021) showed that 1 μM TRIA foliar spray causes a significant increase in leafy vegetative yield of spinach plants under 250 mM NaCl salt stress. The net carbon assimilation rate, quantum yield of photosystem II, and photosynthetic pigmentation were significantly enhanced under both the control and saline conditions, which resulted in increased leafy biomass yield. Sarwar et al. (2021) reported that foliar application of TRIA to cucumber plants caused a significant improvement in fruit yield under 50 mmol L^{-1} soil salinity. TRIA scavenged the ROS by increasing the synthesis of antioxidative enzymes like SOD, CAT, and POD. The accumulation of compatible solutes like proline, glycine, betaine, and phenolic compounds was also increased. Furthermore, the water relations under saline conditions were improved, which in turn enhanced the photosynthetic attribute and ultimately productivity. In another study, Sarwar et al. (2022) reported the role of TRIA in mitigating the ill effects of salinity in pepper plants. The foliar application of 75 μM TRIA considerably improved the vegetative growth characteristics of plants, like plant height, total leaf area, and dry biomass. The antioxidative enzyme synthesis was also upregulated to combat ROS production under ionic stress. Lucia et al. (2022) studied the salinity-alleviating effect of TRIA in sweet corn. The result showed that a 3 μM TRIA foliar spray minimized the deleterious effect of salinity stress in sweet corn. TRIA application increased the plant height, total biomass, photosynthetic characteristics, chlorophyll pigments, NDVI, and osmolyte accumulation under salinity stress. Verma et al. (2023) reported that foliar application of TRIA improved the vegetative growth characteristics like shoot length and root length, fresh weight, and dry weight of *Brassica juncea* under salt stress. The antioxidative defense system and osmotic adjustment were upregulated under the TRIA application. The content of flavonoids, phenolics, and anthocyanins was also increased.

Khedr and Khedr (2024) studied the promotive effect of TRIA under salinity stress in pomegranate. The foliar spray of TRIA at 20 μM concentration showed a very significant increase in fruit quantity and quality. The TRIA application mitigated the harmful effect of salinity and enhanced productivity with increased antioxidant content. Further, the post-harvest storage life was significantly improved. The mineral nutrient content, porphyrin concentration, glycine betaine, proline, total phenolics, total carbohydrates, and so forth, were also increased under TRIA application. Overall, the fruit quality was very significantly improved through TRIA foliar application under control and salinity conditions. Conclusively, these studies showed that TRIA application efficiently regulates salinity-induced oxidative stress through upregulating the antioxidative system and accumulating more osmolyte solutes. These osmolyte solutes maintain leaf water potential and, thereby, normal physiological processes.

10.4.2 DROUGHT STRESS

Drought stress is one of the most complex and major abiotic stresses that severely affect crop productivity. Drought can occur at any stage of crop growth and interfere with normal plant performance, resulting in yield losses. Moisture shortages after seed sowing severely affect seed germination and poor seedling establishment. Vegetative stage drought stress affects major physiological processes like photosynthesis, plant metabolism, transpiration, respiration, and thus poor biomass accumulation. The occurrence of drought stress at the reproductive stage causes a yield reduction through reducing the seed set, poor seed filling, and so forth. The physiological, biochemical, and molecular traits altered during drought stress were briefly described by Saha et al. (2022). Soil moisture depletion led to the synthesis of ROS, which caused damage to the cellular membrane. Disintegration of photosynthetic machinery by oxidative damage results in photoinhibition and ultimately crop yield losses (Chauhan et al., 2023). Further, drought stress also causes leaf water potential depletion, which results in stomatal closure and ultimately reduced transpiration and carboxylation. Nevertheless, plants activate the enzymatic and non-enzymatic antioxidant defense systems to counter the ROS. Drought stress also elicits the synthesis of secondary metabolites to counter excessive photochemical reactions, thereby minimizing ROS production (Choyal et al., 2022; Kumar et al., 2020).

The role of TRIA in imparting drought tolerance is well documented, and its foliar application at low concentrations significantly improves plant growth characteristics under low soil moisture conditions in a variety of plant species. Perveen et al. (2016) showed that the drought tolerance potential of the maize crop was improved under TRIA application. The oxidative stress was effectively managed through the quenching of ROS by enzymatic antioxidants. Alharbi et al. (2021) studied the physio-biochemical and molecular basis of TRIA-induced drought tolerance in rice seedlings. It maintained tissue hydration through the synthesis of soluble sugars and free amino acids. TRIA positively regulated the photosynthetic activity by controlling the photosynthetic efficiency of PS II and minimizing photoinhibition. At the molecular level, it was shown that it induced the expression of aquaporin-related genes (*PIP1,1*, *PIP1,2*, *PIP2,4,* and *PIP2,5*), which positively regulated the root hydraulic conductivity. Ostrowska et al. (2021) reported that foliar sprays of 1 ppm TRIA improved the overall plant health at different growth stages by improving the hydration state of maize under drought stress. The chlorophyll pigmentation, stomatal conductance, and quantum yield of PS II were significantly improved under TRIA application. Ismail and Younis (2021) studied the physiological effects of TRIA application under drought stress in sunflower seedlings. It was found that it induced drought tolerance by enhancing the photosynthesis activity of the seedling. The increase in photosynthetic pigments, the photochemical efficiency of photosystem II, and the increased expression of the Rubisco small subunit cause photosynthetic enhancement, which results in higher biomass under drought stress. The TRIA-induced osmolyte accumulation under drought conditions effectively managed the oxidative stress. El-Beltagi et al. (2022) explored the role of TRIA in the drought tolerance potential of strawberries. The foliar application of 1 ppm TRIA significantly improved the fruit yield per plant under drought conditions. The overall plant growth in terms of plant height, total biomass, and fruit weight was significantly increased. Physiologically, the photosynthetic traits like net photosynthetic trait and stomatal conductance, chlorophyll pigmentation, anti-oxidative characters, and osmoticums were significantly higher under TRIA application. Perveen et al. (2022) reported that foliar application of 1.0 μM TRIA positively regulated the growth characteristics of linseed under drought stress. The vegetative and yield-contributing traits of shoot dry weight, number of fruits per plant, number of seeds per plant, and seed index were significantly increased. Plants with TRIA supplementation effectively managed ROS production through osmotic adjustments. In summary, the above-cited studies showed that the application of TRIA, either foliar or seed priming, efficiently minimized the deleterious effects of drought stress. It could manage photosynthesis, stomatal activity, and leaf tissue hydration under depleted soil moisture.

10.4.3 HEAVY METAL STRESS

Multiple anthropogenic activities, like the indiscriminate use of fertilizers, sludge, sewage, urban runoff, emissions from automobiles, irrigation with contaminated water, and so forth, cause serious problems of heavy metal accumulations in the rhizosphere. The most common heavy metals in agricultural soils are aluminum, cadmium, nickel, chromium, arsenic, and lead. These heavy metals, once taken up by plants at toxic concentrations, interfere with the normal metabolic reactions of plants. They interact with proteins and enzymes and cause the denaturation or inactivation of proteins through substitutions of essential elements. Heavy metals cause oxidative stress in plants, which further denatures essential proteins and enzymes, adversely affects normal physiological processes like photosynthesis and respiration, and induces chlorophyll degradation and necrosis. To counter the adverse effects of heavy metal toxicity, plants also adapt to detoxification mechanisms like the induction of both enzymatic and non-enzymatic antioxidative machinery. Plants also either exclude or compartmentalize the heavy metal ion into cellular organelles.

Further, immobilization and chelation mechanisms also function to limit the uptake of these heavy metals.

A number of studies have revealed the significance of TRIA in alleviating the toxic effects of heavy metals. Karam et al. (2017a) reported that priming coriander seeds with TRIA reversed the growth penalties caused by arsenic (As) stress. The upregulation of both enzymatic and non-enzymatic antioxidant systems was able to counter the oxidative stress caused by arsenic (As). Further, they investigated the biochemical basis of TRIA-induced Cd phytotoxicity tolerance in *Brassica napus* seedlings. It was shown that the ROS content, lipid peroxidation, and lipoxygenase activity were reduced under TRIA pre-treatment. Further, the phytochelatin, ascorbate, and glutathione content were also increased. The upregulation of the ascorbate-glutathione cycle, which includes monodehydroascorbate reductase (MDHAR), dehydroascorbate reductase (DHAR), and glutathione reductase, caused a significant reduction in oxidative stress and thereby enhanced the Cd stress tolerance (Karam et al., 2017b). Nabi et al. (2021) reported that foliar application of TRIA minimized the Ni-instigated repercussions in *Mentha arvensis*. The herbage yield and essential oil content were increased in Ni-affected plants under TRIA application. The menthone and menthyl acetate levels under TRIA application were increased in Ni-stressed plants. TRIA maintained the ultrastructure of the leaves and roots of the Ni-stressed plants. Younis and Ismail (2022) investigated the biochemical and molecular basis of the Ni stress amelioration potential of TRIA in *Zea mays*. TRIA foliar application modulated the hormonal balance under Ni stress. The IAA, salicylic acid, and IAA/IBA ratios were increased in the roots of Ni-stressed plants under TRIA application. The expression of the ABA signaling gene *ZmASR1* was upregulated under TRIA foliar application and caused enhanced Ni phytotoxicity tolerance in roots. ABA is probably involved in the sequestration of heavy metals through the activation of metal-binding proteins (metallothioneins) (Leszczyszyn et al., 2013). Nabi et al. (2022) reported that the TRIA foliar application was able to ameliorate the adverse effects of arsenic (As) toxicity in *Mentha arvensis*. TRIA stimulated vegetative plant growth as well as the induction of free-radicle detoxifying machinery. Sardar et al. (2022) showed that seed priming with TRIA caused a significant reduction in the detrimental impacts of cadmium phytotoxicity in *Coriandrum sativum* seedlings. It improved the efficacy of an antioxidant system consisting of SOD, CAT, and POX, including non-enzymatic antioxidants like phenols, proline, flavonoids, and so forth. The TRIA-primed seedling exhibited more pronounced yield-contributing traits, thus enhancing CD stress tolerance. Umar et al. (2023) explored the Cr stress attenuation potential of TRIA in *Pisum sativum*. Seed priming with 20 μM TRIA caused a significant improvement in seedling vigor, high fresh and dry biomass, higher photosynthetic characteristics, and improved mineral nutrient uptake. It also minimized the Cr uptake by the pea plants and efficiently ameliorated the Cr toxicity. Ahmed et al. (2023a) studied the impact of TRIA on Pb stress amelioration in broccoli. The seeds primed with 20 μM TRIA showed a very prominent effect on Pb tolerance. Photosynthetic traits and antioxidant activity were significantly increased in TRIA-primed seedlings. In continuation of the above report, the same problem in common beans was investigated (Ahmed et al., 2023b). It was found that a 20 μM TRIA concentration was able to efficiently manage the Pb stress. The seedling was able to maintain higher gas exchange attributes like net photosynthetic rate, stomatal conductance, and transpiration rate. Apart from that, the free radicle scavenging activity, proline, phenolics, and soluble protein content were increased under TRIA application. The TRIA-treated maize seedling showed notably higher photosynthesis and thereby more biomass under Pb contamination (Iqbal et al., 2023).

10.5 CONCLUSION

Triacontanol is a plant growth-promoting molecule that modulates the various physiological and biochemical attributes of plant growth at micromolar concentrations. It causes growth and yield stimulation by enhancing chlorophyll pigmentation, photosynthesis, nutrient absorption, and so forth. TRIA increases chlorophyll pigmentation by reducing chlorophyllase enzyme activity. Further, it maintains high photosynthesis through regulation of both photochemistry and carboxylation reactions. It induces rubisco synthesis through upregulation of genes involved in the small subunit of rubisco protein. Further, it maintains the phytohormone balance and acts synergistically with auxin in lateral root development. TRIA also acts as an abiotic stress alleviator and compensates for the growth penalties caused by the adverse effects of salinity, drought, and heavy metal stress. It regulates oxidative stress by enhancing the synthesis of both enzymatic and non-enzymatic antioxidants. In summary,

TRIA is a dynamic growth stimulator that can be employed in general agricultural practice to achieve sustainable crop productivity. However, extensive further research is required to optimize the crop-specific dose for stimulatory effect.

KEYWORDS

- **carbonic anhydrase**
- **phenylalanine ammonia-lyase**
- **plant biostimulant**
- **reactive oxygen species**
- **triacontanol**

REFERENCES

Ahmed, A. A., & AbdulKafoor, A. H. (2023). Response of vegetative growth traits of several soybean cultivars to spraying with the growth regulator triacontanol (TRIA). In *IOP Conference Series: Earth and Environmental Science*, *1213*(1), 012045. https://doi.org/10.1088/1755-1315/1213/1/012045 (accessed on 6th November 2025).

Ahmed, S., Ahmad, M., Sardar, R., & Ismail, M. A. (2023a). Triacontanol priming as a smart strategy to attenuate lead toxicity in *Brassica oleracea* L. *International Journal of Phytoremediation, 25*(9), 1173–1188. https://doi.org/10.1080/15226514.2022.2143478 (accessed on 6th November 2025).

Ahmed, S., Amjad, M., Sardar, R., Siddiqui, M. H., & Irfan, M. (2023b). Seed priming with triacontanol alleviates lead stress in *Phaseolus vulgaris* L. (common bean) through improving nutritional orchestration and morpho-physiological characteristics. *Plants, 12*(8), 1672. https://doi.org/10.3390/plants12081672 (accessed on 6th November 2025).

Alharbi, B. M., Abdulmajeed, A. M., & Hassan, H. (2021). Biochemical and molecular effects induced by triacontanol in acquired tolerance of rice to drought stress. *Genes, 12*(8), 1119. https://doi.org/10.3390/genes12081119 (accessed on 6th November 2025).

Ali, H. M. M., & Perveen, S. (2020). Effect of foliar applied triacontanol on wheat (*Triticum aestivum* L.) under arsenic stress: A study of changes in growth, yield and photosynthetic characteristics. *Physiology and Molecular Biology of Plants, 26*, 1215–1224. https://doi.org/10.1007/s12298-020-00831-0 (accessed on 6th November 2025).

Asadi Karam, E., & Keramat, B. (2017). Foliar spray of triacontanol improves growth by alleviating oxidative damage in coriander under salinity. *Indian Journal of Plant Physiology, 22*, 120–124. https://doi.org/10.1007/s40502-017-0286-z (accessed on 6th November 2025).

Bhandari, S., Bhandari, A., & Shrestha, J. (2021). Effect of different doses of triacontanol on growth and yield of kohlrabi (*Brassica oleracea* L. var. *gongylodes*). *Helion, 7*(10), e08242. https://doi.org/10.1016/j.heliyon.2021.e08242 (accessed on 6th November 2025).

Borowski, E., Blamowski, Z. K., & Michałek, W. (2000). Effects of Tomatex/triacontanol/ on chlorophyll fluorescence and tomato (*Lycopersicon esculentum* Mill.) yields. *Acta Physiologiae Plantarum, 22*, 271–274. https://doi.org/10.1007/s11738-000-0030-5 (accessed on 6th November 2025).

Chen, X., Yuan, H., Chen, R., Zhu, L., Du, B., Weng, Q., & He, G. (2002). Isolation and characterization of triacontanol-regulated genes in rice (*Oryza sativa* L.): Possible role of triacontanol as a plant growth stimulator. *Plant and Cell Physiology, 43*(8), 869–876. https://doi.org/10.1093/pcp/pcf100 (accessed on 6th November 2025).

El-Beltagi, H. S., Ismail, S. A., Ibrahim, N. M., Shehata, W. F., Alkhateeb, A. A., Ghazzawy, H. S., El-Mogy, M. M., & Sayed, E. G. (2022). Unravelling the effect of triacontanol in combating drought stress by improving growth, productivity, and physiological performance in strawberry plants. *Plants, 11*(15), 1913. https://doi.org/10.3390/plants11151913 (accessed on 6th November 2025).

El-Shafey, A. S., Hassan, H. M., & Sayed Ahmad, H. F. (2018). Involvement of triacontanol in regulation of seed germination and seedling growth of *Zea mays* L. *Egyptian Journal of Experimental Biology (Zoology), 14*(2), 375–383. http://doi.org/10.5455/egyjebb.20181002105804 (accessed on 6th November 2025).

He, Y. W., & Loh, C. S. (2002). Induction of early bolting in *Arabidopsis thaliana* by triacontanol, cerium and lanthanum are correlated with increased endogenous concentration of isopentenyl adenosine (iPAdos). *Journal of Experimental Botany, 53*(368), 505–512. https://doi.org/10.1093/jexbot/53.368.505 (accessed on 6th November 2025).

Iqbal, J., Perveen, S., Parveen, A., Saeed, M., Zafar, S., & Iqbal, N. (2023). Interactive effect of salicylic acid and triacontanol to improve lead stress tolerance in maize (*Zea mays* L.). *Arabian Journal of Geosciences, 16*(5), 314. https://doi.org/10.1007/s12517-023-11317-0 (accessed on 6th November 2025).

Ismail, H. A., & Younis, A. A. (2021). Triacontanol foliar spray alleviated drought stress effects by maintaining photosynthesis and cellular redox balance in sunflower seedlings. *Egyptian Academic Journal of Biological Sciences, H. Botany, 12*(2), 103–118. https://doi.org/10.21608/eajbsh.2021.207875 (accessed on 6th November 2025).

Ivanov, A. G., & Angelov, M. N. (1997). Photosynthesis response to triacontanol correlates with increased dynamics of mesophyll protoplast and chloroplast membranes. *Plant Growth Regulation, 21*, 145–152. https://doi.org/10.1023/A:1005790121111 (accessed on 6th November 2025).

Jin, C. D., & Hong, Y. N. (1984). Effects of triacontanol on growth and peroxisomal enzyme activities in radish (*Raphanus sativus* L.) seedlings. *Korean Journal of Botany (Korea R.), 27*(4), 241–251. https://koreascience.kr/article/JAKO198411920127433.pdf (accessed on 6th November 2025).

Karam, E. A., Keramat, B., Asrar, Z., & Mozafari, H. (2017). Study of interaction effect between triacontanol and nitric oxide on alleviating of oxidative stress arsenic toxicity in coriander seedlings. *Journal of Plant Interactions, 12*(1), 14–20. https://doi.org/10.1080/17429145.2016.1267270 (accessed on 6th November 2025).

Karam, E. A., Maresca, V., Sorbo, S., Keramat, B., & Basile, A. (2017). Effects of triacontanol on ascorbate-glutathione cycle in *Brassica napus* L. exposed to cadmium-induced oxidative stress. *Ecotoxicology and Environmental Safety, 144*, 268–274. https://doi.org/10.1016/j.ecoenv.2017.06.035 (accessed on 6th November 2025).

Kathuria, E., Sanadhya, D., Malik, C. P., & Kakralyea, B. L. (2012). Photosynthesis is improved by exogenous TRIA in salt-stressed maize seedlings. *The Journal of Plant Science Research, 28*(2), 239.

Khan, R., Khan, M. M. A., Singh, M., Nasir, S., Naeem, M., Siddiqui, M. H., & Mohammad, F. (2007). Gibberellic acid and triacontanol can ameliorate the opium yield and morphine production in opium poppy (*Papaver somniferum* L.). *Acta Agriculture Scandinavica Section B – Soil and Plant Science, 57*(4), 307–312.

Khan, Z. H., Mohammad, F., & Khan, M. M. A. (2014). Enhancing the growth, yield and production of essential oil and citral in lemongrass by the application of triacontanol. *International Journal of Agricultural Research, 4*, 113–122.

Khedr, E. H., & Khedr, N. (2024). Unraveling the mechanisms behind salt stress alleviation and enhanced fruit storability in pomegranate through triacontanol treatment. *Scientia Horticulture, 325*, 112634. https://doi.org/10.1016/j.scienta.2023.112634 (accessed on 6th November 2025).

Lesniak, A. P., Haug, A., & Ries, S. K. (1989). Stimulation of ATPase activity in barley (*Hordeum vulgare*) root plasma membranes after treatment with triacontanol and calmodulin. *Physiologic Plantarum, 75*(1), 75–80. https://doi.org/10.1111/j.1399-3054.1989.tb02066.x (accessed on 6th November 2025).

Leszczyszyn, O. I., Imam, H. T., & Blindauer, C. A. (2013). Diversity and distribution of plant metallothionein's: A review of structure, properties and functions. *Metallomics, 5*(9), 1146–1169. https://doi.org/10.1039/c3mt00072a (accessed on 6th November 2025).

Li, F., Yang, N., Qian, M., & Gan, L. (2018). Auxin is involved in triacontanol-induced lateral root development in *Arabidopsis thaliana*. *Journal of Nanjing Agricultural University, 41*(3), 473–480. https://doi.org/10.7685/jnau.201709016 (accessed on 6th November 2025).

Li, X., Zhong, Q., Li, Y., Li, G., Ding, Y., Wang, S., Liu, Z., Tang, S., Ding, C., & Chen, L. (2016). Triacontanol reduces transplanting shock in machine-transplanted rice by improving the growth and antioxidant systems. *Frontiers in Plant Science, 7*, 872. https://doi.org/10.3389/fpls.2016.00872 (accessed on 6th November 2025).

Lúcia, S., Mária, T. H., Györgyi, K., Géza, T., Arzu, R. G., & József, Z. (2022). Assessment of 1-triacontanol treatment of sweet corn (*Zea mays* L. convar. *saccharata*) aimed at the improvement of salt tolerance based on a pot experiment. *Acta Universitatis Sapientiae, Agriculture and Environment, 14*, 1–12. https://doi.org/10.2478/ausae-2022-0001 (accessed on 6th November 2025).

Maity, U., Dutta, P., Laya, B., & Chaudhuri, K. M. (2019). Enhancement of green leaf yield in tea [*Camellia sinensis* (L.) O. Kuntze] as a function of pigments by foliar application of triacontanol. *International Journal of Bio-Resource and Stress Management, 10*(4), 451–455. http://dx.doi.org/10.23910/IJBSM/2019.10.4.2019a (accessed on 6th November 2025).

Misra, A., & Srivastava, N. K. (1991). Effect of the triacontanol formulation 'Miraculan' on photosynthesis, growth, nutrient uptake, and essential oil yield of lemongrass

(*Cymbopogon flexuosus* Steud. Watts). *Plant Growth Regulation, 10*, 57–63. https://doi.org/10.1007/BF00035131 (accessed on 6th November 2025).

Morré, D. J., Selldén, G., Zhu, X. Z., & Brightman, A. (1991). Triacontanol stimulates NADH oxidase of soybean hypocotyl plasma membrane. *Plant Science, 79*(1), 31–36. https://doi.org/10.1016/0168-9452(91)90065-G (accessed on 6th November 2025).

Nabi, A., Aftab, T., Masroor, M., Khan, A., & Naeem, M. (2022). Exogenous triacontanol provides tolerance against arsenic-induced toxicity by scavenging ROS and improving morphology and physiological activities of *Mentha arvensis* L. *Environmental Pollution, 295*, 118609. https://doi.org/10.1016/j.envpol.2021.118609 (accessed on 6th November 2025).

Nabi, A., Parwez, R., Aftab, T., Khan, M. M. A., & Naeem, M. (2021). Triacontanol protects *Mentha arvensis* L. from nickel-instigated repercussions by escalating antioxidant machinery, photosynthetic efficiency and maintaining leaf ultrastructure and root morphology. *Journal of Plant Growth Regulation, 40*, 1594–1612. https://doi.org/10.1007/s00344-020-10208-y (accessed on 6th November 2025).

Naeem, M., Ansari, A. A., Aftab, T., Alam, M. M., Khan, M. M. A., & Uddin, M. (2019). Triacontanol improves production of anticancer alkaloids in *Catharanthus roseus* L. *Asian Journal of Pharmaceutical Research and Health Care, 11*(1), 21–27.

Ostrowska, A., Grzesiak, M. T., & Hura, T. (2021). Exogenous application of growth stimulators improves the condition of maize exposed to soil drought. *Acta Physiologiae Plantarum, 43*, 1–9. https://doi.org/10.1007/s11738-021-03232-2 (accessed on 6th November 2025).

Pang, Q., Chen, X., Lv, J., Li, T., Fang, J., & Jia, H. (2020). Triacontanol promotes the fruit development and retards fruit senescence in strawberry: A transcriptome analysis. *Plants, 9*(4), 488. https://doi.org/10.3390/plants9040488 (accessed on 6th November 2025).

Perveen, S., Iqbal, M., Nawaz, A., Parveen, A., & Mahmood, S. (2016). Induction of drought tolerance in *Zea mays* L. by foliar application of triacontanol. *Pakistan Journal of Botany, 48*(3), 907–915. https://www.pakbs.org/pjbot/PDFs/48(3)/08.pdf (accessed on 6th November 2025).

Perveen, S., Parvaiz, M., Shahbaz, M., Saeed, M., & Zafar, S. (2022). Triacontanol positively influences growth, yield, biochemical attributes and antioxidant enzymes of two linseed (*Linum usitatissimum* L.) accessions differing in drought tolerance. *Pakistan Journal of Botany, 54*(3), 843–853. http://dx.doi.org/10.30848/PJB2022-3(2) (accessed on 6th November 2025).

Ramos-Zambrano, E., Juárez-Yáñez, T. E., Tapia-Maruri, D., Camacho-Díaz, B. H., Jiménez-Aparicio, A. R., & Martínez-Ayala, A. L. (2021). Effects of triacontanol and light on stomatal and photochemical responses in *Solanum lycopersicum* L. *Journal of Plant Growth Regulation, 40*(5), 2208–2220. https://doi.org/10.1007/s00344-020-10262-6 (accessed on 6th November 2025).

Sanches, L. G., Santos, A. J. D. S., Carreiro, D. D. A., Cunha, J. G. D., Lobo, J. T., Cavalcante, Í. H., & Paiva Neto, V. B. D. (2023). Biochemical responses in 'Kent' mango grown in Brazilian semi-arid region under different doses of triacontanol. *Revista Brasileira de Engenharia Agrícola e Ambiental, 27*, 309–316. https://doi.org/10.1590/1807-1929/agriambi.v27n5p309-316 (accessed on 6th November 2025).

Sardar, R., Ahmed, S., Akbar, M., Yasin, N. A., & Li, G. (2022). Alleviation of cadmium phytotoxicity in triacontanol treated *Coriandrum sativum* L. by modulation of physiochemical attributes, oxidative stress biomarkers and antioxidative system. *Chemosphere, 295*, 133924. https://doi.org/10.1016/j.chemosphere.2022.133924 (accessed on 6th November 2025).

Sarwar, M., Anjum, S., Alam, M. W., Ali, Q., Ayyub, C. M., Haider, M. S., Ashraf, M. I., & Mahboob, W. (2022). Triacontanol regulates morphological traits and enzymatic activities of salinity affected hot pepper plants. *Scientific Reports, 12*(1), 3736. https://doi.org/10.1038/s41598-022-06516-w (accessed on 6th November 2025).

Sarwar, M., Anjum, S., Ali, Q., Alam, M. W., Haider, M. S., & Mehboob, W. (2021). Triacontanol modulates salt stress tolerance in cucumber by altering the physiological and biochemical status of plant cells. *Scientific Reports, 11*(1), 24504. https://doi.org/10.1038/s41598-021-04174-y (accessed on 6th November 2025).

Shahbaz, M., Noreen, N., & Perveen, S. (2013). Triacontanol modulates photosynthesis and Osmo protectants in canola (*Brassica napus* L.) under saline stress. *Journal of Plant Interactions, 8*(4), 350–359. https://doi.org/10.1080/17429145.2013.764469 (accessed on 6th November 2025).

Soundararajan, M. (2015). Induction of organogenesis in tomato callus (*Lycopersicon esculentum* Mill. cv. PKM-1) using plant growth promoters including triacontanol and antioxidants. *Indian Journal of Fundamental and Applied Life Sciences, 5*(3), 81–87.

Swamy, S. G., Ramanarayan, K., Inamdar, L. S., & Inamdar, S. R. (2009). Triacontanol and jasmonic acid differentially modulate the lipid organization as evidenced by the fluorescent probe behavior and 31P nuclear magnetic resonance shifts in model membranes. *Journal of Membrane Biology, 228*, 165–177. https://doi.org/10.1007/s00232-009-9169-1 (accessed on 6th November 2025).

Tompa, B., & Fodorpataki, L. (2021). Influence of triacontanol and salt stress on the growth and metabolism of spinach. *Acta Universitatis Sapientiae, Agriculture and Environment, 13*(1), 65–76. https://doi.org/10.2478/ausae-2021-0006 (accessed on 6th November 2025).

Umar, I., Ahmed, S., Yasin, N. A., Wahid, A., Alamri, S., Hamid, Y., & Sardar, R. (2023). Role of exogenously applied triacontanol in amelioration of chromium stress in *Pisum sativum* by enhanced nutrition orchestration, growth and proline content. *South African Journal of Botany, 162*, 96–107. https://doi.org/10.1016/j.sajb.2023.09.002 (accessed on 6th November 2025).

Verma, T., Bhardwaj, S., Raza, A., Djalovic, I., Prasad, P. V., & Kapoor, D. (2023). Mitigation of salt stress in Indian mustard (*Brassica juncea* L.) by the application of triacontanol and hydrogen sulfide. *Plant Signaling and Behavior, 18*(1), 2189371. https://doi.org/10.1080/15592324.2023.2189371 (accessed on 6th November 2025).

Waqas, M., Shahzad, R., Khan, A. L., Asaf, S., Kim, Y. H., Kang, S. M., Bilal, S., Hamayun, M., & Lee, I. J. (2016). Salvaging effect of triacontanol on plant growth, thermotolerance, macro-nutrient content, amino acid concentration and modulation of defense hormonal levels under heat stress. *Plant Physiology and Biochemistry, 99*, 118–125. https://doi.org/10.1016/j.plaphy.2015.12.012 (accessed on 6th November 2025).

Xuan, W., & Feng, N. (2000). Effects of triacontanol on seed germination and seedling development of super sweet maize (*Zea mays*). *Fujian Agricultural Science and Technology, (6)*, 7–8.

Yong, W. (2000). Effects of cholesterol and triacontanol on flowering in duckweed. *Zhi Wu Sheng Li Xue Bao = Acta Psychophysiological Sinica, 26*(1), 39–45.

Younis, A. A., & Ismail, H. A. (2019). Triacontanol alleviated nickel toxicity in maize seedling by controlling its uptake and enhancing antioxidant system. *Journal of Advances in Plant Biology, 1*(3), 1–15. https://doi.org/10.14302/issn.2638-4469.japb-19-3051 (accessed on 6th November 2025).

Younis, A., & Ismail, H. (2022). Alleviation of *Zea mays* L. nickel toxicity by triacontanol foliar spray. *Egyptian Journal of Pure and Applied Science, 60*(1), 17–33. https://dx.doi.org/10.21608/ejaps.2022.109286.1015 (accessed on 6th November 2025).

CHAPTER 11

Recent Advancements in Gibberellic Acid (GA3) Production Technologies, Potential Applications, and Role in Horticultural Crop Production

SANCHALI KUNDU,[1] KOUSTAV MONDAL,[1] SANGHATI KUNDU,[2] and SOMANJANA KHATUA[3,*]

[1]*Department of Botany, Krishnagar Government College, Krishnagar, West Bengal, India*

[2]*Department of Botany, Santipur College, Nadia, West Bengal, India*

[3]*Department of Botany, Faculty of Science, University of Allahabad, Prayagraj, Uttar Pradesh, India*

*Corresponding author

ABSTRACT

Gibberellic acids (GA_3) are derived from geranylgeranyl diphosphate (GGDP) and serve a significant role in a number of physiological functions. These diterpene plant growth hormones regulate diverse aspects of growth and development, such as seed germination, stem elongation, dormancy elimination, reproductive development, fruit senescence, leaf senescence, and enzyme induction. Exogenous application of GA_3 has been demonstrated to enhance plant resilience to salinity, chilling, drought, and trace element stresses, mitigating their negative effects by influencing antioxidant enzyme activity as well as reducing intracellular reactive oxygen species (ROS) content. Commercially, GA_3 is widely utilized to improve the phenotypic characteristics, earliness, and productivity of

Plant Biostimulants and Growth Regulators in Sustainable Agriculture.
Deepak Kumar, Sanjay K. Singh, Nutan Kaushik, and Mahaveer P. Sharma (Eds.)

DOI: 10.1201/9781779641458-15

numerous vegetable and ornamental crops. However, its extensive use is hampered by high production costs. The current industrial production of GA_3 relies on methods for submerged fermentation (SmF). Alternatively, solid-state fermentation (SSF) has been explored to potentially increase GA_3 yields by utilizing agro-industrial wastes as substrates, thereby contributing to cost reduction. This review provides insights into the chemical and biological properties of GA_3, its fermentation-based production, and recent advancements in this field.

11.1 INTRODUCTION

Gibberellins, commonly referred to as gibberellic acids (GA_3), gained attention in the 1950s among Western scientists, although their discovery dates back to the 1930s in Japan. Japanese rice farmers observed a fungal disease known as foolish seedling, or bakanae, causing rice plants to grow excessively tall and impeding seed production. Plant pathologists have identified the chemical (gibberellic acid) secreted by the pathogenic fungus *Fusarium fujikuroi*, formerly known as *Gibberella fujikuroi*, as the cause of these symptoms. In the 1930s, Japanese scientists isolated two fungal "compounds," one of which, obtained from *Gibberella*, was named gibberellin A (Gupta and Chakrabarty, 2013; Hedden, 2020). During the 1950s, Japanese efforts led to the isolation of gibberellin A1, A2, A3 (later recognized as gibberellic acid), and A4. Simultaneously, the application of GA_3 restored the normal phenotype of maize dwarf mutants. This discovery spurred further investigations into characterizing new gibberellins in plants, fungi, and bacteria (Castro-Camba et al., 2022).

GA_3, a di-terpenoid tetracyclic carboxylic acid, serves as a natural plant growth regulator (PGR) with diverse applications in agriculture and horticulture (Rodrigues, 2012). Its positive effects include enhanced seed germination, a shortened juvenile phase, leaf expansion, abiotic stress tolerance, stem elongation, early flowering, and fruit development across various crops and ornamental plants (Colebrook et al., 2014; Cornea-Cipcigan et al., 2020). GA_3 finds application in orchards and horticultural industries, influencing processes such as malting, flowering, fruit growth enhancement, response to abiotic stress, and parthenocarpic fruit production (Kataoka et al., 2004; Li et al., 2011; Sharma and Singh, 2009). Despite its significance, the current industrial production of GA_3, primarily sourced from the fungus *G. fujikuroi*, faces challenges due to low yields, resulting in high production costs. Annual GA_3 production is limited, and the cost per kg varies between $150 and $500 USD (Camara et al., 2020). The increasing demand for GA_3 necessitates alternative production methods, with submerged fermentation (SmF) proving economically non-viable (Camara et al., 2018). Recent research focuses on low-cost, environmentally friendly production using substrates such as corn cob, wheat straw, and byproducts of rice, coconut, and soybean (Monrroy, 2022). This review highlights the need for advancements in GA_3 production technologies to make them more accessible, cost-effective, and readily available. By exploring low-energy and environmentally friendly production methods, the aim is to reduce costs, minimize losses, and broaden the applications of GA_3 in agriculture and horticulture. This review aims to contribute to the understanding of current advancements in GA_3 production technologies for improved horticultural practices.

11.2 CHEMICAL PROPERTIES OF GA_3

GA_3 ($C_{19}H_{22}O_6$) stands out as the most prevalent and first structurally characterized gibberellin. Chemically, it is described as a tetracyclic dihydroxy-γ-lactone acid, featuring a double bond at C1–C2, a γ-lactone ring at C-10, and a hydroxyl group at C-13 (Camara et al., 2018; Rodrigues et al., 2012). GA_3 is a white crystalline powder exhibiting a melting point within the range of 233–235°C. It demonstrates solubility in various solvents, including alcohol, acetone, ethyl acetate, and butyl acetate. However, its solubility in petroleum ether, benzene, and chloroform is limited. In water, GA_3 displays low solubility, reaching only 5 g/L. The compound readily undergoes decomposition at higher temperatures, alkaline pH levels, and in aqueous solutions. Specifically, at 20°C, GA_3 has a half-life of approximately 14 days, while at 50°C, its half-life shortens to 2 days (Bruckner and Blechschmidt, 1991). The stability of the molecule is attributed to the presence of a C1-C2 double bond in its chemical structure, enhancing its bioactivity. The effectiveness of GA_3 on plants may be influenced by changes in its structure, which can occur during decomposition. Various physical factors, such as temperature, pH of the solution, and reaction time, can lead to distinct decomposition products. It is noteworthy that alterations in the structure of GA_3 may impact its efficacy in plant applications (Pandya et al., 2023).

11.3 PRODUCTION OF GA_3

12.3.1 GA_3 BIOSYNTHESIS

In both plants and fungi, the synthesis of GA_3 follows the terpenes pathway originating from geranylgeranyl diphosphate (GGDP). Initially, four isoprenoid molecules combine to create a linear structure comprising 20 carbon atoms, referred to as GGDP. Under the influence of *ent*-copalyl diphosphate synthase, this molecule undergoes a transformation, leading to the formation of *ent*-copalyl diphosphate. Subsequently, *ent*-copalyl diphosphate is enzymatically converted into a tetracyclic compound known as *ent*-kaurene, facilitated by the action of *ent*-kaurene synthase (Salazar-Cerezo et al., 2018). In both plants and fungi, the biosynthesis of GA_3 follows a similar pathway from GGDP to *ent*-kaurene, with subsequent divergence in the conversion of GA_{12}-aldehyde. Initiated by *ent*-kaurene oxidase in plants and P450-4 in fungi, C-19 of *ent*-kaurene undergoes sequential oxidation to yield *ent*-kaurenoic acid, subsequently leading to the formation of GA_{12}-aldehyde. In plants, GA_{12}-aldehyde takes two distinct routes: it is initially converted to GA_{12}, and through the action of GA_{20}-oxidase responsible for C19-GA production, it is transformed into GA_9. Simultaneously, GA_{12} undergoes 13-hydroxylation to produce GA_{53}, which is further converted to GA_{20} by C20-oxidase. GA_3-oxidase then catalyzes the conversion of GA_{20} and GA_9 into GA_1 and GA_4, respectively, by adding a 3β-hydroxyl group. Notably, the synthesis of GA_3 diverges among species and is influenced by environmental conditions. In fungi, GA_{12}-aldehyde is 3β-hydroxylated, forming GA_{14}-aldehyde, which undergoes oxidation to yield GA_{14}. GA_{14}, in turn, is converted to GA_4 through C20 oxidation. GA4 represents the initial bioactive molecule, desaturating to form GA_7, which is subsequently converted to GA_3 through 13-hydroxylation. While the biosynthesis pathways in plants and fungi share similarities, from GGDP to *ent*-kaurene, the divergence occurs in the conversion of GA_{12}-aldehyde. This distinction is attributed to the order in which the steps of 3β-hydroxylation and 13-hydroxylation take place in plants and fungi (Camara et al., 2018; Gupta and Chakrabarty, 2013).

11.3.2 INDUSTRIAL PRODUCTION OF GA_3

GA_3, a natural compound produced by plants, fungi, and bacteria, is found in minute amounts within plant tissues, particularly in the vegetative parts, making its extraction challenging and commercially unviable (Keswani et al., 2022; Pandya et al., 2023). Over the years, significant efforts have been invested in optimizing GA_3 production, reducing costs, increasing productivity, and exploring alternative production methods (Peng et al., 2020). Industrial-scale production of GA_3 primarily relies on microbial fermentation, with earlier techniques like liquid surface fermentation (LSF) proving limited in advantages. Currently, *Fusarium moniliforme* is a key player in bulk GA_3 production through fermentation technology. The entire GA_3 production process involves critical parameters in both upstream and downstream stages, especially when utilizing fungal cells. Key considerations include strain selection, strain improvement, inoculum development, utilizing cost-effective raw materials, extraction methods, and recovering GA_3 from the fermentation medium (Pandya et al., 2023). Biologically, GA_3 preparation primarily involves two methods: SmF and solid-state fermentation (SSF). These bioprocesses have been extensively explored for the production of microbial secondary metabolites, showcasing their versatility in GA_3 synthesis (Kumar et al., 2021).

11.3.2.1 SUBMERGED FERMENTATION (SMF)

Since 1935, fermentation has stood as the predominant and widely adopted method for GA_3 production, particularly utilizing microorganisms like *G. fujikuroi* or *F. moniliforme* in SmF. However, this approach has associated high production costs, thereby limiting the broader applications of GA_3 (Monrroy, 2022). While SmF has been practiced with diverse culture media and substrates to enhance production rates (Table 11.1), its yields tend to be relatively low. The purification of the product, diluted within the fermented medium, incurs additional costs and poses challenges in management (Pinheiro et al., 2024). In a study by Cihangir (2002), the utilization of the precursor mevalonic acid with *Aspergillus Niger* in submerged culture resulted in a significant increase in GA_3 production, reaching up to 60 μmol. Durán-Páramo et al. (2004) achieved a production rate of up to 0.60 mg/L/h using the stirred tank reactor (STR) technique with glucose as the carbon source.

TABLE 11.1 Production of Gibberellic Acid (GA_3) Through Different Microorganisms on Various Substrates by Fermentation Under Optimized Conditions

Microorganisms	Substrate	System	Production	Incubation Time	Temperature (°C)	References
Fusarium moniliforme	SmF with immobilized cells, glucose	Flask	680 mg/L	12 days	25	Kahlon and Malhotra (1986)
Gibberella fujikuoroi p-3	SSF	Shake flask	355 mg/kg	7 days	28	Kumar and Lonsane (1987)
Gibberella fujikuroi NRRL 2284	SSF, maize cob particles	250 mL flask	4.8 g/kg	8 days	30	Pastrana et al. (1995)
Gibberella fujikuroi NRRL 2278	SmF	250-mL Erlenmeyer flask	25 mg/mL	5 days	29	Tomasini et al. (1997)
	SSF, cassava flour	Column fermenter	250 mg/kg	36 h		
Gibberella fujikuroi H-984	SmF, glucose and rice flour	Fermenter	1100 mg/L	-	29	Escamilla Silva et al. (1999)
Gibberella fujikuroi CDBB H-984	Glucose and rice meal	3.5-l bioreactor	2.862 g/L	-	30	Escamilla et al. (2000)
Gibberella fujikuroi LPB-06	SSF, coffee husk	250-mL Erlenmeyer flask	492.5 mg/kg	7 days	29	Machado et al. (2002)
Gibberella fujikuroi LPB-06	SSF, coffee husk and cassava bagasse	Packed-bed column bioreactor	0.925 g/kg	6 days	29	Machado et al. (2004)
Gibberella fujikuroi NRRL2284	SmF-fed-batch operation, glucose	Fermenter	1.680 mg/L	4 days	30	Shukla et al. (2005)
Gibberella fujikuroi ATCC 12616	SSF, cassava bagasse and rice hull	Fermenter	1.58 mg/g	8 days	28	Otálvaro et al. (2008)
Gibberella fujikuoroi p-3	SmF	Shake flask	218.4 mg/kg	8 days	28	
Fusarium moniliforme LPB 03	SSF, CP	Erlenmeyer flask	5.9 g /kg	3 days	29	Rodrigues et al. (2009)
Fusarium fujikuroi SG2	SmF, glucose and rice flour	Flask	1.175 mg/L	7 days	–	Uthandi et al. (2010)
Fusarium proliferatum NCIM 1105	SSF, pigeon pea pod	–	6.4–7.8 g/kg	8–10 days	29	Satpute et al. (2010)
	Pea pods		5.7–6.4 g/kg			

TABLE 11.1 *(Continued)*

Microorganisms	Substrate	System	Production	Incubation Time	Temperature (°C)	References
	Corncobs		5.2–6.1 g/kg			
	Sorghum straw		4.1–5.5 g/kg			
Fusarium moniliforme NCIM 1100	SSF, jatropha seed cake	Flask	105 g/kg	4 days	30	Rangaswamy (2012)
Fusarium fujikuroi IMI58289	SmF, glucose	Multi-fermenter system	216 mg/L	10 days	-	Albermann et al. (2013)
Penicillium variable	SSF, banana peel	Flask	31.57 mg/kg	7 days	27	Isa and Mat Don. (2014)
Fusarium moniliforme LPB 03	SSF, CP and soy husk	250 mL Erlenmeyer flask	5.9 g/kg	7 days	29	Rodrigues et al. (2016)
Fusarium fujikuroi CDBB H-984	SmF, corn oil	7 L Batch Stirred Tank Bioreactor	360 mg/L	12 days	29	Rios-iribe et al. (2016)
Fusarium moniliforme LPB 03	SSF, CP	Erlenmeyer flask	946 mg/L	144 h	29	de Oliveira et al. (2017)
		Column bioreactor	837 mg/L	120 h		
	SmF, CP	Erlenmeyer flask	236 mg/L	216 h		
		10 L STR	273 mg/L	96 h		
		1.5 L BCR	203 mg/L	216 h		
	SSSF, CP	Erlenmeyer flask	331 mg/L	240 h		
Paecilomyces sp. ZB	Cow dung	250 mL flask	1.312 g/kg	8 days	37	El-Sheikh et al. (2020)
Gibberella fujikuroi	SSF	600 mL beaker	934.41 mg/kg	7 days	28	Werle et al. (2020)
Gibberella fujikuroi LPB 06	SSSF, a mixture of CP and SH	1.5L BCR	279.6 mg/L	168 h	29	Camara et al. (2020)
		2L STR	205.2 mg/L	96 h		
Gibberella fujikuroi NRRL 2278	SSF, raw rice bran and barley malt residue	400-mL beaker	10.1 g/kg	7 days	28	Pinheiro et al. (2023)

de Oliveira et al. (2017) conducted experiments with *G. fujikuroi* in Erlenmeyer flasks, observing GA_3 synthesis starting at 48 h, with the highest productivity at 120 h recorded at 1.52 mg/L/h. In studies by Omojasola and Adejoro (2018), orange albedo's potential as an agro-industrial waste was explored as a cost-effective substrate for GA_3 production through SmF. The use of *F. moniliforme* ATCC 10052 and *A. niger* CBS 513.88 on the orange albedo substrate yielded promising results, with GA_3 concentrations reaching 5.53 g/L and 6.33 g/L, respectively. These findings underscore the viability of orange albedo as an economical and promising substrate for GA_3 production.

11.3.2.2 SOLID STATE FERMENTATION (SSF)

SSF is a fermentation process that employs a solid matrix containing sufficient moisture to support microbial growth without the presence of additional free water. The solid material can also serve as a nutrient source or a supporting medium infused with a complete nutrient recipe essential for the growth of microorganisms, primarily fungi and molds. SSF aims to replicate the natural environment for these microorganisms, allowing them to thrive optimally in their native habitat (Kumar et al., 2021). This method requires less energy for sterilization, as the lower water activity minimizes the need for extensive sterilization procedures. It is also less susceptible to bacterial contamination. SSF offers numerous advantages, such as higher enzymatic productivity, reduced susceptibility to substrate inhibition, and higher final product concentrations. Environmentally, SSF allows for the utilization of solid agro-industrial wastes as substrates or energy sources in their natural form, contributing to solid waste management and reducing wastewater production (Soccol et al., 2017). The utilization of agro-industrial residues in fermentation as an alternative medium allows for the reuse of these materials as a carbon source, thereby reducing their environmental impact within a biorefinery concept. Additionally, there is evidence suggesting that plant extracts may contain precursors or inducers of gibberellin synthesis, further enhancing GA_3 production (Camara et al., 2020). Other positive aspects of SSF include higher quality and activity of extracts, the absence of the need for organic solvents (which may confer toxicity to extracts), lower capital and operating costs, reduced downstream processing, and minimized stirring requirements (Soccol et al., 2017). In direct comparison with SmF, SSF exhibits significant advantages, particularly in terms of production costs, potentially reducing operating costs by up to 47% (Pinheiro et al., 2024). Consequently, SSF is considered a more efficient technique for the production of secondary metabolites (Kumar et al., 2021).

In a screening of various fungal strains by Kumar and Lonsane (1987), *G. fujikuroi* P-3 emerged as a consistent producer of GA_3 in SSF. Maximum GA_3 production values were 218.4 mg/kg in SmF and 355 mg/kg in SSF, both achieved at different incubation periods (eight days for SmF and seven days for SSF). The absolute amount of GA_3 produced was 1.626 times higher in SSF compared to SmF. Based on the total available carbohydrates in commercial wheat bran (WB), the percent conversion in SmF and SSF was 0.096 and 0.156, respectively. Preliminary cost analysis revealed a net savings of about 60% and 50% on fermentation medium cost and downstream processing expenditure, respectively, compared to the SmF technique in current use. A study by Tomasini et al. (1997) employed sugar cane bagasse as an inert support for SSF, demonstrating excellent growth of *G. fujikuroi*. GA_3 production reached 250 mg/kg of solid medium in 35 hours, while only 23 mg/L was obtained in 120 h of liquid fermentation, highlighting the rapidity of the solid culture system. Rodrigues et al. (2009) conducted a comprehensive study on strain (*F. moniliforme* LPB 03) and substrate (citric pulp) selection for GA_3 production through SSF. Citric pulp extract combined with sucrose was the most effective medium for producing inoculums and GA_3 production by SSF (5.9 g/kg of dry citric pulp after 3 days of fermentation). The same group's further research showed that soy husk and citric pulp's natural composition promotes the synthesis of GA_3 without the requirement for nutritional supplements (Rodrigues et al., 2016). SSF using agricultural residues, such as pigeon pea pod, corncobs, sorghum straw, and pea pod, was explored for GA_3 production. Pigeon pea pods, enriched with nitrogen in SSF, provided excellent support for the growth of *F. proliferatum* NCIM1105. The porous texture of the cellulosic materials in the pigeon pea pod exhibited good water-holding capacity and oxygen retention, making it a suitable substrate for SSF. Optimum conditions for moisture content (70%), temperature (29°C), and pH (4.76) were identified, resulting in optimum GA_3 yield (Satpute et al., 2010). Isa and Mat Don (2014) successfully optimized the culture conditions for maximum GA_3 production by the Penicillium *variable* in a static flask reactor using response surface methodology. The maximum GA_3 yield reached 31.95 mg GA_3/kg substrate under optimized conditions of a 7-day incubation period, 21% v/w inoculum size, and 2% v/w precursor (olive oil) concentration. *Paecilomyces* sp. ZB effectively used cow dung substrate for growth and GA_3 production, showing promise as an alternative for cost-effective GA_3 production (El-Sheikh et al., 2020). In a study by de Oliveira et al. (2017), 7.60 g of GA_3 per gramme of substrate (citric pulp, a waste obtained from orange juice extraction) was produced after 6 days by SSF at 29°C with an initial moisture of 75%. Comparative studies among different fermentation systems showed that the solid-state process yielded the best GA_3 production results. Machado et al. (2002) screened six fungal strains for GA_3 production, and *G. fujikuroi* LPB-6 was identified as the highest producer. In comparison to SmF, GA_3 production was 6.1 times higher in SSF. Optimizing conditions with a saline solution enhanced GA_3 accumulation by 1.7 times, reaching a maximum of 492.5 mg of GA_3/kg of dry substrate under finally optimized conditions. In a subsequent study, Machado et al. (2004) aimed at GA_3 production from *G. fujikuroi* using coffee husk and cassava bagasse in a packed-bed column bioreactor. They reported a production of 0.925 g GA_3/kg of substrate on the 6th day of fermentation, employing cassava bagasse and coffee husk as substrates with 77 wt.% moisture. Otálvaro et al. (2008) investigated GA_3 production from cassava bagasse and rice hull via SSF using *G. fujikuroi*, obtaining 1.58 mg of product per gram of substrate after 8 days of the process. Pinheiro et al. (2024) investigated the usage of barley malt residue (BMR) and raw rice bran (RRB) as substrates for the synthesis of GA_3 through *G. fujikuroi*. Using 30 wt.% RRB and 70 wt.% BMRs for a growth medium containing 70 wt.% moisture was found suitable for the highest yield of GA_3 after seven days of incubation. Higher concentrations of NH_4NO_3 were found to favor GA_3 formation, and kinetic investigation showed increasing GA_3 production, reaching 10.1 g/kg of substrate on the seventh day. These studies collectively highlight the diverse substrates, fermentation conditions, and fungi used in SSF for efficient GA_3 production, demonstrating the versatility and potential of this approach in optimizing the process.

11.3.2.3 SEMI-SOLID-STATE FERMENTATION (SSSF)

A successful GA_3 production process was developed using a SSSF approach, employing a mixture of citric pulp comprising 70% and soy husk comprising 30%, supplemented with sources such as urea and glucose. In a Bubble Column Reactor (BCR), GA_3 production commenced at 48 h, reaching its peak at 168 h with a maximum yield of 279.6 mg/L and a productivity of 1.66 mg/L/h. In STR, production initiated at 24 h, continued to grow until 96 h, and then stabilized. At the point of stabilization, GA_3 production in STR was 205.2 mg/L, with a productivity of 2.14 mg/L/h. While GA_3 production reached higher levels in BCR, greater productivity was achieved in STR. This superior performance is likely attributed to forced aeration and improved agitation, which facilitate the necessary oxygenation and homogenization of the medium (Camara et al., 2020).

11.3.3 ENHANCEMENT OF GA_3 PRODUCTION BY MUTATION BREEDING

The utilization of simvastatin, an HMG-CoA reductase inhibitor, serves as a supportive element in the enrichment method for *F. fujikuroi* strains, contributing to enhanced GA_3 production. This approach proves beneficial for mutagenesis breeding. Additionally, the incorporation of glycerol during the fermentation process has been identified to stimulate GA_3 production. Recent research findings demonstrated that *F. fujikuroi* mutant S109, when cultivated in a 5-L bioreactor with the addition of glycerol, yielded 2.8 g/L of GA_3. This represents a notable 33.3% increase compared to the original strain (Peng et al., 2020).

11.4 POTENTIAL APPLICATIONS OF GA_3

11.4.1 SEED GERMINATION

Seed germination is a crucial developmental process in plants, determining the site of plant growth and ensuring successful plant propagation. Environmental cues, including light, moisture, and temperature, play pivotal roles in regulating seed germination. Among the internal developmental programs responding to these external signals, GA_3 and abscisic acid (ABA) have been extensively studied. ABA and GA_3 exert antagonistic effects on seed germination through intricate signaling crosstalk. ABA delays germination, while GA_3 promotes it. Favorable environmental conditions activate genes encoding ABA-deactivating enzymes and GA_3 biosynthetic enzymes (Kim and Park, 2008). GA_3 influences seed germination by enhancing the growth potential of the embryo and inducing hydrolytic enzymes. During seed germination, embryonic GA_3 release triggers seed coat weakening by stimulating gene expression related to cell expansion and modification, as observed in *Arabidopsis*. GA_3 acts as a natural regulator of germination processes, stimulating the production of hydrolytic enzymes such as α-amylase in the aleurone layer of germinating cereal grains (Gupta and Chakrabarty, 2013). The signaling is regulated by repressors known as DELLA proteins, including REPRESSOR OF GA1-3 (RGA), GA-INSENSITIVE (GAI), and REPRESSOR OF GA1-3-LIKE1/2/3, with RGL2 being a major DELLA factor involved in repressing seed germination (Zhong et al., 2015). Even under favorable conditions, seed germination can be delayed by factors such as high salinity. High salinity reduces GA biosynthesis, leading to the repression of the GA synthetic enzyme gene GA3ox1 (Kim and Park, 2008).

Exogenous application of GA_3 has been shown to facilitate seed germination in various plant species. For instance, in *Fraxinus hupehensis*, GA_3 promotes germination by enhancing soluble sugars and weakening lipolysis (Song et al., 2019). Wang et al. (2018) demonstrated that exogenous GA_3 application facilitates rice seed germination under low-temperature conditions by promoting gibberellin signal transduction and suppressing ABA synthesis. Ge et al. (2023) found that exogenous GA_3 promotes germination and shortens the after-ripening process of *Panax notoginseng* seeds by increasing GA_3 contents. Abdani Nasiri et al. (2018) showed that the application of GA_3 significantly increases seed germination and seedling growth in *Salvia officinalis*. In a recent investigation focused on achieving consistent germination in *Tilia miqueliana* seeds, the evaluation during cold stratification included the analysis of ingredients including sugar, starch, and soluble protein, as well as the activities of enzymes like protease and α-amylase. The combined treatment of magnetically treated water and GA_3 exhibited the highest values for both enzyme activities. These observed changes are indicative of dormancy breakage, potentially leading to accelerated germination (Yao and Shen, 2018).

11.4.2 STEM ELONGATION AND PLANT GROWTH

Enhancing shoot growth and development, crucial agronomic traits with substantial impact on plant yield and architecture, is a key focus in plant science, with GA_3 recognized for its direct influence on these processes (Castro-Camba et al., 2022). In reaction to both darkness and light, GA_3 helps to promote cell division and expansion (Gupta and Chakrabarty, 2013). Similarly, Shan et al. (2021) used the dwarf cultivar Henong 60 (HN60) and the high-stalk cultivar Heinong 48 (HN48) to investigate the underlying process of soybean internode expansion in reaction to GA_3. Their findings indicated that the application of GA_3 to soybean stems, leaves, and roots increased the length of internodes and stems' GA_3 content. This treatment increased the amount of secondary xylem in soybean stems and increased the fraction of pith in primary xylem. The control of stem elongation has been linked to the apical meristem, an essential location for GA_3 production in soybean stems. The importance of GA_3 levels in stems for internode elongation was highlighted by transport experiments, which revealed that GA_3 is carried acropetally via the xylem and phloem in soybean stems. By changing the direction of cellulose microfibrils, GA_3 therapy can affect stem elongation (Sauter et al., 1993). Additionally, when GA_3 was administered to *Arabidopsis*, the percentage of xylem increased significantly (Ragni et al., 2011). While most current research on GA_3 application involves methods such as leaf spraying or seed soaking, understanding its influence on shoot growth remains a focal point (Shan et al., 2021).

11.4.3 FRUIT DEVELOPMENT

Fruit development progresses following fruit set after ovary fertilization, with GA_3 playing a crucial role in inducing fruit setting and growth (Shinozaki et al., 2020). GA_3 significantly impacts fruit quality improvement, especially in berries and various temperate fruit crops such as apple, pear, peach, cherry, apricot, and so forth. (Ghosh and Halder, 2018). In tomatoes, a commercially important fleshy berry-type fruit, the application of GA_3 induces fruit set even in unpollinated ovaries (Fos et al., 2001). Grapes, vital for the international wine and beverage industries, face challenges from diseases that reduce yield, quality, color, and flavors and induce premature ageing (Evers et al., 2010; Smart and Robinson, 1991). GA_3 promotes cell division and enlargement during grape development, influencing growth rate, early sugar uptake, and increasing carbohydrate and water content, particularly in *"Emperatriz"* seedless grapes (Casanova, 2009; Ungsa et al., 2018). In strawberry plants, such as

Gaviota, foliar application of GA_3 has been found to increase strawberry production. Moreover, the application of gibberellins plays a crucial role in overcoming seed rest effects, influencing the physiology of fruit set and development. It also prevents internal browning and frost injury in temperate fruits like apples, peaches, apricots, pears, and cherries, thereby significantly impacting agricultural industries (Ghosh and Halder, 2018). Higher levels of GA_3 induce the size and production of parthenocarpic fruit, and the accumulation of GA_3 is associated with its development (Fos et al., 2000; Kataoka et al., 2003). These insights underscore the importance of GA_3 in enhancing fruit development and quality across diverse fruit crops.

11.4.4 FLOWERING

The development of stamens and anthers is intricately linked to GA_3 signaling, as disruptions in GA_3 signaling pathways lead to male sterility. Various GA-insensitive or -deficient mutants in plant species such as rice, petunia, maize, tomato, barley, and *Arabidopsis* exhibit structural defects in stamens and anthers, often resulting in the formation of sterile flowers (Kwon and Paek, 2016). GA_3 plays a crucial role in regulating flower initiation and development, being indispensable for both male and female fertility rather than the differentiation of floral organs. In cases of severe GA_3 deficiency observed in mutants of *Arabidopsis* and tomato, abnormal stamen development occurs, and extreme GA_3 deficiency can lead to female sterility. Severe GA_3-deficient mutants may exhibit underdeveloped sepals, petals, and pistils, and in some instances, premature abortion of the flower. The application of bioactive GAs or their precursors, such as GA_9, has been shown to restore normal flower development. Notably, *Arabidopsis* stamens require a higher concentration of GA_3 compared to other floral organs, emphasizing their significance as a rich source of GAs, as demonstrated in rice (Gupta and Chakrabarty, 2013). GA_3 has been found to stimulate pistillate flower development in various plants, including castor bean, corn, and hyoscyamus (Pharis and King, 1985).

11.4.5 PHYTOREMEDIATION

Heavy metal contamination has emerged as a significant environmental issue, leading to soil pollution. Phytoremediation, employing plants to extract and accumulate heavy metals in roots, stems, and branches, has shown promise as an effective and eco-friendly solution (Shehata et al., 2019). Phytoextraction, a method within phytoremediation, is recognized for its cost-effectiveness and efficiency (Ji et al., 2015). GA_3, acting as a signaling molecule in plants, plays a crucial role in triggering specific responses to stresses (viz., abiotic and biotic stresses) (Gong et al., 2021). Studies indicate that foliar application of PGRs, including GA_3, can enhance the uptake and translocation of heavy metals in plants, thereby improving the efficiency of phytoremediation. For instance, GA_3 application significantly increased Pb uptake and translocation in *Zea mays*, demonstrating enhanced phytoremediation efficiency for Pb (Hadi et al., 2010). Investigations of *Triticum aestivum* under nickel stress revealed that the combined use of calcium and gibberellin mitigated the negative effects of nickel toxicity by increasing enzyme levels, chlorophyll content, and proline volume in the plant (Siddiqui et al., 2011). GA_3 has also shown promise in improving the phytoremediation of cadmium in *Solanum nigrum*, with a concentration of 1000 mg/L increasing plant biomass and cadmium concentration in the stem (Ji et al., 2015). In the case of *Parthenium hysterophorus*, GA_3 alone increased plant growth and biomass in Cd-contaminated soil, accumulating more than 50% of total Cd in the root (Hadi et al., 2014). Foliar application of GA_3 was found to reduce reactive oxygen species (ROS) generation and enhance Cu phytoextraction in different plant parts of jute (*Corchorus capsularis*) (Saleem et al., 2020). Additionally, GA_3 demonstrated its potential to improve stress tolerance and enhance phytoremediation efficiency in *Brassica juncea* under Cd- and U-contaminated soils (Chen et al., 2020). In a study involving ridge gourd (*Luffa acutangula*) for polycyclic aromatic hydrocarbon removal, foliar application of GA_3 positively influenced phytoextraction of Cu and mitigated levels of oxidative stress in pea cultivars raised on Cu-containing soil. GA_3 has also been investigated in the context of constructed wetland systems using Tifton 85 bermudagrass, demonstrating its influence on plant development and nutrient removal from synthetic municipal wastewater (Araújo et al., 2018). Overall, these findings underscore the potential of GA_3 as a valuable tool for enhancing the effectiveness of phytoremediation processes.

11.4.6 ABIOTIC STRESS TOLERANCE

Modern agriculture confronts various abiotic stresses, including salinity, drought, chilling, and heat, which are major factors affecting crops (Tardieu and Tuberosa, 2010). These stresses contribute to pre- and post-harvest losses, causing a reduction of more than 50% in the average yield of major crops (Bray et al., 2000; Gust et al., 2010). Adverse effects on crops, such as drought, salinity, osmotic stress, excessive ROS generation, heat, cold, irradiation, and light stresses, are well documented (Fahad et al., 2015). Phytohormones play a crucial role in mediating plant responses to abiotic stress, enabling plants to cope with challenging conditions and mitigate stress-induced damage (Skirycz and Inze, 2010). Among these phytohormones, GA_3 actively regulates plant responses to stress (Chakrabarti and Mukherji, 2003).

In the context of Cu stress, GA_3 application significantly alleviated the inhibitory effects of Cu stress on spinach seedling growth. The treatment reversed the negative impacts on biomass, reduced Cu, malondialdehyde, and proline content, increased antioxidant enzyme activities, and restored chlorophyll contents in spinach exposed to Cu-contaminated soil (Gong et al., 2021). Raza Gurmani et al. (2022) demonstrated the potential of GA_3 in enhancing salinity tolerance, particularly in sensitive pea varieties. GA_3 treatment increased chlorophyll content, boosting the net assimilation rate of pea varieties under NaCl stress by elevating antioxidant enzyme activity. Seed priming with GA_3 emerged as an effective strategy to improve wheat performance under salt stress. This approach enhanced plant height, yield, yield-related traits, as well as Ca^{2+} and K^+ concentrations, and transpiration rates, while reducing Na^+ concentrations (Aldesuquy et al., 2001). GA_3 application similarly proved beneficial for maize growth under salt stress, reducing oxidative stress, increasing antioxidant enzyme activities, antioxidant gene expression, and K^+ ion concentration (Shahzad et al., 2021). In addressing heat stress in tomatoes, the exogenous application of 75 mg/L GA_3 demonstrated potential by improving morphological, physiological, and biochemical characteristics (Guo et al., 2022). GA_3 has also shown efficacy in alleviating oxidative stress in wheat under salt stress, enhancing nutrient contents, and improving plant yield. Additionally, GA_3 effectively reduced Cr contents and improved sunflower yield under Cr stress. Its application enhanced growth in Cr-stressed pea plants by improving antioxidant enzymes, countering the toxic effects (Iftikhar et al., 2019). These findings highlight the versatile role of GA_3 in mitigating various abiotic stresses and promoting crop resilience.

11.5 ROLE OF GA_3 IN HORTICULTURAL CROP PRODUCTION

Horticulture, derived from the Latin words hortus (meaning garden) and colere (meaning to cultivate), involves the cultivation of various crops, including ornamental plants for aesthetic appeal, fruits and vegetables for nutrition, and medicinal plants (Janick, 1979). The products of horticultural plants not only serve as foods but also contribute essential vitamins and minerals to the diet. Additionally, secondary metabolites from these plants are often utilized for treating human diseases. In the face of population growth, global environmental changes, and land degradation, horticultural plants are gaining increasing importance (Zhang et al., 2022). In contemporary times, the horticulture industry is a significant force driving the global agricultural economy's growth (Kumar et al., 2011). Recent reports indicate that the global horticultural market reached a value of USD 20.4 billion in 2021, with projections suggesting it could exceed USD 56.5 billion by 2030. This sector also has the potential to connect small-scale farmers to new global markets through innovative contract farming schemes, creating employment opportunities (De Blasis, 2020). Horticultural crop production plays a crucial role in meeting global food nutrition demands, improving food quality, generating employment, and fostering socio-economic development. In India, horticultural crops contribute 30.7% to the agricultural GDP and employ 20% of the labor force. Over the past few decades, India has experienced a notable 4.8% growth in horticultural crop production. Among the five primary PGRs, GA_3 plays a versatile role in modern horticultural practices, serving as a valuable tool for manipulating horticultural systems (Baruah et al., 2023).

11.5.1 EFFECT OF GA_3 ON VEGETABLE CROPS

The production of small-seeded vegetable crops faces challenges such as poor emergence, stand establishment, canopy development, and limited options for effective weed control. To address these issues, the potential use of PGRs, specifically GA_3, was explored to improve early emergence growth rates and potentially synchronize weed germination with control tactics. A controlled-environment study was conducted to assess the impact of GA_3 on several vegetable crops. Results from the study revealed that at 7 days after seeding, the emergence of carrots was significantly higher when treated with 2, 4, 8, 16, or 64 ppm GA_3 compared to untreated seeds.

The total emergence of carrots throughout the study period was 14% greater when seeds were treated with 4 ppm GA_3 compared to untreated seeds. Similarly, the treatment of cabbage with as low as 2 to 4 ppm GA_3 resulted in increased cabbage emergence rates and total plant emergence over the study period compared to untreated seeds (Schuler and Colquhoun, 2022). According to Miceli et al. (2019), the inclusion of 10–6 M GA_3 in the nutrient solution of a hydroponic floating system has been proposed to enhance the growth and quality of lettuce and rocket plants.

11.5.2 EFFECT OF GA_3 ON FRUIT-YIELDING PLANTS

Rabbiteye blueberry, a native species in the southeastern United States, was the subject of a study by Zang et al. (2016) investigating the impact of GA_3 on fruit quality, return bloom, and plant growth. In this study, three representative cultivars were treated with 500 mg/L of GA_3 in a hydroponic floating system. The foliar application of GA_3 significantly increased return bloom across all three cultivars. Compared to untreated control plants, the application of GA_3 to “Powder blue,” “Garden blue,” and “Climax” resulted in a substantial increase in the number of inflorescences in the following year—54.3%, 69.7%, and 60.0%, respectively. GA_3 application also led to improved viable seed count for each fruit weight, leaf area, dry weight, fresh weight, chlorophyll A&B levels, and total chlorophyll content. However, the harvest date and fruit ripening for all three cultivars were delayed in the subsequent year. Despite this delay, GA_3 application did not significantly alter the length/diameter ratio of rabbiteye blueberry fruits. These findings suggest that GA_3 application shows promise in enhancing plant growth and fruit quality in rabbiteye blueberries, albeit with delayed ripening.

In sweet cherry production, fruit cracking, softening, and rapid decay post-harvest pose significant challenges. Usenik et al. (2005) found that the application of GA_3 to sweet cherries resulted in slower fruit ripening, delayed maturity, increased yield, and improved fruit quality. Yildirim and Koyuncu (2010) also observed similar outcomes in the sweet cherry cultivar (namely ‘0900 Ziraat’), where GA_3 treatments delayed the harvest date, increased fruit weight by 10.71% compared to the control, and reduced fruit cracking by 77.80%. GA_3 application influenced fruit color, resulting in brighter and darker red-colored fruits. Studies on other crops, such as tomatoes and apples, have shown positive effects of GA_3 treatment. In tomatoes, GA_3 promotes cell expansion and enlargement, enhancing the shape index. According to the findings of Zhang et al. (2023), self-pollinated apple plants that were sprayed with GA_3 had higher fruit weights, a lower percentage of asymmetric fruits, and external fruits that were shaped and of comparable quality to cross-pollinated fruits.

11.5.3 EFFECT OF GA_3 ON ORNAMENTAL PLANTS

GA_3 is widely employed to boost productivity and enhance phenotypic characteristics in various ornamental crops. The application of GA_3 through spraying has been shown to induce early flowering and enhance the size, fresh weight, and dry weight of flowers in *Chrysanthemum*, as evidenced by studies conducted by Nagarjuna et al. (1988) and Koriesh et al. (1989). Sajid et al. (2016) described that treatment of GA_3 at a level of 250 mg/L caused early flowering and increased the number of branches, leaves, and leaf area. To assess the impact of GA_3, a dual experiment was conducted to examine the influence of GA_3 at different concentrations (50, 100, or 150 mg/L) on seed germination timing as well as the growth and ornamental quality of selected *Cyclamen* sp. Additionally, a greenhouse experiment involving both long and short sunlight photoperiods was conducted to investigate the significance of sunlight exposure in the development of the plants. The results indicated that under short sunlight exposure, seeds germinated earlier compared to those in a long sunlight photoperiod. Plants treated with foliar applications of GA_3 exhibited increased plant height, leaf area, number of leaves, root length, and tuber development. Significant variations were observed in germination percentage, mean germination time, and seedling vigor index with different concentrations of GA_3 supplementation and greenhouse sunlight intensity. In conclusion, the study suggested that the application of GA_3 serves as an economic strategy to enhance germination and promote the growth of selected *Cyclamen* species, leading to improved seedling vigor and ornamental quality (Cornea-Cipcigan et al., 2020).

11.6 CONCLUSION

To sum up, the review article delivers a comprehensive overview of recent strides in GA_3 production technologies, their potential applications, and their pivotal role in horticultural crop production. The exploration of diverse production methods underscores the dynamic nature of GA_3 synthesis, emphasizing the need for sustainable and environmentally friendly approaches. The future outlook involves further refining GA_3 production technologies, potentially through innovative and cost-effective methods. Ongoing research is imperative to unravel the intricate molecular mechanisms governing GA_3-mediated responses across different plant species, facilitating targeted and optimized application strategies. Additionally, there is a focus on comprehending the environmental impact of GA_3 usage and exploring eco-friendly formulations in alignment with broader sustainability goals in agriculture.

GA_3 is pivotal in signaling systems and mechanisms governing seed germination, stem elongation, meristematic tissue development, and the differentiation of floral organs. It serves a critical role in breaking seed dormancy, particularly in cereal grains and various crop species. Notably, mutants with a GA_3 deficit fail to germinate in the absence of exogenous levels of the hormone. Even though our knowledge of gibberellic acids' (GA_3) functions in seed germination, grain development, sex expression, plant growth, and floral development has advanced, the exact mechanisms are still somewhat unknown. Unraveling the transport mechanism of GA_3 is vital for plant species' survival and successful crop production. While significant strides have been made in understanding the roles of GA_3 in horticultural species, there is a notable gap compared to research on rice and *Arabidopsis*. The potential integration of GA_3 into precision farming practices, involving customized application schedules and doses, holds promise for maximizing benefits while minimizing environmental concerns. Collaborative endeavors among researchers, industry stakeholders, and farmers are critical for translating these advancements into practical, field-level applications.

KEYWORDS

- **geranylgeranyl diphosphate**
- **gibberellic acids'**
- **horticultural crops**
- **reactive oxygen species**
- **submerged fermentation**

REFERENCES

Abdani Nasiri, A., Mortazaeinezhad, F., & Taheri, R. (2018). Seed germination of medicinal sage is affected by gibberellic acid, magnetic field and laser irradiation. *Electromagnetic Biology and Medicine, 37*(1), 50–56.

Albermann, S., Elter, T., Teubner, A., Krischke, W., Hirth, T., & Tudzynski, B. (2013). Characterization of novel mutants with an altered gibberellin spectrum in comparison to different wild-type strains of *Fusarium fujikuroi. Applied Microbiology and Biotechnology, 97*, 7779–7790.

Aldesuquy, H., & Ibrahim, A. (2001). Interactive effect of seawater and growth bioregulators on water relations, abscisic acid concentration and yield of wheat plants. *Journal of Agronomy and Crop Science, 187*, 185–193.

Araújo, E. D., Borges, A. C., Dias, N. M., & Ribeiro, D. M. (2018). Effects of gibberellic acid on Tifton 85 bermudagrass (*Cynodon* spp.) in constructed wetland systems. *PLoS One, 13*(10), e0206378. https://doi.org/10.1371/journal.pone.0206378 (accessed on 6th November 2025).

Asahira, T. (1967). Studies on fruit development in tomato. I. Ovule development and content of diffusible auxin in synthetic auxin- and gibberellin-induced parthenocarpy tomato fruits in relation to their development. *Memoirs of the Research Institute for Food Science, Kyoto University, 28*, 47–74. https://cir.nii.ac.jp/crid/1572543025612614144 (accessed on 6th November 2025).

Bandelier, S., Renaud, R., & Durand, A. (1997). Production of gibberellic acid by fed-batch solid state fermentation in an aseptic pilot-scale reactor. *Process Biochemistry, 32*(2), 141–145.

Baruah, K., Tare, K., Sherpa, T. L., & Kaushik, K. (2023). Economics of horticultural production. In *Advances in Horticulture* (Vol. 1, pp. 236–259). Bigfoot Publications.

Bray, E. A., Bailey-Serres, J., & Weretilnyk, E. (2000). Responses to abiotic stress. In Gruissem, W., & Jones, R. (Eds.), *Biochemistry & Molecular Biology of Plants* (pp. 1158–1203). American Society of Plant Physiologists.

Bruckner, B., & Blechschmidt, D. (1991). The gibberellin fermentation. *Critical Reviews in Biotechnology, 192*, 163–192.

Camara, M. C., Vandenberghe, L. P. S., Rodrigues, C., de Oliveira, J., Faulds, C., Bertrand, E., & Soccol, C. R. (2018). Current advances in gibberellic acid (GA3) production, patented technologies and potential applications. *Planta, 248*(5), 1049–1062.

Camara, M. C., Vandenberghe, L. P. S., Sextos, G. C., Tanobe, V. O. A., Magalhães Junior, A. I., & Soccol, C. R. (2020). Alternative methods for gibberellic acid production, recovery and formulation: A case study for product cost reduction. *Bioresource Technology, 309*, 123295. https://doi.org/10.1016/j.biortech.2020.123295 (accessed on 6th November 2025).

Casanova, L., Casanova, R., Moret, A., & Agustí, M. (2009). The application of gibberellic acid increases berry size of "Emperatriz" seedless grape. *Spanish Journal of Agricultural Research, 7*(4), 919–927.

Castro-Camba, R., Sánchez, C., Vidal, N., & Vielba, J. M. (2022). Plant development and crop yield: The role of gibberellins. *Plants (Basel), 11*(19), 2650. https://doi.org/10.3390/plants11192650 (accessed on 6th November 2025).

Chakrabarti, N., & Mukherji, S. (2003). Alleviation of NaCl stress by pretreatment with phytohormones in *Vigna radiata*. *Biologia Plantarum, 46*, 589–594.

Chen, L., Long, C., Wang, D., & Yang, J. (2020). Phytoremediation of cadmium (Cd) and uranium (U) contaminated soils by *Brassica juncea* L. enhanced with exogenous application of plant growth regulators. *Chemosphere, 242*, 125112. https://doi.org/10.1016/j.chemosphere.2019.125112 (accessed on 6th November 2025).

Cihangir, N. (2002). Stimulation of the gibberellic acid synthesis by *Aspergillus niger* in submerged culture using a precursor. *World Journal of Microbiology and Biotechnology, 18*, 727–729.

Colebrook, E. H., Thomas, S. G., Phillips, A. L., & Hedden, P. (2014). The role of gibberellin signaling in plant responses to abiotic stress. *Journal of Experimental Biology, 217*(1), 67–75.

Cornea-Cipcigan, M., Pamfil, D., Sisea, C. R., & Mărgăoan, R. (2020). Gibberellic acid can improve seed germination and ornamental quality of selected *Cyclamen* species grown under short and long days. *Agronomy, 10*(4), 516. https://doi.org/10.3390/agronomy10040516 (accessed on 6th November 2025).

De Blasis, F. (2020). Global horticultural value chains, labor and poverty in Tanzania. *World Development Perspectives, 18*, 100201. https://doi.org/10.1016/j.wdp.2020.100201 (accessed on 6th November 2025).

de Oliveira, J., Rodrigues, C., Vandenberghe, L. P. S., Câmara, M. C., Libardi, N., & Soccol, C. R. (2017). Gibberellic acid production by different fermentation systems using citric pulp as substrate/support. *BioMed Research International, 2017*, 5191046. https://doi.org/10.1155/2017/5191046 (accessed on 6th November 2025).

Durán-Páramo, E., Molina-Jiménez, H., Brito-Arias, M. A., & Robles-Martínez, F. (2004). Gibberellic acid production by free and immobilized cells in different culture systems. *Applied Biochemistry and Biotechnology, 114*, 381–388.

El-Sheikh, M. A., Rajaselvam, J., Abdel-Salam, E. M., Vijayaraghavan, P., Alatar, A. A., & Devadhasan Biji, G. (2020). *Paecilomyces* sp ZB is a cell factory for the production of gibberellic acid using a cheap substrate in solid state fermentation. *Saudi Journal of Biological Sciences, 27*(9), 2431–2438.

Escamilla Silva, E. M., Dendooven, L., Uscanga Reynell, J. A., Monroy Ramírez, A. I., González Alatorre, G., & de la Torre Martínez, M. (1999). Morphological development and gibberellin production by different strains of *Gibberella fujikuroi* in shake flasks and bioreactor. *World Journal of Microbiology and Biotechnology*, *15*, 753–755.

Escamilla, E. M., Dendooven, L., Magaña, I. P., Parra, R., & De La Torre, M. (2000). Optimization of gibberellic acid production by immobilized *Gibberella fujikuroi* mycelium in fluidized bioreactors. *Journal of Biotechnology*, *76*(2, 3), 147–155.

Evers, D., Molitor, D., Rothmeier, M., Behr, M., Fischer, S., & Hoffmann, L. (2010). Efficiency of different strategies for the control of grey mold on grapes including gibberellic acid (Gibb3), leaf removal and/or botrycide treatments. *OENO One*, *44*(3), 151–159.

Fahad, S., Hussain, S., Bano, A., Saud, S., Hassan, S., Shan, D., Khan, F. A., Khan, F., Chen, Y., Wu, C., Tabassum, M. A., Chun, M. X., Afzal, M., Jan, A., Jan, M. T., & Huang, J. (2015). Potential role of phytohormones and plant growth-promoting rhizobacteria in abiotic stresses: consequences for changing environment. *Environmental Science and Pollution Research*, *22*, 4907–4921.

Fos, M., Nuez, F., & García-Martínez, J. L. (2000). The gene *pat-2*, which induces natural parthenocarpy, alters the gibberellin content in unpollinated tomato ovaries. *Plant Physiology*, *122*(2), 471–480.

Fos, M., Proaño, K., Nuez, F., & García-Martínez, J. L. (2001). Role of gibberellins in parthenocarpy fruit development induced by the genetic system *pat-3/pat-4* in tomato. *Physiologic Plantarum*, *111*(4), 545–550.

Ge, N., Jia, J. S., Yang, L., Huang, R. M., Wang, Q. Y., Chen, C., Meng, Z. G., Li, L. G., & Chen, J. W. (2023). Exogenous gibberellic acid shortening after-ripening process and promoting seed germination in a medicinal plant *Panax notoginseng*. *BMC Plant Biology*, *23*(1), 67. https://doi.org/10.1186/s12870-023-04084-3 (accessed on 6th November 2025).

Ghosh, S., & Halder, S. (2018). Effect of different kinds of gibberellin on temperate fruit crops: a review. *Journal of Pharmacognosy and Phytochemistry*, *7*(3), 315–319.

Gong, Q., Li, Z. H., Wang, L., Zhou, J. Y., Kang, Q., & Niu, D. D. (2021). Gibberellic acid application on biomass, oxidative stress response, and photosynthesis in spinach (*Spinacia oleracea* L.) seedlings under copper stress. *Environmental Science and Pollution Research International*, *28*(38), 53594–53604.

Guo, T., Gull, S., Ali, M. M., Yousef, A. F., Ercisli, S., Kalaji, H. M., Telesiński, A., Auriga, A., Wróbel, J., Radwan, N. S., & Ghareeb, R. Y. (2022). Heat stress mitigation in tomato (*Solanum lycopersicum* L.) through foliar application of gibberellic acid. *Scientific Reports*, *12*(1), 11324. https://doi.org/10.1038/s41598-022-15590-z (accessed on 6th November 2025).

Gupta, R., & Chakrabarty, S. K. (2013). Gibberellic acid in plant: still a mystery unresolved. *Plant Signaling & Behavior*, *8*(9), e25504. https://doi.org/10.4161/psb.25504 (accessed on 6th November 2025).

Gust, A. A., Brunner, F., & Nürnberger, T. (2010). Biotechnological concepts for improving plant innate immunity. *Current Opinion in Biotechnology*, *21*(2), 204–210.

Hadi, F., Ali, N., & Ahmad, A. (2014). Enhanced phytoremediation of cadmium-contaminated soil by *Parthenium hysterophorus* plant: effect of gibberellic acid (GA3) and synthetic chelator, alone and in combinations. *Bioremediation Journal*, *18*(1), 46–55.

Hadi, F., Bano, A., & Fuller, M. P. (2010). The improved phytoextraction of lead (Pb) and the growth of maize (*Zea mays* L.): the role of plant growth regulators (GA3 and IAA) and EDTA alone and in combinations. *Chemosphere*, *80*(4), 457–462.

Hedden, P. (2020). The current status of research on gibberellin biosynthesis. *Plant and Cell Physiology*, *61*(11), 1832–1849.

Iftikhar, A., Ali, S., Yasmeen, T., Arif, M. S., Zubair, M., Rizwan, M., Alhaithloul, H. A. S., Alayafi, A. A. M., & Soliman, M. H. (2019). Effect of gibberellic acid on growth, photosynthesis and antioxidant defense system of wheat under zinc oxide nanoparticle stress. *Environmental Pollution*, *254*(Part B), 113109. https://doi.org/10.1016/j.envpol.2019.113109 (accessed on 6th November 2025).

Isa, N. K., & Mat Don, M. (2014). Investigation of the gibberellic acid optimization with a statistical tool from *Penicillium variable* in batch reactor. *Preparative Biochemistry and Biotechnology*, *44*(6), 572–585.

Janick, J. (1979). *Horticultural Science* (3rd ed.). W.H. Freeman and Company.

Javed, T., Ali, M. M., Shabbir, R., Anwar, R., Afzal, I., & Mauro, R. P. (2021). Alleviation of copper-induced stress in pea (*Pisum sativum* L.) through foliar application of gibberellic acid. *Biology (Basel)*, *10*(2), 120. https://doi.org/10.3390/biology10020120 (accessed on 6th November 2025).

Ji, P., Tang, X., Jiang, Y., Tong, Y., Gao, P., & Han, W. (2015). Potential of gibberellic acid 3 (GA3) for enhancing the phytoremediation efficiency of *Solanum nigrum* L. *Bulletin of Environmental Contamination and Toxicology*, *95*(6), 810–814.

Kahlon, S. S., & Malhotra, S. (1986). Production of gibberellic acid by fungal mycelium immobilized in sodium alginate. *Enzyme and Microbial Technology*, *8*, 613–616.

Kataoka, K., Uemachi, A., Nonaka, M., & Yazawa, S. (2004). Effect of endogenous gibberellins in the early stages of fruit growth and development of the 'Severianin' tomato. *Journal of Horticultural Science & Biotechnology*, *79*(1), 54–58.

Keswani, C., Singh, S. P., García-Estrada, C., Mezaache-Aichour, S., Glare, T. R., Borriss, R., Rajput, V. D., Minkina, T. M., Ortiz, A., & Sansinenea, E. (2022). Biosynthesis and beneficial effects of microbial gibberellins on crops for sustainable agriculture. *Journal of Applied Microbiology*, *132*(3), 1597–1615.

Kim, S. G., & Park, C. M. (2008). Gibberellic acid-mediated salt signaling in seed germination. *Plant Signaling & Behavior*, *3*(10), 877–979.

Koriesh, E. M., Dahab, A. M. A., & Ali, E. W. M. (1989). Physiological studies on *Chrysanthemum morifolium* Ramat. Effect of cycocel, gibberellic acid and nucleic acids on vegetative growth and some chemical components. *Assiut Journal of Agricultural Sciences*, *20*(1), 27–41.

Kumar, A., Kumar, S., Singh, D., & Shivjee. (2011). Rural employment diversification in India: Trends, determinants and implications on poverty. *Agricultural Economics Research Review*, *24*, 361–372.

Kumar, P. K., & Lonsane, B. K. (1987). Gibberellic acid by solid state fermentation: Consistent and improved yields. *Biotechnology and Bioengineering*, *30*(2), 267–271.

Kumar, V., Ahluwalia, V., Saran, S., Kumar, J., Patel, A. K., & Singhania, R. R. (2021). Recent developments on solid-state fermentation for production of microbial secondary metabolites: Challenges and solutions. *Bioresource Technology*, *323*, 124566. https://doi.org/10.1016/j.biortech.2020.124566 (accessed on 6th November 2025).

Kwon, C. T., & Paek, N. C. (2016). Gibberellic acid: A key phytohormone for spikelet fertility in rice grain production. *International Journal of Molecular Sciences*, *17*(5), 794. https://doi.org/10.3390/ijms17050794 (accessed on 6th November 2025).

Li, Y. H., Wu, Y. J., Wu, B., Zou, M. H., Zhang, Z., & Sun, G. M. (2011). Exogenous gibberellic acid increases the fruit weight of 'Comte de Paris' pineapple by enlarging flesh cells without negative effects on fruit quality. *Acta Physiologiae Plantarum*, *33*, 1715–1722.

Machado, C. M. M., Oishi, B. O., Pandey, A., & Soccol, C. R. (2004). Kinetics of *Gibberella fujikuroi* growth and gibberellic acid production by solid-state fermentation in a packed-bed column bioreactor. *Biotechnology Progress*, *20*(5), 1449–1453.

Machado, C. M. M., Soccol, C. R., De Oliveira, B. H., & Pandey, A. (2002). Gibberellic acid production by solid-state fermentation in coffee husk. *Applied Biochemistry and Biotechnology*, *102*, 179–191. https://doi.org/10.1385/ABAB:102-103:1-6:179 (accessed on 6th November 2025).

Miceli, A., Moncada, A., Sabatino, L., & Vetrano, F. (2019). Effect of gibberellic acid on growth, yield, and quality of leaf lettuce and rocket grown in a floating system. *Agronomy*, *9*(7), 382. https://doi.org/10.3390/agronomy9070382 (accessed on 6th November 2025).

Monrroy, M., & García, J. R. (2022). Gibberellic acid production from corn cob residues via fermentation with *Aspergillus niger*. *Journal of Chemistry*, *2022*, 1112941. https://doi.org/10.1155/2022/1112941 (accessed on 6th November 2025).

Nagarjuna, B., Reddy, V. P., Rao, M. R., & Reddy, E. N. (1988). Effect of growth regulators and potassium nitrate on growth, flowering and yield of chrysanthemum (*Chrysanthemum indicum* L.). *South Indian Horticulture*, *36*, 136–140.

Omojasola, P. F., & Adejoro, D. O. (2018). Submerged fermentation of orange albedo to produce gibberellic acid using *Fusarium moniliforme* and *Aspergillus niger*. *Jordan Journal of Biological Sciences*, *11*(2), 187–194.

Otálvaro, Á. M., Gutiérrez, G. D., Pierotty, D. A., Parada, F. A., & Algecira, N. A. (2008). Gibberellic acid production by *Gibberella fujikuroi* under solid-state fermentation of cassava bagasse and rice hull. *Journal of Biotechnology*, *136*(Supplement), S371. https://doi.org/10.1016/j.jbiotec.2008.07.853 (accessed on 6th November 2025).

Pandya, J. B., Patani, A. N., Raval, V. H., Rajput, K. N., & Panchal, R. R. (2023). Understanding the fermentation potentiality for gibberellic acid (GA3) production using fungi. *Current Microbiology*, *80*(12), 385. https://doi.org/10.1007/s00284-023-03454-2 (accessed on 6th November 2025).

Pastrana, L. M., Gonzalez, M. P., Pintado, J., & Murado, M. A. (1995). Interactions affecting gibberellic acid production in solid-state culture: a factorial study. *Enzyme and Microbial Technology*, *17*(9), 784–790.

Peng, X. L., Zhao, W. J., Wang, Y. S., Dai, K. L., Cen, Y. K., Liu, Z. Q., & Zheng, Y. G. (2020). Enhancement of gibberellic acid production from *Fusarium fujikuroi* by mutation breeding and glycerol addition. *3 Biotech*, *10*(7), 312. https://doi.org/10.1007/s13205-020-02303-4 (accessed on 6th November 2025).

Pharis, R. P., & King, R. W. (1985). Gibberellins and reproductive development in seed plants. *Annual Review of Plant Physiology*, *36*, 517–568.

Pinheiro, U. V., Wancura, J. H. C., Brondani, M., da Silva, C. M., Mainardi, M. A., Gai, R. M., & Jahn, S. L. (2024). Production of gibberellic acid by solid-state fermentation using wastes from rice processing and brewing industry. *Applied Biochemistry and Biotechnology*, *196*(3), 1493–1508.

Ragni, L., Nieminen, K., Pacheco-Villalobos, D., Sibout, R., Schwechheimer, C., & Hardtke, C. S. (2011). Mobile gibberellin directly stimulates *Arabidopsis* hypocotyl xylem expansion. *Plant Cell*, *23*(4), 1322–1336.

Rangaswamy, V. (2012). Improved production of gibberellic acid by *Fusarium moniliforme*. *Microbiology Research*, *2*(3), 51–55.

Raza Gurmani, A., Wang, X., Rafique, M., Jawad, M., Raza Khan, A., Ullah Khan, Q., Ahmed, R., & Fiaz, S. (2022). Exogenous application of gibberellic acid and silicon to promote salinity tolerance in pea (*Pisum sativum* L.) through Na+ exclusion. *Saudi Journal of Biological Sciences*, *29*(6), 103305. https://doi.org/10.1016/j.sjbs.2022.103305 (accessed on 6th November 2025).

Rios-Iribe, E. Y., Hernández-Calderón, O. M., & Escamilla-Silva, E. M. (2016). Kinetic analysis of the uptake of glucose and corn oil used as carbon sources in batch cultures of *Gibberella fujikuroi*. *World Journal of Microbiology and Biotechnology*, *32*(11), 182. https://doi.org/10.1007/s11274-016-2139-8 (accessed on 6th November 2025).

Rodrigues, C., Vandenberghe, L. P. D. S., Teodoro, J., Oss, J. F., Pandey, A., & Soccol, C. R. (2009). A new alternative to produce gibberellic acid by solid state fermentation. *Brazilian Archives of Biology and Technology*, *52*(SPL.ISS), 181–188.

Rodrigues, C., Vandenberghe, L. P., & Goyzueta, L. D. (2016). Production, extraction and purification of gibberellic acid by solid state fermentation using citric pulp and soy husk. *BAOJ Chemistry*, *2*(2), 014, 1–18.

Rodrigues, C., Vandenberghe, L. P., de Oliveira, J., & Soccol, C. R. (2012). New perspectives of gibberellic acid production: a review. *Critical Reviews in Biotechnology*, *32*(3), 263–273.

Sajid, M., Amin, N., Ahmad, H., & Khan, K. (2016). Effect of gibberellic acid on enhancing flowering time in *Chrysanthemum morifolium*. *Pakistan Journal of Botany*, *48*(2), 477–483.

Salazar-Cerezo, S., Martínez-Montiel, N., García-Sánchez, J., Pérez-Y-Terrón, R., & Martínez-Contreras, R. D. (2018). Gibberellin biosynthesis and metabolism: a convergent route for plants, fungi and bacteria. *Microbiological Research*, *208*, 85–98.

Saleem, M. H., Fahad, S., Adnan, M., Ali, M., Rana, M. S., Kamran, M., Ali, Q., Hashem, I. A., Bhantana, P., Ali, M., & Hussain, R. M. (2020). Foliar application of gibberellic acid endorsed phytoextraction of copper and alleviates oxidative stress in jute (*Corchorus capsularis* L.) plant grown in highly copper-contaminated soil of China. *Environmental Science and Pollution Research International*, *27*(29), 37121–37133.

Satpute, D., Sharma, V., Murarkar, K., Bhotmange, M., & Dharmadhikari, D. (2010). Solid-state fermentation for production of gibberellic acid using agricultural residues. *International Journal of Environmental Pollution*, *43*(1–3), 201–213.

Sauter, M., Seagull, R. W., & Kende, H. (1993). Internodal elongation and orientation of cellulose microfibrils and microtubules in deepwater rice. *Planta*, *190*, 354–362.

Schuler, J., & Colquhoun, J. (2022). Influence of gibberellic acid on vegetable crop and weed emergence. *Weed Technology*, *36*(6), 808–813.

Shahzad, K., Hussain, S., Arfan, M., Hussain, S., Waraich, E. A., Zamir, S., Saddique, M., Rauf, A., Kamal, K. Y., Hano, C., & El-Esawi, M. A. (2021). Exogenously applied gibberellic acid enhances growth and salinity stress tolerance of maize through modulating the morpho-physiological, biochemical and molecular attributes. *Biomolecules*, *11*(7), 1005. https://doi.org/10.3390/biom11071005 (accessed on 6th November 2025).

Shan, F., Zhang, R., Zhang, J., Wang, C., Lyu, X., Xin, T., Yan, C., Dong, S., Ma, C., & Gong, Z. (2021). Study on the regulatory effects of GA3 on soybean internode elongation. *Plants (Basel)*, *10*(8), 1737. https://doi.org/10.3390/plants10081737 (accessed on 6th November 2025).

Sharma, R. R., & Singh, R. (2009). Gibberellic acid influences the production of malformed and button berries, and fruit yield and quality in strawberry (*Fragaria ananassa* Duch.). *Scientia Horticulturae*, *119*, 430–433.

Shehata, S. M., Badawy, R. K., & Aboulsoud, Y. I. E. (2019). Phytoremediation of some heavy metals in contaminated soil. *Bulletin of the National Research Centre*, *43*, 189. https://doi.org/10.1186/s42269-019-0214-7 (accessed on 6th November 2025).

Shinozaki, Y., Beauvoit, B. P., Takahara, M., Hao, S., Ezura, K., Andrieu, M. H., Nishida, K., Mori, K., Suzuki, Y., Kuhara, S., Enomoto, H., Kusano, M., Fukushima, A., Mori, T., Kojima, M., Kobayashi, M., Sakakibara, H., Saito, K., Ohtani, Y., Bénard, C., Prodhomme, D., Gibon, Y., Ezura, H., & Ariizumi, T. (2020). Fruit setting rewires central metabolism via gibberellin cascades. *Proceedings of the National Academy of Sciences of the United States of America*, *117*(38), 23970–23981. https://doi.org/10.1073/pnas.2011859117 (accessed on 6th November 2025).

Shukla, R., Chand, S., & Srivastava, A. K. (2005). Improvement of gibberellic acid production using a model-based fed-batch cultivation of *Gibberella fujikuroi*. *Process Biochemistry*, *40*(6), 2045–2050.

Siddiqui, M. H., Al-Whaibi, M. H., & Basalah, M. O. (2011). Interactive effect of calcium and gibberellin on nickel tolerance in relation to antioxidant systems in *Triticum aestivum* L. *Protoplasma*, *248*(3), 503–511.

Skirycz, A., & Inzé, D. (2010). More from less: plant growth under limited water. *Current Opinion in Biotechnology*, *21*(2), 197–203.

Smart, R., & Robinson, M. (1991). *Sunlight into Wine: A Handbook for Wine Grape Canopy Management*. Winetitles. https://www.scirp.org/reference/referencespapers?referenceid=1749326 (accessed on 6th November 2025).

Soccol, C. R., Da Costa, E. S. F., Letti, L. A. J., Karp, S. G., Woiciechowski, A. L., & De Souza Vandenberghe, L. P. (2017). Recent developments and innovations in solid state fermentation. *Biotechnology Research and Innovation*, *1*(1), 52–71.

Somtrakoon, K., & Chouychai, W. (2022). Gibberellic acid treatment improved pyrene phytoremediation efficiency of ridge gourd (*Luffa acutangula* (L.) Roxb.) in soil. *Soil and Sediment Contamination*, *31*(2), 253–263.

Song, Q., Cheng, S., Chen, Z., Nie, G., Xu, F., Zhang, J., Zhou, M., Zhang, W., Liao, Y., & Ye, J. (2019). Comparative transcriptome analysis revealing the potential mechanism of seed germination stimulated by exogenous gibberellin in *Fraxinus hupehensis*. *BMC Plant Biology*, *19*(1), 199. https://doi.org/10.1186/s12870-019-1801-3 (accessed on 6th November 2025).

Tardieu, F., & Tuberosa, R. (2010). Dissection and modelling of abiotic stress tolerance in plants. *Current Opinion in Plant Biology*, *13*(2), 206–212.

Tomasini, A., Fajardo, C., & Barrios-González, J. (1997). Gibberellic acid production using different solid-state fermentation systems. *World Journal of Microbiology and Biotechnology*, *13*(2), 203–206.

Ungsa, M., Kato, K., Takemura, K., Hori, T., Ohara, H., Ohkawa, K., Matsui, H., & Bukovac, M. J. (2003). Effects of the combination of gibberellic acid and ammonium nitrate on the growth and quality of seedless berries in 'Delaware' grape. *Journal of the Japanese Society for Horticultural Science*, *72*(5), 366–371.

Usenik, V., Kastelec, D., & Štampar, F. (2005). Physicochemical changes of sweet cherry fruits related to application of gibberellic acid. *Food Chemistry*, *90*(4), 663–671.

Uthandi, S., Karthikeyan, S., & Sabarinathan, K. G. (2010). Gibberellic acid production by *Fusarium fujikuroi* SG2. *Journal of Scientific and Industrial Research*, *69*, 211–214.

Wang, Y., Cui, Y., Hu, G., Wang, X., Chen, H., Shi, Q., Xiang, J., Zhang, Y., Zhu, D., & Zhang, Y. (2018). Reduced bioactive gibberellin content in rice seeds under low temperature leads to decreased sugar consumption and low seed germination rates. *Plant Physiology and Biochemistry*, *133*, 1–10.

Werle, L. B., Abaide, E. R., Felin, T. H., Kuhn, K. R., Tres, M. V., Zabot, G. L., Kuhn, R. C., Jahn, S. L., & Mazutti, M. A. (2020). Gibberellic acid production from *Gibberella fujikuroi* using agro-industrial residues. *Biocatalysis and Agricultural Biotechnology*, *25*, 101608. https://doi.org/10.1016/j.bcab.2020.101608 (accessed on 6th November 2025).

Yao, W., & Shen, Y. (2018). Effects of gibberellic acid and magnetically treated water on physiological characteristics of *Tilia miqueliana* seeds. *Canadian Journal of Forest Research*, *48*(5), 554–558.

Yildirim, A. N., & Koyuncu, F. (2010). The effect of gibberellic acid applications on the cracking rate and fruit quality in the '0900 Ziraat' sweet cherry cultivar. *African Journal of Biotechnology*, *9*(38), 6307–6311.

Zang, Y. X., Chun, I. J., Zhang, L. L., Hong, S. B., Zheng, W. W., & Xu, K. (2016). Effect of gibberellic acid application on plant growth attributes, return bloom, and fruit quality of rabbiteye blueberry. *Scientia Horticulture*, *200*, 13–18.

Zhang, J., Cao, Y., Tang, J., He, X., Li, M., Li, C., Ren, X., & Ding, Y. (2023). Physiology and application of gibberellins in postharvest horticultural crops. *Horticulture*, *9*(6), 625. https://doi.org/10.3390/horticulturae9060625 (accessed on 6th November 2025).

Zhang, X., Zhao, B., Sun, Y., & Feng, Y. (2022). Effects of gibberellins on important agronomic traits of horticultural plants. *Frontiers in Plant Science*, *13*, 978223. https://doi.org/10.3389/fpls.2022.978223 (accessed on 6th November 2025).

Zhong, C., Xu, H., Ye, S., Wang, S., Li, L., Zhang, S., & Wang, X. (2015). Gibberellic acid-stimulated *Arabidopsis6* serves as an integrator of gibberellin, abscisic acid, and glucose signaling during seed germination in *Arabidopsis*. *Plant Physiology*, *169*(3), 2288–2303.

CHAPTER 12

Future Prospects of Endophytic Volatile Organic Compounds (VOCs) in Sustainable Agriculture

PERUMALLA SRIKANTH,[1] NUTAN KAUSHIK,[2,*] HEMRAJ CHHIPA,[3] TAKWA MARZOUK,[4] AZUCENA GONZALEZ-COLOMA,[5] and NACEUR DJÉBALI[4]

[1]Amity Institute of Horticulture Studies and Research (AIHSR), Amity University, Noida, Uttar Pradesh, India

[2]Amity Food and Agriculture Foundation, Amity University Uttar Pradesh, Noida, Uttar Pradesh, India

[3]College of Horticulture and Forestry, Agriculture University, Kota, Jhalawar, Rajasthan, India

[4]Laboratory of Bioactive Substances, Center of Biotechnology of Borj Cedria, BP 901 Hammam-Lif 2050, Tunisia

[5]Instituto de Ciencias Agrarias, CSIC, Madrid, Spain

*Corresponding author

ABSTRACT

Endophytic microbes, which live within plant tissues without posing a threat, create bioactive metabolites known as endophytic volatile organic compounds, or VOCs. Endophytes have an edge over other biocontrol agents in that they have the ability to colonize the interior tissues of plants. Because of their many functions in improving plant resilience, development, and health, these volatile organic compounds (VOCs)

Plant Biostimulants and Growth Regulators in Sustainable Agriculture.
Deepak Kumar, Sanjay K. Singh, Nutan Kaushik, and Mahaveer P. Sharma (Eds.)

DOI: 10.1201/9781779641458-16

provide a promising new frontier in sustainable agriculture. Moreover, VOCs have a big impact on the physiology and development of plants. These substances frequently stimulate biomass accumulation overall, root elongation, and seed germination in order to act as growth promoters. It has been determined that VOCs such as acetoin, 2,3-butanediol, and other terpenoids play a major role in these growth-promoting actions. This chapter elucidates the processes by which endophytes defend the plant against diseases and enhance plant tolerance to abiotic stresses such as salinity, drought, and heavy metal contamination.

12.1 INTRODUCTION

The germs that live inside a plant are referred to as endophytic microorganisms. They enter a plant by the seed, leaf, stem, and root and do not harm the host plant (Yadav et al., 2018). Endophytes increase plant health by defending plants against phytopathogens, which in turn helps to improve nutrition by transferring nutrients in both directions and plant development by secreting phytohormones (Andreozzi et al., 2019). Plant-endophyte interaction causes plants to ward off adverse environmental factors such as the presence of heavy metals and dryness (Kushwaha et al., 2019). The concept of the "plant microbiome" has fundamentally altered the situation; thus, it is now necessary to examine the coevolution of plants and their symbionts in order to identify the variables involved in their peaceful coexistence and to identify the advantages of their partnership (Turner et al., 2013). When endophytes are present, plant genes are induced, which offers information about how they affect the host plant (Berendsen et al., 2015). Genome sequencing, comparative genomics, microarrays, next-generation sequencing (NGS), metagenomics, and metatranscriptomics are examples of contemporary "omics"-based methods that could provide detailed information on endophytic lifestyles (Kaul et al., 2016). Plant growth-promoting microbes (PGPM) are well-known for eliciting the first stages of systemic resistance in plants. They do this by generating metabolites, for example, volatile organic compounds (VOCs) and antimicrobials (Enebe et al., 2018). It is noteworthy that research on the application of endophyte metabolites to inducing plant resistance is currently scarce and ought to be further examined in subsequent research.

Endophytes are numerous and, according to research, are found in a wide variety of plants. They gained significance because of their ability to create a large number of bioactive metabolites and enzymes that are relevant to biotechnology (Rajamanikyam et al., 2017). When endophytes are injected into a plant, they typically result in a considerable increase in biomass and contribute to advanced commercial agriculture. Due to their capacity to secrete secondary metabolites, endophytes act as immunosuppressants, antimicrobials, biocontrol agents, and anticancer agents. Moreover, they also secrete antiviral compounds and produce insecticidal products, antidiabetic agents, natural antioxidants, and antibiotics. Endophytes are becoming more and more relevant in biotechnology and industry (Shen et al., 2019). Endophytes extracted from the majority of plants have emerged during the past 20 years as a rich source of natural compounds with a variety of uses, including industrial and agricultural. Poisonous compounds can be replaced with enzymes. They survive best in environments with regular temperatures and a pH of 7. Researchers are starting to identify opportunities in the synthesis of microbial enzymes as the years go by. According to numerous findings at the moment (Sahay et al., 2017), microorganisms isolated from severe environments have enormous biotechnological potential in medicine, agriculture, and industry. In different plant tissues, organs, and intercellular spaces, endophytes, such as bacteria, fungi, and actinomycetes, exist without immediately showing symptoms of disease. Throughout long-term coevolution, they have developed a mutually beneficial interaction with host plants. Endophytes benefit plants by supplying them with nutrients, while plants benefit from endophytes by helping to keep them healthy. Because of their unique biological niches, endophytes within plants may have more advantageous and direct effects on plants than soil microbes. The effects of endophytes on promoting plant growth are well supported by the findings of several research teams (Jia et al., 2016).

Alkaloids, polypeptides, polyketides, terpenoids, and other metabolites having diverse biological functions that are produced by endophytes are extremely important and valuable in many different disciplines, particularly in agriculture and the pharmaceutical industry (Dubey et al., 2020). Due to their potential applications as antibiotics, insecticides, natural antioxidants, anticancer agents, antidiabetic medicines, and other things (Singh et al., 2021), these metabolites have garnered a lot of interest. These bioactive metabolites' primary role in plant health protection is to directly or indirectly support the host plants' ability to withstand biotic and abiotic challenges. For instance, hydrolases secreted by endophytic bacteria can break down the cell walls of pathogens; several antimicrobial chemicals produced by endophytes are widely known for strongly suppressing infections; and phytohormones released by endophytes play an important role. Due to their function as stimulators of plant development, endophytic fungi have been used to improve sustainable agriculture. Based on established mechanisms, such as increased food and water uptake and hormone production in plants, this role increases biotic and abiotic stress resistance. Recent studies have also shown that using bacteria and fungus as biological control agents (BCAs) is beneficial for eradicating plant diseases and increasing crop production. The use of fungal BCAs has significantly increased as a result of their high rate of sexual and asexual reproduction as well as their target specificity (Thambugala et al., 2020).

12.2 TYPES OF ENDOPHYTES

12.2.1 BACTERIAL ENDOPHYTES

Gram-positive and gram-negative bacteria are both included in the category of species known as endophytic bacteria, including *Bacillus*, *Agrobacterium*, *Brevibacterium*, *Pseudomonas*, and others (Sun et al., 2013). According to a recent study by Liu et al., wild rice was home to a wide range of endophytes, with Proteobacteria, Bacteroidetes, and Firmicutes predominating (Liu et al., 2021). The host and environment-related elements, such as the stages of plant growth, a location's climate, and other factors, have an impact on the variety of endophytic bacteria in plants (Afzal et al., 2019) (Table 12.1).

TABLE 12.1 Bacterial Endophytes-Based Bioactive Compounds

Sl. No.	Name of the Endophytic Bacteria	Bioactive Compound	Role in Plant System	References
1.	*Enterobacter*	Methyl ester	Anti-bacterial	Doty et al. (2017)
2.	*Bacillus substilis*	Subtilin	Anti-bacterial	Singh et al. (2017)
3.	*Bacillus substilis*	Terpene	Anti-fungal	Nithya et al. (2011)
4.	*Bacillus pumlus*	–	Anti-bacterial	Aljuraifani et al. (2019)

12.2.2 FUNGAL ENDOPHYTES

Ascomycota, Zygomycota, and Basidiomycota are the three primary groups of endophytic fungi. According to life history characteristics and evolutionary relatedness, endophytic fungi are broadly classified into two main types: (1) the clavicipitaceous endophytes colonizing in certain grasses and (2) angiosperms, conifers, ferns, and nonvascular plants' asymptomatic tissues include non-clavicipitaceous endophytes. It is generally recognized that endophytic fungi produce a variety of bioactive substances for the creation of anti-inflammatory, antioxidant, anti-fibrotic, and antiviral drugs. The fungus endophyte *Pestalotiopsis microspora* contains isopestacin, which possesses strong antioxidant capabilities (Terhonen et al., 2019) (Table 12.2).

TABLE 12.2 Fungal Endophytes-Based Bioactive Compounds

Sl. No.	Name of the Endophytic Fungi	Bioactive Compound	Role in Plant System	References
1.	*Pestalotiopsis mangiferae*	Isopectin	Anti-fungal	Subban et al. (2013)
2.	*Rhizopus* sp.	–	Anti-fungal	Pai and Chandra (2018)
3.	*Phompsis* sp.	Steroids	Anti-fungal	Wu et al. (2013)
4.	*Fusarium* sp.	lipopeptides	Anti-fungal	Pai and Chandra (2018)

12.2.3 ACTINOMYCETES ENDOPHYTES

Endophytic actinomycetes are typically isolated from many different kinds of plants, especially mangroves and therapeutic plants in tropical rain forests (Qin et al., 2010). In general, there are more endophytic actinomycetic species present in plant roots as compared to other parts. The two prominent species, *Streptomyces* and *Micromonospora*, have been recognized as important sources of antibiotics and other bioactive compounds, such as Munumbicin D and oronamycin, which are synthesized through *Streptomyces* NRRL 30562 and *Streptomyces* sp. MSU-2110, respectively. There are still several species of endophytic microbes that need to be investigated and characterized (Zin et al., 2017) (Table 12.3).

TABLE 12.3 Actinomycetes Endophytes-Based Bioactive Compounds

Sl. No.	Name of the Endophytic Actinomycetes	Bioactive Compound	Role in Plant System	References
1.	*Streptomyces* sp.	Steroids	Anti-bacterial	Bae et al. (2015)
2.	*Streptomyces* sp.	Alkaloids	Anti-fungal	Zhang et al. (2014)
3.	*Streptomyces* sp.	Alkaloids	Anti-fungal	Singh and Dubey (2018)
4.	*Streptomyces* sp.	Tannins	Anti-fungal	Singh and Dubey (2018)

12.3 VARIOUS ANTI-MICROBIAL ENDOPHYTE STUDIES

Numerous studies have found that endophytes that lacked a specific component showed antibacterial properties, some of which were considerable. Five of the 63 endophytic fungi that were isolated from the rhizomes of the traditional medicine plant *Paris polyphylla* var. *yunnanensis* demonstrated potent antibacterial activity against *Staphylococcus haemolyticus* and *Bacillus subtilis* (Gram-positive) as well as the Gram-negative bacteria *Escherichia coli*, *Xanthomonas vesicatoria*, and *Agrobacterium tumefaci* (Zhao et al., 2008). No specific chemical was identified in 353 endophytic fungi that were attached to medicinal plants. *Botrytis cinerea* and *Alternaria solani*'s growth inhibition percentages were 11.3% and 15.9%, respectively (Jiang et al., 2006). Phongpaichit et al. (2006) collected endophytic fungal fermentation broths from Garcinia plants and used the agar diffusion method to investigate the antibacterial activity against *Candida albicans*, *Staphylococcus aureus*, and *Cryptococcus neoformans*. According to the findings, out of the 377 total strains, 70 (18.6%) exhibited antibacterial activity against at least one harmful microbe. Many endophyte studies are still in the screening phase, and it is necessary to separate the active component.

Antimicrobial ingredients can be tracked via bio-autography and thin-layer chromatography (TLC). On the TLC plate, the crude extract should be put before being separated on the plate using an appropriate liquid solvent. A TLC plate with separated components is injected with the tested microorganisms (bacteria or fungal spores), and the microbial cultivation is then carried out for a few days in a wet environment. Tracking antimicrobial compounds was possible by looking at the TLC plate's inhibitory zone. In the end, procedures including column chromatography, liquid preparation chromatography, and precipitation TLC can be employed to separate the active components (Han et al., 2008). According to Tuntiwachwuttikul et al. (2008), the minimal inhibitory concentration (MIC) or 50% inhibitory concentration (IC_{50}) values of an antibiotic component serve as indicators of its efficacy. Methyl thiazolyl tetrazolium (MTT) or Triphenyltetrazolium chloride (TTC) techniques are used to investigate tracked antibacterial substances (Abe and Matsuki, 2000).

12.4 CLASSIFICATION OF BIOACTIVE COMPOUNDS OF ENDOPHYTES

Among the many bioactive metabolites that have been isolated from endophytes and reported in recent years are aliphatic compounds, flavonoids, terpenoids, steroids, phenols, alkaloids, quinones, and peptides (Figure 12.1).

12.4.1 ALKALOID COMPOUNDS

In endophytes, alkaloids are relatively common secondary metabolites, and some of them have antibacterial properties. A culture of an endophytic *C. globosum* obtained from Ginkgo biloba leaves revealed the presence of chaetoglobosins C (2) and A (1). These two metabolites were demonstrated to be antibacterial in the agar diffusion method against *Mucor miehei*; however, subsequent studies indicate that both of them may be dangerous to mammals due to their substantial toxicity to brine shrimp larvae. Pyrrocidines B (4) and A (3), two recently identified antibiotics, were produced by the endophyte *Acremonium zeae* in maize, which showed substantial antifungal activity against *Aspergillus flavus* and *Fusarium verticillioides*. These interactions and their relevance to the aflatoxin and fumonisin contamination of preharvest maize have been explained. *A. zeae* interacts with other endophytes. The endophytic fungus *Phomopsis* sp. in *Garcinia dulcis* produced phomoenamide (5), which had a MIC of 6.25 mg/mL against Mycobacterium TB. The metabolite had a sizable antibacterial effect. Numerous endophyte-produced alkaloids have been found to have strong antibacterial properties; nevertheless, more research is needed (Qin et al., 2009).

12.4.2 PEPTIDE COMPOUNDS

Numerous endophyte-produced peptides exhibited notable antibacterial properties. *Acremonium* sp. in Taxus baccata produced leuesnostatin A (6), which has antibacterial effects against *Pythium ultimum*. In *Tripterigium wiflordii*, endophytic *Cryptosporiopsis quercina* developed a new peptide called Cryptocandin (7) that has antibacterial activity against *C. albicans*. Echinocandins were a collection of peptides that were discovered to have antimicrobial properties in endophytic *Pezicula* sp. and *Cryptosporiopsis* sp. in *Fagus sylvatica* and *Pinus sylvestris*. Two novel antibacterial components, cyclo(Pro-Tyr) (13) and cyclo(Pro-Thr) (12), were created by the fermentation broth of the endophytic fungus *Penicillium* sp. that was isolated from the mangrove plant *Acrostichum aureum* (Cui et al., 2008).

12.4.3 STEROID COMPOUNDS

On Gomera Island, the endophytic fungus *Nodulisporium* sp. was found in Juniperus cedre, and the antimicrobial substances 5a, 8a-epidioxyergosterol (15), and ergosterol (14) were recovered from the culture extract (Dai et al., 2006). 3b-hydroxy-ergosta-5-ene (16), 3-oxo-ergosta-4, 6, 8 (14), and 22-tetraene are four antibacterial steroids that were isolated within *Artemisia annua* from *Colletotrichum* sp. 3b, 5a-dihydroxy-6bacetoxy-ergosta-7, 22-diene (18) and 3b, 5a-dihydroxy-6b-phenylacetyloxy-ergosta-7, 22-diene (19) demonstrated fungistatic activities for the crop pathogenic fungi *Phytophthora capisici* and *Rhizoctonia cerealis* (Lu et al., 2000).

12.4.4 TERPENOID COMPOUNDS

The primary terpenoids isolated from endophytes are sesquiterpenes, diterpenoids, and triterpenoids. Up to this point, some sesquiterpenes have been shown to have antimicrobial properties. Four novel diterpenoid antibiotics were discovered in endophytes: periconicin A (22), guanancastepene A (20), guanacastepene (21), and perieoniein (Kim et al., 2004). An endophytic mycoflora, *Phomopsis cassiae,* from *Cassia spectabilis,* was used as the source material for the bioassay-guided fractionation that produced five cadinane sesquiterpenes (24–28) derivatives. When tested against *Cladosporium sphaerospermum* and *C. cladosporioides*, compound 28 was shown to be the most effective (Silva et al., 2006).

12.4.5 PHENOLIC COMPOUNDS

Many endophytic cultures that originated from a variety of host plants have yielded phenol and phenolic acids. *Pestalotiopsis adusta*, an endophytic plant, yielded two novel antibiotics, pestalachloride A (29) and B (30), which showed notable antifungal efficacy against three plant diseases (*Gibberella zeae*, *Fusarium culmorum*, and *Verticillium aiboatrum*) (Li et al., 2008). A group of phenolic acids (numbered 31–33) were recovered by Hoffman et al. (2008) from the culture broth of the endophyte *Phoma* sp. from *Saurauia scaberrinae* (Guinea plant). *Rhizoctonium solani*, *P. ultimum*, *S. aureus*, and *Sclerotinia sclerotiorum* were among the Gram-positive bacteria that the compounds showed antibacterial activity against.

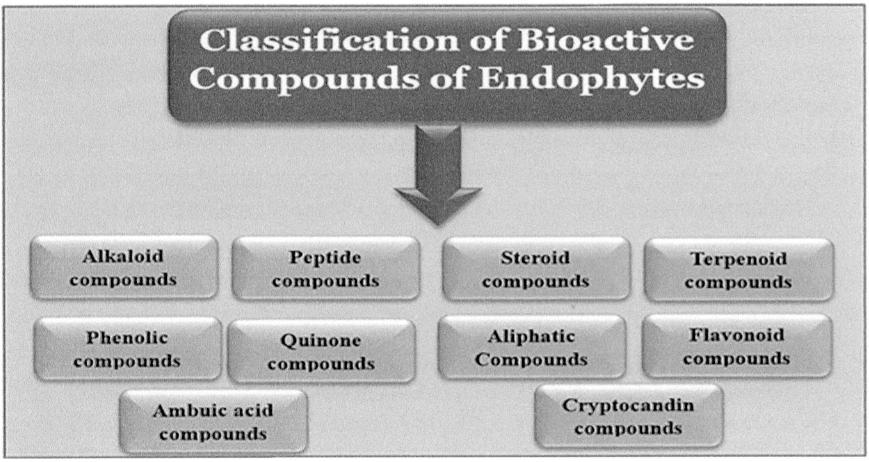

FIGURE 12.1 Classification of bioactive compounds of endophytes.

12.4.6 QUINONE COMPOUNDS

An endophyte called *Ampelomyces* sp. was identified from the therapeutic plant *Urospermum picroides* and produced 3-O-methylalaternin (36) and altersolanol A (37) that showed antimicrobial activity against the Gram-positive pathogens *Enterococcus faecalis*, *S. epidermidis*, and *S. aureus* at 12.5 and 12.5–25 mg/mL as MIC values, respectively. Altersolanol A's antibacterial effect is most likely not a result of its cytotoxic properties but rather because the substance functions as a bacterial membrane electron acceptor, which prevents bacterial development (Aly et al., 2008). Three novel spiroketals (38, 39, and 40), isolated from the recently discovered endophytic fungus Edenia gomezpompae, were initially reported to exhibit considerable growth inhibition (Macías-Rubalcava et al., 2008) (Figure 12.2).

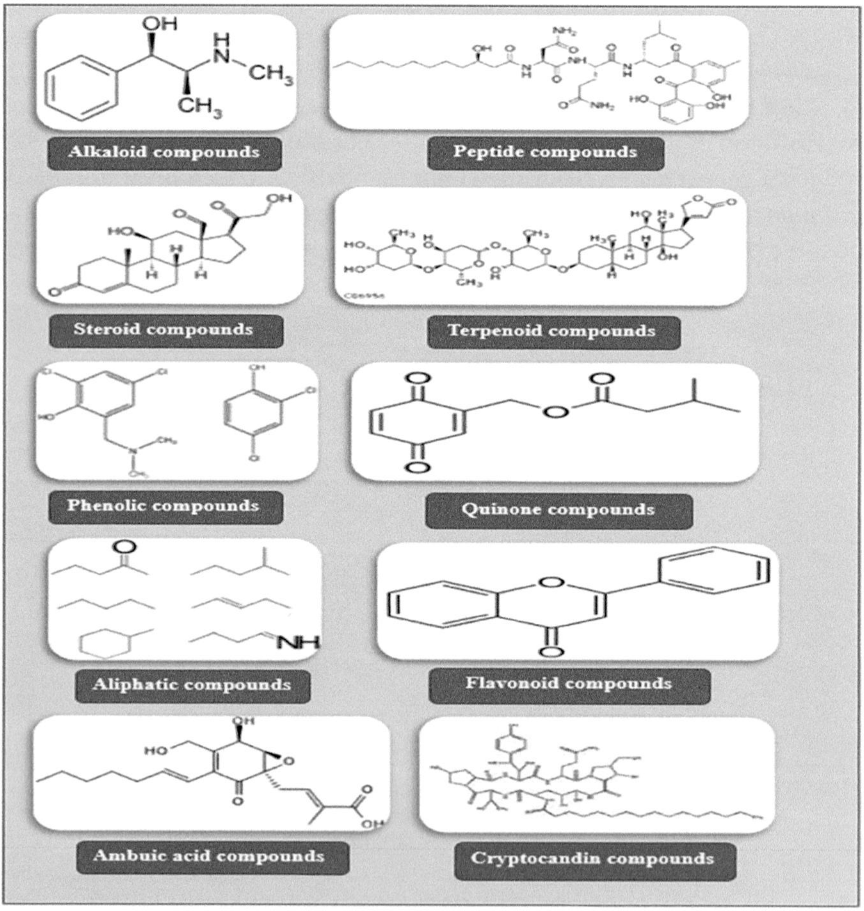

FIGURE 12.2 Different structures of bioactive compounds.

12.4.7 ALIPHATIC COMPOUNDS

Aliphatic chemicals, which are typically found in endophyte culture samples, frequently exhibit biological activity. Two aliphatic compounds with antifungal properties, chaetomugilin A (41) and D (42), were discovered in the endophytic fungus *C. globosum* that was isolated from Ginkgo biloba (Qin et al., 2009). A mangrove endophytic fungus called *Phomopsis* sp. produced cytosporones B (43) and C (44). They prevented the growth of the two mycoflora, *F. oxysporum* and *C. albicans,* with a MIC value of 32 to 64 mg/mL. Similarly, a metabolite called Brefeldin A (45) was also extracted from the endophyte *Penicillium* sp. (host: *Panax ginseng* root). This substance is a special fungal metabolite with a 13-membered lactone ring that exhibits strong antibacterial action (Wang et al., 2006).

12.4.8 FLAVONOID COMPOUNDS

Endophytic mycoflora *Nodulisporium* sp. from *Juniperus virginiana* (red cedar) contained antimicrobial flavonoid derivatives (46–48). These three derivatives can be employed as lead compounds whose functionalities can be enhanced through the application of synthetic chemistry (Dai et al., 2006).

12.4.9 AMBUIC ACID COMPOUNDS

Ambuic acid is a substance that is both anti-fungal and anti-oomycete. Species of the endophytic fungus *Pestalotiopsis microspora* have been discovered in numerous tropical rainforests across the world. It is running against a number of *Pythium ultimum* species and *Fusarium* species (Li et al., 2001). Additionally, according to Gary Strobel's personal communication, it seems to be a quorum-sensing inhibitor. *Colletotrichum* sp., isolated from *Artemisia annua*, is the source of the chemical known as colletonoic acid (Bills et al., 2002). According to Hussain et al. (2014), it exhibits strong antibacterial, antifungal, and anti-algal properties. Similarly, the endophytic fungus *Colletotrichum gloeosporioides* (host: *A. mongolica*) produces colletotric acid as a metabolite. It exhibits antibacterial activity against both the fungus *Helminthosporium sativum* and bacteria (Zou et al., 2000). *Gibberella fujikuroi*, a plant pathogenic fungus, is highly susceptible to Cordycepsidone A, which was isolated from the endophytic fungus *Cordyceps dipterigena* (Varughese et al., 2012).

13.4.10 CRYPTOCANDIN COMPOUNDS

A lipopeptide known as cryptocandin was discovered in the fungus *Cryptosporiopsis quercina*, which is frequently connected to European hardwood species. It is effective against plant-pathogenic fungi, including *Sclerotinia sclerotiorum* and *Botrytis cinerea*, among others (Strobel et al., 1999). It shares a relationship with echinocandins and pneumocandins, two well-known antimycotic substances. According to Li et al. (2000), *Cr. quercina* produces cryptocin, which has the strongest antifungal effects against *Pyricularia oryzae*. Additionally, it exhibits significant efficacy against the phytopathogens *Phytophthora cinnamon, Pythium ultimum,* and *Phytophthora citrophthora,* as well as the fungus *Geotrichum candidum, Fusarium oxysporum*, *Sclerotiorum* sp., and *Rhizoctonia solarum* (Li et al., 2000).

12.5 VARIOUS ENDOPHYTE METABOLITES IN SUSTAINABLE AGRICULTURE

12.5.1 ANTI-MICROBIAL VOLATILE COMPOUNDS

Although the subject of volatile antibiotics would have been a great match in the preceding section, it has been given its own heading since specialized methods are needed to identify these substances. Air contact between the target pathogen and the volatile-producing fungus allows for the detection of their anti-microbial action. The chemical profiles produced by multivariate data analysis using NMR or LC-HRMS can then be combined to separate and identify the volatile mixture (Wu et al., 2015a). The endophytic fungus *Muscodor albus* (host plant: *Cinnamomum zeylanicum*) developed a mixture of at least 28 volatile chemicals that were effective in inhibiting and killing particular other fungi and bacteria (Worapong et al., 2001). None of the chemicals were fatal, but a few of them had some inhibitory effect on the test fungus and bacteria. However, it was observed that they had a potent synergistic effect, and taken together, they killed a wide variety of bacteria and fungi that were harmful to plants and people. There were found to be five different classes of compounds: lipids, alcohols, ketones, esters, and acids. The initial research revealed that the most effective class of inhibitory chemicals was found to be esters (such as isoamyl acetate), which was the most physiologically active molecule (Strobel et al., 2001). Seven new *Muscodor albus* strains were subsequently discovered, each of which produces a unique combination of volatiles. The esters generated by the original *Muscodor albus* sample were not present in any of the subsequent isolates, though. They collectively create 50 different volatiles with a diverse spectrum of biological activities, including the ability to suppress the growth of Gram-positive and Gram-negative bacteria as well as oomycetes and plant-pathogenic fungi (Ezra et al., 2004). In the Bolivian Amazon Basin, the *Ananas ananassoides* (wild pineapple plant) was associated with an endophytic fungus known as *Muscodor crispans*. A combination of antifungal and antibacterial volatile organic chemicals is produced by the fungus. When tested against the volatile mixture of *Muscodor crispans*, *Pythium ultimum* was shown to be the fungal and oomycete plant pathogen with the greatest sensitivity. Additional sensitive plant fungal and oomycete infections include *Verticillium dahliae*, *Alternaria helianthi*, *Fusarium culmorum*, *Phytophtora cinnamomi*, *Sclerotinia sclerotiorum*, *Phytophtora palmivora*, *Botrytis cinerea*, *Rhizoctonia solani*, and *Fusarium oxysporum*. Moreover, *Xanthomonas axonopodis* pv. *citri*, a plant-pathogenic bacterium, showed signs of sensitivity (Mitchell et al., 2010).

An endophytic fungus on *Eucryphia cordifolia* called *Gliocladium* sp. was found to be an unstable generator of antibiotics. It was particularly interesting that annulene, a former rocket propellant, was shown to be its most prevalent volatile inhibitor fuel that was just recently identified as a naturally occurring substance made by an endophytic fungus (Stinson et al., 2003). At least 28 VOCs that are highly promising for post-harvest management are produced by the endophytic fungus *Daldinia concentrica*, which was discovered in an olive tree in Israel. These volatiles prevent the rotting of raisins, plums, and other dried fruits. Additionally, they defend grapes against *Botrytis cinerea*, tomatoes and oranges against *Penicillium digitatum*, and peanuts against *Aspergillus Niger*. Artificial mixes of specific volatiles show considerable promise for use in the agriculture and food industries.

12.5.2 ANTI-INSECT VOLATILE COMPOUNDS

The link between synthetic pesticides and ecological harm has prompted research to find strong, focused, and secure substitutes (Strobel and Daisy, 2003). Biopesticides have thus far only made up a small portion of the pesticide business, but their influence is growing. It is known that a number of endophytes create secondary metabolites with insect-repelling characteristics. Indole diterpenes, known as nodulisporic acids, have strong insecticidal effects on blowfly larvae. They work by opening glutamate-gated chloride channels in insects. *Nodulisporium* sp., an endophyte found in *Bontia daphnoides* plants, produced the first nodulisporic compounds. Since this discovery, efforts have been directed towards finding other *Nodulisporium* spp. and other sources of more potent nodulisporic acid analogues (Demain et al., 2000). Some Epichloe grass combinations produce peramine, a feeding deterrent with pyrrolopyrazine for the Argentine stem weevil insect problem (Johnson et al., 2013a). Epichloe endophytes have been reported to produce yet another class of alkaloids, known as indole-diterpenes. As the main tremorgenic mycotoxin responsible for ryegrass staggers, Lolitrem B, which is specific to Epichloe, is a notorious example; it also inhibits the development and growth of ASW larvae. According to the information in the section "Fungal endophytes for transmitting pest and disease," due to the broad-spectrum insect resistance of strain AR37, the only Epichloe strain known to produce these compounds, epoxy-janthitrems, another type of indole diterpene, probably possess some pesticidal properties (Fletcher et al., 1999).

12.5.3 PLANT GROWTH PROMOTION BY ENDOPHYTIC METABOLITES

The promotion of plant growth can increase host plants' resistance to various stresses, and this is known as one of the tactics used by plants in attack by a pathogen response (Kuldau et al., 2008). Endophytes and their byproducts are well known for promoting plant growth. On the one hand, endophytes significantly increase plants' ability to absorb and use nutrients including potassium (K), phosphorus (P), and nitrogen (N). Specifically, endophytic diazotrophic bacteria are connected to nitrogen fixation in gramineous crops, which promotes host growth and disease resistance. According to studies on lodgepole pine seedlings, *Paenibacillus polymyxa* P2b-2R helped maize seedlings obtain 30% of their foliar nitrogen from the atmosphere and grow 52% longer (Puri et al., 2016). Endophytes, on the other hand, have also been found to hasten plant growth by secreting chemicals like auxin, ethylene, gibberellin, and cytokinin. The growth and development of host plants can be regulated and aided by the secondary metabolites produced by endophytic bacteria of the *Staphylococcus, Azotobacter*, and *Azospirillum* genera. From tissue samples of 15 different tea cultivars, Shan et al. (2018) recovered 46 actinomycetes, the majority of which were able to generate IAA. Different plants have been discovered to have a greater prevalence of growth-promoting endophytes (Borah et al., 2020). Therefore, it is thought that endophytes' support of plant growth can indirectly shield host plants from diseases.

Growth regulators and phytohormones are typically chemically synthesized or extracted from plants in the agriculture sector. Microbial fermentation has been regarded as a more practical and effective strategy to boost output and reduce plant metabolite manufacturing expenses. Few products have been mass produced commercially, despite multiple publications demonstrating success in plant metabolite production from endophytes. The question of whether host plants or endophytes produce bioactive metabolites must also be addressed. Plant-endophyte interaction processes are not fully understood. The native plant-endophyte network is likewise disturbed when endophytes are cultivated *in vitro* apart from the host. Further research is required to clarify the proportional contributions of the "micro-guests" and the "host plants" to the synthesis of particular metabolites.

12.5.4 AGRICULTURAL USE OF ENDOPHTES AND THEIR METABOLITES AS NOVEL BIOLOGICAL CONTROL AGENTS (BCAS)

Potential BCAs are endophytes, which can colonize a plant without causing disease symptoms and have the capacity to create large amounts of bioactive metabolites (Silva et al., 2019). They are suitable alternatives to agrochemicals for reducing environmental impact and ensuring a sustainable agricultural system. In recent years, numerous studies on the use of endophytes for the biological management of plant diseases have been conducted. In a previous study, the use of *Epichloë typhina*, a fungus isolated from Timothy grass, decreased the incidence of *Cladosporium phlei*-caused illness in host plants. This was the first instance of the use of fungi as endophytes to biocontrol foliar cereal diseases (O'Hanlon et al., 2012). It's noteworthy to note that a specific endophyte may have the ability to act as a BCA, and that the same species may also help host plants thrive and better withstand stress. Recently, several biocontrol programs in agriculture have used other microbes, including *Streptomyces* spp., *Bacillus* spp., *Pseudomonas* spp., *Pantoea* spp., and *Enterobacter* spp. (Larran et al., 2016). In contemporary agriculture, there is currently no widespread tendency to exploit and market endophytes and their metabolites as BACs.

Endophytes are an abundant source of numerous useful compounds. On the one hand, plants can be grown and disease resistance can be increased by using natural active chemicals including alkaloids, antibiotics, and PHs. These compounds can be directly used in agricultural production after separation and extraction. On the other side, plant endophytes' secondary metabolites can be used as the active ingredients of green insecticides (biopesticides). A recent study found that only 38% of the bioactive compounds isolated from soil microorganisms had ever been discovered, compared to 51% of those isolated from endophyte fungi. Therefore, a promising route to creating new biopesticides is to thoroughly explore endophyte metabolites with biopesticide activity (Wen et al., 2022). Agribusiness greatly benefits from the secondary metabolites produced by plant endophytes in a variety of harsh settings, including cold, drought, and high latitude. Therefore, by selecting endophytes from particular ecological habitats, specific active compounds can be found. The identification, extraction, and isolation of these bioresources have all been the subject of extensive research. There are currently just a few papers on the structures of lead compounds, but it is anticipated that this will change quickly (Suryanarayanan et al., 2005). The potential toxicity of these anticipated products to humans, animals, and plants is what worries us the most. In actuality, eukaryotes include both plants and animals. The fact that endophytes form a symbiotic relationship with the host makes them a part of the core microbiome of plants. According to studies, endophytes produce the majority of antimicrobial compounds that are eco-friendly, harmful to phytopathogens, and safe for both humans and plants. Plants naturally serve as selection mechanisms for physiologically active chemicals that are not poisonous. Overall, there are several benefits and tremendous promise for using metabolites from plant endophytes as avenues for novel BCAs in plant disease control (Kim et al., 2016).

12.5.5 PHYTOHORMONE PRODUCTION

Endophytes create phytohormones that alter the shape and structure of plants and promote their growth. Endophytes have advanced in the field of agricultural sustainability as a result of this quality. The technique utilized by rhizobacteria to promote plant growth is just like the method that endophytes use in the generation of phytohormones in the host plant. By secreting ethylene, gibberellic acid, and auxins (i.e., IAA), they aid in the non-leguminous plants' growth and protection (Dutta et al., 2014; Khan et al., 2014).

The plant hormone (PH) indole acetic acid (IAA) stimulates plant cell division, differentiation, and extension; seed and tuber germination; root and xylem development; lateral initiation; rate of vegetative growth; adventitious root formation; as well as pigment and metabolite biosynthesis; responses to gravity, light, and fluorescence; photosynthesis; and resistance to environmental stresses. The physiological processes mentioned above can occasionally be slowed down by IAA released by bacteria that promote plant growth by altering the amount of auxin secreted by the plant. Additionally, IAA made by endophytic bacteria has the power to lengthen and increase the surface area of roots, allowing plants to more easily acquire soil-derived nutrients. Moreover, auxin (indole-3-acetic acid; IAA) production increases the size of bacteria's cell walls and their ability to secrete exudates, as well as providing additional resources for the growth of other beneficial bacteria that are already present. Therefore, it is acknowledged that the primary effector molecule in phytostimulation, disease, and plant-microbe interactions is the IAA generated by endophytic bacteria. Numerous studies have shown that endophytic actinomycetes also generate substances that aid in plant growth, such as IAA, which is known to promote the development and lengthening of adventitious roots in plants (Shimizu et al., 2011).

12.5.6 SIDEROPHORE PRODUCTION

Endophytes can produce siderophores, which are tiny molecules that can chelate iron and make it available for plant use while depriving pathogens of iron. Some of the siderophores that endophytes are known to produce, such as the hydroxymate, phenolate, and/or catecholate kinds, can provide biocontrol functions (Rajkumar et al., 2010). Additionally, siderophores, which aid in fixing nitrogen since diazotrophic organisms need Fe^{2+} and Mo factors for growth, improve the iron-deficient plant nitrogenase's production and activity. There is a wealth of literature that supports endophytes' ability to kill insects. Some endophytes strengthen the endodermal cell wall to prevent pests from penetrating the stele. Others produce secondary metabolites that kill insects. Although some harmful metabolites can be linked to endophytes, pyrrolizidine alkaloids, ergovalin, peramine ergot alkaloid, and pyrrolopyrazine alkaloid are just a few of these metabolites (Gao et al., 2010).

In the case of bacteria that stimulate plant growth, Fe^{2+} is converted in the bacterial membrane to a Fe^{3+}-siderophore complex, which is then brought into the cell by endophytes via a gating mechanism. Soluble metal concentration increases when siderophores adhere to the metal surface (Rajkumar et al., 2010). After the presence of heavy metal pollutants has been reduced, plants use a variety of processes to absorb iron from bacterial siderophores, such as ligand exchange or the direct absorption of siderophore-Fe complexes with the help of iron chelates. *Pseudomonas* strain GRP3, a siderophore-producing endophyte, was tested on *Vigna radiata* for iron nutrition, and the results showed significant improvement in chlorophyll (a&b) content and a decrease in chlorotic and iron mineral deficiency symptoms (Sharma et al., 2003). Similarly, several endophytic actinomyces, such as *Streptomyces* sp. UKCW/B, *Streptomyces* sp. GMKU 3100, *Nocardia* sp., and *Streptomyces* sp. mhcr0816, have been reported to produce siderophores (Singh and Dubey, 2018).

12.6 ROLE OF ENDOPHYTIC METABOLITES FOR MANAGING BIOTIC AND ABIOTIC STRESSES IN CROPS

12.6.1 BIOTIC STRESS MANAGEMENT

Endophytes have been shown to have the ability to inhibit phytopathogens through antagonistic activity. It is now understood that endophytes contribute to the induction of ISR (induced systematic resistance) in plants against phytopathogens. According to studies by Salam et al. (2017), foliar endophytes are also known to control the host's genetic expression, which affects the plant's physiological responses and defense mechanisms. Particularly salicylic and jasmonic acids are recognized to be crucial during plant stress responses against phytopathogens. (Khare et al., 2016). Through the SA and JA pathways, gibberellin-producing endophytes are known to increase resistance to the attack of phytopathogens and insects. According to Kavroulakis et al. (2007), *Fusarium solani* induces ISR against *Septoria lycopersici* (a tomato foliar disease) by causing the production of genes associated with pathogenesis in root tissues. *Colletotrichum tropicale*, a foliar endophytic fungus, was inoculated into Theobroma cacao, which displayed a decrease in Phytophthora infection (Mejia et al., 2008).

Arabidopsis thaliana, *T. cacao*, and other host plants showed disease resistance after *C. tropicale* infection due to the induction of numerous ethylene (ET) defense system components as well as a number of additional signaling genes (Meja et al., 2014). Numerous VOCs produced by endophytic bacteria have been shown to have broad-spectrum antibacterial efficacy against phytopathogenic bacteria, fungi, and nematodes. In a study, Sheoran et al. (2015) reported that black pepper-associated endophytic *Pseudomonas putida* BP25 inhibits various phytopathogens such as *Phytophthora capsici, Pythium myriotylum, Gibberella moniliformis, Rhizoctonia solani*, *Athelia rolfsii, Colletotrichum gloeosporioides*, and the plant parasitic nematode *Radopholus similis* by several volatile substances. Rhizobium that produces siderophores has a key role in the biocontrol of phytopathogens, according to Arora et al. Numerous crops are affected by *Macrophomina phaseolina*, which causes charcoal rot. It is known that plants produce phytoalexins, which are low-molecular-weight antimicrobial compounds that contain a variety of metabolite groups, including terpenoids. (Gao et al., 2010). In a 2009 study, Yong et al. found that the endophytic fungus *Fusarium* spp. increased the concentration of terpenoids and the development of *Euphorbia pekinensis*. Comparative studies conducted by a number of researchers revealed that endophytes-containing plants exhibit high lamina density and cellulose content, which results in tougher leaves and lower herbivory rates, particularly by leaf-cutting ants (Estrada et al., 2013) (Table 12.4).

TABLE 12.4 Endophytic Microorganisms and Their Roles in Biotic Stress Management

Sl. No.	Name of the Endophyte	Role of the Endophyte	Bioactive Compound	References
1.	*Bacillus* sp.	Plant growth promotion	lipopeptides	Hassan et al. (2017)
2.	*Aspergillus* sp.	Anti-bacterial	Alkaloids	Sun et al. (2018)
3.	*Penicillium* sp.	Anti-bacterial and Phosphorus solubilizing	Terpenoids	Yao et al. (2014)
4.	*Pseudomonas* sp.	Plant growth promotion	Organocopper	Pham et al. (2017)
5.	*Serratia* sp.	Plant growth promotion	Prodigiosin	Tavares et al. (2018)

12.7 DIFFERENT ROLES OF BIOACTIVE COMPOUNDS OF ENDOPHYTES IN SUSTAINABLE AGRICULTURE

12.7.1 ABIOTIC STRESS MANAGEMENT

Abiotic stressors like salt, dryness, and extremely high temperatures threaten agroecosystems, which are seriously threatened by oxidative stress and heavy metal toxicity (Khare and Arora, 2015). The induction and expression of stress-responsive genes, the production of scavenger molecules like ROS, and the synthesis of antistress metabolites are a few examples of the molecular strategies used by endophytes to increase stress tolerance in plants (Lata et al., 2018). Phytohormones are essential for plants' ability to withstand abiotic stress (Wani et al., 2016). Abscisic acid (ABA), a PH, controls stomatal closure and plant development, which helps plants resist osmotic and other abiotic stressors (Waqas et al., 2012). The presence of advantageous microbes in plants' endospheres influences ABA production and ABA-mediated signaling pathways, which may boost plant development under salt stress conditions. Recently, halo-tolerant *Dietzia natronolimnaea*, the source of salinity, modulated an ABA-signaling cascade. The overexpression of the TaABARE (ABA-responsive gene) and TaOPR1 (12-oxophytodienoate reductase 1) genes in wheat plants served as proof of tolerance (Ilangumaran and Smith, 2017), and Peskan-Berghofer et al. (2015) showed that ABA is necessary for the establishment of the mutualistic symbiosis between the roots of *A. thaliana* and the helpful fungus *Piriformospora indica*. By inoculating rice with endophytic *Trichoderma harzianum*, it has been reported that the overexpression of the genes for aquaporin, dehydrin, and malonialdehyde helps to mitigate abiotic stressors (Pandey et al., 2016c).

According to studies (Mishra et al., 2017), endophytic bacteria may play a role in the decrease of metal phytotoxicity by extracellular precipitation, intracellular accumulation, sequestration, or biotransformation of harmful metal ions into less toxic or non-toxic forms. Endophytes can also inhibit heavy metal-induced oxidative damage by controlling the activity of plant antioxidant enzymes and lipid peroxidation. According to Madhaiyan et al. (2007), inoculating tomato plants with *Methylobacterium oryzae* and *Burkholderia* sp. endophytic bacteria reduced the toxicity and buildup of Ni and Cd, as well as subsequent translocation from roots to shoots. It has been demonstrated that a variety of endophytic Proteobacteria shield plants from the severe effects of herbicides by assisting in their metabolism (Ngigi et al., 2012). According to Germaine et al. (2006), plants with *Pseudomonas* endophyte did not accumulate any pesticide in their tissues after being exposed to 2,4-dichlorophenoxyacetic acid. Numerous herbicide-resistance and metabolizing genes are found in *Pseudomonas punonensis* D1-6's genome, indicating that this endophytic bacterium contributes to the host's herbicide resistance (Lafi et al., 2017).

12.8 CONCLUSION

This presented chapter makes an effort to understand the various defense mechanisms that endophytes employ to keep plants healthy for sustainable agriculture. Endophytic bacteria use a variety of direct and indirect ways of action to promote the plant and hasten plant growth. The main advantage of utilizing such advantageous microorganisms in agriculture is to reduce the usage of various agrochemicals, such as pesticides, chemical fertilizers, other synthetic chemicals, and so forth. This would increase agricultural productivity and sustainability. Even now, endophytes can be quite helpful. Under harsh situations like drought, salt, and waterlogging, these metabolites can also enhance plant nutrient absorption and provide insecticidal and pest control activities. When considered as a whole, the unique bioactive substances released by endophytes, notably endophytic actinomycetes, may significantly help to address current and upcoming issues in agriculture, the environment, and medicine. Last but not least, the use of metagenomics in conjunction with next-generation sequencing technologies is anticipated to unlock the vast untapped reservoir of antimicrobials released by as-yet uncultivated endophytic microorganisms.

KEYWORDS

- **next-generation sequencing**
- **plant growth-promoting microbes**
- **sustainable agriculture**
- **thin-layer chromatography**
- **volatile organic compounds**

REFERENCES

Abe, K., & Matsuki, N. (2000). Measurement of cellular 3-(4, 5-dimethylthiazol-2-yl)-2, 5-diphenyltetrazolium bromide (MTT) reduction activity and lactate dehydrogenase release using MTT. *Neuroscience Research*, *38*, 325–329.

Afzal, I., Shinwari, Z. K., Sikandar, S., & Shahzad, S. (2019). Plant beneficial endophytic bacteria: Mechanisms, diversity, host range and genetic determinants. *Microbiological Research*, *221*, 36–49.

Aljuraifani, A., Aldosary, S., & Ababutain, I. (2019). *In vitro* antimicrobial activity of endophytes, isolated from *Moringa peregrina* growing in Eastern region of Saudi Arabia. *National Academy Science Letters*, *42*, 75–80.

Aly, A. H., Edrada-Ebel, R., Wray, V., Müller, W. E. G., Kozytska, S., & Hentschel, U. (2008). Bioactive metabolites from the endophytic fungus *Ampelomyces* sp. isolated from the medicinal plant *Urospermum picroides*. *Phytochemistry*, *69*, 1716–1725.

Andreozzi, A., Prieto, P., Mercado-Blanco, J., Monaco, S., Zampieri, E., & Romano, S. (2019). Efficient colonization of the endophytes *Herbaspirillum huttiense* RCA24 and *Enterobacter cloacae* RCA25 influences the physiological parameters of *Oryza sativa* L. cv. Baldo rice. *Environmental Microbiology*, *21*, 3489–3504.

Bae, M., Chung, B., Oh, K. B., Shin, J., & Oh, D. C. (2015). Hormaomycins B and C: new antibiotic cyclic depsipeptides from a marine mudflat-derived *Streptomyces* sp. *Marine Drugs*, *13*, 5187–5200.

Berendsen, R. L., Verk, M. C., Stringlis, I. A., Zamioudis, C., Tommassen, J., & Pieterse, C. M. (2015). Unearthing the genomes of plant-beneficial *Pseudomonas* model strains WCS358, WCS374 and WCS417. *BMC Genomics*, *16*, 539. https://doi.org/10.1186/s12864-015-1632-z (accessed on 6th November 2025).

Bills, G., Dombrowski, A., & Peláez, F. (2002). Recent and future discoveries of pharmacologically active metabolites from tropical fungi. In R. Watling, J. C. Frankland, & A. M. Ainsworth (Eds.), *Tropical Mycology: Micromycetes* (Vol. 2, pp. 165–194). CABI Publishing.

Borah, A., & Thakur, D. (2020). Phylogenetic and functional characterization of culturable endophytic actinobacteria associated with *Camellia* spp. for growth promotion in commercial tea cultivars. *Frontiers in Microbiology*, *11*, 318. https://doi.org/10.3389/fmicb.2020.00318 (accessed on 6th November 2025).

Cui, H. B., Mei, W. L., Miao, C. D., Lin, H. P., Hong, K., & Dai, H. F. (2008). Antibacterial constituents from the endophytic fungus *Penicillium* sp. 0935030 of mangrove plant *Acrostichum aureum*. *Chemical Journal of Chinese Universities*, *33*, 407–410.

Dai, J. Q., Krohn, K., Flörke, U., Draeger, S., Schulz, B., & Kiss-Szikszai, A. (2006). Metabolites from the endophytic fungus *Nodulisporium* sp. from *Juniperus cedre*. *European Journal of Organic Chemistry*, 3498–3506.

De Silva, N. I., Brooks, S., Lumyong, S., & Hyde, K. D. (2019). Use of endophytes as biocontrol agents. *Fungal Biology Reviews*, *33*, 133–148.

Demain, A. L., Wrigley, S. K., Hayes, M. A., & Thomas, R. (2000). Microbial natural products: a past with a future. In *Biodiversity: New Leads for Pharmaceutical and Agrochemical Industries* (pp. 3–16). The Royal Society of Chemistry.

Doty, S. L., Freeman, J. L., Cohu, C. M., Burken, J. G., Firrincieli, A., & Simon, A. (2017). Enhanced degradation of TCE on a superfund site using endophyte assisted poplar tree phytoremediation. *Environmental Science & Technology*, *51*(17), 10050–10058. https://doi.org/10.1021/acs.est.7b01504 (accessed on 6th November 2025).

Dubey, A., Malla, M. A., Kumar, A., Dayanandan, S., & Khan, M. L. (2020). Plants endophytes: Unveiling hidden agenda for bioprospecting toward sustainable agriculture. *Critical Reviews in Biotechnology*, *40*, 1210–1231.

Dutta, D., Puzari, K. C., Gogoi, R., & Dutta, P. (2014). Endophytes: Exploitation as a tool in plant protection. *Brazilian Archives of Biology and Technology*, *57*, 621–629.

Enebe, M. C., & Babalola, O. O. (2018). The impact of microbes in the orchestration of plants' resistance to biotic stress: A disease management approach. *Applied Microbiology and Biotechnology*, *103*, 9–25.

Estrada, C., Wcislo, W. T., & Van Bael, S. A. (2013). Symbiotic fungi alter plant chemistry that discourages leaf-cutting ants. *New Phytologist*, *198*, 241–251.

Ezra, D., Hess, W., & Strobel, G. A. (2004). New endophytic isolates of *Muscodor albus*, a volatile-antibiotic-producing fungus. *Microbiology*, *150*, 4023–4031.

Fletcher, L. R., Sutherland, B. L., & Fletcher, C. G. (1999). The impact of endophyte on the health and productivity of sheep grazing ryegrass-based pastures. In D. R. Woodfield & C. Matthew (Eds.), *Ryegrass Endophyte: An Essential New Zealand Symbiosis* (Research and Practice Series No. 7, pp. 11–17). New Zealand Grassland Association.

Gao, D., & Tao, Y. (2012). Current molecular biologic techniques for characterizing environmental microbial community. *Frontiers of Environmental Science & Engineering*, *6*, 82–97.

Gao, F. K., Dai, C. C., & Liu, X. Z. (2010). Mechanisms of fungal endophytes in plant protection against pathogens. *African Journal of Microbiology Research*, *4*, 1346–1351.

Germaine, K., Liu, X., Cabellos, G., Hogan, J., Ryan, D., & Dowling, D. N. (2006). Bacterial endophyte-enhanced phytoremediation of the organochlorine herbicide 2,4-dichlorophenoxyacetic acid. *FEMS Microbiology Ecology*, *57*, 302–310.

Han, Z., Mei, W. L., Cui, H. B., Zeng, Y. B., & Lin, H. P., & Hong, K. (2008). Antibacterial constituents from the endophytic fungus *Penicillium* sp. of mangrove plant *Cerbera manghas*. *Chemical Journal of Chinese Universities*, *29*, 749–752.

Hassan, S. E. D. (2017). Plant growth-promoting activities for bacterial and fungal endophytes isolated from medicinal plant of *Teucrium polium* L. *Journal of Advanced Research*, *8*, 687–695.

Hoffman, A. M., Mayer, S. G., Strobel, G. A., Hess, W. M., Sovocool, G. W., Grange, A. H., & Kelley-Swift, E. G. (2008). Purification, identification and activity of phomodione, a furandione from an endophytic *Phoma* species. *Phytochemistry*, *69*, 1049–1056.

Hussain, H., Root, N., & Jabeen, F. (2014). Seimatoric acid and colletonoic acid: Two new compounds from the endophytic fungi, *Seimatosporium* sp. and *Colletotrichum* sp. *Chinese Chemical Letters*, *25*, 1577–1579.

Ilangumaran, G., & Smith, D. L. (2017). Plant growth promoting rhizobacteria in amelioration of salinity stress: A systems biology perspective. *Frontiers in Plant Science*, *8*, 1768. https://doi.org/10.3389/fpls.2017.01768 (accessed on 6th November 2025).

Jia, M., Chen, L., Xin, H. L., Zheng, C. J., Rahman, K., Han, T., & Qin, L. P. (2016). A friendly relationship between endophytic fungi and medicinal plants: A systematic review. *Frontiers in Microbiology*, *7*, 906. https://doi.org/10.3389/fmicb.2016.00906 (accessed on 6th November 2025).

Jiang, S., Chen, D. J., Ge, M., Zhang, F., & Cai, Y. F. (2006). Studies on the antibacterial activities of endophytic fungi isolated from pharmaceutical plants. *Pharmaceutical Biotechnology*, *13*(5), 351–354.

Johnson, L. J., De Bonth, A. C. M., & Briggs, L. R. (2013a). The exploitation of epichloae endophytes for agricultural benefit. *Fungal Diversity*, *60*, 171–188.

Kaul, S., Sharma, T., & Dhar, M. K. (2016). "Omics" tools for better understanding the plant–endophyte interactions. *Frontiers in Plant Science*, *7*, 955. https://doi.org/10.3389/fpls.2016.00955 (accessed on 6th November 2025).

Kavroulakis, N. S., Zervakis, G. I., Ehaliotis, C., Haralampidis, K., & Papadopoulou, K. K. (2007). Role of ethylene in the protection of tomato plants against soil-borne fungal pathogens conferred by an endophytic *Fusarium solani* strain. *Journal of Experimental Botany*, *58*, 3853–3864.

Khan, A. L., Waqas, M., Kang, S. M., Al-Harrasi, A., Hussain, J., & Al-Rawahi, A. (2014). Bacterial endophyte *Sphingomonas* sp. LK11 produces gibberellins and IAA and promotes tomato plant growth. *Journal of Microbiology*, *52*, 689–695.

Khare, E. K., & Arora, N. K. (2015). Effects of soil environment on field efficacy of microbial inoculants. In *Plant Microbes Symbiosis: Applied Facets* (pp. 353–381). Springer.

Khare, E., Kim, K. M., & Lee, K. J. (2016). Rice *OsPBL1* (*Oryza sativa Arabidopsis PBS1-LIKE 1*) enhanced defense of *Arabidopsis* against *Pseudomonas syringae* DC3000. *European Journal of Plant Pathology*, *146*, 901–910.

Kim, J., Woo, H. R., & Nam, H. G. (2016). Toward systems understanding of leaf senescence: An integrated multi-omics perspective on leaf senescence research. *Molecular Plant*, *9*, 813–825.

Kim, S., Sin, D. S., Lee, T., & Oh, J. B. (2004). Periconicins: Two new fusicoccane diterpenes produced by an endophytic fungus *Periconia* sp. with antibacterial activity. *Journal of Natural Products*, *67*, 448–450.

Kuldau, G., & Bacon, C. (2008). Clavicipitaceous endophytes: Their ability to enhance resistance of grasses to multiple stresses. *Biological Control*, *46*, 57–71.

Kushwaha, P., Kashyap, P. L., Kuppusamy, P., Srivastava, A. K., & Tiwari, R. K. (2019). Functional characterization of endophytic *Bacilli* from pearl millet (*Pennisetum glaucum*) and their possible role in multiple stress tolerance. *Plant Biosystems*, *154*(4), 503–514. https://doi.org/10.1080/11263504.2019.1651773 (accessed on 6th November 2025).

Lafi, F. F., AlBladi, M. L., Salem, N. M., AlBanna, L., Alam, I., & Bajic, V. B. (2017). Draft genome sequence of the plant growth–promoting *Pseudomonas punonensis* strain D1-6 isolated from the desert plant *Erodium hirtum* in Jordan. *Genome Announcements*, *5*, e1437–e1416. http://dx.doi.org/10.1128/genomea.01437-16 (accessed on 6th November 2025).

Larran, S., Simon, M. R., Moreno, M. V., Siurana, M. S., & Perelló, A. (2016). Endophytes from wheat as biocontrol agents against tan spot disease. *Biological Control*, *92*, 17–23.

Lata, R., Chowdhury, S., Gond, S. K., & White, J. F. (2018). Induction of abiotic stress tolerance in plants by endophytic microbes. *Letters in Applied Microbiology*, *66*, 268–276.

Li, E. W., Jiang, L. H., Guo, L. D., Zhang, H., & Che, Y. S. (2008). Pestalachlorides A–C, antifungal metabolites from the plant endophytic fungus *Pestalotiopsis adusta*. *Bioorganic & Medicinal Chemistry*, *16*, 7894–7899.

Li, J. Y., Harper, J. K., & Grant, D. M. (2001). Ambuic acid, a highly functionalized cyclohexanone with antifungal activity from *Pestalotiopsis* spp. and *Monochaetia* sp. *Phytochemistry*, *56*, 463–468.

Li, J. Y., Strobel, G., & Harper, J. (2000). Cryptocin, a potent tetramic acid antimycotic from the endophytic fungus *Cryptosporiopsis cf. quercina*. *Organic Letters*, *2*, 767–770.

Liu, L. H., Yuan, T., Zhang, J. Y., Tang, G. X., Lü, H., Zhao, H. M., Li, H., Li, Y. W., Mo, C. H., Tan, Z. Y., & Cai, Q. Y. (2021). Diversity of endophytic bacteria in wild rice (*Oryza meridionalis*) and potential for promoting plant growth and degrading phthalates. *Science of the Total Environment*, *806*(Part 1), 150310. https://doi.org/10.1016/j.scitotenv.2021.150310 (accessed on 6th November 2025).

Lu, H., Zou, W. X., Meng, J. C., Hu, J., & Tan, R. X. (2000). New bioactive metabolites produced by *Colletotrichum* sp., an endophytic fungus in *Artemisia annua*. *Plant Science*, *151*, 67–73.

Macías-Rubalcava, M. L., Hernández-Bautista, B. E., Jiménez-Estrada, M., González, M. C., Glenn, A. E., Hanlin, R. T., Hernández-Ortega, S., Saucedo-García, A., Muria-González, J. M., & Anaya, A. L. (2008). Naphthoquinone spiroketal with allelochemical activity from the newly discovered endophytic fungus *Edenia gomezpompae*. *Phytochemistry*, *69*(5), 1185–1196. https://doi.org/10.1016/j.phytochem.2007.12.006 (accessed on 6th November 2025).

Madhaiyan, M., Poonguzhali, S., & Sa, T. (2007). Metal tolerating methylotrophic bacteria reduces nickel and cadmium toxicity and promotes plant growth of tomato (*Lycopersicon esculentum* L.). *Chemosphere*, *69*, 220–228.

Mejía, L. C., Herre, E. A., Sparks, J. P., Winter, K., García, M. N., & Van Bael, S. A. (2014). Pervasive effects of a dominant foliar endophytic fungus on host genetic and phenotypic expression in a tropical tree. *Frontiers in Microbiology*, *5*, 479. https://doi.org/10.3389/fmicb.2014.00479 (accessed on 6th November 2025).

Mejia, L. C., Rojas, E. I., Maynard, Z., Van Bael, S. A., Arnold, E. A., & Hebbar, P. (2018). Endophytic fungi as biocontrol agents of *Theobroma cacao* pathogens. *Biological Control*, *46*, 4–14.

Mishra, J., Singh, R., & Arora, N. K. (2017). Alleviation of heavy metal stress in plants and remediation of soil by rhizosphere microorganisms. *Frontiers in Microbiology*, *8*, 1706. https://doi.org/10.3389/fmicb.2017.01706 (accessed on 6th November 2025).

Mitchell, A. M., Strobel, G. A., & Moore, E. (2010). Volatile antimicrobials from *Muscodor crispans*, a novel endophytic fungus. *Microbiology*, *156*, 270–277.

Ngigi, A. N., Getenga, Z. M., Boga, H. I., & Ndalut, P. K. (2012). Biodegradation of S-triazine herbicide atrazine by *Enterobacter cloacae* and *Burkholderia cepacia* sp. from long-term treated sugarcane-cultivated soils in Kenya. *Journal of Environmental Science and Health, Part B*, *47*(8), 769–778. https://doi.org/10.1080/03601234.2012.676364 (accessed on 6th November 2025).

Nithya, K., & Muthumary, J. (2019). Growth studies of *Colletotrichum gloeosporioides* (Penz.) Sacc., a taxol-producing endophytic fungus from *Plumeria acutifolia*. *Indian Journal of Science and Technology*, *2*(11), 14–19.

O'Hanlon, K. A., Knorr, K., Jørgensen, L. N., Nicolaisen, M., & Boelt, B. (2012). Exploring the potential of symbiotic fungal endophytes in cereal disease suppression. *Biological Control*, *63*, 69–78.

Pai, G., & Chandra, M. (2017). Antimicrobial activity of endophytic fungi isolated from ethnomedicinal plant *Phyllanthus reticulatus* Poir. *International Journal of Engineering Science Invention*, *7*, 40–46.

Pandey, V., Ansari, M. W., Tula, S., Yadav, S., Sahoo, R. K., & Shukla, N. (2016c). Dose-dependent response of *Trichoderma harzianum* in improving drought tolerance in rice genotypes. *Planta*, *243*, 1251–1264.

Peskan-Berghofer, T., Vilches-Barro, A., Muller, T. M., Glawischnig, E., Reichelt, M., & Gershenzon, J. (2015). Sustained exposure to abscisic acid enhances the colonization potential of the mutualist fungus *Piriformospora indica* on *Arabidopsis thaliana* roots. *New Phytologist*, *208*, 873–886. https://doi.org/10.1111/nph.13504 (accessed on 6th November 2025).

Pham, V. T., Rediers, H., Ghequire, M. G., Nguyen, H. H., De Mot, R., & Vanderleyden, J. (2017). The plant growth-promoting effect of the nitrogen-fixing endophyte *Pseudomonas stutzeri* A15. *Archives of Microbiology*, *199*, 513–517.

Phongpaichit, S., Rungjindamai, N., Rukachaisirikul, V., & Sakayaroj, J. (2006). Antimicrobial activity in cultures of endophytic fungi isolated from *Garcinia* species. *FEMS Microbiology Letters*, *48*, 367–372.

Puri, A., Padda, K. P., & Chanway, C. P. (2016). Seedling growth promotion and nitrogen fixation by a bacterial endophyte *Paenibacillus polymyxa* P2b-2R and its GFP derivative in corn in a long-term trial. *Symbiosis*, *69*, 123–129.

Qin, J. C., Zhang, Y. M., Gao, J. M., Bai, M. S., Yang, S. X., & Laatsch, H. (2009). Bioactive metabolites produced by *Chaetomium globosum*, an endophytic fungus isolated from *Ginkgo biloba*. *Bioorganic & Medicinal Chemistry Letters*, *19*(6), 1572–1574.

Qin, S., Xing, K., Jiang, J. H., Xu, L. H., & Li, W. J. (2010). Biodiversity, bioactive natural products and biotechnological potential of plant-associated endophytic actinobacteria. *Applied Microbiology and Biotechnology*, *89*, 457–473.

Rajamanikyam, M., Vadlapudi, V., & Upadhyayula, S. M. (2017). Endophytic fungi as novel resources of natural therapeutics. *Brazilian Archives of Biology and Technology*, *60*, e17160542. https://doi.org/10.1590/1678-4324-2017160542 (accessed on 6th November 2025).

Rajkumar, M. A. N., Prasad, M. N. V., & Freitas, H. (2010). Potential of siderophore-producing bacteria for improving heavy metal phytoextraction. *Trends in Biotechnology*, *28*, 142–149.

Sahay, H., Yadav, A. N., Singh, A. K., Singh, S., Kaushik, R., & Saxena, A. K. (2017). Hot springs of Indian Himalayas: potential sources of microbial diversity and thermostable hydrolytic enzymes. *3 Biotech*, *7*, 118. https://doi.org/10.1007/s13205-017-0762-1 (accessed on 6th November 2025).

Salam, N., Khieu, T. N., Liu, M. J., Vu, T. T., Chu-Ky, S., & Quach, N. T. (2017). Endophytic actinobacteria associated with *Dracaena cochinchinensis* Lour.: Isolation, diversity, and their cytotoxic activities. *BioMed Research International*, *2017*, 1308563. https://doi.org/10.1155/2017/1308563 (accessed on 6th November 2025).

Shan, W., Zhou, Y., Liu, H., & Yu, X. (2018). Endophytic actinomycetes from tea plants (*Camellia sinensis*): Isolation, abundance, antimicrobial, and plant-growth-promoting activities. *BioMed Research International*, *2018*, 1470305. https://doi.org/10.1155/2018/1470305 (accessed on 6th November 2025).

Shen, F. T., Yen, J. H., Liao, C. S., Chen, W. C., & Chao, Y. T. (2019). Screening of rice endophytic biofertilizers with fungicide tolerance and plant growth promoting characteristics. *Sustainability*, *11*(4), 1133. https://doi.org/10.3390/su11041133 (accessed on 6th November 2025).

Sheoran, N., Valiya Nadakkakath, A., Munjal, V., Kundu, A., Subaharan, K., & Venugopal, V. (2015). Genetic analysis of plant endophytic *Pseudomonas putida* BP25 and chemo-profiling of its antimicrobial volatile organic compounds. *Microbiological Research*, *173*, 66–78. https://doi.org/10.1016/j.micres.2015.02.001 (accessed on 6th November 2025).

Shimizu, M. (2011). Endophytic actinomycetes: biocontrol agents and growth promoters. In D. K. Maheshwari (Ed.), *Bacteria in Agrobiology: Plant Growth Responses* (pp. 201–220). Springer.

Silva, G. H., Teles, H. L., & Zanardi, L. M. (2006). Cadinene sesquiterpenoids of *Phomopsis cassiae*, an endophytic fungus associated with *Cassia spectabilis* (Leguminosae). *Phytochemistry*, *67*, 1964–1969.

Singh, A., Singh, D., Kharwar, R., White, J., & Gond, S. (2021). Fungal endophytes as efficient sources of plant-derived bioactive compounds and their prospective applications in natural product drug discovery: Insights, avenues, and challenges. *Microorganisms*, *9*(1), 197. https://doi.org/10.3390/microorganisms9010197 (accessed on 6th November 2025).

Singh, M., Kumar, A., Singh, R., & Pandey, K. D. (2017). Endophytic bacteria: a new source of bioactive compounds. *3 Biotech*, *7*, 315. https://doi.org/10.1016/B978-0-12-818734-0.00008-5 (accessed on 6th November 2025).

Singh, R., & Dubey, A. K. (2018). Diversity and applications of endophytic actinobacteria of plants in special and other ecological niches. *Frontiers in Microbiology*, *9*, 1767. https://doi.org/10.3389/fmicb.2018.01767 (accessed on 6th November 2025).

Stinson, M., Ezra, D., & Hess, W. M. (2003). An endophytic *Gliocladium* sp. of *Eucryphia cordifolia* producing selective volatile antimicrobial compounds. *Plant Science*, *165*, 913–922.

Strobel, G. A., Dirkse, E., & Sears, J. (2001). Volatile antimicrobials from *Muscodor albus*, a novel endophytic fungus. *Microbiology*, *147*, 2943–2950.

Strobel, G., & Daisy, B. (2003). Bioprospecting for microbial endophytes and their natural products. *Microbiology and Molecular Biology Reviews*, *67*(4), 491–502. https://journals.asm.org/doi/10.1128/mmbr.67.4.491-502.2003 (accessed on 6th November 2025).

Strobel, S. A., & Strobel, G. A. (2007). Plant endophytes as a platform for discovery-based undergraduate science education. *Nature Chemical Biology*, *3*, 356–359.

Subban, K., Subramani, R., & Johnpaul, M. (2013). A novel antibacterial and antifungal phenolic compound from the endophytic fungus *Pestalotiopsis mangiferae*. *Natural Product Research*, *27*, 1445–1449.

Sun, B. T., Akutse, K. S., Xia, X. F., Chen, J. H., Ai, X., & Tang, Y. (2018). Endophytic effects of *Aspergillus oryzae* on radish (*Raphanus sativus*) and its herbivore, *Plutella xylostella*. *Planta*, *248*, 705–714.

Sun, H., He, Y., Xiao, Q., Ye, R., & Tian, Y. (2013). Isolation, characterization, and antimicrobial activity of endophytic bacteria from *Polygonum cuspidatum*. *African Journal of Microbiology Research*, *7*(16), 1496–1504.

Suryanarayanan, T. S., Wittlinger, S. K., & Faeth, S. H. (2005). Endophytic fungi associated with cacti in Arizona. *Mycological Research*, *109*, 635–639.

Tavares, M., Nascimento, F., Glick, B., & Rossi, M. (2018). The expression of an exogenous ACC deaminase by the endophyte *Serratia grimesii* BXF1 promotes the early nodulation and growth of common bean. *Letters in Applied Microbiology*, *66*, 252–259.

Terhonen, E., Blumenstein, K., Kovalchuk, A., & Asiegbu, F. O. (2019). Forest tree microbiomes and associated fungal endophytes: Functional roles and impact on forest health. *Forests*, *10*(1), 42. https://doi.org/10.3390/f10010042 (accessed on 6th November 2025).

Thambugala, K. M., Daranagama, D. A., Phillips, A. J., Kannangara, S. D., & Promputtha, I. (2020). Fungi vs. fungi in biocontrol: An overview of fungal antagonists applied against fungal plant pathogens. *Frontiers in Cellular and Infection Microbiology*, *10*, 604923. https://doi.org/10.3389/fcimb.2020.604923 (accessed on 6th November 2025).

Tuntiwachwuttikul, P., Taechowisan, T., Wanbanjob, A., Thadaniti, S., Taylor, W. C., & Lansai, A. D. (2008). Secondary metabolites from *Streptomyces* sp. SUC1. *Tetrahedron*, *64*, 7583–7586.

Turner, T. R., James, E. K., & Poole, P. S. (2013). The plant microbiome. *Genome Biology*, *14*, 209. https://doi.org/10.1186/gb-2013-14-6-209 (accessed on 6th November 2025).

Varughese, T., Rios, N., & Higginbotham, S. (2012). Antifungal depsidone metabolites from *Cordyceps dipterigena*, an endophytic fungus antagonistic to the phytopathogen *Gibberella fujikuroi*. *Tetrahedron Letters*, *53*(13), 1624–1626. https://doi.org/10.1016/j.tetlet.2012.01.076 (accessed on 6th November 2025).

Wang, F. W., Jiao, R. H., Cheng, A. B., Tan, S. H., & Song, Y. C. (2006). Antimicrobial potentials of endophytic fungi residing in *Quercus variabilis* and brefeldin A obtained from *Cladosporium* sp. *World Journal of Microbiology and Biotechnology*, *23*, 79–83.

Wani, S. H., Kumar, V., Shriram, V., & Sah, S. K. (2016). Phytohormones and their metabolic engineering for abiotic stress tolerance in crop plants. *Crop Journal*, *4*, 162–176.

Waqas, M., Khan, A. L., Kamran, M., Hamayun, M., Kang, S. M., & Kim, Y. H. (2012). Endophytic fungi produce gibberellins and indoleacetic acid and promotes host-plant growth during stress. *Molecules*, *17*, 10754–10773.

Wen, J., Okyere, S. K., Wang, S., Wang, J., Xie, L., Ran, Y., & Hu, Y. (2022). Endophytic fungi: An effective alternative source of plant-derived bioactive compounds for pharmacological studies. *Journal of Fungi*, *8*(2), 205. https://doi.org/10.3390/jof8020205 (accessed on 6th November 2025).

Worapong, J., Strobel, G., & Ford, E. J. (2001). *Muscodor albus* anam. gen. et sp. nov., an endophyte from *Cinnamomum zeylanicum*. *Mycotoxin*, *79*, 67–79.

Wu, C., Kim, H. K., & Van Wezel, G. P. (2015a). Metabolomics in the natural products field – a gateway to novel antibiotics. *Drug Discovery Today: Technologies*, *13*, 11–17. https://doi.org/10.1016/j.ddtec.2015.01.004 (accessed on 6th November 2025).

Wu, S. H., Huang, R., Miao, C. P., & Chen, Y. W. (2013). Two new steroids from an endophytic fungus *Phomopsis* sp. *Chemistry & Biodiversity*, *10*, 1276–1283.

Yadav, A. N. (2018). Biodiversity and biotechnological applications of host-specific endophytic fungi for sustainable agriculture and allied sectors. *Acta Scientific Microbiology*, *1*(5), 1–5. https://actascientific.com/ASMI/pdf/ASMI-01-0044.pdf (accessed on 6th November 2025).

Yao, M., Liu, J., Jin, S., Jiao, J., Gai, Q., & Wei, Z. (2014). A novel biotransformation of astragalosides to astragaloside IV with the deacetylation of fungal endophyte *Penicillium canescens*. *Process Biochemistry*, *49*, 807–812.

Zhang, J., Wang, J. D., Liu, C. X., Yuan, J. H., Wang, X. J., & Xiang, W. S. (2014). A new prenylated indole derivative from endophytic actinobacteria *Streptomyces* sp. neau-D50. *Natural Product Research*, *28*, 431–437.

Zhao, J. L., Cai, X. Y., Li, J., Zhang, Y., Peng, Y., & Zhou, L. G. (2008). Endophytic fungi from rhizomes of *Paris polyphylla* var. *yunnanensis* and their antibacterial activity. *Journal of Biotechnology*, *136* (Supplement), S609. https://doi.org/10.1016/j.jbiotec.2008.07.1411 (accessed on 6th November 2025).

Zin, N. M., Remali, J., Nasrom, M. N., Ishak, S. A., Baba, M. S., & Jalil, J. (2017). Bioactive compounds fractionated from endophyte *Streptomyces* SUK 08 with promising *ex-vivo* antimalarial activity. *Asian Pacific Journal of Tropical Biomedicine*, *7*, 1062–1066.

Zou, W. X., Meng, J. C., & Lu, H. (2000). Metabolites of *Colletotrichum gloeosporioides*, an endophytic fungus in *Artemisia mongolica*. *Journal of Natural Products*, *63*(11), 1529–1530. https://doi.org/10.1021/np000204t (accessed on 6th November 2025).

CHAPTER 13

The Role of Plant Growth Regulators in Abiotic Stress Tolerance in Crops

MOHD. SAMAMA,[1] PRIYANK VYAS,[1] NAVIN KUMAR,[1]
PALLAVI JOSHI,[2,*] and PAYAL GUPTA[1,*]

[1]*Department of Biotechnology, Graphic Era University, Dehradun, Uttarakhand, India*

[2]*School of Agriculture, Graphic Era Hill University, Dehradun, Uttarakhand, India*

*Corresponding authors

ABSTRACT

Abiotic stressors that affect plants, such as high salt, drought, and extreme temperatures, can have a negative impact on their growth and crop yield. Plants have evolved a number of coping mechanisms, including the use of plant growth regulators (PGRs), to deal with such challenges. PGRs are compounds with synthetic or natural origins that can enhance the growth of plants as well as abiotic stress resistance in plants. Numerous studies have been conducted on how PGRs respond to abiotic stress tolerance, and the results point to the fact that different PGRs may have varying degrees of success in helping plants cope with abiotic challenges. For instance, stomatal closure is known to be triggered by the administration of abscisic acid (ABA), a PGR that has been extensively investigated. Plants are subjected to many abiotic stresses, for example, high salinity, drought, and extreme temperatures, which can severely affect their growth and crop yield. To monitor such stresses, plants have incorporated various strategies, including the use of PGRs. PGRs are substances of synthetic or natural

Plant Biostimulants and Growth Regulators in Sustainable Agriculture.
Deepak Kumar, Sanjay K. Singh, Nutan Kaushik, and Mahaveer P. Sharma (Eds.)

DOI: 10.1201/9781779641458-17

origin that can enhance plant growth and play a major role in limiting water loss from the plant during drought stress. Similarly, gibberellic acid (GA) increases the root length and growth rate, helping plants get hold of more resources in a drought-prone environment. Ethylene is another PGR that plays a critical role in plant growth under stressful conditions. Ethylene helps to induce tolerance to temperature extremes by regulating the gene expression responsible for the cold stress response. Therefore, judicious use of the various PGRs has a positive impact on plant health and crop production.

13.1 INTRODUCTION

Plant growth regulators (PGRs) are natural or synthetic organic compounds that have the power to regulate plant growth and development in both positive and negative ways at low concentrations. PGRs work by impacting metabolism, gene expression, and various signaling pathways (Rademacher, 2015). PGRs act as a signal molecule in intercellular communication and participate in cell differentiation, growth, and division. They are synthesized by plants inside the cell but are not tissue-localized and thus move wherever they are required. Five major classes of PGRs include auxins, gibberellins, abscisic acid (ABA), ethylene, and cytokinins. Mostly, auxin and gibberellin are responsible for growth and cell elongation-related activities, while cytokinin is responsible for cell division. Ethylene is a PGR gas and is required for fruit ripening. Lastly, ABA is a sesquiterpene and controls senescence and abscission in plants (Fahad et al., 2015).

A plant undergoes a variety of stresses throughout its life cycle, like low and high water, temperature, oxidative stress, heavy metal stress, and so forth. These stresses impact the physiology and development of plants and, thus, productivity, which creates trouble for farmers (Bray, 2000). Plants have developed and acquired various ways to overcome stress mediated through PGRs. The plants mount a complex, integrated series of responses based on the level, type, and location of stress. A complete understanding of these intertwined responses is crucial to accelerating the PGR application for commercial benefits. All PGRs are chemically and structurally different from each other and have the ability to protect plants from any kind of stress that they encounter during their lives.

During plants' stress tolerance activity, the concentration of endogenous PGRs gets modulated, which helps the plant do its best. The advances made in molecular and genetics fields have expanded our understanding of PGRs' role in abiotic stress tolerance, and the same has been applied to making crops more resilient to abiotic stress. These enhancements in traits have been done either by modifying their genes or by employing the PGRs exogenously. The abiotic stress response in crop plants and the overall impact of PGRs on control are complex and described here in detail. The chapter will cover all aspects of PGRs and their application to crop productivity improvement comprehensively.

13.2 ABIOTIC STRESS

Abiotic stresses are the harmful/stressful conditions created by abiotic factors like temperature, water, rays, and so forth, that affect the normal development of plants. These factors are external, temporary/seasonal, and adverse effects may get reversed with changes in the environment. The conditions include temperature variation, excessive salinity, lack of water or too much water (drought or flooding), insufficient nutrients, and high levels of metal concentration in plants (Figure 13.1). These factors can have a negative effect on the health and growth of plants (Ghosh and Majee, 2022).

13.2.1 COLD STRESS

Cold stress in plants is a commonly observed problem occurring at low temperatures, typically ranging from 0 to 15°C. This stressor negatively affects various aspects of plant physiology, including seed germination, ATP generation, and photosynthesis, leading to a decrease in crop production. Every plant experiences a change in temperature in its life cycle, and plants know how to sustain this change without impacting plant growth. To combat the detrimental effects of cold stress, PGRs play a crucial role in maintaining the biochemical and physiological attributes of plants. One such regulator, salicylic acid (SA), has been found to have significant impacts under cold stress conditions. SA improves stomatal movement, enhances the water-holding capacity of plants, and boosts enzymatic activity, thereby aiding plants in coping with cold stress (Saleem et al., 2021).

13.2.2 HEAT STRESS

The major loss in crop production can be attributed to the heat stress experienced by plants when exposed to high temperatures, specifically those exceeding 40°C. A significant impact on rice crops occurs during the flowering phase, where exposure to temperatures as high as 41°C for a duration of 4 hours can lead to complete sterility (Jha et al., 2014). To enhance rice crop production, PGRs have been found to be effective in increasing the soluble sugar and protein content in pollen, thereby promoting successful pollen fertilization (Fahad et al., 2015). Another factor influenced by heat stress is the production of ethylene, an important plant growth hormone and regulator. Ethylene (ETH), a gaseous plant hormone (PH), has both time- and dose-dependent effects on plants during heat stress. While a low concentration of ETH activates defense signaling in plants, a high concentration of ETH inhibits processes in cucumber and wheat during post-harvest storage. ETH production during heat stress has been associated with leaf senescence and reduced pollen germination, fruit set, spike fertility, and grain weight. However, these effects may serve as a survival strategy to maintain optimal progeny under heat-stress conditions (Poór et al., 2022). But, at 40°C, suppression of ethylene synthesis has been observed in *Lycopersicon esculentum*. Not all plants respond to heat stress in the same way, as some species, like creeping bent grass, show no change in ET concentration at 35°C, indicating variations in heat-tolerant responses among different plant species.

13.2.3 HEAVY METAL STRESS

Heavy metal (HM) stress in plants occurs when there is an elevated concentration of non-essential metals, such as arsenic (As), lead (Pb), cadmium (Cd), and mercury (Hg). These metals enter the plant and disrupt biochemical and metabolic processes (Sharma et al., 2020). The impact of heavy metals on plants is particularly evident in photosynthesis, as these metals hinder enzymatic activity at the cellular and molecular levels. In addition, hydrogen peroxide (H_2O_2) plays a crucial role in the transportation of PGRs throughout the plant. The interaction between auxin and hydrogen peroxide has been observed in *Arabidopsis*, where it regulates primary root gravitropism and moderates lateral root growth

(Nazir et al., 2020). HM toxicity negatively impacts plant growth and development, accumulating in both aboveground and underground plant parts. High concentrations of certain HMs, such as cadmium, copper, lead, and nickel, can delay seed germination and reduce root and shoot biomass. HMs affect enzymatic activity, inhibiting germination and growth processes. Oxidative stress, decreased enzyme activity, and protein damage are observed under HM exposure (Sytar et al., 2019). Different plant species respond variably to different HMs, and the plant's reaction to HM stress depends on the concentration and type of HM, as well as interactions with other stresses.

13.2.4 SALINITY STRESS

Salinity stress in plants arises from the presence of soluble salts in the underlying groundwater (Pitman and Läuchli, 2002). Soil salinity is a global problem that affects around 30% of irrigated land and 6% of the total land area. This has resulted in a significant loss of around 12 billion US dollars in agricultural production. The agricultural lands affected by salinity cover an area of more than 831 million hectares worldwide. The increasing occurrence of salinity in arable lands highlights the importance of

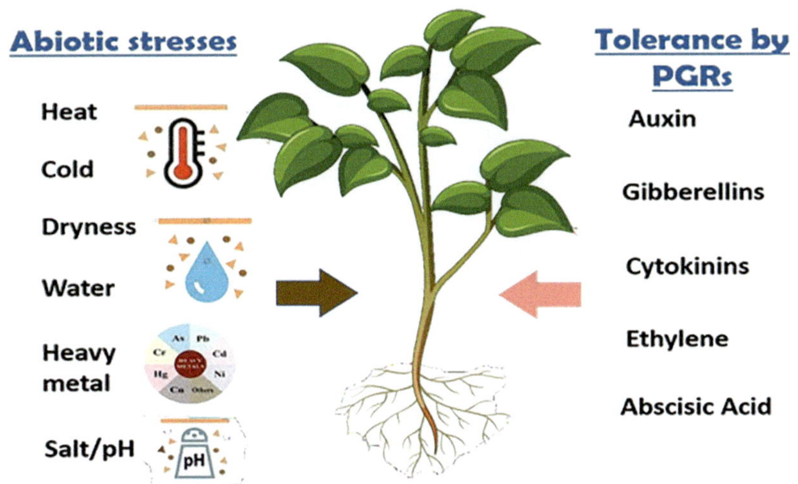

FIGURE 13.1 The diagrammatic representation of abiotic stresses that plants encounter and the different PGRs to mitigate the stress and promote growth.

understanding how plants tolerate and adapt to these conditions to maintain crop productivity. Salinity stress negatively impacts plant growth and development, leading to reduced photosynthesis, respiration, and protein synthesis in sensitive species. It also triggers the excessive production of harmful reactive oxygen species (ROS) like superoxide anion, hydrogen peroxide, and hydroxyl radicals, particularly in the chloroplasts and mitochondria of plants (Fahad et al., 2015). The oxidative damage caused by salinity can be detrimental. This stress condition negatively impacts plant development and growth by causing the accumulation of salt ions (Na^+ and Cl^-) within plant tissues, leading to disruptions in vital metabolic processes (Quamruzzaman et al., 2021).

13.3 PLANT GROWTH REGULATORS (PGRS)

PGRs are synthetic organic compounds made artificially outside the plant body that help in the maintenance of the morphological and physiological characteristics of plants. They function like natural PHs but are known as exogenous PHs. These are frequently employed in agriculture, horticulture, and viticulture to promote growth under unfavorable conditions (such as a short growing season, poor soil fertility, and disease) and to avoid fruit drop (Farman et al., 2019). Based on their role in plant growth, one category of PGRs that promote plant growth and are known as plant growth promotors includes auxins, gibberellins, and cytokinins. Another group of PGRs that inhibit plant growth and are referred to as plant growth inhibitors includes ethylene and ABA. They also work as biostimulants (BS) and bio-inhibitors to activate and block the production of specific enzymes inside the plant cells. These are also useful in the cloning of cells in tissue culture (Halmann, 1990). The action of PGRs in crops depends on many factors, such as environmental conditions, the growth phase of the plant, and crop type (Soni et al., 2022). PGRs and plant growth hormones perform a similar function inside the plant, although their sites of synthesis and chemical behavior are different. Plant growth hormones, or phytohormones, are organic compounds present naturally in a plant other than nutrition, synthesized in the meristematic region, that move to the different parts of plants and control cell division, cell elongation, and cell differentiation and influence plants' responses to environmental stress.

13.4 ROLE OF PLANT GROWTH REGULATORS IN ABIOTIC STRESS MITIGATION

13.4.1 AUXINS

Auxins are low-molecular-weight organic compounds that play important roles in plant cell division, cell elongation, and cell expansion. They mainly occur in roots and stem tips through metabolism and are transported to elongation regions (George et al., 2008). Auxin interacts with salicylic acid and ABA to regulate plant growth under abiotic stress. They are commercially available in liquid and powder form. The liquid form is absorbed faster than the powder form. There are two types of auxins: natural and synthetic. Here, natural auxins include indole-3-acetic acid (IAA), indole-3-propionic acid (IPrA), phenylacetic acid, and indole-3-butyric acid (Figure 13.2). These molecules are either synthesized by plants or provided by commensal bacteria, but synthetic auxins (2,4-dichlorophenoxyacetic acid, α-naphthaleneacetic acid, and 2-methoxy-3,6-dichlorobenzoic acid) are made of and are believed to be: Due to their presence in plant tissues, the most potent do not oxidize.

An auxin named 2,4-dichlorophenoxyacetic acid is the most commonly used herbicide worldwide (Castro et al., 2017; Woodward and Bartel, 2005). Auxin biosynthesis does not occur through a single pathway. However, biosynthetic pathways can describe the synthesis of molecular groups with similar structures, such as IAA, 4-Cl-IAA, IBA, and IPrA. Furthermore, biosynthetic pathways vary among plant species. The major biosynthetic pathway of IAA in plants involves tryptophan biosynthesis, followed by the deamination of amino acids to IPA. IPA is decarboxylated to form IAld and indole-3-acetaldehyde, after which it is oxidized to IAA by specific dehydrogenases. Even when free, IAA is the biologically active form of the hormone, and in plants, most auxins exist in a covalently bound state that is thought to be hormonally inactive. In addition, various substances have been developed and evaluated that mimic or modify the physiological effects of auxin, such as 4-Cl-IAA. 4-Cl-IAA is a more potent PH than IAA because it is metabolized more slowly than IAA.

Auxin regulates the expression of several transcription factors, including the ARF (auxin response factor), AUX/IAA (auxin/indole-3-acetic acid), and GH3 (Gretchen Hagen 3) genes. The ARF transcription

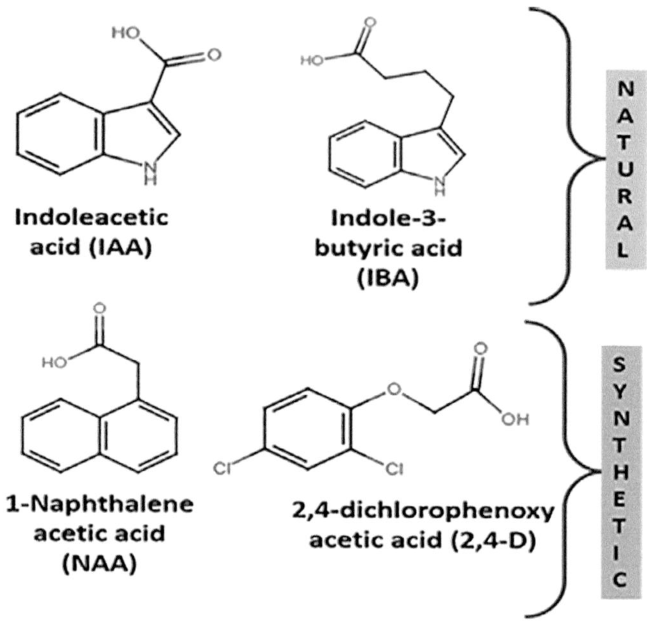

FIGURE 13.2 Chemical structure of natural and synthetic auxins

factor family is a key mediator of auxin signaling. ARFs regulate the expression of downstream genes by binding to auxin response elements (AuxREs) in their promoters (He et al., 2018). ARF activity is regulated by the AUX/IAA protein family, which acts as repressors of ARF activity. Auxin promotes AUX/IAA degradation, leading to ARF activation and subsequent changes in gene expression.

13.4.1.1 RESPONSE OF AUXIN TO ABIOTIC STRESS

Plant roots sense soil moisture and respond through auxin-regulated hydrotropism. Auxin, together with ABA, promotes root hair growth through two opposing phenomena: gravitropism and hydrotropism, which occur simultaneously in the soil. Roots always extend toward moisture in the soil and are primarily controlled by ABA signaling but can also be overridden by auxin-mediated gravitropism. ABA was found to regulate the growth of both primary roots and root hairs by promoting auxin passage, as observed in *Arabidopsis* and rice plants (Muhammad et al., 2022). In addition, it keeps the root elongation process active and increases proton secretion by activating plasma membrane attachment. The root structure of *Arabidopsis* plants is altered by a hydrolytic enzyme known as IAA-Ala-resistance-3 (IAR3) under osmotic conditions. IAR3 may also generate free auxin through the hydrolysis of indoleacetic acid (IAA) in roots during drought. Therefore, to survive in such a harsh environment, IAR3 is thought to act as a bioactivator of auxin formation and stimulate lateral root formation in water-deficient soils (Bielach et al., 2017; Salehin et al., 2019). The YUCCA gene responsible for auxin synthesis in *A. thaliana* was also reported in potato plants and found to be responsible for regulating auxin production and thus the drought tolerance response (Cheol et al., 2013).

Low-salinity conditions reduced elongation of the lateral root; however, they increased the size of the lateral root, in sharp contrast to the high-salinity conditions that hindered root development (Zou et al., 2022). Auxins are important mediators in the regulation of salt stress and work in concert with other signaling molecules, such as PHs. Under saline conditions, the auxin signaling genes AXR1, AXR4, IAR4, and TIR1 promoted plant lateral root growth (Van Zelm et al., 2020).

13.4.2 GIBBERELLINS

Gibberellin (GBR) is a natural chemical that helps plants grow. These chemicals, including tetracyclic and diterpenoid compounds, are mostly inactive and serve as building blocks for synthesizing active forms of GBR (Salazar-Cerezo et al., 2018). Gibberellic acid, also known as GBR, was first discovered in Japan in the 1930s, when fungi caused the rice plants to grow too tall and produce few seeds. This molecule is secreted by the fungus *Gibberella dichroi*, which he later identified as GBR. Structurally, GBR consists of four isoprene units, forming a kaurene cycle. A total of 136 GBRs have been identified in higher plant fungi and bacteria, but only a few are active. One of the most active GBRs is GA_3, often produced industrially for agricultural and horticultural purposes. Interestingly, mutant rice and wheat cultivars have been engineered with genes that alter GBR biosynthesis, resulting in more grain production without adversely affecting other aspects of plant development. A semi-dwarf plant was born. GBR production in plants is regulated by various environmental factors such as temperature, light, water, nutrient availability, and abiotic and biotic stresses.

The GBR production process has been elucidated at the molecular level by studying plants, fungi, and bacteria (Salazar-Cerezo et al., 2018). GBR is made by different enzymes in different parts of the cell. Gibberellin synthesis occurs through the role of terpene cyclases (Hedden and Thomas, 2012). This process begins with a molecule called GGPP, which, through a series of reactions involving enzymes such as CPS (ent-copalyl diphosphate synthase) and KS (ent-kaurene synthase), produces ent-kaurene (Figure 13.3). Ent-kaurene is then converted to GA_{12} by the enzymes eKO and KAO. Finally, GA_{12} is converted to bioactive GA (GA1 or GA4) by the enzymes GA20ox (GA20 oxidase) and GA3ox (GA3 oxidase) (Jiang and Asami, 2018; Salazar-Cerezo et al., 2018).

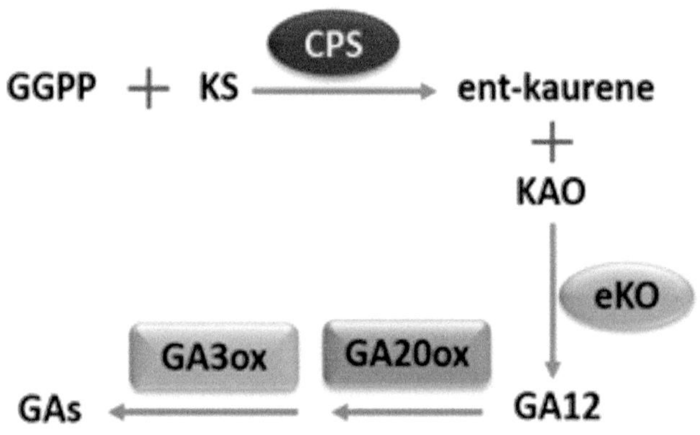

FIGURE 13.3 A schematic representation of biosynthesis of gibberellins.

Gibberellin regulates various developmental processes and the expression of several transcription factors, including DELLA proteins, GAMYB (Gibberellin MYB) transcription factors, and PIF (Phytochrome Interacting Factor) proteins. DELLA proteins are key negative regulators of gibberellin signaling. They act by inhibiting the activity of GAMYB transcription factors. Gibberellin promotes DELLA protein degradation, leading to GAMYB activation and subsequent changes in gene expression. PIF proteins are also involved in gibberellin signaling. They act downstream of GAMYB transcription factors and regulate the expression of genes involved in stem elongation and other developmental processes.

13.4.2.1 RESPONSE OF GIBBERELLIN TO ABIOTIC STRESS

Gibberellins are PHs essential for various processes related to plant development and growth. The importance of gibberellins in plant responses to abiotic challenges such as cold, salinity, and drought has also been demonstrated (Castro-Camba et al., 2022). Many plants alter gibberellin production and signaling pathways in response to abiotic stressors. For example, exogenous gibberellins have been shown to help some plant species cope with salt stress. Increased gibberellin concentrations have also been shown to reduce drought stress in maize plants by promoting photosynthesis, water uptake, and stomatal closure (Shahzad et al., 2021). In response to cold stress, rice reduces gibberellin biosynthesis, and this reduction in gibberellin levels helps plants conserve energy and stunt growth to withstand stress. Overall, the ways in which plants respond to abiotic stressors are complex and diverse (Colebrook et al., 2014). Under flooding stress, plants modulate their energy metabolism, phytohormone biosynthesis, photosynthesis, and respiration (Zhou et al., 2020). Rice plants often adopt an 'escape strategy' during long submerge conditions where they promote shoot elongation (Nagai et al., 2022). This elongation is triggered by ethylene response factor (ERF) proteins; SNORKEL1 and SNORKEL2 via ethylene signaling, as during submerged conditions these genes induce gibberellin production in plants and hence shoot elongation (Hattori et al., 2009). Gibberellins are just one of many hormones and signaling molecules involved in abiotic stress response. However, one of the most important steps in creating plants that are more tolerant to abiotic challenges is understanding how gibberellins interact with other signaling pathways in response to stress. And this interaction/inter-signaling is explained to an extent in some reviews.

13.4.3 CYTOKININS

Cytokinins are chemical compounds dispersed in plants to perform a number of functions (deferring leaf senescence and promoting cell division) (Sakakibara et al., 2006). There are around 20 natural plant cytokinins, but the precise mechanism of cytokinin activity isn't completely clear. They are, for the most part, utilized in plant tissue culture (Schäfer et al., 2015). It acts directly on protein synthesis, which is significant for cell division, cell growth, and responding to various stimuli such as water availability, light, nutrients, abiotic stressors, and biotic stressors (Farman et al., 2019). Synthetic or natural cytokinins included in plants have been observed to mobilize nutrients to the specific area of activity. On occasion, cytokinin application is driven by a reduction in protein and chlorophyll degradation, and in the case of dormant buds, cytokinin application results in the early development of buds. In the vegetative stages of a plant, cytokinin plays a more prominent role, and their effects are quantifiable and discernible within the formative organization when connected. For example, during root formation, root development is improved, whereas during stem development, stem growth is observed. Cytokinins also control plant stress, and the amount of cytokinin present in a plant can serve as a marker of the level of plant stress. Plants growing in highly saline soil conditions often have very high levels of cytokinins.

Cytokinins are normally occurring adenine subordinates that contain a variable side chain at the 6-position in purine, an isoprenoid chain, and a fragrant ring. The subordinate and adjusted shapes of cytokinin contain an exocyclic nitrogen molecule. Cytokinin, moreover, forms glycosidic conjugates that encourage cytokinin transport, secure the particle from debasement, and induce reversible/irreversible inactivation (Werner et al., 2001). These conjugates can have an engineered beginning that is urea-based. In angiosperms, the roots are the essential location of cytokinin biosynthesis, resulting in high concentrations of endogenous cytokinins. The development of cytokinins between soil and roots impacts the rate of their biosynthesis and digestion systems within the root and shoot of the plant. Exogenous take-up of cytokinins through the root has also appeared to impact the plant digestion system (Stirk and van Staden, 2010). Broadly, cytokinin biosynthesis is administered by three key chemicals: adenosine phosphate isopentenyltransferase (IPT), cytochrome P450 monooxygenase 735A (CYP735A), and Lonely-Guy phosphoribohydrolase (LOG). IPT catalyzes the arrangement of isopentenyl adenine (iP) nucleotides, particularly iP riboside 5'-diphosphate (iPRDP) and iP riboside 5'-triphosphate (iPRTP), from dimethyl diphosphate (DMAPP) and adenine nucleotides by particular utilization of ATP or ADP (Kakimoto, 2003). The chemical CYP735A acts on the prenyl side chain conclusion of iP ribonucleotides, where it encourages the transformation of iP nucleotides to trans-Zeatin (tZ) nucleotides. These tZ nucleotides are inert and, afterward, get changed into a dynamic shape by the activity of LOG.

Cytokinins also act by regulating the expression of various genes and transcription factors. Cytokinin-regulated genes include those involved in cell division, cell differentiation, and stress responses. It also regulates the expression of several transcription factors, including ARR (*Arabidopsis* Response Regulator) proteins and type-B ARRs. ARR proteins are key mediators of cytokinin signaling. They act by binding to and regulating the expression of downstream genes. Type-B ARRs are a subclass of ARR proteins that function as transcription factors. They regulate the expression of cytokinin-responsive genes by binding to cis-acting elements in their promoters.

13.4.3.1 ROLE OF CYTOKININS IN STRESS RESPONSE

Modulation of photoperiod, resilience to stresses, and homeostasis are some of the strategies that cytokinin imparts in plants to overcome abiotic stress. Cytokinin signaling activates in response to any change in nutrients, osmotic condition, temperature, and chemoattractants (Mandal et al., 2022). Cytokinin plays a multifaceted role in both plant development and abiotic stress response via modulating its level (Dobra et al., 2010; Zwack and Rashotte, 2015). The cytokinin signaling pathway works via a phosphor-relay cascade and regulates the expression of genes responsible for plant stress adaptation (Pavlů et al., 2018; Thu et al., 2017). Different photoprotective pathways were found to be disturbed in cytokinin-deficient plants, illustrating the significance of the hormone in protecting plants from extreme light. As mentioned earlier, all phytohormones work in an interconnected manner; cytokinin interacts with ABA, ethylene, jasmonate, and so forth, during stress tolerance in plants (Artner and Benkova, 2019; Skalak et al., 2021).

Under nutrient stress, the root system of plants is first to respond, and cytokinin plays a role because nutritional deficiency modulates cytokinin metabolism as well as signaling genes (Bielach et al., 2012; Pavlů et al., 2018). In *A. thaliana* mutants, phosphate deprivation activated cytokinin signaling (Franco-Zorrilla et al., 2002). Moreover, cytokinin is also responsible for plant responses to sulfur and iron acquisition and iron homeostasis (Maruyama-Nakashita et al., 2004; Séguéla et al., 2008). Under arsenic stress, cytokinins respond by making plants more resistant to arsenate, a prevalent form of arsenic (Mohan et al., 2016). Inadequate light is also a stress to plants, and here also cytokinins play a role in rescuing plants from stress. However, plants themselves have developed some defensive mechanisms: cyclic electron transport, light-rescuing movement of chloroplasts and leaves, and the emission of excess light energy as heat (Takahashi and Badger, 2011). Alone, these defensive mechanisms are not sufficient to protect plants from light stress and need the involvement of phytohormones (Yamamoto, 2016). Any drop in cytokinin level results in increased photoinhibition and reduced photoprotection in plants due to a considerable drop in D1 protein, a component of PSII (Cortleven et al., 2014). Furthermore, reduced cytokinin also impacts the antioxidant system of plants, which highlights the importance of cytokinin in the light stress response. Under drought conditions, the expression of *A. thaliana* CK biosynthesis IPT gene family members (GFMs), particularly AtIPT3 and AtIPT9, was downregulated (Ghosh et al., 2018). This *IPT* family involved in cytokinin biosynthesis is widely distributed among crop plants (Hai et al., 2020). IPT expression gets modulated under both low- and high-water conditions.

13.4.4 ETHYLENE

Ethylene (ETH) is an unsaturated hydrocarbon that plays a part in plant development and senescence. It has a vital part within the distinctive physiological forms included within trimester aging (Iqbal et al., 2017). ETH occurs in all plants; in any case, its concentration shifts and depends on a number of factors, such as the environment and the formative arrangement of the plant (Pierik et al., 2006). During the total life cycle of a plant, the concentration of ETH can change by at least 100-fold. Ethylene was one of the primary chemicals to be recognized as a common development controller and influencer of plant advancement. Its history dates back to the early 19th century, when the Russian understudy, Neljubov, to begin with, portrayed the three-fold development reaction of pea seedlings using ETH. These reactions included decreased stature, overstated hypocotyl snare, and restraint of root development. In spite of the fact that ETH was the primary phytohormone to be chemically distinguished, it picked up full recognition, as it were, after the advent of gas chromatography within the first half of the 20th century (Iqbal et al., 2013). ETH is a gas and diffuses quickly from generation locales, which can make its detection difficult (Grierson, 2012; Jiang and Asami, 2018). One of the special characteristics of ETH is its capacity to interact with other phytohormones at the molecular level. For instance, it interacts with AUXs and ABA to advance leaf development, and this interaction, in some way or another, affects ETH generation. In addition, the quick metabolite for ETH biosynthesis, cytokinin, plays a noteworthy part in leaf start through the upkeep of S-adenosyl-methionine (SAM) (Iqbal et al., 2017).

There are three ways in which plants can deliver ETH: the activated methyl cycle, the S-methyl methionine cycle, and the methionine or Yang cycle. Ethylene is made by a basic process that includes three enzymatic responses. The primary response makes SAM by breaking down methionine with a chemical called SAM synthetase. The second response produces 1-aminocyclopropane-1-carboxylic acid (ACC) with the assistance of a chemical called ACC synthase (ACS). The third and last response oxidizes ACC to ethylene with the assistance of a protein called ACC oxidase (ACO) (Pattyn et al., 2021).

13.4.4.1 RESPONSE OF ETHYLENE TO ABIOTIC STRESS

In abiotic stress conditions, the enzymes (ACC synthase and ACC oxidase) catalyzing the conversion of methionine to ethylene get activated in order to fulfill the ethylene requirement of the plant. French bean, orange, avocado, faba, and numerous other crop plants have shown increased ethylene generation in response to drought conditions (Poór et al., 2022). While the plants' response to flooding stress is also mediated by ethylene and other PGRs, in high-water conditions, the lack of oxygen within submerged roots causes the ACC enzymes to synthesize ethylene, where the enzymes are transported upward within the plant. The production of ethylene creates nastic motions and encourages aerenchyma generation. Under salt stress, ethylene also gets impacted, which raises Na/K homeostasis and makes a difference in advancing salt resilience. Ethylene plays a pivotal role in salt resistance by keeping K^+ in roots and shoots instead of causing Na to drop. Further, ethylene also affects the ROS in plants in response to saline stress by promoting the antioxidant system, which elevates nitrate and sulphate assimilation (Riyazuddin et al., 2020).

Ethylene accumulation also increases the plant's cold tolerance ability; α-methylcyclopropene is an inhibitor of ethylene, and its addition reduced the cold tolerance ability of the tomato plant, while ethephon increased it (Chen et al., 2022). Stress severity in plants is directly proportional to ethylene concentration in the plant; the theory is applicable to mild exposure of the plant to Cd concentration (Schellingen et al., 2014). Chelation is one of the most successful strategies plants adopt to overcome HM stress, followed by vacuolar sequestration in the cytosol. Major chelators belong to the classes of metallothionein, glutathione, phytochelatin, nicotinamine, organic acids, and so forth (Anjum et al., 2015). Besides, the metal-specific response of plants is different and needs to be investigated. For more detailed information on the HM response of plants, the review can be studied (Keunen et al., 2016). Among various HMs, Cd is the most toxic inorganic ion to plants that initiates ethylene production in *Hordeum vulgare*, *Lycopersicon esculentum*, *P. sativum*, *Brassica juncea*, and so forth (Keunen et al., 2016). In view of mechanistic information, in *Nicotiana glutinosa*, the expression of ACO1 and ACO3 genes is induced in the presence of Cu. These genes increase ethylene production (Ruduś et al., 2013).

13.4.5 ABSCISIC ACID

Abscisic acid (ABA) could be a PGR that also acts as a signaling particle and directs different aspects of plant development and growth, including developing life development, seed dormancy, germination, cell division, elongation, and flower induction (Chen et al., 2020). ABA is a small sesquiterpene and has numerous vital parts to play in plants' reaction to environmental stresses caused by components such as drought, salinity, cold, pathogen assault, and UV radiation. ABA was, to begin with, found in the early 1960s during studies related to natural product abscission and was at first called "abscisine II" (Cardoso et al., 2020; Ohkuma and Smith, 1963). It is a 15-carbon compound with a place for terpenoids, and its structure has both trans and cis isomers depending on the introduction of the carboxyl group at carbon-2. It also has S (+) or R (-) enantiomers, which are determined by the symmetric carbon atom at the 1' position within the ring. The characteristic form of ABA is (S)-cis-ABA, whereas commercially accessible ABA could be a blend of the S and R shapes in generally rising to sums (Mehrotra et al., 2014). Light can change the (S)-cis-abscisic acid and (R)-cis-abscisic isomers into the biologically inactive shapes (S)-2-trans-ABA and (R)-2-trans-ABA, individually. Both (S)-cis- and (R)-cis-shapes are dynamic in prolonged ABA reactions, such as seed development, but (S)-cis-Abscisic is stronger in quick ABA reactions like stomatal closure. Abscisic is present in all plants, including green growth and terrestrial plants, some phytopathogenic fungi, microbes, and metazoans, including sea sponges and humans (Jiang and Asami, 2018).

There are two ways that ABA can be made: (i) by parasites straightforwardly from farnesyl pyrophosphate (Figure 13.4); and (ii) by plants through a pathway that includes the cleavage of a β-carotene precursor. This transformation is impacted by the dry-season push. The plant pathway is thought to be the most likely, and it involves the transformation of zeaxanthin to violaxanthin by an enzyme called zeaxanthin epoxidase within the chloroplast. It is usually taken after several other steps that result in the production of neoxanthin. Within the nearness of oxygen, xanthoxin is made, and after that, it is changed over into ABA by an arrangement of responses that take place within the cytoplasm (Finkelstein 2013; Xiong and Zhu 2003). There are also a few ways to make ABA and similar molecules through chemical synthesis.

13.4.5.1 RESPONSE OF ABSCISIC ACID (ABA) TO ABIOTIC STRESS

HM stress itself is a complicated stress because it causes harmfulness, disturbs ionic adjustment, the digestion system, and produces oxidative stretch as well (Ghori et al., 2019; Vishwakarma et al., 2017). Plants have some intrinsic stress tolerance mechanisms but cannot avoid/prevent the stress. However, to boost the stress reaction, exogenous application of ABA moves forward the HM stress mitigation in plants by making strides in the osmolyte collection as well as activating the antioxidant pathways (Kumar et al., 2022; Tang et al., 2020). ABA moreover intervenes in the translocation of HM from root to shoot and simultaneously controls the closing of stomata, reduction of transpiration, and restricts HM development to other parts (Saradevi et al., 2017). In response to HM, plants (*O. sativa, S. tuberosum, T. aestivum*, etc.) themselves activate their endogenous ABA biosynthesis (Kumar et al., 2022). ABA in turn facilitates the expression of metal chelating proteins; the transcript level of PCS1 was increased in response to Cd (Stroinski et al., 2013). Furthermore, ABA moreover interacts with other PGRs and mediates HM mitigating response together with them too.

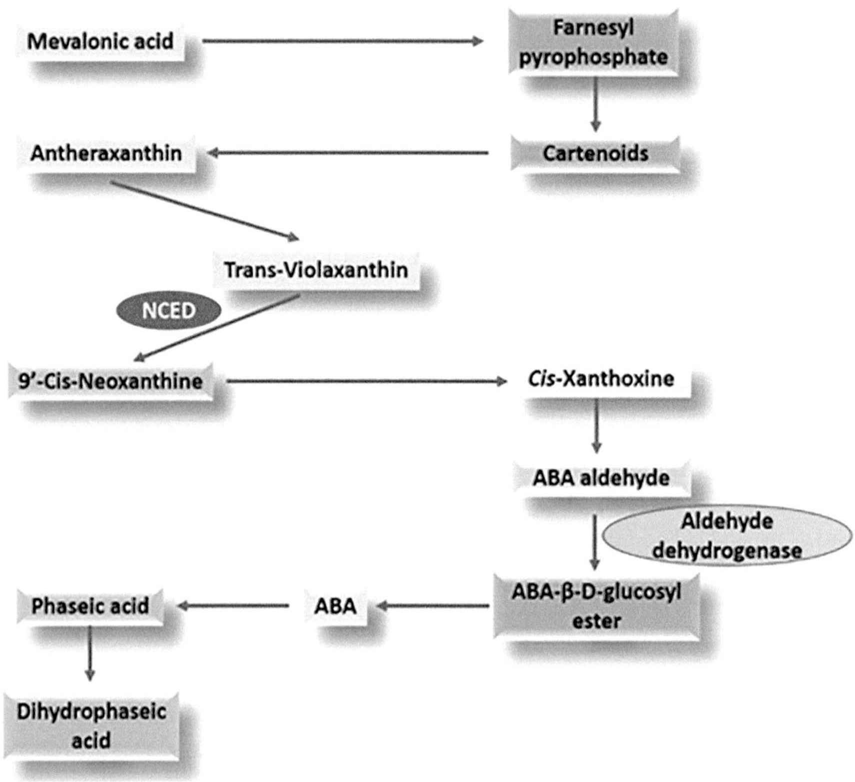

FIGURE 13.4 A schematic representation of the biosynthesis of abscisic acid.

Drought stress significantly impacts ABA levels in plants, as does the transcript level of ABA1, AAO3, 9-cis-epoxycartenoid (NCED3), and ABA3 (Dar et al., 2017; Kim et al., 2010). The transcriptional factors that control their expression include ABF, MYC MYB, NAC, ERF, bZIP, and DREB/CBF (Verma et al., 2016). ABA promotes stomatal closure to control transpiration under drought stress (Kim et al., 2010). The transport of ABA is regulated by the multidrug and toxic compound extrusion (MATE) transporter family, detoxification efflux carrier (AtDTX50) in *A. thaliana* (Zhang et al., 2014). The further ABA-mediated response is facilitated by ABA receptors (PYR, RCAR) in drought stress (Saavedra et al., 2010). Further, the ABA responsive element binding protein maintains photosynthesis in crop plants. However, for all these ABA-mediated gene regulations, histone acetylation is crucial (Sridha and Wu, 2006; Zhang et al., 2008). Conclusively, ABA is an important PGR in regulating abiotic stress response in plants.

13.5 CONCLUSION

Plants are one of the oldest inhabitants on earth and highly evolved in terms of acquisition of stress tolerance mechanisms. They are well adapted to varied environments ranging from high water to low water tolerance, from dry arid soil to water lodging, and saline soil. But sometimes, due to weather change and global warming, plants face some environmental stress different from regular stress. At this point, it becomes necessary for plants to respond to the stress for their survival. The development and growth of plants are intricately regulated by a variety of physiological and biochemical mechanisms. Among these mechanisms, plant growth promoters play a crucial role in orchestrating plant growth and development. Notably, the plant growth promoters: auxin, cytokinins, gibberellins, and ABA are recognized as key regulators in this context. PGRs have played a vital role in rescuing plants from stress and maintaining the needed physiological activity of plants. A cocktail of PGRs can also be developed for betterment and crop enhancement. This kind of exogenous addition will reduce the chances of crop failure.

KEYWORDS

- **abscisic acid**
- **gibberellic acid**
- **indoleacetic acid**
- **plant growth regulators**
- **reactive oxygen species**

REFERENCES

Anjum, N. A., Hasanuzzaman, M., Hossain, M. A., Thangavel, P., Roychoudhury, A., Gill, S. S., Rodrigo, M. A., Adam, V., Fujita, M., Kizek, R., Duarte, A. C., Pereira, E., & Ahmad, I. (2015). Jacks of metal/metalloid chelation trade in plants—an overview. *Frontiers in Plant Science*, *6*, 192. https://doi.org/10.3389/fpls.2015.00192 (accessed on 6th November 2025).

Artner, C., & Benkova, E. (2019). Ethylene and cytokinin: Partners in root growth regulation. *Molecular Plant*, *12*, 1312–1314.

Bielach, A., Hrtyan, M., & Tognetti, V. B. (2017). Plants under stress: involvement of auxin and cytokinin. *International Journal of Molecular Sciences*, *18*(7), 1427. https://doi.org/10.3390/ijms18071427 (accessed on 6th November 2025).

Bielach, A., Podlešáková, K., Marhavý, P., Duclercq, J., Cuesta, C., Müller, B., Grunewald, W., Tarkowski, P., & Benková, E. (2012). Spatiotemporal regulation of lateral root organogenesis in *Arabidopsis* by cytokinin. *Plant Cell*, *24*(10), 3967–3981.

Bray, E. A. (2000). Responses to abiotic stresses. *Biochemistry and Molecular Biology of Plants*, 1158–1203.

Castro-Camba, R., Sánchez, C., Vidal, N., & Vielba, J. M. (2022). Interactions of gibberellins with phytohormones and their role in stress responses. *Horticulture*, *8*(3), 241. https://doi.org/10.3390/horticulturae8030241 (accessed on 6th November 2025).

Chen, H., Bullock, D. A., Alonso, J. M., & Stepanova, A. N. (2022). To fight or to grow: The balancing role of ethylene in plant abiotic stress responses. *Plants*, *11*(1), 33. https://doi.org/10.3390/plants11010033 (accessed on 6th November 2025).

Cheol Park, H., Cha, J. Y., & Yun, D. J. (2013). Roles of YUCCAs in auxin biosynthesis and drought stress responses in plants. *Plant Signaling & Behavior*, *8*, 337–349.

Colebrook, E. H., Thomas, S. G., Phillips, A. L., & Hedden, P. (2014). The role of gibberellin signaling in plant responses to abiotic stress. *Journal of Experimental Biology*, *217*, 67–75.

Cortleven, A., Nitschke, S., Klaumünzer, M., AbdElgawad, H., Asard, H., Grimm, B., Riefler, M., & Schmülling, T. (2014). A novel protective function for cytokinin in the light stress response is mediated by the *Arabidopsis histidine kinase2* and *Arabidopsis histidine kinase3* receptors. *Plant Physiology*, *164*(3), 1470–1483.

Dar, N. A., Amin, I., Wani, W., Wani, S. A., Shikari, A. B., Wani, S. H., & Masoodi, K. Z. (2017). Abscisic acid: A key regulator of abiotic stress tolerance in plants. *Plant Gene*, *11*, 106–111.

Dobra, J., Motyka, V., Dobrev, P., Malbeck, J., Prasil, I. T., Haisel, D., et al. (2010). Comparison of hormonal responses to heat, drought and combined stress in tobacco plants with elevated proline content. *Journal of Plant Physiology*, *167*, 1360–1370.

Fahad, S., Hussain, S., Matloob, A., Khan, F. A., Khaliq, A., Saud, S., Hassan, S., Shah, D., Khan, F., Ullah, N., Faiq, M., Khan, M. F., Tareen, A. K., Khan, A., Ullah, A., Ullah, N., & Huang, J. (2015). Phytohormones and plant responses to salinity stress: a review. *Plant Growth Regulation*, *75*, 391–404. https://doi.org/10.1007/s10725-014-0013-y (accessed on 6th November 2025).

Farman, S., Mushtaq, A., & Azeem, M. W. (2019). Plant growth regulators (PGRs) and their applications: A review. *International Journal of Chemical and Biochemical Sciences*, *15*, 94–103.

Franco-Zorrilla, J. M., Martin, A. C., Solano, R., Rubio, V., Leyva, A., & Paz-Ares, J. (2002). Mutations at *CRE1* impair cytokinin-induced repression of phosphate starvation responses in *Arabidopsis*. *Plant Journal*, *32*, 353–360.

Ghosh, A., Shah, M. N. A., Jui, Z. S., Saha, S., Fariha, K. A., & Islam, T. (2018). Evolutionary variation and expression profiling of isopentenyl transferase gene family in *Arabidopsis thaliana* L. and *Oryza sativa* L. *Plant Gene*, *15*, 15–27.

Ghosh, S., & Majee, M. (2022). Protein *l*-isoAspartyl Methyltransferase (PIMT) and antioxidants in plants. *Vitamins & Hormones*, *121*, 413–432.

Hai, N. N., Chuong, N. N., Tu, N. H. C., Kisiala, A., Hoang, X. L. T., & Thao, N. P. (2020). Role and regulation of cytokinin's in plant response to drought stress. *Plants (Basel)*, *9*(4), 422. https://doi.org/10.3390/plants9040422 (accessed on 6th November 2025).

Halmann, M. (1990). Synthetic plant growth regulators. *Advances in Agronomy*, *43*, 47–105.

Hattori, Y., Nagai, K., Furukawa, S., Song, X. J., Kawano, R., Sakakibara, H., Wu, J., Matsumoto, T., Yoshimura, A., Kitano, H., Matsuoka, M., Mori, H., & Ashikari, M. (2009). The ethylene response factors *SNORKEL1* and *SNORKEL2* allow rice to adapt to deep water. *Nature*, *460*, 1026–1030. https://doi.org/10.1038/nature08258 (accessed on 6th November 2025).

He, M., He, C. Q., & Ding, N. Z. (2018). Abiotic stresses: general defenses of land plants and chances for engineering multi-stress tolerance. *Frontiers in Plant Science*, *9*, 1771. https://doi.org/10.3389/fpls.2018.01771 (accessed on 6th November 2025).

Jha, U. C., Bohra, A., & Singh, N. P. (2014). Heat stress in crop plants: its nature, impacts and integrated breeding strategies to improve heat tolerance. *Plant Breeding*, *133*, 679–701.

Keunen, E., Schellingen, K., Vangronsveld, J., & Cuypers, A. (2016). Ethylene and metal stress: small molecule, big impact. *Frontiers in Plant Science*, *7*, 23. https://doi.org/10.3389/fpls.2016.00023 (accessed on 6th November 2025).

Kim, T. H., Böhmer, M., Hu, H., Nishimura, N., & Schroeder, J. I. (2010). Guard cell signal transduction network: advances in understanding abscisic acid, CO_2, and Ca^{2+} signaling. *Annual Review of Plant Biology*, *61*, 561–591.

Mandal, S., Ghorai, M., Anand, U., Samanta, D., Kant, N., Mishra, T., et al. (2022). Cytokinin and abiotic stress tolerance—What has been accomplished and the way forward? *Frontiers in Genetics*, *13*, 943025. https://doi.org/10.3389/fgene.2022.943025 (accessed on 6th November 2025).

Maruyama-Nakashita, A., Nakamura, Y., Yamaya, T., & Takahashi, H. (2004). A novel regulatory pathway of sulfate uptake in *Arabidopsis* roots: implication of *CRE1/WOL/AHK4*-mediated cytokinin-dependent regulation. *Plant Journal*, *38*, 779–789.

Mohan, T. C., Castrillo, G., Navarro, C., Zarco-Fernández, S., Ramireddy, E., Mateo, C., Zamarreño, A. M., Paz-Ares, J., Muñoz, R., García-Mina, J. M., Hernández, L. E., Schmülling, T., & Leyva, A. (2016). Cytokinin determines thiol-mediated arsenic tolerance and accumulation. *Plant Physiology*, *171*(2), 1418–1426.

Muhammad Aslam, M., Waseem, M., Jakada, B. H., Okal, E. J., Lei, Z., Saqib, H. S. A., Yuan, W., Xu, W., & Zhang, Q. (2022). Mechanisms of abscisic acid-mediated drought stress responses in plants. *International Journal of Molecular Sciences*, *23*(3), 1084. https://doi.org/10.3390/ijms23031084 (accessed on 6th November 2025).

Nagai, K., Kurokawa, Y., Mori, Y., Minami, A., Reuscher, S., Wu, J., Matsumoto, T., & Ashikari, M. (2022). *SNORKEL* genes relating to flood tolerance were pseudogenization in normal cultivated rice. *Plants (Basel)*, *11*(3), 376. https://doi.org/10.3390/plants11030376 (accessed on 6th November 2025).

Nazir, F., Fariduddin, Q., & Khan, T. A. (2020). Hydrogen peroxide as a signaling molecule in plants and its crosstalk with other plant growth regulators under heavy metal stress. *Chemosphere*, *252*, 126486. https://doi.org/10.1016/j.chemosphere.2020.126486 (accessed on 6th November 2025).

Pavlů, J., Novák, J., Koukalová, V., Luklová, M., Brzobohatý, B., & Černý, M. (2018). Cytokinin at the crossroads of abiotic stress signaling pathways. *International Journal of Molecular Sciences*, *19*(8), 2450. https://doi.org/10.3390/ijms19082450 (accessed on 6th November 2025).

Pitman, M. G., & Läuchli, A. (2002). Global impact of salinity and agricultural ecosystems. In A. Läuchli & U. Lüttge (Eds.), *Salinity: Environment–Plants–Molecules* (pp. 3–20). Springer, Dordrecht. https://doi.org/10.1007/0-306-48155-3_1 (accessed on 6th November 2025).

Poór, P., Nawaz, K., Gupta, R., Ashfaque, F., & Khan, M. I. R. (2022). Ethylene involvement in the regulation of heat stress tolerance in plants. *Plant Cell Reports*, *41*, 675–698. https://doi.org/10.1007/s00299-021-02675-8 (accessed on 6th November 2025).

Quamruzzaman, M., Manik, S. M. N., Shabala, S., & Zhou, M. (2021). Improving performance of salt-grown crops by exogenous application of plant growth regulators. *Biomolecules*, *11*(6), 788. https://doi.org/10.3390/biom11060788 (accessed on 6th November 2025).

Rademacher, W. (2015). Plant growth regulators: backgrounds and uses in plant production. *Journal of Plant Growth Regulation*, *34*, 845–872.

Riyazuddin, R., Verma, R., Singh, K., Nisha, N., Keisham, M., Bhati, K. K., Kim, S. T., & Gupta, R. (2020). Ethylene: A master regulator of salinity stress tolerance in plants. *Biomolecules*, *10*(6), 959. https://doi.org/10.3390/biom10060959 (accessed on 6th November 2025).

Ruduś, I., Sasiak, M., & Kępczyński, J. (2013). Regulation of ethylene biosynthesis at the level of 1-aminocyclopropane-1-carboxylate oxidase (ACO) gene. *Acta Physiologiae Plantarum*, *35*, 295–307.

Saavedra, X., Modrego, A., Rodríguez, D., González-García, M. P., Sanz, L., Nicolás, G., & Lorenzo, O. (2010). The nuclear interactor *PYL8/RCAR3* of *Fagus sylvatica FsPP2C1* is a positive regulator of abscisic acid signaling in seeds and stress. *Plant Physiology*, *152*(1), 133–150. https://doi.org/10.1104/pp.109.146381 (accessed on 6th November 2025).

Saleem, M., Fariduddin, Q., & Janda, T. (2021). Multifaceted role of salicylic acid in combating cold stress in plants: A review. *Journal of Plant Growth Regulation*, *40*, 464–485.

Salehin, M., Li, B., Tang, M., Katz, E., Song, L., Ecker, J. R., Kliebenstein, D. J., & Estelle, M. (2019). Auxin-sensitive Aux/IAA proteins mediate drought tolerance in *Arabidopsis* by regulating glucosinolate levels. *Nature Communications*, *10*, 4021. https://doi.org/10.1038/s41467-019-12002-1 (accessed on 6th November 2025).

Schellingen, K., Van Der Straeten, D., Vandenbussche, F., Prinsen, E., Remans, T., Vangronsveld, J., & Cuypers, A. (2014). Cadmium-induced ethylene production and responses in *Arabidopsis thaliana* rely on *ACS2* and *ACS6* gene expression. *BMC Plant Biology*, *14*, 214. https://doi.org/10.1186/s12870-014-0214-6 (accessed on 6th November 2025).

Séguéla, M., Briat, J., Vert, G., & Curie, C. (2008). Cytokinin's negatively regulate the root iron uptake machinery in *Arabidopsis* through a growth-dependent pathway. *Plant Journal*, *55*(2), 289–300. https://doi.org/10.1111/j.1365-313X.2008.03502.x (accessed on 6th November 2025).

Shahzad, K., Hussain, S., Arfan, M., Hussain, S., Waraich, E. A., Zamir, S., Saddique, M., Rauf, A., Kamal, K. Y., Hano, C., & El-Esawi, M. A. (2021). Exogenously applied gibberellic acid enhances growth and salinity stress tolerance of maize through modulating the morpho-physiological, biochemical and molecular attributes. *Biomolecules*, *11*(7), 1005. https://doi.org/10.3390/biom11071005 (accessed on 6th November 2025).

Sharma, A., Kumar, V., Shahzad, B., Ramakrishnan, M., Singh Sidhu, G. P., Bali, A. S., et al. (2020). Photosynthetic response of plants under different abiotic stresses: A review. *Journal of Plant Growth Regulation*, *39*, 509–531. https://doi.org/10.1007/s00344-019-10018-x (accessed on 6th November 2025).

Skalak, J., Nicolas, K. L., Vankova, R., & Hejatko, J. (2021). Signal integration in plant abiotic stress responses via multistep phosphorelay signaling. *Frontiers in Plant Science*, *12*, 644823. https://doi.org/10.3389/fpls.2021.644823 (accessed on 6th November 2025).

Soni, S., Vishwakarma, G., Singh, S. C., Kumar, S., Singh, R. K., Awasthi, P., & Gangwar, V. (2022). Commercial use of plant growth regulators in horticultural crops: An overview. *Pharma Innovation Journal*, *11*, 112–119.

Sridha, S., & Wu, K. (2006). Identification of *AtHD2C* as a novel regulator of abscisic acid responses in *Arabidopsis*. *Plant Journal*, *46*, 124–133.

Sytar, O., Kumari, P., Yadav, S., Brestic, M., & Rastogi, A. (2019). Phytohormone priming: regulator for heavy metal stress in plants. *Journal of Plant Growth Regulation*, *38*, 739–752.

Takahashi, S., & Badger, M. R. (2011). Photoprotection in plants: A new light on photosystem II damage. *Trends in Plant Science*, *16*, 53–60.

Thu, N. B. A., Hoang, X. L. T., Truc, M. T., Sulieman, S., Thao, N. P., & Tran, L. P. (2017). Cytokinin signaling in plant response to abiotic stresses. In *Mechanism of Plant Hormone Signaling under Stress* (Vol. 1, pp. 71–100). https://doi.org/10.1002/9781118889022.ch4 (accessed on 6th November 2025).

Van Zelm, E., Zhang, Y., & Testerink, C. (2020). Salt tolerance mechanisms of plants. *Annual Review of Plant Biology*, *71*, 403–433.

Verma, V., Ravindran, P., & Kumar, P. P. (2016). Plant hormone-mediated regulation of stress responses. *BMC Plant Biology*, *16*, 86. https://doi.org/10.1186/s12870-016-0771-y (accessed on 6th November 2025).

Yamamoto, Y. (2016). Quality control of photosystem II: The mechanisms for avoidance and tolerance of light and heat stresses are closely linked to membrane fluidity of the thylakoids. *Frontiers in Plant Science*, *7*, 1136. https://doi.org/10.3389/fpls.2016.01136 (accessed on 6th November 2025).

Zhang, H., Zhu, H., Pan, Y., Yu, Y., Luan, S., & Li, L. (2014). A DTX/MATE-type transporter facilitates abscisic acid efflux and modulates ABA sensitivity and drought tolerance in *Arabidopsis*. *Molecular Plant*, *7*, 1522–1532.

Zhang, X., Wollenweber, B., Jiang, D., Liu, F., & Zhao, J. (2008). Water deficits and heat shock effects on photosynthesis of a transgenic *Arabidopsis thaliana* constitutively expressing *ABP9*, a bZIP transcription factor. *Journal of Experimental Botany*, *59*, 839–848.

Zhou, W., Chen, F., Meng, Y., Chandrasekaran, U., Luo, X., Yang, W., & Shu, K. (2020). Plant waterlogging/flooding stress responses: From seed germination to maturation. *Plant Physiology and Biochemistry*, *148*, 228–236.

Zou, Y., Zhang, Y., & Testerink, C. (2022). Root dynamic growth strategies in response to salinity. *Plant, Cell & Environment*, *45*, 695–704.

Zwack, P. J., & Rashotte, A. M. (2015). Interactions between cytokinin signaling and abiotic stress responses. *Journal of Experimental Botany*, *66*, 4863–4871.

PART V

Nanomaterials in Agriculture

CHAPTER 14

Biogenic Synthesis of Nanoparticles and Nanomaterials as Plant Biostimulants and Their Role in Sustainable Crop Production

SWAPNAJA GULAWANI,[1] NUSRAT IQBAL,[2] SAURABH DUBEY,[3] and PRATIBHA SRIVASTAVA[1,*]

[1]Bioprospecting Group, Agharkar Research Institute, Pune, Maharashtra, India

[2]Formulation Division, Institute of Pesticide Formulation Technology, Udyog Vihar, Gurugram, Haryana, India

[3]RS Formulation Development and Consultant, Bhiwadi, Rajasthan, India

*Corresponding author

ABSTRACT

Natural resources are treasures of valuable natural substances with tremendous bioprospecting potential utilized by anthropogenic activities. Anthropogenic action generates harmful and beneficial implications for living beings and the environment. Concerning agriculture, the escalating population forced humans to use agrochemicals to fulfil the food demand, resulting in an ample food supply with adverse side effects on health and the environment. Researchers opted for suitable alternatives, such as biofertilizers and nanofertilizers, to overcome these limitations. The supply of substantial amounts of biofertilizers for the mass production of crops is also challenging to fulfil, and metal-based nanomaterials adhere to

Plant Biostimulants and Growth Regulators in Sustainable Agriculture.
Deepak Kumar, Sanjay K. Singh, Nutan Kaushik, and Mahaveer P. Sharma (Eds.)

DOI: 10.1201/9781779641458-19

toxicity issues. In light of these problems, another remarkable field came into the limelight: nanoparticles (NPs) containing biostimulants (BS) for increasing crop production and healthy plant growth and development. In this area, various microorganisms, such as algae, fungi, bacteria, yeast, viruses, and plants, are explored for the biogenic synthesis of NPs as BS in agriculture. These BS can be beneficial for acting on plant pathogens, controlling nutrient uptake, and improving soil quality. The current chapter describes the correlation between nanomaterials and agriculture, the biogenic synthesis of NPs with applications, and using nanomaterials as a stimulant for enhanced crop production.

14.1 INTRODUCTION

Sustainable food production is an elementary need for any country and has become a matter of concern due to the escalating population (Vermeir et al., 2020). Crop production has to undergo a series of processes, from seed sowing to harvesting. During the entire process, it has to go through several environmental conditions, which may be favorable or unfavorable due to natural calamities and anthropogenic activities. A country's economy has a direct relationship with agriculture. Although agrochemicals have enormously affected crop yield in the past several decades, they were encountered due to their adverse effects on human health (Mandal et al., 2022). Simultaneously, intensive agriculture practices, frequent use of agrochemicals, environmental pollution, and disruption of soil biodiversity significantly impact food quality (Tripathi et al., 2020). Therefore, modern agriculture is directed toward eco-friendly and safer methods for the production of crops using biotechnology.

Technology advancement has brought several features that have helped agriculture sectors (Yadav et al., 2023). Biofertilizers have been used for organic farming, providing an environment-friendly alternative to chemical-based agriculture with improved crop yield and soil quality (Mahmud et al., 2021). The intervention of nanotechnology in biofertilizers has been strongly supported in resolving the problems associated with agriculture and the supply of the world's growing demand for food in an eco-friendly manner (Ralya et al., 2018). Nanoparticles (NPs) of metals, semimetals, and nonmetals, carbon nanomaterials (carbon nanotubes), organic materials (such as chitosan, cyclodextrin, and dendrimers), and other nanomaterials (such as nanoclays) having at least one dimension less than 100 nm enhance surface area (Ramsden, 2005). These features generate value addition for them, different from their own bulk material of the same composition in various bioprospecting applications such as energy, pharmaceuticals, medical appliances, and agriculture (Jackson et al., 2017; Mitchell et al., 2020; Prasad et al., 2017; Serrano et al., 2009).

Biological approaches for the synthesis of nano-agriproducts have become an exciting area of research, with broad applications from nanopesticides, nanoinsecticides, nanoherbicides, nanobactericides and nanofungicides, bio-conjugated nanocomplexes, and nanoemulsions emerging therefrom, which are helpful in combating different biotic and abiotic stresses (Giri et al., 2023). The materials are more sustainable and target-oriented than conventional agrochemicals. Plants must be treated with biostimulants (BS) for enhanced growth. The widely defined PBs are organic substances that, when applied to plants or the soil rhizosphere, stimulate the natural process to facilitate nutrient uptake, abiotic stress-bearing potential, and crop quality, regardless of their nutrient content (du Jardin, 2015). According to their nature, they are categorized into two broad groups: (i) non-microbial BS (chitosan, humic, and fulvic acids, protein hydrolysates, phosphites, seaweed extracts, and silicon) and (ii) microbial BS (arbuscular mycorrhizal fungi, plant growth-promoting rhizobacteria, and yeast) (Colla and Rouphael, 2015). Due to their excellent solubility and multiplication capability in the rhizosphere, BS could be conducive to agriculture in combating environmental stress and enhancing crop yields.

14.2 AGRICULTURE AND NANOTECHNOLOGY

Agriculture contributes uniquely to the country's economy and acts as a bedrock by providing food to every nation. There are continuous efforts to develop sustainable agro-formulations to enhance qualitative crop productivity with reduced risk associated with conducting agriculture under varying climate conditions (Dethier et al., 2012; Sekhon et al., 2014). Nanotechnology offers several intriguing applications. Encapsulating beneficial microorganisms, including fungi or bacteria, can increase the accessibility of nitrogen, phosphorus, and potassium in the root area, thereby promoting the health of the plant's roots (Dasgupta et al., 2015; Neme et al., 2021). Nanotechnology not only provides advanced methods in agriculture applications, but it also boosts the country's economy by reducing the foreign exchange involved in the import of fertilizers. As mentioned below, there are several ways to prepare and apply nanofertilizers in agriculture.

- By encapsulating beneficial microbes, such as bacteria and fungi, which can improve plant root health through nutrient uptake, including nitrogen, phosphorous, and potassium in the root area.
- Nanoscale fertilizers (nutrient-loaded nanoparticles).
- Nanoscale additives (addition of nano-sized additives in traditional fertilizers).
- Nanoscale fertilizers (coating of nanoparticles on traditional fertilizers).

Urea, ammonia, various synthetic fertilizers, and plant wastes have all been used to create nanoscale fertilizers. The process of producing a nano-sized N fertilizer includes urea being deposited on calcium cyanamide (Fraceto et al., 2017; Kottegoda et al., 2011). Nanofertilizer action is also influenced by soil rhizosphere and plant growth, as well as the form, content, and concentration of nanoparticles utilized (Fellet et al., 2021; Zelleke et al., 2019). Nanofertilizers, due to their nanosized particles, can be easily absorbed by plants and are also helpful in generating disease- and drought-resistant crops (Wang et al., 2023). Various reactions, including reduction, oxidation, and recombination, take place to deliver the proper nutrients to plants (Elemike et al., 2019). Natural sources, such as microbes and plants, are explored for the biosynthesis of nanoparticles (NPs), which can act as stimulants for plant growth and nutrient uptake (Goel et al., 2023). This chapter will provide a detailed biogenic synthesis of NPs and their applications in agriculture.

14.3 SYNTHESIS OF NANOPARTICLES

An attractive feature of NPs is their size in the 1–100 nm range, which provides astonishing physical, chemical, and biological features due to their surface-to-volume ratio compared to their bulk counterparts. Metals such as silver, titanium, zinc, copper, and gold gained popularity after their conversion into nanomaterials by utilizing several methods (Figure 14.1), such as physical, chemical (sputtering, sol-gel, electrochemical, ball milling, and laser ablation), and biological, with numerous varieties of applications (Amulya et al., 2020; Roy et al., 2021).

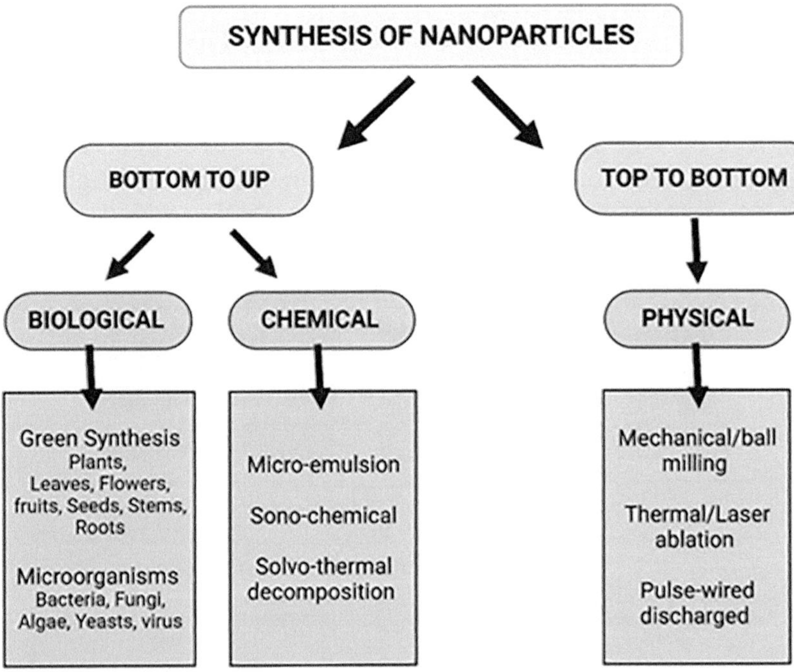

FIGURE 14.1 Representation of various synthetic methods for nanoparticle preparation.

Synthesis of NPs using greener methods is more fascinating than physical and chemical methods because of enhanced advantages, production of ultra-fine NPs, fast synthesis, reduced toxicity, and involvement of less energy (Batool et al., 2021; Jasim et al., 2017; Sarkar et al., 2020;). The process or development of NPs can determine whether they are inorganic, organic, natural, or biogenic. The field of nanoscience refers to the production of NPs utilizing microbes or plants for biogenic or green synthesis.

14.3.1 BIOGENIC SYNTHESIS OF NANOPARTICLES BY MICROBES

The synthesis of NPs using microbes is an eco-friendly procedure. It can generate NPs with large specific surface area, size, multiple shapes, structures, morphology, and enhanced catalytic activity, with improved interactions between the enzymes and metal salts due to the bacterial carrier matrix (Zhang et al., 2011). Many microbes, including bacteria, fungi, actinomycetes, and yeast (Figure 14.2), have been considered biological factories (Das et al., 2021) and have shown potential for synthesizing NPs intracellularly or extracellularly (Fariq et al., 2017).

Haefeli et al. (1984) and Klaus et al. (1999) reported the first synthesis of AgNPs by *Pseudomonas stutzeri* AG 259, a silver mine bacterium that reduces silver ions to small silver atoms, that is, AgNP. The

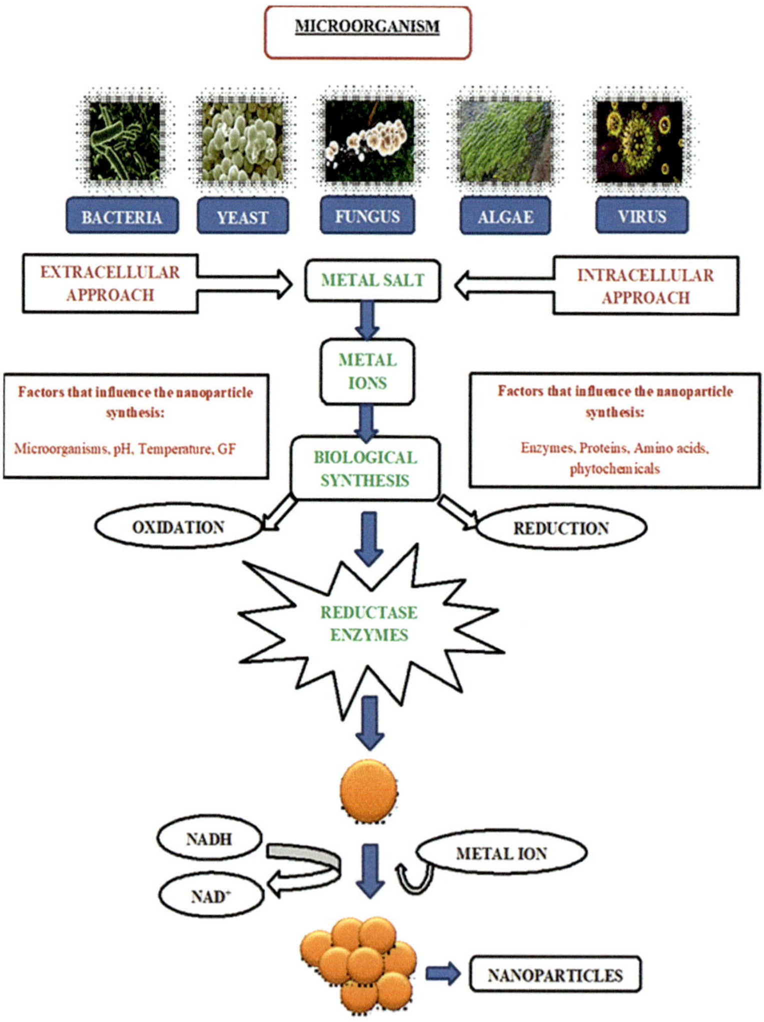

FIGURE 14.2 Diagrammatic representation with the mechanistic approach for the synthesis of nanoparticles using microorganisms.

above report acts as an ignition engine for other researchers to explore several microorganisms for the biosynthesis of AgNPs. Interestingly, microorganisms are considered better targets for synthesizing metal and metal oxide NPs, such as palladium, silver, copper, gold, titanium oxides, and zinc oxides. Nanoscale materials have many forms, such as nanorods, nanowires, nanoconjugates, and nanotubes (Albanese et al., 2012). However, these biogenic synthesized NPs have become more suitable alternatives to resolve the limitations of conventional methods (Singh et al., 2016).

The process of NP synthesis using microorganisms is mediated by a cascade of reactions that occur due to the different biomolecules present in microbes. Therefore, they are independent of any capping or stabilizing agents during synthesis, resulting in more stable NPs with reduced size compared to the conventional technique (Sintubin et al., 2012). However, microbial communities can synthesize metallic NPs through two main mechanisms: intracellular and extracellular (Ghosh et al., 2021). Among these two procedures, extracellular synthesis methods have received extensive support in the field of nanobiotechnology because they are devoid of necessary steps for the intracellular restoration of NPs, such as sonication to break down the microbial cell wall, several centrifugations, and washing steps required to obtain NPs (Jain et al., 2011).

14.3.2 BIOGENIC SYNTHESIS OF NANOPARTICLES USING BACTERIA

A class of non-pathogenic organisms known as plant growth-promoting rhizobacteria (PGPRs) are found mainly in the rhizosphere and are linked to the roots of legume plants. Using several processes, PGPRs can increase crop output because they have the ability to transform nitrogen from the atmosphere into a form that plants can use. *Bacillus*, *Xanthobacter*, *Azospirillum*, *Azomonas*, *Rhizobium*, and *Pseudomonas* fall under the category of nitrogen-fixing bacteria and fix atmospheric nitrogen either symbiotically or non-symbiotically (Nayana et al., 2020). The extreme use of conventional agrochemicals for prolonged duration has generated fatal effects on all natural life forms, which alerted the researchers' attention to PGPRs in the world for improving and sustainable crop productivity (Yadav et al., 2013). Moreover, the eco-friendliness and frugal effectiveness of PGPRs make them a suitable alternative in agricultural areas with enhanced nutrient efficiency to handle crop pathogens. Researchers also explored

PGPRs as biofertilizers, phytostimulants, biocontrol agents (Vasant et al., 2023), and for NP synthesis by intracellular and extracellular processes (Figure 14.3). In the intracellular synthesis process, bacterial cells are involved in a particular ion-transport system, which plays a crucial role in the microbial cell wall during the biosynthesis of metallic nanomaterials. Similarly, in an extracellular mechanism, microbes also secrete various chemical enzymes, such as reductase, for the reduction of metal ions to fine NPs (Hulkoti and Taranath, 2014). In the intracellular process, negatively charged bacterial cell walls attract positive metal ions for the synthesis of NPs due to the presence of electrostatic interactions (Lahiri et al., 2021).

Bacteria have an extraordinary potential for removing heavy metal ions due to their easy management and high growth rates. Because of their unique potential, they are considered nanofactories and one of the most remarkable choices for NP synthesis. However, bacteria need specific parameters, such as oxygen, temperature, and incubation periods, to optimize their growth and management (Ayesha, 2017). Besides, bacteria can also easily biomineralize metal ions and modify the genetic engineering materials within microorganisms (Fernandes et al., 2010). Bacteria can easily cope with stressful conditions such as extracellular accumulation, efflux pumps, reductions of metal ions, and intracellular sequestration due to their natural defense system (Bahrulolum et al., 2021).

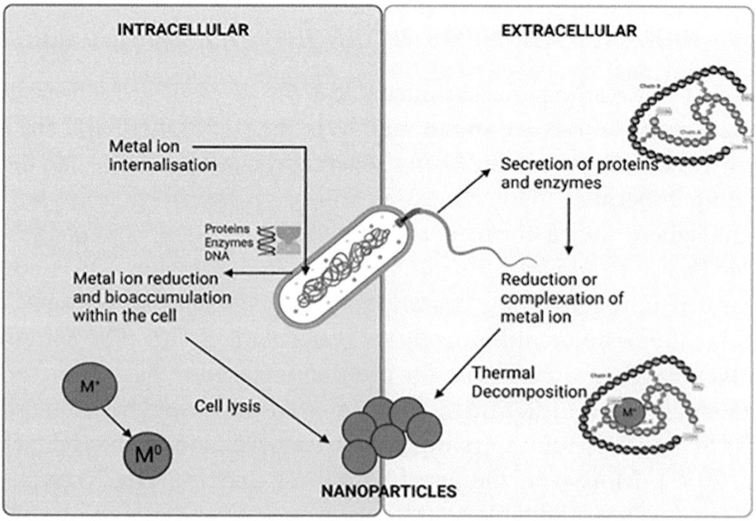

FIGURE 14.3 Representation of the mechanism involved in the biosynthesis of nanoparticles through bacteria.

In a greener approach using microorganisms, scientists use bacterial biomass or cell extracts to prepare several NPs for various applications. Therefore, bacteria can be acclaimed as one of the biofactories for the synthesis of NPs such as copper, silver, magnesium, gold, calcium sulfide, platinum, and cadmium (Langke et al., 2007). Synthesized NPs find extensive application, primarily in the sectors of agriculture and biomedicine. Recently, *Bacillus brevis* was used to synthesize silver nanoparticles (AgNPs), which showed outstanding antimicrobial properties against multidrug-resistant strains of *Salmonella typhi* and *Staphylococcus aureus* (Saravanan et al., 2018). Using an intracellular production method, *Bacillus* sp. and *Pseudomonas stutzeri* both produced silver NPs (Alsamhary, 2020; Rajora et al., 2016). Additionally, it was found that *Pseudomonas aeruginosa* strains were capable of producing gold NPs of varied sizes (Husseiny et al., 2007). Additionally, other bacterial strains have been used to produce bimetallic NPs like Ag-Au (Koyatti et al., 2022). Around the world, significant advancements in PGPR biofertilizer technology have been made. It has been demonstrated that PGPRs are highly potent and useful microorganisms for improving soil fertility and crop productivity (Upadhyaya et al., 2023). Table 14.1 lists the many bacterial species that are used in the synthesis of metal NPs.

14.3.3 BIOGENIC SYNTHESIS OF NANOPARTICLES BY FUNGI

Fungi are one of the critical resources for the biogenic production of metallic NPs because of their easy handling, high tolerance, and capacity to accumulate metal. Fungi are also large producers of extracellular proteins, which produce stable NPs with enhanced bioreduction and generate biocompatible NPs. The leverages of fungi are related to ample biomass production compared to other microorganisms and don't require any further procedures to extract filtrate (Sonawane et al., 2022). Additionally, many fungal species have a rapid growth rate, making it relatively easy to cultivate and maintain them in lab conditions. The easiest-to-find eukaryotic cells in our ecosystem come from fungi, which are the best primary source. Apart from being a food source, fungi also function as preservatives, decomposers, symbiosis partners, and biogeochemical cycling processes in environmental habitats. They are biodiverse and can be easily isolated, cultured, and maintained at a laboratory scale without complex instrumentation. The mycelial mass of fungi can sustain high agitation and pressure; therefore, it

TABLE 14.1 A brief Overview of Some of the Bacterial Species that are Employed in the Production of Metal Nanoparticles and Their Uses

Bacteria	Nanoparticles	Size	Shape	Mechanism	References
Actinobacter	Ag, Au	13.2, 63.8	Spherical	Intracellular and Extracellular	Jain et al. (2022)
Acinetobacter	Au	19	Spherical, triangular-polyhedral	Extracellular	Behrulolum et al. (2012); Wadhwani et al. (2016)
Azospirillum brasilense	Se, Au	25–80, 5–50	Spherical	Extracellular	Kupryashina et al. (2013); Tugarova et al. (2020); Zhang et al. (2023)
Bacillus subtilis	Fe_3O_4	12–34	Spherical	Extracellular	Jubran et al. (2020); Khan et al. (2022)
Bacillus siamensis	Ag	25–50	Spherical	Extracellular	Ibrahim et al. (2019)
Bacillus cereus	FeO	43–159	Hexagonal	Extracellular	Ahmed et al. (2021); García-Ovando et al. (2022)
Cupriavidus sp.	Ag	10–50	Spherical	Extracellular	Ameen et al. (2020)
Gluconacetobacter liquefaciens IFO 12388	Au	10–30	Spherical	Intracellular	Liu et al. (2020)
Klebsiella pneumoniae	Cu	22–44	Spherical	Extracellular	Noman et al. (2021)
Nocardiopsis flavascens	Ag	5–50	Spherical	Extracellular	Ranjani et al. (2018)
Pseudomonas chlororaphis	CdS	6.7 nm	Spherical	Extracellular	Ashengroph et al. (2020)
Pseudomonas fluorescens	Au, Ag	5–50	Spherical	Extracellular	Ahsan (2020); Rajsree and Suman (2012)
Pseudomonas stutzeri	Mg^{2+}	11.9–43.29	Spherical	Intracellular	Sidkey et al. (2020)
Streptomyces griseoplanus	Ag	19.5–20.9	Spherical	Extracellular	Vijayabharathi et al. (2018)
Mycobacterium sp.	Au	5–55	Spherical	Extracellular	Camas et al. (2018); Nadeem et al. (2023)
Serratia sp.	Ag	1–100	Circular, triangular, rod shaped	Extracellular	Akl et al. (2020)
Sinomonas mesophila	Ag	4–50	Spherical	Extracellular	Manikprabhu et al. (2016)
Shewanella loihica	Cu	10–16	Spherical	Extracellular	Lv et al. (2018)

is more suitable for the large-scale generation of NPs (Anjum et al., 2023). By controlling fungal metabolism, it is possible to produce NPs with the appropriate form and size. This can be done by optimizing physicochemical culture parameters, including time, pH, biomass amount, and temperature (Ameen, 2022). According to Hyde et al. (2019), fungi are remarkably adept at breaking down extracellular food. To achieve this, they secrete specific enzymes that hydrolyze complex materials into simpler molecules, which are subsequently taken up and used as a source of energy. Incorporating the function of reducing enzymes, fungi can also create metallic NPs both intracellularly and extracellularly (Figure 14.4).

As an example of the intracellular synthesis of NPs in fungi, when gold NPs are created inside fungi, metal gold ions are linked to the fungal surfaces by electrostatic interactions, generating reversed metal ion charges above the surface of the fungal cell wall (Srivastava and Bhargava, 2022). The enzymes reduce the metal ions and transform them into nano-shapes and, further, into the finest NPs. During extracellular mechanisms, the silver NPs are formed by reducing silver ions. The enzyme nitrate reductase, found in the fungus cytoplasm (Figure 14.4), takes part in the production of fungal NPs by intracellular and extracellular mechanisms (Jain et al., 2011). These operations are carried out with the aid of enzymes.

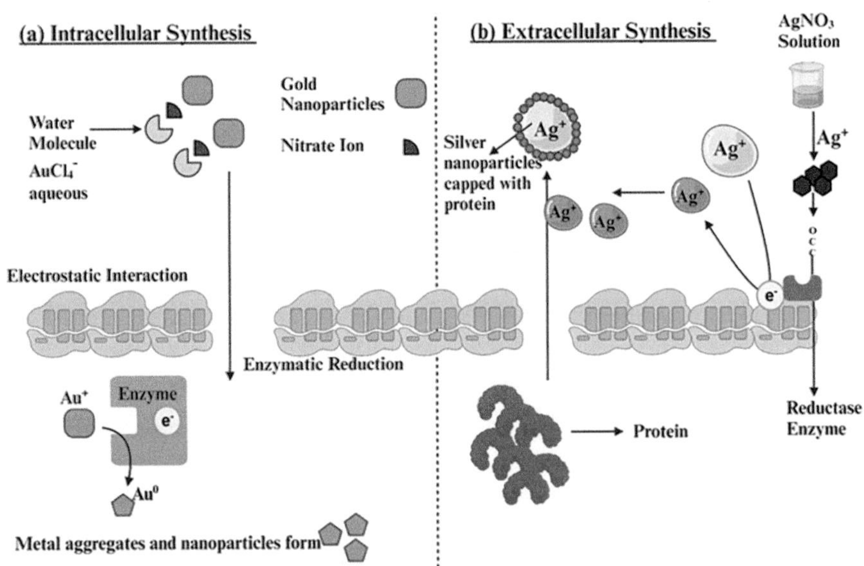

FIGURE 14.4 Representation of intracellular and extracellular synthesis of nanoparticles in fungi.

Over the last two decades, a great deal of fungal species has been studied, including *Aspergillus niger*, *Coriolus versicolor*, *Aspergillus fumigates*, *Colletotrichum* sp., *Fusarium semitectum*, *Trichoderma asperellum*, *Fusarium salani*, *Penicillium brevicompactum*, *Fusarium oxysporum*, and so forth, for the biogenic synthesis of NPs both extracellularly and intracellularly (Ranjani et al., 2020; Singh et al., 2013). Synthesized NPs have a size range of 5–200 nm and spherical to hexagonal shapes. Researchers have focused more on the extracellular manufacturing of NPs using fungal species since the NPs are not attached to any cells. The typical fungus species Ascomycetes, Basidiomycetes, and Phycomycetes have been employed successfully for the green production of NPs. A list of the several fungus species from various phyla that were employed to create biocompatible metal or metal oxide NPs is mentioned in Table 14.2.

14.3.4 BIOGENIC SYNTHESIS OF NANOPARTICLES USING ALGAE

Algae are the simplest photosynthetic organisms in the ecosystem, having ecological and environmental significance, and exist under different environmental conditions such as seawater, freshwater, and on surfaces of wet rocks (Ramanan et al., 2016; Shalaby, 2011). They may be unicellular or multicellular and categorized into six distinct groups, such as microalgae, macroalgae, brown algae, red algae, blue-green algae, and green algae. They have broader bioprospecting potential in fields such as agriculture, medicine, cosmetics, aquaculture, and pharmaceutics. Besides, algae have been utilized in industry for valuable products such as biofuels and natural coloring (Kothari et al., 2017; Maghzian et al., 2023). Different groups of algae are explored for the synthesis of NPs, such as *Phaeophyceae*, *Chlorophyceae*, *Rhodophyceae, Cyanophyceae*, and many other euglenoids and diatoms (Mukherjee et al., 2021; Sharma et al., 2016). Algae can absorb metal ions and reduce the metal ions to NPs; hence, they are considered potential candidates for NP synthesis. Algae are also of prime interest for NP synthesis due to their affordability, ease of handling, low-temperature need, and lesser toxicity (Chaudhary et al., 2020; Sharma et al., 2019). Like other biological methods, algae are one of the best natural commodities; hence, they are considered biofactories for NP synthesis because their biomass is useful for NP biosynthesis in both forms, that is, live biomass and dead dried biomass (Khan et al., 2022).

TABLE 14.2 List of Various Fungi Employed for the Green Synthesis of Nanoparticles

Fungi	Nanoparticle	Size	Shape	Application	References
Aspergillus niger	ZnO	53–69	Spherical	Antimicrobial	Daniel et al. (2023); Kalpana et al. (2018)
Aspergillus sydwoii	Ag	24	cubic	antifungal	Wang et al. (2021)
Beauveria bassiana	Ag	10–50	Triangular, Circular	Nanopesticides	Santos et al. (2021); Sayed and Kim (2018)
Candida glabrata	Ag	2–15	Spherical	Antimicrobial	Jalal et al. (2018)
Cladosporium cladosporioides	Au	60	Round	Antioxidant, antidiabetic, anti-Alzheimer	Balaji et al. (2008); Joshi et al. (2017); Popli et al. (2018)
Fusarium nygamai	Mn^{2+}	11.90	Round	Anti-oxidant, antimicrobial	El-Ansari et al. (2023)
Fusarium oxysporum	Ag	21–37	Spherical	Medicine and agriculture	Karbasian et al. (2008)
Fusarium solani	Ag	5–35	Spherical	Biomedical applications	Ingle et al. (2009); Rai et al. (2021)
Fusarium scripi	Ag	2–20	Spherical	Extracellular	Adebayo (2021)
Fusarium semitectum	Ag	60	Spherical	Biomedical applications	Basavraja et al. (2008)
Macrophanina phaseolina	Cu, Au, Ag/AgCl	5–30	Spherical	Crop protection	Spagnoletti et al. (2019); Sreedharan et al. (2019)
Monascus purpureus	Se	6.5	Spherical	Antibacterial	Sun et al. (2021)
Pleurotus ostreatus	Au	10–30	Spherical	Antimicrobial	Bawadekji et al. (2018); Wang et al. (2017)
Phenerochaete chrysosporium	Ag, Au, Pd	39–40	Spherical-oval	Antibacterial	Huang et al. (2020); Liu et al. (2022); Sanghi et al. (2011); Tarver et al. (2019)
Penicillium potonicum	Ag	10–15	Spherical	Antibacterial	Neethu et al. (2018); Singh et al. (2014)
Trichoderma longibrachiatum	Ag	10	Spherical	Antifungal	Elamawi et al. (2018); Omran et al. (2019)
Trichoderma harzianum	Au	32–44	Spherical	Antifungal	Guilger et al. (2017)

Similarly, the synthesis of NPs from algal species follows an extracellular or intracellular mechanism at a particular concentration, which depends on the types of algae species used (Dahoumane et al., 2014; Gursoy et al., 2021). In intracellular processes, algal metabolites such as NADH or NADPH reduce metallic ions through various synthesis processes such as respiration, nitrogen fixation, and photosynthesis to produce metallic NPs. In extracellular processes, the biomolecules present in algae, such as proteins, polysaccharides, reducing agent sugars, peptides, and pigments, reduce the metal ions as well as catalyze metal ions for the production of NPs (LewisOscar et al., 2016; Shankar, 2016 (Figure 14.5; Table 14.3). However, compared to other biological techniques, algae-mediated NP synthesis is faster.

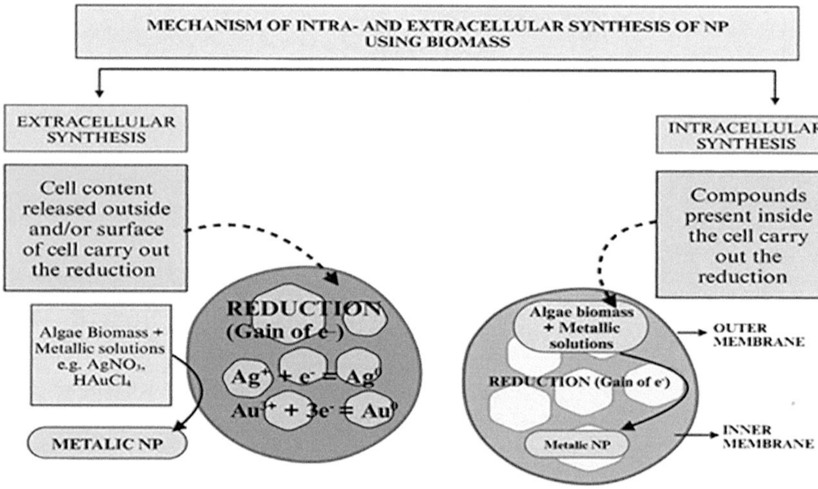

FIGURE 14.5 Schematic mode of action for the biogenic synthesis of nanoparticles using algae.

14.3.5 BIOGENIC SYNTHESIS OF NANOPARTICLES USING YEASTS

One of the unicellular microorganisms, yeasts, is classified as either Ascomycetes, like *Saccharomyces* and *Candida*, or *Basidiomycetes*, like *Filobasidiella* and *Rhodotorula* (Tao et al., 2023). Reproduction in yeast occurs through an asymmetric cell division process known as budding. Modern applications of yeasts include the manufacture of heterologous chemicals, single-cell proteins (SCP), and their employment in the

TABLE 14.3 List of Algal Species Reported for the Biogenic Synthesis of Nanoparticles

Algae	NPs	Size (nm)	Shape	Pathway	Application	References
Bifurcaria bifurcata	CuO	5–45	Spherical	Extracellular	Antimicrobial	Abboud et al. (2014)
Fucus vesiculosus	Au	20–50	Spherical	Extracellular	Antitumor	Mata et al. (2009); Pinto et al. (2020)
Gracilaria edulis	Ag and ZnO	55–99 66–95	Spherical Rod shape	Extracellular	Anticancer	Priyadarshani et al. (2014)
Gelidium corneum	Ag	20–40	Spherical	Extracellular	Antimicrobial	Öztürk et al. (2020)
Desmodesmus sp.	Ag	15–30	Spherical	Intracellular	Dye	Dağlıoğlu and Öztürk (2019)
Kappaphycus alvarezii	Au	12.5–40	Spherical	Extracellular	Antibacterial	Khan et al. (2022); Rajasulochana et al. (2010)
Sargassum incisifolium	Ag, Au	20	Spherical	Extracellular	Antimicrobial	Mmola et al. (2016)
Sargassum wightii	Ag	8–27	Spherical	Extracellular	Antibacterial	Govindaraju et al. (2009)
Sargassum wightii	Au	8–12	Spherical	Extracellular	catalytic	Singaravelu et al. (2007)
Spirulina platensis	Au	20–30	Spherical	Extracellular	Antioxidant and antibacterial	Hamza et al. (2018); Kalabegishvili et al. (2012)
Tetraselmis kochinensis	Au	5–35	Spherical, triangular	Intracellular	biomedical	Senapati et al. (2012)
Turbinia ornata	Ag	8.5	Spherical	Extracellular	Antimicrobial	Anuluxan et al. (2022)
Turbinaria conaides	Au, Ag	5–20	Spherical	Extracellular	Antimicrofouling	Rajeshkumar et al. (2013); Vijayan et al., (2014)

biofuels industry, in addition to their traditional use in the creation of a variety of fermented foods like alcoholic beverages and bakery goods (Arevalo-Villena et al., 2017; Segal-Kischinevzky et al., 2020). Recently, yeast has also received significant interest in a wide range of industrial uses because of its swift growth and development. Additionally, yeast is also considered for the greener synthesis of NPs, which is boosting in industries because they are prime ingredients and already have applications in producing cultivated crops such as bread, wine, and beer (Zheng et al., 2021). Yeast has several valuable properties, including being simple to manage in lab settings, growing more quickly in harsh environments by utilizing nutrients, temperature, and pH, and being highly cost-effective for manufacturing NPs. Consequently, due to their distinctive qualities, yeast is viewed as a more relevant source of microorganisms for NP synthesis (Skalickova et al., 2017).

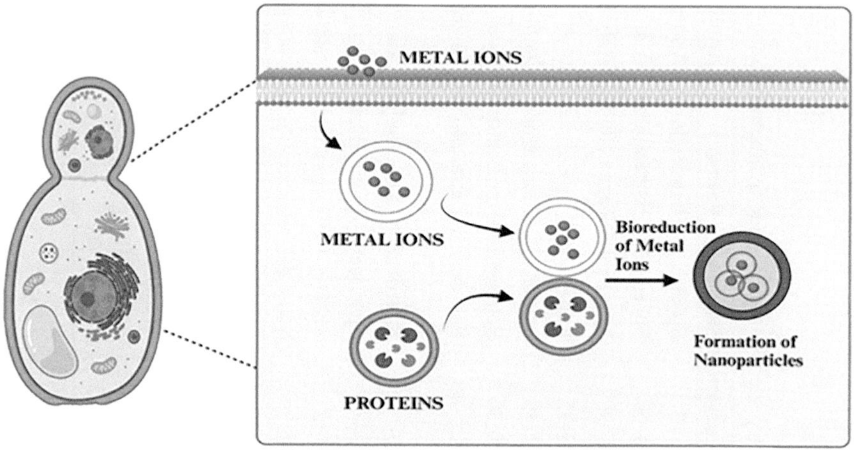

FIGURE 14.6 Diagrammatic representation of the eco-friendly synthesis of nanoparticles using yeast.

Yeasts are essential in agriculture as biological control agents (BCAs), cures, and environmental quality indicators. The adaptation ability of yeast in difficult ecological circumstances, such as a wide range of temperatures and pH, and highly concentrated organic and inorganic contaminants, grows readily on economical media. Yeasts can naturally absorb enormous amounts of potentially dangerous metal ions from the environment and store them in large quantities. Metals can be mobilized, immobilized, or transformed, among other detoxifying techniques, in response to this environmental stress. These bioremediation processes of yeasts can play key roles in the biogenic synthesis of NPs (Tarekegn et al., 2020). The stress caused by the presence of metal ions stimulates a metabolic series of chemical reactions to generate stress-reducing molecules such as phytochelatin synthase and glutathione that act as reducing agents. These compounds interact with metal ions such as cadmium, zinc, silver, selenium, gold, nickel, copper, and so forth, and reduce them to the respective NPs. Add-on mechanisms involved in this process include the activity of membrane-bound oxidoreductases and quinones (Mendoza-Cozati et al., 2005). Metal ion adsorption causes a rise in pH and the immediate activation of pH-sensitive oxidoreductases, which serve as reducing and stabilizing factors for the production of NP. The syntheses of NPs can be either intracellularly or extracellularly, depending on the type of yeast species.

However, the mechanism behind the synthesis of NPs by yeast is ambiguous to date. In this context, Salvadori et al. (2014) elucidated a possible mechanism for the synthesis of NPs. The extracellular processes of synthesis demonstrated the interaction of metal cations and the amide groups present in the yeast cell wall. Hence, bioreduction is possible due to the different extracellular enzymes present in the yeast cell wall. Additionally, in the intracellular process, the amide groups and cations in the yeast cell wall interacted electrostatically to produce NPs through the process of bio-reduction, in which enzymes reduce the ions, collect the metal ions that result, and simultaneously produce ultra-finely synthesized NPs from the yeast cells (Figure 14.6). Hence, yeast is also entrusted as a biological factory for the production of NPs (Salvadori et al., 2017). Table 14.4 summarizes the synthesis of NPs of different shapes and sizes using yeast as a biological factory.

For the production of AgNPs, a variety of yeast species have been employed, including *Saccharomyces cerevisiae*, *Saccharomyces boulardii* (Korbekandi et al., 2014), *Candida utilis* NCIM 3469 (Waghmare, 2015), silver-tolerant yeast strain MKY3 (Kowshik et al., 2002), and a marine yeast strain called *Yarrowia lipolytica*. Elahian et al. (2002) explored a genetically modified strain of *Pichia pastoris* for AgNP biosynthesis. It has been demonstrated that the yeast *Pichia jadinii* (formerly *Candida utilis*), which was isolated from a metal-rich waste, produces AuNPs from the metal (Mariekie and Anthony, 2006). Agnihotri et al. (2009) also

TABLE 14.4 List of Yeast Species Reported for the Biogenic Synthesis of Nanoparticles

Yeast	NPs	Size (nm)	Shape	Biosynthetic Pathway	Application	References
Magnusiomyces ingens LHF1	Se	70–90	Spherical and quasispheroidal	Intracellular	bioremediation	Lian et al. (2019)
Pichia jadinii	Au	100	Spherical	Intracellular	–	Gericke and Punches (2006)
Candida glabrata	CdS, Se	2	Spherical	Intracellular	Antimicrobial	Bafghi et al. (2022)
Rhodotorula mucilaginosa	Cu	10.5	Spherical	Intracellular	Antifungal	Cunha et al. (2018)
Saccharomyces cerevisiae	Ag	16.07	Oval	Extracellular	Antibacterial	Olobayotan et al. (2019)
Schizosaccharomyces pombe	Cds	1–1.5	Hexagonal	Intracellular	Diode	Kowshik et al. (2002)
Torulopsis sp.	PbS	2–5	Spherical	Intracellular	Antibacterial	Kowshik et al. (2002)
Yeast strain MKY3	Ag	2–5	Hexagonal	Extracellular	Antimicrobial	Kowshik et al. (2003)
Yarrowia lipolytica	Au	2–5	Spherical	Extracellular	Antibiofilm	Pimprikar et al. (2002)

reported a description of the green synthesis of AuNPs using the tropical yeast *Yarrowia lipolytica*. Additionally, it has been shown that acid mine drainage-identified extremophilic yeasts can generate AuNPs and AgNPs (Mourato et al., 2011). Other NPs, including copper nanoparticles (CuNPs) and palladium nanoparticles (PdNPs), have also been reported to be biosynthesized using *Saccharomyces cerevisiae* (do Nascimento et al., 2018). It has been reported that *Cryptococcus laurentii* and *Rhodotorula glutinis*, two epiphytic yeasts isolated from apple peels, have the potential to be used as an effective nanofungicide against phytopathogenic fungi that cause postharvest diseases in pome fruits. Fernandez et al. (2016) demonstrated the antifungal activity of AgNPs synthesized using these two epiphytic yeasts. The synthesis of NPs utilizing these two epiphytic yeasts has substantial benefits in agroecosystems since they are generally recognized as safe (GRAS) microorganisms and are considered to be innocuous (Sukhwal, 2016).

14.3.6 BIOSYNTHESIS OF NANOPARTICLES USING VIRUSES

Viruses are basically considered infectious pathogens that are ubiquitous and spread very quickly in nature. Viral diseases have been under study for more than 100 years, beginning with the identification of the plant pathogen *tobacco mosaic virus* (Zaitlin, 1998), and more than 5000 viruses have been observed till today (Scholthof et al., 2011). More recently, scenarios have been modified, and scientists from various disciplines have directed viruses as useful tools for numerous beneficial applications (Pokorski and Steinmetz, 2011). In the 1950s, bacteriophages were used as cloning vectors and expression platforms and became the root cause of bacteriophage therapy (Clark and March, 2006; Marks and Sharp, 2000). In the beginning of the 1970s, efforts were made to make virus-like particles (VLPs) as vaccines against viral diseases (Garcea and Gissmann, 2004; Grgacic and Anderson, 2006; Ludwig and Wagner, 2007). In another exciting way, viruses are used as scaffolds to fabricate the green synthesis of metalized NPs and nanotube structures using mineralization techniques (Lee et al., 2020; Shenton et al., 1999). Attention has focused on these techniques recently, leading to the development of mineralized nanotubes and nanowires and their applications in batteries and data storage devices (Lee et al., 2009).

Despite being infectious, they are also considered potential candidates for NP synthesis due to their uniformity in size and geometry, as well as their characteristic surface properties (Lee et al., 2009). Like other microorganisms, when viruses are exposed to metal ions, they contribute to the formation of nanomaterials on the external and internal surfaces through different interactions with the amino acids of their viral cells (Chen et al., 2018). Multiple groups of researchers have proposed that viruses do not constitute pathogens exclusively, but they are also used as biofactories for the production of NPs for various industrial applications (Blum et al., 2005; Carballal et al., 2023; Singh et al., 2006). Thus, as Table 14.5 illustrates, viruses clearly have the ability to synthesize a wide variety of NPs with varying sizes and shapes.

14.3.7 BIOGENIC SYNTHESIS OF NANOMATERIALS USING PLANTS

In continuation with other natural organisms, plants can also synthesize organic acids through root exudation. During excretion, several useful plants can synthesize organic acids through root exudation, just like other natural organisms, such as citrate, oxalate, and lactate, dissolve mineral oxides to release minerals that are excreted by the roots of young plants, which can attract beneficial microorganisms that colonize the roots of the plants (Almeida et al., 2020; Seisenbaeva et al., 2013). According to a study by Groenke et al. (2012) utilizing lactate, TiO_2 NPs exhibit a great affinity for the cell wall components of both plant and microbial cells. This information is useful for understanding the biological plant-microorganism interaction. For instance, a plant extract from the roots, leaves, flowers, seeds, or fruits can be used in place of the reducing agent employed in conventional chemical procedures. These extracts can either be aqueous or organic; aqueous extracts have been employed more frequently in the synthesis of NPs than organic extracts, which are typically generated using ethanol as a solvent. It is important to understand that biological processes have the potential to create NPs of all sorts, sizes, and shapes. These characteristics could serve as the foundation for a number of an NP's antimicrobial activity-related modes of action. The utilization of plant extracts is a cost-effective, non-toxic, and environmentally friendly synthesis technique, similar to other biological techniques (Figure 14.7).

TABLE 14.5 List of Viruses Reporting the Biogenic Synthesis of Nanoparticles

Virus	NPs	Size (nm)	Shape	Biosynthetic Pathway	Application	References
M13 bacteriophage	ZnS, CdS	3–5, 20	Hexagonal Cubical	Extracellular	Fluorescent, agriculture imaging	Huang et al. (2011); Lai et al. (2021); Mao et al. (2003)
Tobacco mosaic virus (TMV)	CdS, PbS, Si, Fe_2O_3	5–50, 30, 3, 2–22	Tubular Prismatic Tubular rod	Intracellular and Extracellular	Anticancer	Bruckman et al. (2016); Love et al. (2014); Shenton et al. (1999)
Bacteriophage	Au	20–100	Rectangular, rhomboidal, hexagonal, triangular, and sphere shapes	Extracellular	Anti-biofilm	Ahiwale et al. (2017); Paczesny and Bielec (2020)
CPMV (Cowpea mosaic virus)	Fe, Ni, Co, Pt, NiFe and CoPt	35	No shape mentioned	Intracellular and Extracellular		Aljabali et al. (2010); Balke and Zeltins (2019)

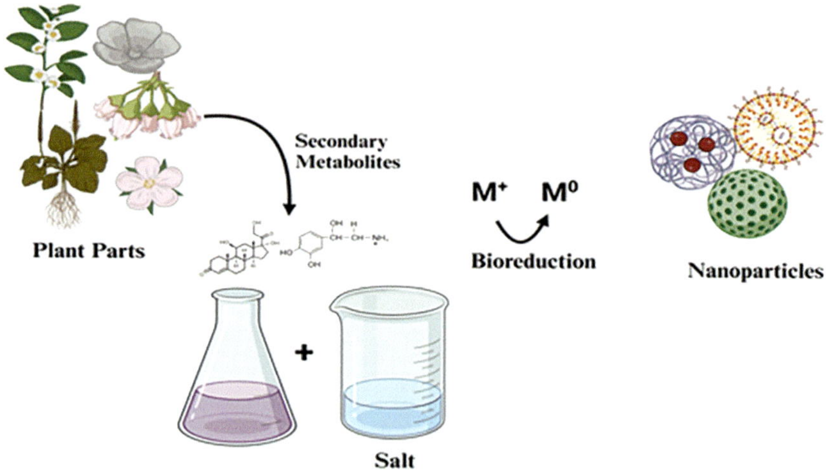

FIGURE 14.7 Green synthesis of nanoparticles using plants as bioresources.

The biodiversity of plants has tremendously increased, leading researchers to explore several plants for eco-friendly synthesis of NPs. Some of the plants are listed in Table 14.6. *Rosa sp.*, *Achillea wilhelmsii, Piper sarmentosum, Mentha piperita, Ipomoea digitata, Pelargonium graveolens, Azadirachta indica, Cymbopogon flexuosus, Tamarindus indica, Aloe vera, Coriandrum sativum, Cinnamomum camphora, Capsicum annuum, Gliricidia sepium, Datura metel, Jatropha curcas, Paullinia pinnata, Zingiber officinale, Terminalia chebula, Camellia sinensis, Gardenia jasminoides, Moringa oleifera,* and *Calendula officinalis*. The NPs generated from plant extracts and ionic compounds, such as Ag, Au, Cu, Se, Zn, Ti, and Fe, have shown excellent antimicrobial properties against several phytopathogens (Rai and Bai, 2011). Thus, these NPs possess the potential to be used in agriculture (Table 14.6).

14.4 PLANT PROBIOTICS AS BIOSTIMULANTS

BS, such as plant probiotics (PP), can be considered a distinct group of microbial strains that could act as biofertilizers that influence the growth of plants via direct and indirect mechanisms (Rai et al., 2023). The endophytomicrobiome and the rhizomicrobiome, two distinct hubs of the microbial community, are found in both the inner regions

TABLE 14.6 List of the Plants Used for the Biogenesis of Nanoparticles

Plant Name	Part	NPs	Size (nm)	Shape	Application	References
Achillea wilhelmsii	leaves	Ag	38	Spherical	Antibacterial	Asariha et al. (2020)
Annona muricata	leaves	Au	25	Spherical	Antifungal	Folorunso et al. (2019)
Aloe vera	leaves	Se	50	Spherical	Antifungal	Fardsadegh and Jafarizadeh-Malmiri (2019)
Acalypha indica	leaves	Ag	50	Spherical	Antifungal	Krishnarag et al. (2012)
Azadirachta indica	leaves	Ag,	60	Spherical	Antimicrobial	Roy et al. (2019)
Camellia sinensis	leaves	Fe, Cu, Ag	42–60, 26–40, 10–20	Spherical	Antifungal	Ali et al. (2022); Asghar et al. (2018)
Calendula officinalis	leaves	Ag	35.7	Spherical	Antimicrobial	Balciunaitiene et al. (2022)
Capsicum annuum	leaves	Ag, Au	16	Spherical	Antimicrobial	Lomelí-Rosales et al. (2022)
Carica papaya	leaves	CuO	80	Spherical	Antimicrobial	Arsene et al. (2023); Chen et al. (2019)
Cinnamomum camphora	leaves	Ag	19.57	Spherical or ellipsoidal	Antimicrobial	Li et al. (2021)
Coriandrum sativum	leaves	Ag	12	Spherical	Antimicrobial	Ashraf et al. (2019)
Cymbopogon citratus	leaves	Ag	47	Spherical	Antimicrobial	Rakin-Uz-Zaman et al. (2022)
Datura metel	flower	Ag	25–70	Spherical	Antibacterial	Fatimah et al. (2023)
Eucalyptus robusta	leaves	Fe_2O_3	8	Spherical	Antimicrobial	Vitta et al. (2020)
Gliricidia sepium	leaves	Ag	10–50	Spherical	Antimicrobial	Raut et al. (2009)
Gloriosa superba	leaves	CeO_2	5	Spherical	Antimicrobial	Arumugam et al. (2015)
Ipomoea digitata	flower	Ag	20	Face-centered cubic	Antimicrobial	Varadavenkatesan et al. (2018)
Jatropha curcas	leaves	Ag	43.67	Spherical	Antimicrobial	Chauhan et al. (2016)
Mentha piperita	leaves	Ag	–	–	Antimicrobial	Harsha et al. (2022)

TABLE 14.6 *(Continued)*

Plant Name	Part	NPs	Size (nm)	Shape	Application	References
Moringa oleifera	leaves	Ag	38	Spherical	Antimicrobial	Mohammed and Hawar (2022)
Ocimum sanctum	leaves	Cu, Ag	25, 55	Cylindrical, Elliptical, Spherical	Antifungal	Gautam et al. (2023); Shende et al. (2016)
Psidium guajava L	leaves	TiO_2	32.58	Spherical	Antimicrobial	Santoshkumar et al. (2014)
Piper nigrum	leaves	Ag	7–50	Spherical	Antimicrobial	Paulkumar et al. (2014)
Punica granatum L	leaves	ZnO	50–100	Spherical	Antifungal	Mishra and Sharma (2015)
Solanum tuberosum	leaves	Ag	39–80	Spherical	Antifungal	Almadiy and Nenaah (2018)
Solanum lycopersicum	leaves	Ag	35–50	Spherical	Antimicrobial	Santiago et al. (2019)
Trigonella foenum-graecum	leaves	TiO_2	20–90	Spherical	Antimicrobial	Subhapriya and Gomathipriya (2018)
Withania somnifera	leaves	Se	45–90	Spherical	Antimicrobial	Alagesan and Venugopal (2019)

of plant tissues and the rhizosphere (Jiang et al., 2022; Pandey et al., 2022). These microbiomes have a variety of characteristics, including the solubility of nutrients (Khan et al., 2022), fixing nitrogen (Abdelkhalek et al., 2022), ammonia (Upadhyay et al., 2022), PH (i.e., IAA) synthesis (Nazli et al., 2020), and killing of plant pathogens (Mathur et al., 2019; Reddy et al., 2022). Due to these features, PPs possess plant growth-stimulating capabilities. Recent developments in science and technology have enlightened the role of nanomaterials in the agricultural field (Neme et al., 2021). Nanomaterials are also reported to improve nutrient uptake potential in plants and increase crop yield (Guleria et al., 2022). Therefore, the combination of BS with compatible NPs has become a promising area of research in the agriculture sector for providing eco-friendly agriculture. This method of effective fertilization can replace conventional fertilization since it is more effective at utilizing resources, releases nutrients sustainably and under control, increases crop output with a moderate amount of fertilizer, and has minimal negative effects on the soil. Additionally, the utilization of NMs and PPs is economically viable and can contribute to environmental sustainability. The "concoction" of NMs and PPs, according to the literature, can be regarded as a "nanobiofertilizer (NBF)" since it has the advantages of both substances, contributes to the delayed and targeted release of nutrients, increases nutrient use effectiveness, and causes a sizable rise in crop output (Kumari and Singh, 2020).

Due to its wide range of plant growth-stimulating characteristics, including the solubilization of nutrients, nitrogen fixation, production of siderophores, EPS, PHs, and anti-pathogenic compounds, the microbial component of this cocktail benefits the plant system. The effective microbial component leads to improvements in soil fertility, functional enzymatic activity, soil microbial biomass, organic carbon content, and NPK content. On the other hand, the second and most powerful segment, "NMs," maximizes the benefits and supports plant growth through the efficient and regulated release of nutrients, control over the soil's ability to fix nutrients, an enhancement in the bioavailability of nutrients to plants, a greater tolerance of plants to environmental stress, and protection of plants from pests. Plants can be treated with nanomaterials and plant probiotics in a variety of ways, including soil application, foliar spray, seed treatment, and seedling therapy. According to the experiments chosen, advances in nanotechnology have also enabled the encapsulation of plant probiotic strains in suitable nanomaterials (Akhtar et al., 2022; Panichikkal et al., 2019, 2021) or the inclusion complex of both NMs and PPs in the appropriate carrier (Moradi et al., 2022). This idea maintains the microbiological component's effectiveness and shelf life, as well as the steady and controlled supply of both NMs and PPs. Through the use of NPs, this two-pronged approach immediately boosts nutrient availability while also promoting plant development through efficient microbial treatment. The agricultural industry might be significantly changed by the application of such an efficient dose combination of NMs and PPs, which would eventually lead to benefits for sustainable agricultural productivity and food consumption (Agri et al., 2022; Kumari et al., 2021). When NM and PP are used together, as an efficient but two-way strategy, they have a positive influence on food crops in several ways, including enhanced crop production, a decrease in the negative effects of environmental stress, an increase in soil fertility, and a strengthening of the bioeconomy.

14.5 SIGNIFICANT ASPECTS OF BIOGENIC NANOMATERIALS FOR PLANT GROWTH AND DEVELOPMENT

Several mechanisms have been investigated to understand the role of NMs on plant growth and development under different environmental conditions (Upadhyay et al., 2023). During multiple high-throughput experiments, the effects of NMs on particular gene expression patterns—whether downregulated or upregulated—during plant production and growth, including photosynthesis, seed germination, and biotic and abiotic stress adaptation, were taken into consideration. Because ZnO-NPs were applied, many photosynthesis-related genes (such as "CHLI," "LHCa/b," and "RSSU") were expressed (Mardi et al., 2022). The impact of silica NPs was envisaged during wheat cultivation under field conditions with upregulation of several genes related to sugar metabolism (α-glucosidase, SUS, SPC), PHs (TIR1 for IAA; PP2C; SnRK2 and ABF for abscisic acid, PYR/PYL), and chlorophyll metabolism ("POR"; "CHLH," and "CAO") (Li et al., 2023). By activating the flowering induction gene SFT, foliar application of Mn ferrite NMs caused early blooming in tomatoes. Similar research was observed with the overexpression of genes involved in the production of gibberellin (GA20ox3, SIGAST; and GA20ox2) (Yue et al., 2022). Ferredoxin-NADP reductase (PetH) was majorly expressed when CeO_2 NPs were applied; however, there was a downregulation of photosystem II lipoprotein (Psb27) under different concentrations of CeO_2 NPs (Hong et al., 2023). In *Brassica chinensis* L., these types of genes are engaged in the photosynthetic process. The regulation and expression of abiotic stress-responsive genes is one of the most crucial strategies for plant survival during stressful conditions (Sahil et al., 2021). In *Solanum lycopersicum*, chitosan NPs increased the expression of genes HsfA1a, SlAREB1, LeNCED, and LePIP1, which are responsive to drought (Mohamed and Abdel-Hakeem, 2023). The application of ZnO-NPs to wheat increased the expression of genes associated with drought resistance, including DREB2, CAT1, and P5CS, which are responsible for proline biopreparation, dehydration-responsive element-binding proteins and catalase activity, respectively (Raeisi et al., 2022). Recently, Subotic et al. (2022) found that exposure of tomatoes to a nanomaterial called hyper-harmonized hydroxyl-modified fullerene (3HFWC) increased the aquaporin gene expression (PIP1;3, PIP2;4, and PIP1;5) involved in the transportation of water and solutes across the plant membrane. Alkaloids are one of the nitrogen-containing phytochemicals that have a justifiable role in plants under stress and dry conditions. Therefore, their biosynthesis is urged during stressful conditions (Amirifar et al., 2022). Additional nanomaterials can increase the rate of phytochemical production in plants. According to Ali et al. (2021), the expression of key manufacturing genes for the biosynthesis of alkaloids, including STS (strictosidine synthase), GS (geissoschizine synthase), PRX1 (peroxidase 1), and DAT (deacetylvindoline-4-O-acetyltransferase), was upregulated during drought stress conditions in *Catharanthus roseus* on application of chitosan NMs. Rahmatizadeh et al. (2021) demonstrated the impact of nano-SiO_2 as a potential mediator, inducing the expression of "LeNRAMP3" and "LeFER" in tomato plants. The overexpression of these genes may improve the nutritional status of tomato plants under Cd stress, proving that the LeFER transporter is crucial in depreciating the negative effects of Cd stress. Additionally, Table 14.7 shows how multiple genes linked to different functions were subsequently upregulated in response to NMs.

TABLE 14.7 List of the Nanomaterials That are Useful for Plant Growth and Development

Plant Name	Nanoparticle	Agricultural Effect	References
Tomato (*Solanum lycopersicum* L.)	Se	Increased crop production with better tomato fruit quality; Enhanced solubility of solid substance; Upregulation of antioxidant enzymes such as POX, CAT, and PPO under salinity stress	Saffan et al. (2022)
Maize	Cu	High grain yield and supportive in plant growth	Van Nguyen et al. (2022)
Maize	Zn	Increased crop yield and enriched extract of maize cultivation in Zn-scarce soils	Azam et al. (2022)
Wheat (*Triticum aestivum* L.)	Cu	DNA damage and DNA Methylation in NaCl stress condition	Hosseinpour et al. (2022)
Potato (*Solanum tuberosum*)	Nanochitosan	Preventing *Ralstonia solanacearum*-induced bacterial wilt	Khairy et al. (2022)
Vigna radiata L.	Nanourea	Decrease in nitrate (NO_3-N) leaching remarkable increase in the protein quantity; phenolic content and ROS scavenging activity;	Sharma et al. (2022)
		Qualitative morphological growth as well biomass of crop	
Spinach	Nanogypsum	Depreciation of salinity-sodality effects and enhanced growth of spinach in sodic-saline soil	Salama et al. (2022)
Fenugreek	Nanophosphorus	Better adaptation to irrigation stress condition,	Abou-Sreea et al. (2022)
		Upsurged plant growth and production by increasing water efficient intake, osmoregulatory molecules (specifically, soluble sugars and proline) with upregulation of antioxidant enzymes	

14.6 FUTURE DIRECTIONS

Science and technology always provide solutions for emerging problems and continuously grow in sustainable directions. The advantage of nanotechnology with BS can be intensified in the future with an understanding of several essential aspects mentioned below:

1. **Tailored Nanoparticle Design:** Future research will likely focus on designing NPs with specific characteristics and functionalities tailored to support the unique needs of different crops and agricultural conditions. This could be possible with the precise engineering of nanomaterials for improved nutrient delivery, stress resilience, and disease management.
2. **Targeted Delivery Systems:** Scientists are likely to explore highly precise and targeted delivery systems for biostimulant NPs. It could involve the use of smart nanocarriers that release BS in response to plant needs or environmental conditions.
3. **Integration with Precision Agriculture:** The future of BS lies in their integration with precision agriculture technologies. It could involve using data-driven approaches to optimize biostimulant timing, dosage, and application for maximum crop yield and quality.
4. **Sustainability and Eco-Friendly Nanomaterials:** Researchers will prioritize the development of sustainable, eco-friendly nanomaterials. These materials should be biodegradable, minimize environmental impacts, and align with organic farming practices, ensuring long-term agricultural sustainability.
5. **Synergistic Approaches:** Investigating the synergistic effects of combining nanomaterial-based BS with biofertilizers will be a focus. This integrated approach can provide comprehensive solutions for nutrient management and plant health.
6. **Microbial-Based BS:** BS derived from beneficial microorganisms, such as mycorrhizal fungi or PGPB, may gain more attention. These microbe-based BS can enhance nutrient uptake and overall plant health.
7. **Customization for Crop Varieties:** Tailoring nanomaterial-based BS to specific crop varieties and their unique requirements is a promising avenue. This approach can maximize the benefits of BS in various agricultural contexts.
8. **Regulatory Frameworks:** With the increasing use of nanomaterials, regulatory frameworks are likely to evolve to ensure product safety, efficacy, and proper labeling. This will provide guidance to farmers and manufacturers.

9. **Nanoparticle Characterization:** Advancements in characterizing biostimulant NPs are essential to understanding their behavior, interactions, and long-term effects in diverse agroecosystems.
10. **Demonstrating Economic Viability:** Future research may focus on demonstrating the economic viability of biostimulant use. Cost-benefit analyses and case studies can provide insights into the practicality of incorporating BS into farming practices.
11. **Education and Outreach:** Increasing awareness and providing education to farmers about the benefits and proper use of nano-material-based BS will be crucial. Extension services and farmer training programs can play a pivotal role.
12. **Global Collaboration:** Given that agriculture is a global concern, international collaboration and knowledge sharing in biostimulant research can accelerate progress and address global food security challenges.

In summary, the future scope of BS in agriculture involves cutting-edge research in nanotechnology, biogenic synthesis, and precision agriculture (Figure 14.8), with pesticide-free crops. By harnessing the potential of BS, we can improve crop production, enhance plant health, and contribute to sustainable and environmentally responsible agricultural practices.

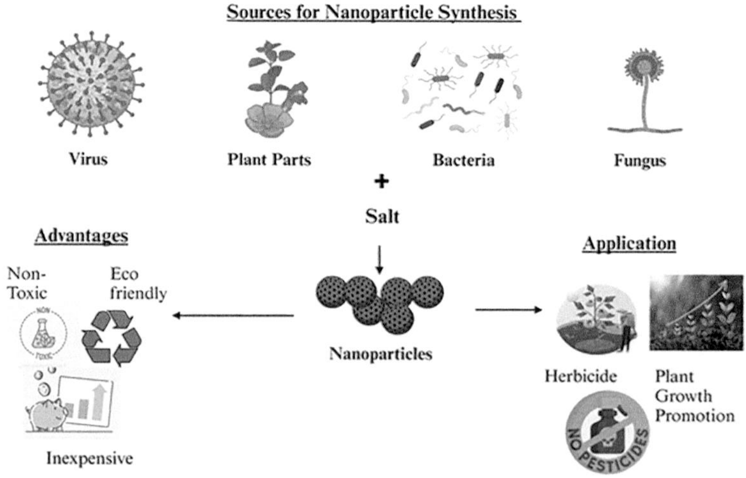

FIGURE 14.8 Diagrammatic representation showing the advantages of biogenic nanoparticles in agriculture.

14.7 CONCLUSION

The synthesis of NPs and nanomaterials through biogenic means, particularly using plants, holds immense promise in the realm of sustainable agriculture. These tiny structures, often derived from natural sources, exhibit remarkable properties that can revolutionize crop production while minimizing environmental impact. Moreover, NPs synthesized via biogenic routes exhibit unique characteristics such as enhanced stability, biocompatibility, and tailored functionality. As PBs, nanomaterials can improve nutrient uptake, water retention, and stress tolerance in crops, thereby enhancing yield, quality, and resilience in the face of changing environmental conditions. Furthermore, the use of biogenic NPs aligns with the principles of sustainable agriculture by promoting resource efficiency, reducing chemical inputs, and minimizing environmental pollution. By harnessing the natural processes of plants, this approach offers a sustainable solution to the challenges facing modern agriculture, including food security, climate change, and soil degradation. In conclusion, biogenic synthesis of NPs and nanomaterials holds great promise as PBs for sustainable crop production. By harnessing the natural processes of plants, these nanostructures offer innovative solutions to enhance agricultural productivity while minimizing environmental impact. Moving forward, concerted efforts are required to overcome technical, regulatory, and ethical challenges to realize the full potential of nanotechnology in agriculture and pave the way towards a more sustainable future.

KEYWORDS

- **nanoparticles**
- **biostimulants**
- **biogenic synthesis**
- **single-cell proteins**
- **nanobiofertilizer**

REFERENCES

Abboud, Y., Saffaj, T., Chagraoui, A., Abboud, Y., Saffaj, T., Chagraoui, A., El Bouari, A., Brouzi, K., Tanane, O., & Ihssane, B. (2014). Biosynthesis, characterization and antimicrobial activity of copper oxide nanoparticles (CONPs) produced using brown alga extract (*Bifurcaria bifurcata*). *Applied Nanoscience*, *4*, 571–576.

Abdelkhalek, A., El-Gendi, H., Al-Askar, A. A., Maresca, V., Moawad, H., Elsharkawy, M. M., Younes, H. A., & Behiry, S. I. (2022). Enhancing systemic resistance in faba bean (*Vicia faba* L.) to bean yellow mosaic virus via soil application and foliar spray of nitrogen-fixing *Rhizobium leguminosarum* bv. *viciae* strain 33504-Alex1. *Frontiers in Plant Science*, *13*, 933498. https://doi.org/10.3389/fpls.2022.933498 (accessed on 6th November 2025).

Abou-Sreea, A. I. B., Kamal, M., El Sowfy, D. M., Rady, M. M., Mohamed, G. F., Al-Dhumri, S. A., Al-Harbi, M. S., & Abdou, N. M. (2022). Small-sized nanophosphors has a positive impact on the performance of fenugreek plants under soil-water deficit stress: A case study under field conditions. *Biology*, *11*(1), 115. https://doi.org/10.3390/biology11010115 (accessed on 6th November 2025).

Adebayo, E. A., Azeez, M. A., Alao, M. B., Oke, A., & Aina, D. A. (2021). Fungi as veritable tool in current advances in nanobiotechnology. *Helion*, *7*(11), e08480. https://doi.org/10.1016/j.heliyon.2021.e08480 (accessed on 6th November 2025).

Agnihotri, M., Joshi, S., Kumar, A. R., Zinjarde, S., & Kulkarni, S. (2009). Biosynthesis of gold nanoparticles by the tropical marine yeast *Yarrowia lipolytica* NCIM 3589. *Materials Letters*, *63*, 1231–1234.

Agri, U., Chaudhary, P., Sharma, A., & Kukreti, B. (2022). Physiological response of maize plants and its rhizosphere microbiome under the influence of potential bioinoculants and nano chitosan. *Plant and Soil*, *474*, 451–468. https://doi.org/10.1007/s12088-017-0640-x (accessed on 6th November 2025).

Ahiwale, S. S., Bankar, A. V., Tagunde, S., et al. (2017). A bacteriophage mediated gold nanoparticles synthesis and their anti-biofilm activity. *Indian Journal of Microbiology*, *57*(2), 188–194. https://doi.org/10.1007/s12088-017-0640-x (accessed on 6th November 2025).

Ahmed, T., Noman, M., Manzoor, N., Shahid, M., Abdullah, M., Ali, L., Wang, G., Hashem, A., Al-Arjani, A. B. F., Alqarawi, A. A., Abd_Allah, E. F., & Li, B. (2021). Nanoparticle-based amelioration of drought stress and cadmium toxicity in rice via triggering the stress responsive genetic mechanisms and nutrient acquisition. *Ecotoxicology and Environmental Safety*, *209*, 111829. https://doi.org/10.1016/j.ecoenv.2020.111829 (accessed on 6th November 2025).

Ahsan, T. (2020). Bio fabrication of silver nanoparticles from *Pseudomonas fluorescens* to control tobacco mosaic virus. *Egyptian Journal of Biological Pest Control*, *30*, 66. https://doi.org/10.1186/s41938-020-00268-3 (accessed on 6th November 2025).

Akhtar, N., Ilyas, N., Meraj, T. A., Pour-Aboughadareh, A., Sayyed, R. Z., Mashwani, Z. U., & Poczai, P. (2022). Improvement of plant responses by nano biofertilizer: A step towards sustainable agriculture. *Nanomaterials (Basel)*, *12*(6), 965. https://doi.org/10.3390/nano12060965 (accessed on 6th November 2025).

Akl, B., Nadar, M., & El-Saadony, M. T. (2020). Biosynthesis of silver nanoparticles by *Serratia marcescens* sp. *sakuensis* and its antibacterial application against some pathogenic bacteria. *Journal of Applied Chemistry and Biotechnology*, *11*(1), 1–8. https://jacb.journals.ekb.eg/article_76656.html (accessed on 6th November 2025).

Alagesan, V., & Venugopal, S. (2019). Green synthesis of selenium nanoparticle using leaves extract of *Withania somnifera* and its biological applications and photocatalytic activities. *Bio Nanoscience*, *9*, 105–116.

Albanese, A., Tang, P. S., & Chan, W. C. (2012). The effect of nanoparticle size, shape, and surface chemistry on biological systems. *Annual Review of Biomedical Engineering*, *14*, 1–16.

Ali, E. F., El-Shehawi, A. M., Ibrahim, O. H. M., Abdul-Hafeez, E. Y., Moussa, M. M., & Hassan, F. A. S. (2021). A vital role of chitosan nanoparticles in improvisation the drought stress tolerance in *Catharanthus roseus* (L.) through biochemical and gene expression modulation. *Plant Physiology and Biochemistry*, *161*, 166–175.

Ali, S. G., Jalal, M., Ahmad, H., Sharma, D., Ahmad, A., Umar, K., & Khan, H. M. (2022). Green synthesis of silver nanoparticles from *Camellia sinensis* and its antimicrobial and antibiofilm effect against clinical isolates. *Materials (Basel)*, *15*(19), 6978. https://doi.org/10.3390/ma15196978 (accessed on 6th November 2025).

Aljabali, A. A., Barclay, J. E., Lomonossoff, G. P., & Evans, D. J. (2010). Virus templated metallic nanoparticles. *Nanoscale*, *2*, 2596–2600.

Almadiy, A. A., & Nenaah, G. E. (2018). Ecofriendly synthesis of silver nanoparticles using potato steroidal alkaloids and their activity against phytopathogenic fungi. *Brazilian Archives of Biology and Technology*, *61*, 1–14.

Almeida, D. S., Delai, L. B., Christine, A. C., Sawaya, H. F., & Rosolem, C. A. (2020). Exudation of organic acid anions by tropical grasses in response to low phosphorous availability. *Scientific Reports*, *10*, 16955. https://doi.org/10.1038/s41598-020-73398-1 (accessed on 6th November 2025).

Alsamhary, K. I. (2020). Eco-friendly synthesis of silver nanoparticles by *Bacillus subtilis* and their antibacterial activity. *Saudi Journal of Biological Sciences*, *27*, 2185–2191.

Ameen, F. (2022). Optimization of the synthesis of fungus-mediated bi-metallic Ag-Cu nanoparticles. *Applied Sciences*, *12*(3), 1384. https://doi.org/10.3390/app12031384 (accessed on 6th November 2025).

Ameen, F., Al Yahya, S., Govarthanan, M., Aljahdali, N., Al-Enazi, N., Alsamhary, K. I., Alshehri, W. A., Alwakeel, S. S., & Alharbi, S. A. (2020). Soil bacteria *Cupriavidus* sp. mediates the extracellular synthesis of antibacterial silver nanoparticles. *Journal of Molecular Structure*, *1202*, 127233. https://doi.org/10.1016/j.molstruc.2019.127233 (accessed on 6th November 2025).

Amirifar, A., Hemati, A., Asgari Lajayer, B., Pandey, J., & Astatkie, T. (2022). Impact of various environmental factors on the biosynthesis of alkaloids in medicinal plants. In T. Aftab (Ed.), *Environmental challenges and medicinal plants. Environmental challenges and solutions* (pp. 229–248). Springer, Cham. https://doi.org/10.1007/978-3-030-92050-0_9 (accessed on 6th November 2025).

Amulya, M. A. S., Nagaswarupa, H. P., Kumar, M. R. A., Ravikumar, C. R., & Kusuma, K. B. (2020). Sono chemical synthesis of $MnFe_2O_4$ nanoparticles and their electrochemical and photocatalytic properties. *Journal of Physics and Chemistry of Solids*, *148*, 109661. https://doi.org/10.1016/j.jpcs.2020.109661 (accessed on 6th November 2025).

Anjum, S., Vyas, A., & Sofi, T. (2023). Fungi-mediated synthesis of nanoparticles: characterization process and agricultural applications. *Journal of the Science of Food and Agriculture*, *103*, 4727–4741.

Anuluxan, S., Thavaranjit, A. C., Prabagar, S., et al. (2022). Synthesis of silver nanoparticles from *Turbinaria ornata* and its antibacterial activity against water contaminating bacteria. *Chemical Papers*, *76*(4), 2365–2374.

Apte, M., Girme, G., Nair, R., Bankar, A., Kumar, A. R., & Zinzarde, S. (2013). Melanin mediated synthesis of gold nanoparticles by *Yarrowia lipolytica*. *Materials Letters*, *95*, 149–152.

Arevalo-Villena, M., Briones-Perez, A., Corbo, M. R., Sinigaglia, M., & Bevilacqua, A. (2017). Biotechnological application of yeasts in food science: starter culture, probiotics, and enzyme production. *Journal of Applied Microbiology*, *123*(6), 1360–1372.

Arsene, M. M. J., Viktorovna, P. I., Alla, M., Mariya, M., Davares, A. K. L., Carime, B. Z., Anatolievna, G. O., Vyacheslavovna, Y. N., Vladimirovna, Z. A., Andreevna, S. L., Aleksandrovna, V. E., Alekseevich, B. L., Nikolaïevna, B. M., Parfait, K., & Andrey, V. (2023). Antimicrobial activity of phytofabricated silver nanoparticles using *Carica papaya* L. against Gram-negative bacteria. *Veterinary World*, *16*(6), 1301–1311.

Arumugam, A., Karthikeyan, C., Syedahamed, A., Hameed, H., Gopinatha, K., Gowria, S., & Karthika, V. (2015). Synthesis of cerium oxide nanoparticles using *Gloriosa superba* L. leaf extract and their structural, optical and antibacterial properties. *Materials Science and Engineering: C*, *49*, 408–415.

Asariha, M., Chahardoli, A., Karimi, N., Gholamhosseinpour, M., Khoshroo, A., Nemati, H., Shokoohinia, Y., & Fattahi, A. (2020). Green synthesis and structural characterization of gold nanoparticles from *Achillea wilhelmsii* leaf infusion and *in vitro* evaluation. *Bulletin of Materials Science*, *43*, 57. https://doi.org/10.1007/s12034-019-2005-z (accessed on 6th November 2025).

Asghar, M. A., Zahir, E., Shahid, S. M., Khan, M. N., Iqbal, J., & Walker, G. (2018). Iron, copper and silver nanoparticles: green synthesis using green and black tea leaves extracts and evaluation of antibacterial, antifungal and aflatoxin B1 adsorption activity. *LWT–Food Science and Technology*, *90*, 98–107.

Ashengroph, M., Khaledi, A., & Bolbanabad, E. M. (2020). Extracellular biosynthesis of cadmium sulphide quantum dot using cell-free extract of *Pseudomonas chlororaphis* CHR05 and its antibacterial activity. *Process Biochemistry*, *89*, 63–70.

Ashraf, A., Zafar, S., Zahid, K., Shah, M. S., Al-Ghanim, K. H., Al-Misned, F., & Mahboob, S. (2019). Synthesis, characterization, and antibacterial potential of silver nanoparticles synthesized from *Coriandrum sativum* L. *Journal of Nanomaterials*, *12*, 275–281.

Ayesha, A. (2017). Bacterial synthesis and nanoparticles. *Nano Science & Nano Technology: An Indian Journal*, *11*(2), 1–30.

Azam, M., Bhatti, H. N., Khan, A., Zafar, L., & Iqbal, M. (2022). Zinc oxide nano-fertilizer application (foliar and soil) effect on the growth, photosynthetic pigments and antioxidant system of maize cultivar. *Biocatalysis and Agricultural Biotechnology*, *42*, 102343. https://doi.org/10.1016/j.bcab.2022.102343 (accessed on 6th November 2025).

Bafghi, M. H., Zarrinfar, H., Darroudi, M., Zargar, M., & Nazari, R. (2022). Green synthesis of selenium nanoparticles and evaluate their effect on the expression of *ERG3*, *ERG11* and *FKS1* antifungal resistance genes in *Candida albicans* and *Candida glabrata*. *Letters in Applied Microbiology*, *74*(5), 809–819. https://doi.org/10.1111/lam.13667 (accessed on 6th November 2025).

Bahrulolum, H., Nooraei, S., Javanshir, N., Tarrahimorfrad, H., Mirbagheri, V. S., Eastom, A. J., & Ahmadian, G. (2021). Green synthesis of metal nanoparticles using microorganism and their application in the agrifood sector. *Journal of Nanobiotechnology*, *19*, 86. https://doi.org/10.1186/s12951-021-00834-3 (accessed on 6th November 2025).

Balaji, D. S., Basavaraja, S., Deshpande, R., Mahesh, D. B., Prabhakar, B. K., & Venkataraman, A. (2009). Extracellular synthesis of functionalized silver nanoparticles by strains of *Cladosporium cladosporioides* fungus. *Colloids and Surfaces B: Bio Interfaces*, *68*, 88–92.

Balciunaitiene, A., Puzeryte, V., Radenkoys, V., Krasnova, I., Memvanga, P. B., Viskelis, P., Streimikyte, P., & Viskelis, J. (2022). Sustainable-green synthesis of silver nanoparticles using aqueous *Hyssopus officinalis* and *Calendula officinalis* extracts and their antioxidant and antibacterial activities. *Molecules*, *27*(22), 7700. https://doi.org/10.3390/molecules27227700 (accessed on 6th November 2025).

Balke, I., & Zeltins, A. (2019). Use of plant viruses and virus-like particles for the creation of novel vaccines. *Advanced Drug Delivery Reviews*, *145*, 119–129.

Basavaraja, S., Balaji, S., Lagashetty, A., Rajasab, A. H., & Venkataraman, A. (2008). Extracellular biosynthesis of silver nanoparticles using the fungus *Fusarium semitectum*. *Materials Research Bulletin*, *43*, 1164–1170.

Batool, S. U., Javed, B., Sohail, M., Zehra, S. S., Mashwani, Z. U. R., Raja, N. I., Khan, T., AlHaithloul, H. A. S., Alghanem, S. M., Al-Mushhin, A. A. M., Hashem, M., & Alamri, S. (2021). Exogenous applications of bio-fabricated silver nanoparticles to improve biochemical, antioxidant, fatty acid and secondary metabolite contents of sunflower. *Nanomaterials (Basel)*, *11*(7), 1750. https://doi.org/10.3390/nano11071750 (accessed on 6th November 2025).

Bawadekji, A., Oueslati, A., Ali, A., & Basha, J. (2018). Biosynthesis of gold nanoparticles using *Pleurotus ostreatus* (Jacq. ex. Fr.) Kummer extract and their antibacterial and antifungal activities. *Journal of Applied Environmental and Biological Sciences*, *9*, 142–147.

Blum, A. S., Soto, C. M., Wilson, C. D., Brower, T. L., Pollack, S. K., Schull, T. L., Chatterji, A., Lin, T., Johnson, J. E., Amsinck, C., Franzon, P., Shashidhar, R., & Ratna, B. R. (2005). An engineered virus as a scaffold for three-dimensional self-assembly on the nanoscale. *Small*, *1*, 702–706.

Bruckman, M. A., Czapar, A. E., VanMeter, A., Randolph, L. N., & Steinmetz, N. F. (2016). Tobacco mosaic virus-based protein nanoparticles and nanorods for chemotherapy delivery targeting breast cancer. *Journal of Controlled Release*, *231*, 103–113.

Camas, M., Sazak, A., & Kyeremeh, K. (2018). Extracellular synthesis and characterization of gold nanoparticles using *Mycobacterium* sp. BRS2A-AR2 isolated from the aerial roots of the Ghanaian mangrove plant, *Rhizophora racemosa*. *Indian Journal of Microbiology*, *58*, 214–221.

Carballal, R. C., Gutierrez Rivas, C., & Fuentes, M. C. (2023). Viral protein nanoparticles (Part 1): Pharmaceutical characteristics. *European Journal of Pharmaceutical Sciences*, *187*, 106460. https://doi.org/10.1016/j.ejps.2023.106460 (accessed on 6th November 2025).

Chaudhary, R., Nawaz, K., Khan, A. K., Hano, C., Abbasi, B. H., & Anjum, S. (2020). An overview of the algae-mediated biosynthesis of nanoparticles and their biomedical applications. *Biomolecules*, *10*(11), 1498. https://doi.org/10.3390/biom10111498 (accessed on 6th November 2025).

Chauhan, N., Tyagi, A. K., Kumar, P., & Malik, A. (2016). Antibacterial potential of *Jatropha curcas* synthesized silver nanoparticles against food borne pathogens. *Frontiers in Microbiology*, *7*, 1748. https://doi.org/10.3389/fmicb.2016.01748 (accessed on 6th November 2025).

Chen, C. C., Stark, M., Baikoghli, M., & Cheng, R. H. (2018). Surface functionalization of hepatitis E virus nanoparticles using chemical conjugation methods. *Journal of Visualized Experiments*, *135*, 57020. https://www.jove.com/t/57020/surface-functionalization-hepatitis-e-virus-nanoparticles-using (accessed on 6th November 2025).

Chen, J., Mao, S., Xu, Z., & Ding, W. (2019). Various antibacterial mechanisms of biosynthesized copper oxide nanoparticles against soil borne *Ralstonia solanacearum*. *RSC Advances*, *9*, 3788–3799.

Clark, J. R., & March, J. B. (2006). Bacteriophages and biotechnology: Vaccines, gene therapy and antibacterials. *Trends in Biotechnology*, *24*, 212–218.

Colla, G., & Rouphael, Y. (2015). Bio stimulants in horticulture. *Scientia Horticulture*, *196*, 1–2. https://doi.org/10.1016/j.scienta.2015.10.044 (accessed on 6th November 2025).

Cunha, F. A., Cunha, M. D. C. S. O., da Frota, S. M., Mallmann, E. J. J., Freire, T. M., Costa, L. S., Paula, A. J., Menezes, E. A., & Fechine, P. B. A. (2018). Biogenic synthesis of multifunctional silver nanoparticles from *Rhodotorula glutinis* and *Rhodotorula mucilaginosa*: Antifungal, catalytic and cytotoxicity activities. *World Journal of Microbiology and Biotechnology*, *34*(9), 127. https://doi.org/10.1007/s11274-018-2514-8 (accessed on 6th November 2025).

Dağlıoğlu, Y., & Öztürk, B. Y. (2019). A novel intracellular synthesis of silver nanoparticles using *Desmodesmus* sp. (Scenedesmaceae): Different methods of pigment change. *Rendiconti Lincei. Scienze Fisiche e Naturali*, *30*, 611–621.

Dahoumane, S. A., Wijesekera, K., Filipe, C. D., et al. (2014). Stoichiometrically controlled production of bimetallic gold–silver alloy colloids using micro-alga cultures. *Journal of Colloid and Interface Science*, *416*, 67–72. https://doi.org/10.1016/j.jcis.2013.10.048 (accessed on 6th November 2025).

Daniel, A. I., Keyster, M., & Klein, A. (2023). Biogenic zinc oxide nanoparticles: A viable agricultural tool to control pathogenic fungi and its potential effects on soil and plants. *Science of the Total Environment*, *897*, 165483. https://doi.org/10.1016/j.scitotenv.2023.165483 (accessed on 6th November 2025).

Das, B. S., Das, A., Mishra, A., & Arakha, M. (2021). Microbial cells as biological factory for nanoparticle synthesis. *Frontiers of Materials Science*, *15*, 177–191.

Dasgupta, N., Ranjan, S., Mundekkad, D., Ramalingam, C., Shanker, R., & Kumar, A. (2015). Nanotechnology in agro-food: From field to plate. *Food Research International*, *69*, 381–400.

Dethier, J. J., & Effenberger, A. (2012). Agriculture and development: A brief review of the literature. *Economic Systems*, *36*, 175–205.

do Nascimento, J. M., de Oliveira, J. D., de Lima Rizzo, A. C., & Leite, S. G. F. (2018). Biogenic production of copper nanoparticles by *Saccharomyces cerevisiae*. *Journal of Bio Nanoscience*, *12*, 689–693.

Du Jardin, P. (2015). Plant bio stimulants: Definition, concept, main categories and regulation. *Scientia Horticulture*, *196*, 3–14.

Elahian, F., Heidari, R., & Charghan, V. R. (2020). Genetically modified *Pichia pastoris*, a powerful resistant factory for gold and palladium bioleaching and nanostructure heavy metal biosynthesis. *Artificial Cells, Nanomedicine, and Biotechnology*, *48*, 259–265.

Elamawi, R. M., Al-Harbi, R. E., & Hendi, A. A. (2018). Biosynthesis and characterization of silver nanoparticles using *Trichoderma longibrachiatum* and their effect on phytopathogenic fungi. *Egyptian Journal of Biological Pest Control*, *28*, 28. https://doi.org/10.1186/s41938-018-0028-1 (accessed on 6th November 2025).

El-Ansary, A. E., Omran, A. A., Mohamed, H. I., & El-Mahdy, O. M. (2023). Green synthesized silver nanoparticles mediated by *Fusarium nygamai* isolate AJTYC1: Characterizations, antioxidant, antimicrobial, anticancer, and photocatalytic activities and cytogenetic effects. *Environmental Science and Pollution Research International*, *30*, 100477–100499.

Elemike, E. E., Uzoh, I. M., Onwudiwe, D. C., & Babalola, O. O. (2019). The role of nanotechnology in the fortification of plant nutrients and improvement of crop production. *Applied Sciences*, *9*(3), 499. https://doi.org/10.3390/app9030499 (accessed on 6th November 2025).

Fardsadegh, B., & Jafarizadeh-Malmiri, H. (2019). *Aloe vera* leaf extract mediated green synthesis of selenium nanoparticles and assessment of their *in vitro* antimicrobial activity against spoilage fungi and pathogenic bacteria strains. *Green Processing and Synthesis*, *8*, 399–407.

Fariq, A., Khan, T., & Yasmin, A. (2017). Microbial synthesis of nanoparticles and their potential in biomedicine. *Journal of Applied Biomedicine*, *15*, 241–248.

Fatimah, I., Hidayat, H., Nugroho, B., & Husein, S. (2023). Green synthesis of silver nanoparticles using *Datura metel* flower extract assisted by ultrasound method and its antibacterial activity. *Recent Patents on Nanotechnology*, *17*, 68–73.

Fellet, G., Pilotto, L., Marchiol, L., & Braidot, E. (2021). Tools for nano-enabled agriculture: Calcium phosphate, silicon and chitosan nanostructure. *Agronomy*, *11*(6), 1239. https://doi.org/10.3390/agronomy11061239 (accessed on 6th November 2025).

Fernandes, R., Roy, V., Wu, H. C., & Bentley, W. E. (2010). Engineered biological nano factories trigger quorum sensing response in targeted bacteria. *Nature Nanotechnology*, *5*, 213–217.

Fernandez, J. G., Fernandez-Baldo, M. A., Berni, E., Cami, G., Duran, N., Raba, J., & Sanz, M. I. (2016). Production of silver nanoparticles using yeasts and evaluation of their antifungal activity against phytopathogenic fungi. *Process Biochemistry*, *51*(9), 1306–1313.

Folorunso, A., Akintelu, S., Oyebamiji, A. K., Ajayi, S., Abiola, B., Abdusalam, I., & Morakinyo, A. (2019). Biosynthesis, characterization and antimicrobial activity of gold nanoparticles from leaf extracts of *Annona muricata*. *Journal of Nanostructure in Chemistry*, *9*, 111–117. https://doi.org/10.1007/s40097-019-0301-1 (accessed on 6th November 2025).

Fraceto, L. F., Grillo, R., de Medeiros, G. A., Scognamiglio, V., Rea, G., & Bartolucci, C. (2016). Nanotechnology in agriculture: Which innovation potential does it have? *Frontiers in Environmental Science*, *4*, 20. https://doi.org/10.3389/fenvs.2016.00020 (accessed on 6th November 2025).

Garcea, R. L., & Gissmann, L. (2004). Virus-like particles as vaccines and vessels for the delivery of small molecules. *Current Opinion in Biotechnology*, *15*, 513–517.

García-Ovando, A. E., Piña, J. E. R., Naranjo, E. U. E., Chávez, J. A. C., & Esquivel, K. (2022). Biosynthesized nanoparticles and implications by their use in crops: Effects over physiology, action mechanisms, plant stress response and toxicity. *Plant Stress*, *6*, 100109. https://doi.org/10.1016/j.stress.2022.100109 (accessed on 6th November 2025).

Gautam, D., Dolma, K. G., Khandelwal, B., Gupta, M., Singh, M., Mahboob, T., et al. (2023). Green synthesis of silver nanoparticles using *Ocimum sanctum* Linn. and its antibacterial activity against multidrug resistant *Acinetobacter baumannii*. *PeerJ*, *11*, e15590. https://doi.org/10.7717/peerj.15590 (accessed on 6th November 2025).

Gericke, M., & Pinches, A. (2006). Biological synthesis of metal nanoparticles. *Gold Bulletin*, *83*, 132–140.

Ghosh, S., Ahmad, R., Banerjee, K., Alajmi, M. F., & Rahman, S. (2021). Mechanistic aspects of microbe-mediated nanoparticle synthesis. *Frontiers in Microbiology*, *12*, 626834. https://doi.org/10.3389/fmicb.2021.638068 (accessed on 6th November 2025).

Ghosh, S., Shah, S., & Webster, T. J. (2021). Recent trends in fungal biosynthesis of nanoparticles. In *Fungi Bio Prospects in Sustainable Agriculture, Environment and Nanotechnology. Fungal Metabolites and Nanotechnology* (Vol. 3, pp. 403–452).

Giri, V. P., Shukla, P., Tripathi, A., Verma, P., Kumar, N., Pandey, S., Dimkpa, C. O., & Mishra, A. (2023). A review of sustainable use of biogenic nanoscale agro-materials to enhance stress tolerance and nutritional values of plants. *Plants*, *12*(4), 815. https://doi.org/10.3390/plants12040815 (accessed on 6th November 2025).

Goel, S., Kumar, R., & Beniwal, V. (2023). Biogenic synthesis of nanoparticles for sustainable crop production: A review. *Letters in Applied Nano Bioscience*, *12*(1), 10. https://doi.org/10.33263/LIANBS121.010 (accessed on 6th November 2025).

Govindaraju, K., Kiruthiga, V., Kumar, V. G., & Singaravelu, G. (2009). Extracellular synthesis of silver nanoparticles by a marine alga, *Sargassum wightii* Grevilli and their antibacterial effects. *Journal of Nanoscience and Nanotechnology*, *9*, 5497–5501.

Grgacic, E. V., & Anderson, D. A. (2006). Virus-like particles: Passport to immune recognition. *Methods*, *40*, 60–65.

Groenke, N., Seisenbaeva, G. A., Kaminskyy, V., Zhivotovsky, B., Kost, B., & Kessler, V. G. (2012). Structural characterization, solution stability, and potential health and environmental effects of the Nano-TiO_2 bioencapsulation matrix and the model product of its biodegradation TiBALDH. *RSC Advances*, *2*, 4228–4235.

Guilger, M., Pasquoto-Stigliani, T., Bilesky-Jose, N., Grillo, R., Abhiash, P. C., Fraceto, F., & Lima, R. D. (2017). Biogenic silver nanoparticles based on *Trichoderma harzianum*: Synthesis, characterization, toxicity evaluation and biological activity. *Scientific Reports*, *7*, 44421. https://doi.org/10.1038/srep44421 (accessed on 6th November 2025).

Guleria, G., Thakur, S., Shandilya, M., Sharma, S., Thakur, S., & Kalia, S. (2022). Nanotechnology for sustainable agro-food systems: The need and role of nanoparticles in protecting plants and improving crop productivity. *Plant Physiology and Biochemistry*, *194*, 533–549.

Gursoy, N., Ozturk, B. Y., & Dag, L. (2021). Synthesis of intracellular and extracellular gold nanoparticles with a green machine and its antifungal activity. *Turkish Journal of Biology*, *45*, 196–213.

Haefeli, C., Franklin, C., & Hardy, K. (1984). Plasmid-determined silver resistance in *Pseudomonas stutzeri* isolated from a silver mine. *Journal of Bacteriology*, *158*, 389–392.

Hamza, W. T., Taha, H. M., & Abouelkheir, S. S. (2018). *Spirulina platensis* synthesized gold nanoparticles (AuNPs) as an antibacterial and antioxidant agent. *Egyptian Journal of Experimental Biology (Biology)*, *14*, 385–391.

Harsha, L., Jain, R. K., & Prasad, A. S. (2022). Synthesis, characterization, and antimicrobial activity of silver nanoparticles derived from *Mentha × piperita + Ocimum tenuiflorum*: An *in vitro* study. *Journal of Advanced Pharmaceutical Technology & Research*, *13*(Supplement), S272–S276.

Hong, J., Jia, S., Wang, C., Li, Y., He, F., & Gardea-Torresdey, J. L. (2023). Transcriptome reveals the exposure effects of CeO_2 nanoparticles on *Pakchoi* (*Brassica chinensis* L.) photosynthesis. *Journal of Hazardous Materials*, *444*(Part A), 130427. https://doi.org/10.1016/j.jhazmat.2022.130427 (accessed on 6th November 2025).

Hosseinpour, A., Ilhan, E., Özkan, G., Öztürk, H. İ., Haliloglu, K., & Cinisli, K. T. (2022). Plant growth-promoting bacteria (PGPBs) and copper (II) oxide (CuO) nanoparticle ameliorates DNA damage and DNA methylation in wheat (*Triticum aestivum* L.) exposed to NaCl stress. *Journal of Plant Biochemistry and Biotechnology*, *31*, 751–764.

Huang, X., Stein, B. D., Cheng, H., Malyutin, A., Tsvetkova, I. B., Baxter, D. V., Remmes, N. B., Verchot, J., Kao, C., Bronstein, L. M., & Dragnea, B. (2011). Magnetic virus-like nanoparticles in *N. benthamiana* plants: A new paradigm for environmental and agronomic biotechnological research. *ACS Nano*, *5*(5), 4037–4045.

Huang, Z., Zeng, Z., Song, Z., Chen, A., Zeng, G., Xiao, R., He, K., Yuan, L., Li, H., & Chen, G. (2020). Antimicrobial efficacy and mechanisms of silver nanoparticles against *Phanerochaete chrysosporium* in the presence of common electrolytes and humic acid. *Journal of Hazardous Materials*, *383*, 121153. https://doi.org/10.1016/j.jhazmat.2019.121153 (accessed on 6th November 2025).

Hulkoti, N. I., & Taranath, T. C. (2014). Biosynthesis of nanoparticles using microbes—A review. *Colloids and Surfaces B: Bio Interfaces*, *121*, 474–483.

Husseiny, M. I., Abd El-Aziz, M., Badr, Y., & Mahmoud, M. A. (2007). Biosynthesis of gold nanoparticles using *Pseudomonas aeruginosa*. *Spectrochimica Acta Part A: Molecular and Biomolecular Spectroscopy*, *67*, 1003–1006.

Hyde, K. D., Xu, J., Rapior, S., Jeewon, R., Lumyong, S., Niego, A. G. C., et al. (2019). The amazing potential of fungi: 50 ways we can exploit fungi industrially. *Fungal Diversity*, *97*, 1–136. https://doi.org/10.1007/s13225-019-00430-9 (accessed on 6th November 2025).

Ibrahim, E., Fouad, H., Zhang, M., Zhang, Y., Qiu, W., Yan, C., Li, B., Mo, J., & Chen, J. (2019). Biosynthesis of silver nanoparticles using endophytic bacteria and their role in inhibition of rice pathogenic bacteria and plant growth promoting bacteria. *RSC Advances*, *9*, 29293–29299.

Ingle, A., Rai, M., Gade, A., & Bawaskar, M. (2009). *Fusarium solani*: A novel biological agent for the extracellular synthesis of silver nanoparticles. *Journal of Nanoparticle Research*, *11*, 2079–2085.

Jackson, T. C., Patani, B. O., & Ekpa, D. E. (2017). Nanotechnology in diagnostic: A review. *Advances in Nanoparticles*, *6*, 93–102.

Jain, N., Bhargava, A., Majumdar, S., Tarafdar, J. C., & Panwar, J. (2011). Extracellular biosynthesis and characterization of silver nanoparticles using *Aspergillus flavus* NJP08: A mechanism perspective. *Nanoscale*, *3*, 635–641.

Jain, S., Gupta, I., Walia, P., & Swami, S. (2022). Application of actinobacteria in agriculture, nanotechnology, and bioremediation. In *Actinobacterial-Diversity, Application and Medical Aspects* (pp. 1–29). IntechOpen. https://www.intechopen.com/chapters/83340 (accessed on 6th November 2025).

Jalal, M., Ansari, M. A., Alzohairy, M. A., Ali, S. G., Khan, H. M., Almatroudi, A., & Raees, K. (2018). Biosynthesis of silver nanoparticles from oropharyngeal *Candida glabrata* isolates and their antimicrobial activity against clinical strains of bacteria and fungi. *Nanomaterials*, *8*(8), 586. https://doi.org/10.3390/nano8080586 (accessed on 6th November 2025).

Jasim, B., Thomas, R., Mathew, J., & Radhakrishnan, E. K. (2017). Plant growth and diosgenin enhancement effect of silver nanoparticles in Fenugreek (*Trigonella foenum-graecum* L.). *Saudi Pharmaceutical Journal*, *25*, 443–447.

Jiang, G., Zhang, Y., Gan, G., Li, W., Wan, W., Jiang, Y., Yang, T., Zhang, Y., Xu, Y., Wang, Y., Shen, Q., Wei, Z., & Dini-Andreote, F. (2022). Exploring rhizo-microbiome transplants as a tool for protective plant-microbiome manipulation. *ISME Communications*, *2*, 10. https://doi.org/10.1038/s43705-022-00094-8 (accessed on 6th November 2025).

Joshi, C. G., Danagoudar, A., Poyya, J., Kudva, A. K., & Dhananjaya, B. L. (2017). Biogenic synthesis of gold nanoparticles by marine endophytic fungus-*Cladosporium cladosporioides* isolated from seaweed and evaluation of their antioxidant and antimicrobial properties. *Process Biochemistry*, *63*, 137–144.

Jubran, A. S., Al-Zamely, O. M., & Al-Ammar, M. H. (2020). A study of iron oxide nanoparticles synthesis by using bacteria. *International Journal of Pharmaceutical Quality Assurance*, *11*, 1–8.

Kalabegishvili, T., Kirkesali, E., Rcheulishvili, A., et al. (2012). Synthesis of gold nanoparticles by blue-green algae *Spirulina platensis*. *Journal of Applied Microbiology and Biotechnology, Advanced Science, Engineering and Medicine*, *4*, 1–10. http://www1.jinr.ru/Preprints/2012/031(E14-2012-31).pdf (accessed on 6th November 2025).

Kalpana, V. N., Kataru, B. A. S., Sravani, N., Vigneshwari, T., Panneerselvam, A., & Rajeswari, V. D. (2018). Biosynthesis of zinc oxide nanoparticles using culture filtrates of *Aspergillus niger*: Antimicrobial textiles and dye degradation studies. *Open Nano*, *3*, 48–55.

Karbasian, M., Atyabi, S. M., Siadat, S. D., Momen, S. B., & Norouzian, D. (2008). Optimizing nano-silver formation by *Fusarium oxysporum* PTCC 5115 employing response surface methodology. *American Journal of Agricultural and Biological Sciences*, *3*, 433–437.

Kashyap, A. S., Manzar, N., Rajawat, M. V. S., Kesharwani, A. K., Singh, R. P., Dubey, S. C., Pattanayak, D., Dhar, S., Lal, S. K., & Singh, D. (2021). Screening and biocontrol potential of rhizobacteria native to Gangetic plains and hilly regions to induce systemic resistance and promote plant growth in chilly against bacterial wilt disease. *Plants*, *10*(10), 2125. https://doi.org/10.3390/plants10102125 (accessed on 6th November 2025).

Khairy, A. M., Tohamy, M. R. A., Zayed, M. A., Mahmoud, S. F., El-Tahan, A. M., El-Saadony, M. T., & Mesiha, P. K. (2022). Eco-friendly application of nano-chitosan for controlling potato and tomato bacterial wilt. *Saudi Journal of Biological Sciences*, *29*(4), 2199–2209.

Khan, A., Singh, J., Upadhayay, V. K., Singh, A. V., & Shah, S. (2019). Microbial biofortification: A green technology through plant growth promoting microorganisms. In S. Shah, V. Venkatramanan, & R. Prasad (Eds.), *Sustainable Green Technologies for Environmental Management* (pp. 255–269). Springer, Singapore. https://doi.org/10.1007/978-981-13-2772-8_13 (accessed on 6th November 2025).

Khan, F., Shahid, A., Zhu, H., Wang, N., Javed, M. R., Ahmad, N., Xu, J., & Alam, M. A., & Mehmood, M. A. (2022). Prospects of algae-based green synthesis of nanoparticles for environmental applications. *Chemosphere*, *293*, 133571. https://doi.org/10.1016/j.chemosphere.2022.133571 (accessed on 6th November 2025).

Khan, M. S., Ranjani, S., & Hemalatha, S. (2022). Synthesis and characterization of *Kappaphycus alvarezii* derived silver nanoparticles and determination of antibacterial activity. *Materials Chemistry and Physics*, *282*, 125985. https://doi.org/10.1016/j.matchemphys.2022.125985 (accessed on 6th November 2025).

Khan, S., Akhtar, N., Rehman, S. U., Shujah, S., Rha, E. S., & Jamil, M. (2022). *Bacillus subtilis* synthesized iron nanoparticles (Fe_3O_4NPs) induced metabolic and anti-oxidative response in rice (*Oryza sativa*) under arsenic stress. *Toxics*, *10*(10), 618. https://doi.org/10.3390/toxics10100618 (accessed on 6th November 2025).

Klaus, T., Joerger, R., Olsson, E., & Granqvist, C. G. (1999). Silver-based crystalline nanoparticles, microbially fabricated. *Proceedings of the National Academy of Sciences of the United States of America*, *96*, 13611–13614.

Korbekandi, H., Jouneghani, R. M., Mohseni, S., Pourhossein, M., & Iravani, S. (2014). Synthesis of silver nanoparticle using biotransformation of *Saccharomyces boulardii*. *Green Processing and Synthesis*, *3*, 271–277.

Kothari, R., Pandey, A., Ahmad, S., Kumar, A., Pathak, V. V., & Tyagi, V. V. (2017). Microalgal cultivation for value-added products: A critical enviro-economical assessment. *3 Biotech*, *7*(4), 243. https://doi.org/10.1007/s13205-017-0812-8 (accessed on 6th November 2025).

Kottegoda, N., Munaweera, I., Madusanka, N., & Karunaratne, V. (2011). A green slow-release fertilizer composition based on urea-modified hydroxyapatite nanoparticles encapsulated wood. *Current Science*, *101*, 73–78.

Kowshik, M., Ashtaputre, S., Kharrazi, S., Vogel, W., Urban, J., Kulkarni, S. K., & Paknikar, K. M. (2003). Extracellular synthesis of silver nanoparticles by a silver-tolerant yeast strain MKY3. *Nanotechnology*, *14*, 95–100.

Kowshik, M., Deshmukh, N., Vogel, W., Urban, J., Kulkarni, S. K., & Paknikar, K. M. (2002). Microbial synthesis of semiconductor CdS nanoparticles, their characterization, and their use in the fabrication of an ideal diode. *Biotechnology and Bioengineering*, *78*(5), 583–588. https://doi.org/10.1002/bit.10233 (accessed on 6th November 2025).

Kowshik, M., Vogel, W., Urban, J., Kulkarni, S. K., & Paknikar, K. M. (2002). Microbial synthesis of semiconductor PbS nano crystallites. *Advanced Materials*, *14*, 815–818.

Koyatti, R., Kudle, K. R., Nagati, V., Merugu, R., & Padigya, P. R. M. (2022). Extracellular synthesis of mono and bimetallic nanocomposites from novel strains of *Rhodopseudomonas palustris* and evaluation of their biomedical applications. *Macromolecular Symposia*, *402*(1), 2100378. https://doi.org/10.1002/masy.202100378 (accessed on 6th November 2025).

Krishnaraj, C., Ramachandran, R., Mohan, K., & Kalaichelvan, P. T. (2012). Optimization for rapid synthesis of silver nanoparticles and its effect on phytopathogenic fungi. *Spectrochimica Acta Part A: Molecular and Biomolecular Spectroscopy*, *93*, 95–99.

Kumari, H., Khati, P., Gangola, S., Chaudhary, P., & Sharma, A. (2021). Performance of plant growth promontory rhizobacteria on maize and soil characteristics under the influence of TiO_2 nanoparticles. *Pantnagar Journal of Research*, *19*, 28–39.

Kumari, R., & Singh, D. P. (2020). Nano-biofertilizer: An emerging eco-friendly approach for sustainable agriculture. *Proceedings of the National Academy of Sciences, India Section B: Biological Sciences*, *90*, 733–741.

Kuprashina, M., Vetchinkina, E. P., Burov, A., Punomareva, E. G., & Nikitina, V. E. (2013). Biosynthesis of gold nanoparticles by *Azospirillum brasilense*. *Mikrobiologiia*, *82*, 833–840.

Lahiri, D., Nag, M., Sheikh, H. I., Sarkar, T., Edinur, H. A., Pati, S., & Ray, R. R. (2021). Microbiologically-synthesized nanoparticles and role in silencing the biofilm signaling cascade. *Frontiers in Microbiology*, *12*, 636588. https://doi.org/10.3389/fmicb.2021.636588 (accessed on 6th November 2025).

Lai, J. Y., Inoue, N., Oo, C. W., Kawasaki, H., & Lim, T. S. (2021). One-step synthesis of M13 phage-based nanoparticle and their fluorescence properties. *RSC Advances*, *11*, 1367–1375.

Latif, M., Bukhari, S. A. H., Alrajhi, A. A., Alotaibi, F. S., Ahmad, M., Shahzad, A. N., Dewidar, A. Z., & Mattar, M. A. (2022). Inducing drought tolerance in wheat through exopolysaccharide-producing rhizobacteria. *Agronomy*, *12*(5), 1140. https://doi.org/10.3390/agronomy12051140 (accessed on 6th November 2025).

Lee, K. J., Pussepitiyalage, V. B., Lee, Y. H., Loesch-Fries, L. S., Harris, M. T., Hemmati, S., & Solomon, K. V. (2020). Engineering tobacco mosaic virus and its virus-like particles for synthesis of bio templated nanomaterials. *Biotechnology Journal*, *16*(4), 2000311. https://doi.org/10.1002/biot.202000311 (accessed on 6th November 2025).

Lee, L. A., Niu, Z., & Wang, Q. (2009). Viruses and virus-like protein assemblies—Chemically programmable nanoscale building blocks. *Nano Research*, *2*, 349–364.

Lee, Y. J., Yi, H., Kim, W. J., Kang, K., Yun, D. S., Strano, M. S., Ceder, G., & Belcher, A. M. (2009). Fabricating genetically engineered high-power lithium-ion batteries using multiple virus genes. *Science*, *324*, 1051–1055. https://www.science.org/doi/10.1126/science.1171541 (accessed on 6th November 2025).

Lengke, M. F., Fleet, M. E., & Southam, G. (2007). Biosynthesis of silver nanoparticles by filamentous cyanobacteria from a silver(I) nitrate complex. *Langmuir*, *23*, 2694–2699.

LewisOscar, F., Vismaya, S., Arunkumar, M., et al. (2016). Algal nanoparticles: Synthesis and biotechnological potentials. In N. Thajuddin & D. Dhanasekaran (Eds.), *Algae—Organisms for Imminent Biotechnology* (Vol. 7, pp. 157–182). https://doi.org/10.5772/62909 (accessed on 6th November 2025).

Li, W., Qu, F., Chen, Y., Sun, Y., Zhang, J., Xie, G., You, Q., & Xu, H. (2021). Antimicrobial activity of silver nanoparticles synthesized by the leaf extract of *Cinnamomum camphora*. *Biochemical Engineering Journal*, *172*, 108050. https://doi.org/10.1016/j.bej.2021.108050 (accessed on 6th November 2025).

Li, Y., Xi, K., Liu, X., Han, S., Han, X., Li, G., Yang, L., Ma, D., Fang, Z., Gong, S., Yin, J., & Zhu, Y. (2023). Silica nanoparticles promote wheat growth by mediating hormones and sugar metabolism. *Journal of Nanobiotechnology*, *21*(1), 2. https://doi.org/10.1186/s12951-022-01753-7 (accessed on 6th November 2025).

Lian, S., Diko, C. S., Yan, Y., Li, Z., Zhang, H., Ma, Q., & Qu, Y. (2019). Characterization of biogenic selenium nanoparticles derived from cell-free extracts of a novel yeast *Magnusiomyces ingens*. *3 Biotech*, *9*, 221. https://doi.org/10.1007/s13205-019-1748-y (accessed on 6th November 2025).

Liu, R., Guo, Z., Li, J., Wu, F., Cui, K., Cheng, P., Chen, Y., Ding, Y., Cui, M., & Wu, Z. (2022). Mechanism of *Phanerochaete chrysosporium*-mediated tetracycline hydrochloride removal promoted by low-dose silver nanoparticles. *Journal of Environmental Chemical Engineering*, *10*(3), 107474. https://doi.org/10.1016/j.jece.2022.107474 (accessed on 6th November 2025).

Liu, Y., Perumalsamy, H., Kang, C. H., Kim, S. H., Hwang, J. S., Koh, S. C., Yi, T. H., & Kim, Y. J. (2020). Intracellular synthesis of gold nanoparticles by *Gluconacetobacter liquefaciens* for delivery of peptide CopA3 and ginsenoside and anti-inflammatory effect on lipopolysaccharide-activated macrophages. *Artificial Cells, Nanomedicine, and Biotechnology*, *48*, 777–788.

Lomelí-Rosales, D. A., Zamudio-Ojeda, A., Reyes-Maldonado, O. K., López-Reyes, M. E., Basulto-Padilla, G. C., Lopez-Naranjo, E. J., Zuñiga-Mayo, V. M., & Velázquez-Juárez, G. (2022). Green synthesis of gold and silver nanoparticles using leaf extract of *Capsicum chinense* plant. *Molecules*, *27*(5), 1692. https://doi.org/10.3390/molecules27051692 (accessed on 6th November 2025).

Love, A. J., Makarov, V., Yaminsky, I., Kalinina, N. O., & Taliansky, M. E. (2014). The use of tobacco mosaic virus and cowpea mosaic virus for the production of novel nanometals. *Virology*, *449*, 133–139.

Lu, S., Zheng, F., Wen, L., He, Y., Wang, D., Wu, M., & Wang, B. (2021). Yeast engineering technologies and their applications to the food industry. *Food Biotechnology*, *35*(3), 252–271. https://doi.org/10.1080/08905436.2021.1942037 (accessed on 6th November 2025).

Ludwig, C., & Wagner, R. (2007). Virus-like particles—universal molecular toolboxes. *Current Opinion in Biotechnology*, *18*, 537–545.

Lv, Q., Zhang, B., Xing, X., Zhao, Y., Cai, R., Wang, W., & Gu, Q. (2018). Biosynthesis of copper nanoparticles using *Shewanella loihica*. *Journal of Hazardous Materials*, *347*, 141–149.

Maghzian, A., Aslani, A., Zahedi, R., & Yaghoubi, M. (2023). How to effectively produce value-added products from microalgae? *Renewable Energy*, *204*, 262–276.

Mahmud, A. L., Upadhyay, S. K., Srivastava, A. K., & Bhojiya, A. A. (2021). Biofertilizers: A nexus between soil fertility and crop productivity under abiotic stress. *Current Research in Environmental Sustainability*, *3*, 100063. https://doi.org/10.1016/j.crsust.2021.100063 (accessed on 6th November 2025).

Mandal, A., Sarkar, B., Mandal, S., Vithanage, M., Patra, A. K., & Manna, M. C. (2020). Impact of agrochemicals on soil health. In *Agrochemicals Detection, Treatment and Remediation: Pesticides and Chemical Fertilizers* (pp. 161–187). https://doi.org/10.1016/B978-0-08-103017-2.00007-6 (accessed on 6th November 2025).

Manikprabhu, D., Cheng, J., Chen, W., Sunkara, A. K., Mane, S. B., Kumar, R., Das, M., Hozzein, W. N., Duan, Y. Q., & Li, W. J. (2016). Sunlight mediated synthesis of silver nanoparticles by a novel actinobacterium (*Sinomonas mesophila* MPKL 26) and its antimicrobial activity against multi-drug-resistant *Staphylococcus aureus*. *Journal of Photochemistry and Photobiology B: Biology*, *158*, 202–205. https://doi.org/10.1016/j.jphotobiol.2016.01.018 (accessed on 6th November 2025).

Mao, C., Flynn, C. E., Hayhurst, A., Sweeney, R., Qi, J., Georgiou, G., Iverson, B., & Belcher, A. M. (2003). Viral assembly of oriented quantum dot nanowires. *Proceedings of the National Academy of Sciences of the United States of America*, *100*, 6946–6951. https://doi.org/10.1073/pnas.0832310100 (accessed on 6th November 2025).

Mardi, A., Mohajjel Shoja, H., & Mohajel Kazemi, E. (2022). Comparative study of growth responses, photosynthetic pigment content, and gene expression pattern in tobacco plants treated with ZnO nano and ZnO bulk particles. *Journal of Nanoparticle Research*, *24*, 208. https://doi.org/10.1007/s11051-022-05583-4 (accessed on 6th November 2025).

Mariekie, G., & Anthony, P. (2006). Microbial production of gold nanoparticles. *Gold Bulletin*, *39*, 22–28.

Marks, T., & Sharp, R. (2000). Bacteriophages and biotechnology: A review. *Journal of Chemical Technology and Biotechnology*, *75*, 6–17.

Mata, Y., Torres, E., Blazquez, M., Ballester, A., González, F., & Muñoz, J. (2009). Gold (III) biosorption and bio reduction with the brown alga *Fucus vesiculosus*. *Journal of Hazardous Materials*, *166*, 612–618.

Mathur, A., Koul, A., & Hattewar, J. (2019). Plant growth-promoting rhizobacteria (PGPRs): Significant revolutionary tools for achieving long-term sustainability and combating the biotic stress caused by the attack of pathogens affecting crops in agriculture. In R. Sayyed (Ed.), *Plant Growth Promoting Rhizobacteria for Sustainable Stress Management. Microorganisms for Sustainability* (Vol. 13, pp. 379–388). Springer, Singapore. https://doi.org/10.1007/978-981-13-6986-5_15 (accessed on 6th November 2025).

Mendoza-Cozatl, D., Loza-Tavera, H., Hernandez-Navarro, A., & Moreno-Sanchez, R. (2005). Sulfur assimilation and glutathione metabolic metabolism under cadmium stress in yeast, protists and plants. *FEMS Microbiology Reviews*, *29*, 653–671.

Mishra, V., & Sharma, R. (2015). Green synthesis of zinc oxide nanoparticles using fresh peels extract of *Punica granatum* and its antimicrobial activities. *International Journal of Pharmaceutical Research and Health Sciences*, *3*, 694–699.

Mitchell, M. J., Billingsley, M. M., Haley, R. M., Wechsler, M. E., Peppas, N. A., & Langer, R. (2021). Engineering precision nanoparticles for drug discovery. *Nature Reviews Drug Discovery*, *20*, 101–124.

Mmola, M., Rose-Hill, M. L., & Durrell, K. (2016). Enhanced antimicrobial and anticancer activity of silver and gold nanoparticles synthesized using *Sargassum incisifolium* aqueous extract. *Molecules*, *21*(12), 1633. https://doi.org/10.3390/molecules21121633 (accessed on 6th November 2025).

Mohamed, N. G., & Abdel-Hakeem, M. A. (2023). Chitosan nanoparticles enhance drought tolerance in tomatoes (*Solanum lycopersicum* L.) via gene expression modulation. *Plant Gene*, *34*, 100406. https://doi.org/10.1016/j.plgene.2023.100406 (accessed on 6th November 2025).

Mohammed, G. F., & Hawar, S. N. (2022). Green biosynthesis of silver nanoparticles from *Moringa oleifera* leaves and its antimicrobial and cytotoxicity activities. *International Journal of Biomaterials*, *2022*, 4136641. https://doi.org/10.1155/2022/4136641 (accessed on 6th November 2025).

Moradi Pour, M., Saberi Riseh, R., & Skorik, Y. A. (2022). Sodium alginate-gelatin nano formulations for encapsulation of *Bacillus velezensis* and their use for biological control of pistachio gummosis. *Materials*, *15*(6), 2114. https://doi.org/10.3390/ma15062114 (accessed on 6th November 2025).

Mourato, A., Gadanho, M., Lino, A. R., & Tenreiro, R. (2011). Biosynthesis of crystalline silver and gold nanoparticles by extremophilic yeasts. *Bioinorganic Chemistry and Applications*, *2011*, 546074. https://doi.org/10.1155/2011/546074 (accessed on 6th November 2025).

Mukherjee, A., Sarkar, D., & Sasmal, S. (2021). A review of green synthesis of metal nanoparticles using algae. *Frontiers in Microbiology*, *12*, 693899. https://doi.org/10.3389/fmicb.2021.693899 (accessed on 6th November 2025).

Mushtaq, Z., Nazir, A., Asghar, H. N., & Zahir, Z. A. (2022). Interactive effect of siderophore-producing bacteria and L-tryptophan on physiology, tuber characteristics, yield, and iron concentration of potato. *Potato Research*, *65*, 1015–1027.

Nadeem, M., Pervez, L., Khan, A. M., Burton, R. A., Ullah, S., Nadhman, A., & Celli, J. (2023). Microbial-mediated synthesis of gold nanoparticles—current insights and future vistas. *Gold Bulletin*, *56*, 69–81.

Nayana, A. R., Joseph, B. J., Jose, A., & Radhakrishnan, E. K. (2020). Nanotechnological advances with PGPR applications. In S. Hayat, J. Pichtel, M. Faizan, & Q. Fariduddin (Eds.), *Sustainable Agriculture Reviews 41* (pp. 163–180). Springer, Cham. https://doi.org/10.1007/978-3-030-33996-8_9 (accessed on 6th November 2025).

Nazli, F., Mustafa, A., Ahmad, M., Hussain, A., Jamil, M., Wang, X., Shakeel, Q., Imtiaz, M., & El-Esawi, M. A. (2020). A review on practical application and potentials of phytohormone-producing plant growth-promoting rhizobacteria for inducing heavy metal tolerance in crops. *Sustainability*, *12*(21), 9056. https://doi.org/10.3390/su12219056 (accessed on 6th November 2025).

Neethu, S., Midhun, S. J., Sunil, M. A., Soumya, S., Radhakrishnan, E. K., & Jyothis, M. (2018). Efficient visible light induced synthesis of silver nanoparticles by *Penicillium polonicum* ARA 10 isolated from *Chetomorpha antennina* and its antibacterial efficacy against *Salmonella enterica* serovar Typhimurium. *Journal of Photochemistry and Photobiology B: Biology*, *180*, 175–185.

Neme, K., Nafady, A., Uddin, S., & Tola, Y. B. (2021). Application of nanotechnology in agriculture, postharvest loss reduction and food processing: Food security implication and challenges. *Helion*, *7*(12), e08539. https://doi.org/10.1016/j.heliyon.2021.e08539 (accessed on 6th November 2025).

Noman, M., Ahmed, T., Shahid, M., Niazi, M. B. K., Qasim, M., Kouadri, F., Abdulmajeed, A. M., Alghanem, S. M., Ahmad, N., Zafar, M., & Shehbaj, A. (2021). Biogenic copper nanoparticles produced by using *Klebsiella pneumoniae* strain NST2 curtailed salt stress effects in maize by modulating the cellular oxidative repair mechanism. *Ecotoxicology and Environmental Safety*, *217*, 112264. https://doi.org/10.1016/j.ecoenv.2021.112264 (accessed on 6th November 2025).

Olobayotan, I., & Akin-Osanaiye, B. (2019). Biosynthesis of silver nanoparticles using baker's yeast, *Saccharomyces cerevisiae*, and its antibacterial activities. *Access Microbiology*, *1*, 526. https://doi.org/10.1099/acmi.ac2019.po0316 (accessed on 6th November 2025).

Omran, B. A., Nassar, H. N., Younis, S. A., Fatthallah, N. A., Hamdy, A., El-Shatoury, E. H., & El-Gendy, N. S. (2019). Physiochemical properties of *Trichoderma longibrachiatum* DSMZ 16517-synthesized silver nanoparticles for the mitigation of halotolerant sulphate-reducing bacteria. *Journal of Applied Microbiology*, *126*, 138–154.

Öztürk, B. Y., Gürsu, B. Y., & Dağ, İ. (2020). Antibiofilm and antimicrobial activities of green synthesized silver nanoparticles using marine red algae *Gelidium corneum*. *Process Biochemistry*, *89*, 208–219.

Paczesny, J., & Bielec, K. (2020). Application of bacteriophages in nanotechnology. *Nanomaterials*, *10*(10), 1944. https://doi.org/10.3390/nano10101944 (accessed on 6th November 2025).

Pandey, S. S., Jain, R., Bhardwaj, P., Thakur, A., Kumari, M., Bhushan, S., & Kumar, S. (2022). Plant probiotics–endophytes pivotal to plant health. *Microbiological Research*, *263*, 127148. https://doi.org/10.1016/j.micres.2022.127148 (accessed on 6th November 2025).

Panichikkal, J., Thomas, R., John, J. C., & Radhakrishnan, E. K. (2019). Biogenic gold nanoparticle supplementation to plant beneficial *Pseudomonas monteilii* was found to enhance its plant probiotic effect. *Current Microbiology*, *76*, 503–509.

Paulkumar, K., Gnanajobitha, G., Vanaja, M., Rajeshkumar, S., Malarkodi, C., Pandian, K., & Annadurai, G. (2014). *Piper nigrum* leaf and stem assisted green synthesis of silver nanoparticles and evaluation of its antibacterial activity against agricultural plant pathogens. *The Scientific World Journal*, *2014*, 829894. https://doi.org/10.1155/2014/829894 (accessed on 6th November 2025).

Pighi, L., Pumpel, T., & Schinner, F. (1989). Selective accumulation of silver by fungi. *Biotechnology Letters*, *11*, 275–280.

Pimprikar, P. S., Joshi, S. S., Kumar, A. R., Zinjarde, S. S., & Kulkarni, S. K. (2009). Influence of biomass and gold salt concentration on nanoparticle synthesis by the tropical marine yeast *Yarrowia lipolytica* NCIM 3589. *Colloids and Surfaces B: Bio Interfaces*, *74*, 309–316.

Pinto, R. J. B., Bispo, D., Vilela, C., Botas, A. M. P., Ferreira, R. A. S., Menezes, A. C., Campos, F., Oliveira, H., Abreu, M. H., Santos, S. A. O., & Freire, C. S. R. (2020). One-minute synthesis of size-controlled fucoidan-gold nano systems: Antitumoral activity and dark field imaging. *Materials (Basel)*, *13*(5), 1076. https://doi.org/10.3390/ma13051076 (accessed on 6th November 2025).

Pokorski, J. K., & Steinmetz, N. F. (2011). The art of engineering viral nanoparticles. *Molecular Pharmaceutics*, *8*, 29–44.

Popli, D., Anil, V., Subramanyam, A. B., Namratha, M. N., Rajitha, V. R., Rao, S. R., Rai, R. V., & Govindappa, M. (2017). Endophytic fungi, *Cladosporium* species-mediated synthesis of silver nanoparticles possessing *in-vitro* antioxidant, anti-diabetic and anti-Alzheimer activity. *Journal of Microscopy and Ultrastructure*, *46*, 676–683.

Prasad, R., Bhattacharya, A., & Nguyen, Q. D. (2017). Nanotechnology in sustainable agriculture: Recent developments, challenges and perspectives. *Frontiers in Microbiology*, *8*, 1014. https://doi.org/10.3389/fmicb.2017.01014 (accessed on 6th November 2025).

Priyadharshini, R. I., Prasannaraj, G., Geetha, N., & Venkatachalam, P. (2014). Microwave mediated extracellular synthesis of metallic silver and zinc oxide nanoparticles using macro-algae (*Gracilaria edulis*) extracts and its anticancer activity against human PC3 cell lines. *Applied Biochemistry and Biotechnology*, *174*(8), 2777–2790.

Raeisi Sadati, S. Y., Jahanbakhsh Godehkahriz, S., Ebadi, A., & Sedghi, M. (2022). Zinc oxide nanoparticles enhance drought tolerance in wheat via physio-biochemical changes and stress genes expression. *Iranian Journal of Biotechnology*, *20*(1), 12–24.

Rahmatizadeh, R., Jamei, R., Arvin, M. J., & Rezanejad, F. (2021). Upregulation of *LeNRAMP3* and *LeFER* genes in *Solanum lycopersicum* confers its cadmium tolerance. *Russian Journal of Plant Physiology*, *68*, S92–S102.

Rai, M., Bonde, S., Golinska, P., Trzcinska-Wencel, J., Gade, A., Abd-Elsalam, K. A., Shende, S., Gaikwad, S., & Ingle, A. P. (2021). *Fusarium* as a novel fungus for the synthesis of nanoparticles: Mechanism and applications. *Journal of Fungi*, *7*(2), 139. https://doi.org/10.3390/jof7020139 (accessed on 6th November 2025).

Rai, R. V., & Bai, J. A. (2011). Nanoparticles and their potential application as antimicrobials. In A. Méndez-Vilas (Ed.), *Science Against Microbial Pathogens: Communicating Current Research and Technological Advances* (pp. 197–209). Formatex Research Center, Badajoz, Spain.

Rai, S., Solanki, M. K., Solanki, A. C., & Samal, S. (2023). Microbial endophytes as probiotics for the plant health: An overview. In M. K. Solanki, M. K. Yadav, B. P. Singh, & V. K. Gupta (Eds.), *Microbial Endophytes and Plant Growth* (pp. 269–281). Academic Press.

Rajasree, R. S. R., & Suman, T. Y. (2012). Extracellular biosynthesis of gold nanoparticles using a Gram-negative bacterium *Pseudomonas fluorescens*. *Asian Pacific Journal of Tropical Biomedicine*, *2*, S796–S799.

Rajasulochana, P., Dhamotharan, R., Murugakoothan, P., Murugesan, S., & Krishnamoorthy, P. (2010). Biosynthesis and characterization of gold nanoparticles using the alga *Kappaphycus alvarezii*. *International Journal of Nanoscience*, *9*(5), 511–516.

Rajeshkumar, S., Malarkodi, C., Gnanajobitha, G., Paulkumar, K., Vanaja, M., Kannan, C., & Annadurai, G. (2013). Seaweed-mediated synthesis of gold nanoparticles using *Turbinaria conoides* and its characterization. *Journal of Nanostructure in Chemistry*, *3*, 44. https://doi.org/10.1186/2193-8865-3-44 (accessed on 6th November 2025).

Rajora, N., Kaushik, S., Jyoti, A., & Kothari, S. L. (2016). Rapid synthesis of silver nanoparticles by *Pseudomonas stutzeri* isolated from textile soil under optimized conditions and evaluation of their antimicrobial and cytotoxicity properties. *IET Nanobiotechnology*, *10*(6), 367–373. https://doi.org/10.1049/iet-nbt.2015.0107 (accessed on 6th November 2025).

Rakib-Uz-Zaman, S. M., Apu, E. H., Muntasir, M. N., Mowna, S. A., Khanom, M. G., Jahan, S. S., Akter, N., Khan, M. A. R., Shuborna, N. S., Shams, S. M., & Khan, K. (2022). Biosynthesis of silver nanoparticles from *Cymbopogon citratus* leaf extract and evaluation of their antimicrobial properties. *Challenges*, *13*(1), 18. https://doi.org/10.3390/challe13010018 (accessed on 6th November 2025).

Raliya, R., Saharan, V., Dimkpa, C., & Biswas, P. (2018). Nano fertilizer for precision and sustainable agriculture: Current state and future perspectives. *Journal of Agricultural and Food Chemistry*, *66*, 6487–6453.

Ramanan, R., Kim, B. H., Cho, D. H., Oh, H. M., & Kim, H. S. (2016). Algae-bacteria interactions: Evolution, ecology, and emerging applications. *Biotechnology Advances*, *34*, 14–29.

Ramsden, J. J. (2005). What is nanotechnology? *Nanotechnology Perceptions*, *1*, 3–17.

Ranjani, A., Gopinath, P. M., Ananth, S., Narchonai, G., Santhanam, P., Thajuddin, N., & Dhanasekaran, D. (2018). Multidimensional dose–response toxicity exploration of silver nanoparticles from *Nocardiopsis flavascens* RD30. *Applied Nanoscience*, *8*, 699–713.

Ranjani, S., Shariq, A. M., Adnan, M., Senthil-Kumar, N., Ruckmani, K., & Hemalatha, S. (2020). Synthesis, characterization and applications of endophytic fungal nanoparticles. *Inorganic and Nano-Metal Chemistry*, *51*(2), 280–287.

Raut, R., Lakkakula, J., Kolekar, N., Mendhulkar, V. D., & Sahebrao, K. (2009). Phytosynthesis of silver nanoparticle using *Gliricidia sepium* (Jacq.). *Current Nanoscience*, *5*, 112–117.

Reddy, E. C., Reddy, G. S., Goudar, V., Sriramula, A., Swarnalatha, G. V., Al Tawaha, A. R. M., & Sayyed, R. Z. (2022). Hydrolytic enzyme-producing plant growth-promoting rhizobacteria (PGPR) in plant growth promotion and biocontrol. In R. Z. Sayyed & V. G. Uarrota (Eds.), *Secondary Metabolites and Volatiles of PGPR in Plant-Growth Promotion* (pp. 303–312). Springer, Cham. https://doi.org/10.1007/978-3-031-07559-9_15 (accessed on 6th November 2025).

Roy, A., Elzaki, A., Tirth, V., Kajoak, S., Osman, H., Algahtani, A., Islam, S., Faizo, N. L., Khandaker, M. U., Islam, M. N., Emran, T. B., & Bilal, M. (2021). Biological synthesis of nano catalysts and their applications. *Catalysts*, *11*(12), 1494. https://doi.org/10.3390/catal11121494 (accessed on 6th November 2025).

Roy, P., Das, B., Mohanty, A., & Mohapatra, S. (2017). Green synthesis of silver nanoparticles using *Azadirachta indica* leaf extract and its antimicrobial study. *Applied Nanoscience*, *7*, 843–850.

Saffan, M. M., Koriem, M. A., El-Henawy, A., El-Mahdy, S., El-Ramady, H., Elbehiry, F., Omara, A. E. D., Beyoumi, Y., Badgar, K., & Prokisch, J. (2022). Sustainable production of tomato plants (*Solanum lycopersicum* L.) under low-quality irrigation water as affected by bio-nano fertilizers of selenium and copper. *Sustainability*, *14*(6), 3236. https://doi.org/10.3390/su14063236 (accessed on 6th November 2025).

Sahil, K., Sahil Keshan, R., Patra, A., Mehta, S., Abdelmotelb, K. F., Lavale, S. A., Chaudhary, M., Aggarwal, S. K., & Chattopadhyay, A. (2021). Expression and regulation of stress-responsive genes in plants under harsh environmental conditions. In A. Husen (Ed.), *Harsh Environment and Plant Resilience* (pp. 25–44). Springer, Cham. https://doi.org/10.1007/978-3-030-65912-7_2 (accessed on 6th November 2025).

Salama, A. M., Abd El-Halim, A. E. H. A., Ibrahim, M. M., Aiad, M. A., & El-Shal, R. M. (2022). Amendment with nanoparticulate gypsum enhances spinach growth in saline-sodic soil. *Journal of Soil Science and Plant Nutrition*, *22*, 3377–3385.

Salvadori, M. R., Ando, R. A., Nascimento, C. A. O., & Corrêa, B. (2017). Dead biomass of Amazon yeast: A new insight into bioremediation and recovery of silver by intracellular synthesis of nanoparticles. *Journal of Environmental Science and Health Part A*, *52*(11), 1112–1120. https://doi.org/10.1080/10934529.2017.1340754 (accessed on 6th November 2025).

Salvadori, M. R., Ando, R. A., Oller do Nascimento, C. A., & Corrêa, B. (2014). Intracellular biosynthesis and removal of copper nanoparticles by dead biomass of yeast isolated from the wastewater of a mine in the Brazilian Amazonia. *PLoS ONE*, *9*(1), e87968. https://doi.org/10.1371/journal.pone.0087968 (accessed on 6th November 2025).

Sanghi, R., Verma, P., & Puri, S. (2011). Enzymatic formation of gold nanoparticles using *Phenerochaete chrysosporium*. *Advances in Chemical Engineering and Science*, *1*, 154–162.

Santhoshkumar, T., Rahuman, A. A., Jayaseelan, C., Rajakumar, G., Marimuthu, S., Kirthi, A. V., Velayutham, K., Thomas, J., Venkatesan, J., & Kim, S. K. (2014). Green synthesis of titanium dioxide nanoparticles using *Psidium guajava* extract and its antibacterial and antioxidant properties. *Asian Pacific Journal of Tropical Medicine*, *7*(12), 968–976.

Santiago, T. R., Bonatto, C., Rossato, M., Lopes, C., Mizubuti, E., & Silva, L. (2019). Green synthesis of silver nanoparticles using tomato leaf extract and their entrapment in chitosan nanoparticles to control bacterial wilt. *Journal of the Science of Food and Agriculture*, *99*, 4248–4259.

Santos, T. C., Silva, T. M., Cardosco, J. C., Albuquerque-Júnior, R. L. C., Zielinska, A., Souto, E. B., Severino, P., & Mendonça, M. D. C. (2021). Biosynthesis of silver nanoparticles mediated by entomopathogenic fungi: Antimicrobial resistance, nano pesticides, and toxicity. *Antibiotics*, *10*(7), 852.

Saravanan, M., Barik, S. K., Mubarakali, D., Prakash, P., & Pugazhendhi, A. (2018). Synthesis of silver nanoparticles from *Bacillus brevis* (NCIM2533) and their antibacterial activity against pathogenic bacteria. *Microbial Pathogenesis*, *116*, 221–226.

Sarkar, J., Chakraborty, N., Chatterjee, A., Bhattacharjee, A., Dasgupta, D., & Acharya, K. (2020). Green synthesized copper oxide nanoparticles ameliorate defense and antioxidant enzymes in *Lens culinaris*. *Nanomaterials*, *10*(2), 19136. https://doi.org/10.3390/nano10020312 (accessed on 6th November 2025).

Sayed, A. M. M., & Kim, S. (2018). Myco-silver nanoparticles synthesized using *Beauveria bassiana* and *Metahizium brunnem* as a smart pest control. *Egyptian Journal of Plant Protection Research Institute*, *1*, 1–18.

Scholthof, K. B. G., Adkins, S., Czosnek, H., Palukaitis, P., Jacquot, E., Hohn, T., Hohn, B., Saunders, K., Candresse, T., Ahlquist, P., Hemenway, C., & Foster, G. D. (2011). Top 10 plant viruses in molecular plant pathology. *Molecular Plant Pathology*, *12*(9), 938–954.

Segal-Kischinevzky, C., Romero-Aguilar, L., Alcaraz, L. D., López-Ortiz, G., Martínez-Castillo, B., Torres-Ramírez, N., Sandoval, G., & González, J. (2020). Yeast inhabiting extreme environments and their biotechnological applications. *Microorganisms*, *10*(4), 794. https://doi.org/10.3390/microorganisms10040794 (accessed on 6th November 2025).

Seisenbaeva, G. A., Daniel, G., Nedelec, J. M., & Kessler, V. G. (2013). Solution equilibrium behind the room-temperature synthesis of nanocrystalline titanium dioxide. *Nanoscale*, *5*, 3330–3336.

Sekhon, B. S., Manjunatha, S. B., Biradar, D. P., & Aladakatti, Y. R. (2014). Nanotechnology and its applications in agriculture: A review. *Journal of Farm Sciences*, *29*, 31–53.

Senapati, S., Syed, A., Moeez, S., Kumar, A., & Ahmad, A. (2012). Intracellular synthesis of gold nanoparticles using alga *Tetraselmis kochinensis*. *Materials Letters*, *79*, 116–118.

Serrano, E., Rus, G., & Garcia-Martinez, J. (2009). Nanotechnology for sustainable energy. *Renewable and Sustainable Energy Reviews*, *13*, 2373–2384.

Shalaby, E. A. (2011). Algae as promising organisms for environment and health. *Plant Signaling & Behavior*, *6*, 1338–1350.

Shankar, P. D., Shobana, S., Karuppusamy, I., Pugazhendhi, A., Ramkumar, V. S., Arvindnarayan, S., & Kumar, G. (2016). A review on the biosynthesis of metallic nanoparticles (gold and silver) using biocomponents of microalgae: Formation mechanism and applications. *Enzyme and Microbial Technology*, *95*, 28–44.

Sharma, A., Kumar, S., & Singh, R. (2022). Synthesis and characterization of a novel slow-release nanourea/chitosan nanocomposite and its effect on *Vigna radiata* L. *Environmental Science: Nano*, *9*, 4177–4189.

Sharma, A., Sharma, S., Sharma, K., Chetri, S. P. K., Vashishtha, A., Singh, P., Kumar, R., Rathi, B., & Agrawal, V. (2016). Algae as crucial organisms in advancing nanotechnology: A systematic review. *Journal of Applied Phycology*, *28*, 1759–1774.

Sharma, D., Kanchi, S., & Bisetty, K. (2019). Biogenic synthesis of nanoparticles: A review. *Arabian Journal of Chemistry*, *12*, 3576–3600.

Shende, S. S., Gaikwad, N. D., & Bansod, S. D. (2016). Synthesis and evaluation of antimicrobial potential of copper nanoparticle against agriculturally important phytopathogens. *International Journal of Biological Research*, *1*, 41–47.

Shenton, W., Douglas, T., Young, M., Stubbs, G., & Mann, S. (1999). Inorganic-organic nanotube composites from template mineralization of tobacco mosaic virus. *Advanced Materials*, *11*, 253–265.

Sidkey, N. M., Arafa, R. A., Moustafa, Y. M. M. M., Morsi, R. E., & El Hateir, M. M. (2020). Biosynthesis of Mg and Mn intracellular nanoparticles via extremo-metallotolerant *Pseudomonas stutzeri* B4 Mg/W and *Fusarium nygamai* F4 Mn/S. *Journal of Microbiology, Biotechnology and Food Sciences*, *9*, 1181–1187.

Singaravelu, G., Arockiamary, J. S., Kumar, V. G., & Govindaraju, K. (2007). A novel extracellular synthesis of monodisperse gold nanoparticles using marine alga, *Sargassum wightii* Greville. *Colloids and Surfaces B: Bio Interfaces*, *57*(1), 97–101.

Singh, D., Rathod, V., Ninaganagouda, S., Hiremath, J., Singh, A. K., & Mather, J. (2014). Optimization and characterization of silver nanoparticle by endophytic fungi *Penicillium* sp. isolated from *Curcuma longa* (turmeric) and application studies against MDR *E. coli* and *S. aureus*. *Bioinorganic Chemistry and Applications*, *2014*, 408021. https://doi.org/10.1155/2014/408021 (accessed on 6th November 2025).

Singh, D., Rathod, V., Ninganagouda, S., Herimath, J., & Kulkarni, P. (2013). Biosynthesis of silver nanoparticle by endophytic fungi *Penicillium* sp. isolated from *Curcuma longa* (turmeric) and its antibacterial activity against pathogenic Gram-negative bacteria. *Journal of Pharmaceutical Research*, *7*, 448–453. https://doi.org/10.1016/j.jopr.2013.06.003 (accessed on 6th November 2025).

Singh, P., Gonzalez, M. J., & Manchester, M. (2006). Viruses and their uses in nanotechnology. *Drug Development Research*, *67*, 23–41.

Singh, P., Kim, Y. J., Zhang, D., & Yang, D. C. (2016). Biological synthesis of nanoparticles from plants and microorganisms. *Trends in Biotechnology*, *34*, 588–599.

Sintubin, L., Verstraete, W., & Boon, N. (2012). Biologically produced nano silver: Current state and future perspectives. *Biotechnology and Bioengineering, 109*, 2422–2436.

Skalickova, S., Boron, M., & Sochor, J. (2017). Nanoparticles biosynthesized by yeast: A review of their applications. *Kvasný Průmysl, 63*(6), 290–292.

Sonawane, H., Shelke, D., Chambhare, M., Dixit, N., Math, S., Sen, S., Borah, S. N., Islam, N. F., Joshi, S. J., Yousaf, B., Rinklebe, J., & Sarma, H. (2022). Fungi-derived agriculturally important nanoparticles and their application in crop stress management–prospects and environmental risks. *Environmental Research, 212*, 113543. https://doi.org/10.1016/j.envres.2022.113543 (accessed on 6th November 2025).

Spagnoletti, F. N., Spedalieri, C., Kronberg, F., & Giacometti, R. (2019). Extracellular biosynthesis of bactericidal Ag/AgCl nanoparticles for crop protection using the fungus *Macrophomina phaseolina*. *Journal of Environmental Management, 231*, 457–466.

Sreedharan, S. M., Gupta, S., Saxena, A. K., & Singh, R. (2019). *Macrophomina phaseolina*: Micro based biorefinery for gold nanoparticle production. *Annals of Microbiology, 69*, 435–455.

Srinivasa, C., Kavitha, G. C., Pallavi, C., Shivamallu, C., Sushma, P., Kollur, S. P., Aiyaz, M., Shukla, A. K., Murali, M., & Ansari, M. A. (2021). Role of viruses in nanoparticles synthesis. In Ansari, M. A., & Rehman, S. (Eds.), *Microbial Nanotechnology: Green Synthesis and Applications* (pp. 103–119). Springer, Singapore. https://doi.org/10.1007/978-981-16-1923-6_6J (accessed on 6th November 2025).

Srivastava, S., & Bhargava, A. (2022). Biological synthesis of nanoparticles: Fungi. In *Green Nanoparticles: The Future of Nanobiotechnology* (pp. 101–137). Springer, Singapore. https://doi.org/10.1007/978-981-16-7106-7_6 (accessed on 6th November 2025).

Subhapriya, S., & Gomathipriya, P. (2018). Green synthesis of titanium dioxide (TiO_2) nanoparticles by *Trigonella foenum-graecum* extract and its antimicrobial properties. *Microbial Pathogenesis, 116*, 215–220.

Subotić, A., Jevremović, S., Milošević, S., Trifunović-Momčilov, M., Đurić, M., & Koruga, Đ. (2022). Physiological response, oxidative stress assessment and aquaporin genes expression of cherry tomato (*Solanum lycopersicum* L.) exposed to hyper-harmonized fullerene water complex. *Plants, 11*(21), 2810. https://doi.org/10.3390/plants11212810 (accessed on 6th November 2025).

Sukhwal, A., Jain, D., Joshi, A., Rawal, P., & Kushwaha, H. S. (2016). Biosynthesized silver nanoparticles using aqueous leaf extract of *Tagetes patula* L. and evaluation of their antifungal activity against phytopathogenic fungi. *IET Nanobiotechnology, 11*, 531–537.

Sun, Y., Shi, Y., Jia, H., Ding, H., Yue, T., & Yuan, Y. (2021). Biosynthesis of selenium nanoparticles of *Monascus purpureus* and their inhibition to *Alicyclobacillus acidoterrestris*. *Food Control, 130*, 108366. https://doi.org/10.1016/j.foodcont.2021.108366 (accessed on 6th November 2025).

Tao, Z., Yuan, H., Liu, M., Zhang, S., Liu, H., Jiang, Y., Huang, D., & Wang, T. (2023). Yeast extract: Characteristics, production, applications and future perspectives. *Journal of Microbiology and Biotechnology, 33*(2), 151–166.

Tarekgen, M. M., Salilih, F. Z., Ishetu, A. I., & Yilidiz, F. (2020). Microbes used as a tool for bioremediation of heavy metal from the environment. *Food Science and Technology, 6*(1), 1783174. https://doi.org/10.1080/23311932.2020.1783174 (accessed on 6th November 2025).

Tarver, S., Gray, D., Loponov, K., Das, D. B., Sun, T., & Sotenko, M. (2019). Biomineralization of Pd nanoparticles using *Phenerochaete chrysosporium* as a sustainable approach to turn platinum group metals (PGMs) wastes into catalyst. *International Biodeterioration & Biodegradation, 143*, 104724. https://doi.org/10.1016/j.ibiod.2019.104724 (accessed on 6th November 2025).

Tripathi, S., Srivastava, P., Devi, R. S., & Bhadouria, R. (2020). Influence of synthetic fertilizers and pesticides on soil health and soil microbiology. In Prasad, M. N. V. (Ed.), *Agrochemicals Detection, Treatment and Remediation: Pesticides and Chemical Fertilizers* (pp. 25–54). Butterworth Heinemann, Elsevier, Ltd. https://doi.org/10.1016/B978-0-08-103017-2.00002-7 (accessed on 6th November 2025).

Tugarova, A. V., Mamchencova, P. V., Khanadeev, V. A., & Kamnev, A. A. (2020). Selenite reduction by the *Azospirillum brasilense*, synthesis of extracellular selenium nanoparticles and their characterization. *New Biotechnology, 58*, 17–24.

Upadhayay, V. K., Singh, A. V., Khan, A., Singh, J., Pareek, N., & Raghav, A. (2022b). FE-SEM/EDX based zinc mobilization analysis of *Burkholderia cepacia* and *Pantoea rodasii* and their functional annotation in crop productivity, soil quality, and zinc biofortification of paddy. *Frontiers in Microbiology, 13*, 852192. https://doi.org/10.3389/fmicb.2022.852192 (accessed on 6th November 2025).

Upadhyay, V. K., Chitara, M. K., Mishra, D., Jha, M. N., Jaiswal, A., Kumari, G., Ghosh, S., Patel, V. K., Naitam, M. G., Singh, A. K., Pareek, N., Taj, G., Maithani, D., Kumar, A., Dasila, H., & Sharma, A. (2023). Synergistic impact of nanomaterials and plant probiotics in agriculture: A tale of two-way strategy for long term sustainability. *Frontiers in Microbiology, 14*, 1133968. https://doi.org/10.3389/fmicb.2023.1133968 (accessed on 6th November 2025).

Van Nguyen, D., Nguyen, H. M., Le, N. T., Nguyen, K. H., Nguyen, H. T., Le, H. M., Nguyen, A. T., Dinh, N. T. T., Hoang, S. A., & Ha, C. V. (2022). Copper nanoparticle application enhances plant growth and grain yield in maize under drought stress conditions. *Journal of Plant Growth Regulation, 41*, 364–375.

Varadavenkatesan, T., Selvaraj, R., & Vinayagam, R. (2018). Dye degradation and antibacterial activity of green synthesized Ag-NPs using *Ipomoea digitata* Linn. flower extract. *International Journal of Science and Environment and Technology, 140*, 35–45.

Vasant, G., Bhatt, S., Raghav, R., & Joshi, P. J. (2023). Revitalization of PGPR through integrating nanotechnology for sustainable development in agriculture. In *Plant-Microbe Interaction–Recent Advances in Molecular and Biochemical Approaches, Volume 2: Agriculture Aspects of Microbiome Leading to Plant Defense* (pp. 227–248). Academic Press. https://doi.org/10.1016/B978-0-323-91876-3.00009-9 (accessed on 6th November 2025).

Vermier, I., Weitzers, B., Houwer, J. D., Geuens, M., Slabbinck, H., Spruyt, A., Van Kerckhove, A., Van Lippevelde, W., De Steur, H., & Verbeke, W. (2020). Environmentally sustainable food consumption: A review and research agenda from a goal directed perspective. *Frontiers in Psychology, 11*, 1603. https://doi.org/10.3389/fpsyg.2020.01603 (accessed on 6th November 2025).

Vetchinkina, E., Loshchinina, E., Kupryashina, M., Burov, A., & Pylaev, T. (2018). Green synthesis of nanoparticles with extracellular and intracellular extracts of basidiomycetes. *PeerJ, 6*, e5237. https://doi.org/10.7717/peerj.5237 (accessed on 6th November 2025).

Vijayabharathi, R., Arumugam, S., & Gopalkrishnan, S. (2018). Extracellular biosynthesis of silver nanoparticles using *Streptomyces griseoplanus* SAI-25 and its antifungal activity against *Macrophomina phaseolina*, the charcoal rot pathogen of sorghum. *Biocatalysis and Agricultural Biotechnology, 14*, 166–171.

Vijayan, S. R., Santhiyagu, P., Singamuthu, M., Ahila, N. K., Jayaraman, R., & Ethiraj, K. (2014). Synthesis and characterization of silver and gold nanoparticles using aqueous extract of seaweed, *Turbinaria conoides*, and their antimicrofouling activity. *The Scientific World Journal, 2014*, 938272. https://doi.org/10.1155/2014/938272 (accessed on 6th November 2025).

Vitta, Y., Figueroa, M., Calderon, M., & Ciangherotti, C. (2020). Synthesis of iron nanoparticles from aqueous extract of *Eucalyptus robusta* Sm and evaluation of antioxidant and antimicrobial activity. *Materials Science for Energy Technologies, 3*, 97–103.

Wadhwani, S. A., Shedbalkar, U. U., Singh, R., Vashisth, P., Pruthi, V., & Chopade, B. A. (2016). Kinetics of synthesis of gold nanoparticles by *Acinetobacter* sp. SW30 isolated from environment. *Indian Journal of Microbiology, 56*, 439–444.

Waghmare, S. R., Mulla, M. N., Marathe, S. R., & Sonawane, K. D. (2015). Ecofriendly production of silver nanoparticles using *Candida utilis* and its mechanistic action against pathogenic microorganisms. *3 Biotech, 5*, 33–38. https://doi.org/10.1007/s13205-014-0196-y (accessed on 6th November 2025).

Wang, D., Xue, B., Wang, L., Zhang, Y., Liu, L., & Zhou, Y. (2021). Fungus-mediated green synthesis of nano-silver using *Aspergillus sydowii* and its antifungal/antiproliferative activities. *Scientific Reports, 11*(1), 10356. https://doi.org/10.1038/s41598-021-89854-5 (accessed on 6th November 2025).

Wang, W., Yao, F., Wang, D., Zhou, X., Wang, X., Zhao, C., Fang, M., & Cai, J. (2017). Biosynthesis of gold nanoparticles using *Pleurotus ostreatus* extract with their electrochemical activity of detection of carbenazin in vegetable. *International Journal of Electrochemical Science, 12*, 842–851.

Wang, X., Xie, H., Wang, P., & Yin, H. (2023). Nanoparticles in plants: Uptake, transport and physiological activity in leaf and root. *Materials (Basel), 16*(8), 3097. https://doi.org/10.3390/ma16083097 (accessed on 6th November 2025).

Yadav, A., Theivasanthi, T., Paul, P. K., & Upadhyay, K. C. (2013). Extracellular biosynthesis of silver nanoparticles from plant growth promoting rhizobacteria *Pseudomonas* sp. *International Journal of Current Microbiology and Applied Sciences, 4*, 1057–1068.

Yadav, N., Garg, V. P., Chillar, A. K., & Rana, J. S. (2023). Recent advances in nanotechnology for the improvement of conventional agricultural systems: A review. *Plant Nano Biology, 4*, 100032. https://doi.org/10.1016/j.plana.2023.100032 (accessed on 6th November 2025).

Yong, E. (2012). Yeast suggests speedy start for multicellular life. *Nature*. https://doi.org/10.1038/nature.2012.9810 (accessed on 6th November 2025).

Yue, L., Feng, Y., Ma, C., Wang, C., Chen, F., Cao, X., Wang, J., White, J. C., Wang, Z., & Xing, B. (2022). Molecular mechanisms of early flowering in tomatoes induced by manganese ferrite ($MnFe_2O_4$) nanomaterials. *ACS Nano, 16*(4), 5636–5646.

Zaitlin, M. (1998). The discovery of the causal agent of the tobacco mosaic disease. In Kung, S. D., & Yang, S. F. (Eds.), *Discoveries in Plant Biology* (pp. 105–110). World Publishing.

Zelleke, G., Agegnehu, G., Abera, D., & Rashid, S. (2019). Fertilizer and soil fertility potential in Ethiopia. *Gates Open Research, 3*, 482. https://doi.org/10.21955/gatesopenres.1115635.1 (accessed on 6th November 2025).

Zhang, T., Qi, M., Wu, Q., Xiang, P., Tang, D., & Li, Q. (2023). Recent research progress on the synthesis and biological effects of selenium nanoparticles. *Frontiers in Nutrition,*

10, 1183487. https://doi.org/10.3389/fnut.2023.1183487 (accessed on 6th November 2025).

Zhang, X., Yan, S., Tyagi, R. D., & Surampalli, R. Y. (2011). Synthesis of nanoparticles by microorganisms and their application in enhancing microbiological reaction rates. *Chemosphere, 82*, 489–494.

CHAPTER 15

Nano-Biofertilizers in Agriculture: A Sustainable Method for Increasing Crop Yields

SHANIYA ZEHRA,[1] PRABHA RAI KALAL,[1] MAHAVEER P. SHARMA,[2] AND ANJANA JAJOO[1,3,*]

[1]*School of Life Sciences, Devi Ahilya University, Indore, Madhya Pradesh, India*

[2]*ICAR–National Soybean Research Institute, Indore, Madhya Pradesh, India*

*Corresponding author

ABSTRACT

Globally, the rising human population has raised serious concerns about agriculture productivity and sustainability. Current agriculture uses large amounts of chemical fertilizers, which increases the risk of environmental pollution, ultimately jeopardizing the ecosystem's sustainability. One new area of science that is relevant to sustainable agriculture is nanotechnology, which offers innovative solutions to maximize crop yield while minimizing negative effects on the environment. Nanomaterials ensure precise and site-target delivery of nutrients, pesticides, and fertilizers directly to the plants due to their higher surface area and smaller size (<100 nm), reducing waste and optimizing resource use. Recently, in the agricultural landscape, biofertilizers have been efficiently used as biostimulants (BS) and plant growth promoters. Nano-biofertilizers (NBF) are

Plant Biostimulants and Growth Regulators in Sustainable Agriculture.
Deepak Kumar, Sanjay K. Singh, Nutan Kaushik, and Mahaveer P. Sharma (Eds.)

DOI: 10.1201/9781779641458-20

a blend of nanomaterials and biofertilizers that show great promise as a more economical and environmentally beneficial option to conventional agrochemical methods. The synthesis and application of nanobiofertilizer is a useful step towards smart fertilizer that enhances plant growth and augments crop yield. However, the literature on the formulation and application of nano-biofertilizers at the commercial level is limited. This book chapter will discuss the significance of NBF, procedures to prepare NBF, and its applications in overcoming the challenges of conventional agronomic practices. It will also discuss limitations and future strategies to promote the use of nanobiofertilizer in sustainable agriculture.

15.1 INTRODUCTION

The 20th-century Green Revolution marked a significant increase in global food production. This revolution was mainly driven by two key advances: the use of chemical inputs to improve crop yields and advanced genetic modifications involving plant breeding. Fertilization has played a crucial role in modern agriculture, enhancing agricultural efficiency and ensuring higher-quality crop yields. In the last two decades, there has been an exponential increase in consumption of chemical fertilizers worldwide (Savci, 2012). Agrochemicals work best when applied at the recommended dose and help overcome nutrient deficiencies to increase crop growth. Nevertheless, the benefits linked to chemical fertilizers come at a high cost to the environment. High-dose chemical fertilizers have been applied frequently, increasing their concentration in cropland and causing irreversible damage to the soil's structure (Solanki et al., 2015). Moreover, a significant amount of these agrochemicals enters the food chain, inducing heritable mutations in consumers and affecting future generations (Solanki et al., 2015). Also, plants can only utilize half of the chemical fertilizers that are applied because of their larger particle size and chemical composition, which results in significant waste. This not only results in economic losses but also makes agriculture less sustainable (Sharma and Singhvi, 2017). Due to this, to make agriculture sustainable, it is necessary to explore eco-friendly and cost-effective practices. Nanotechnology is playing a crucial role in the current era of agriculture as it offers innovative solutions to enhance crop productivity while minimizing environmental impacts. Using nanomaterials in fertilizers, the so-called "nano-fertilizers," is a relatively new technology and is considered a viable solution for enhancing fertilizer use efficiency, mitigating nutrient deficiencies, and thus improving crop yields. Nano-fertilizers are also being used as bio-stimulants as an alternative to harmful agrochemicals, which stimulate plant growth and also provide protection against various biotic and abiotic stresses (Sharma et al., 2019). Nanotechnology offers precise and effective delivery of nutrients and fertilizers directly to the plants, reducing waste and thus optimizing resource use. In recent times, there has been a growing trend towards the utilization of smart fertilizers known as nano-biofertilizers (NBF). These innovative products are a result of advances in nano-biotechnology and are created by combining nanoparticles (NPs) with biofertilizers (Tekam et al., 2023). As a result, NBF combines the beneficial properties of both NPs and biofertilizers into a single, integrated solution. NBFs excel in their ability to release nutrients to plants gradually and in a controlled, slow-release manner. As a result, they reduce the reliance on chemical fertilizers, proving to be both cost-effective and environmentally friendly. Efforts should be directed towards researching and developing innovative combinations of nano-biofertilizers to enable their widespread commercialization, making them accessible to farmers with convenient storage and application methods. The large-scale commercialization of nano-biofertilizers is still in its early stages. This chapter primarily focuses on the importance, optimization, and formulation of nano-biofertilizers, explaining their modes of action as well, all geared towards promoting sustainable agriculture in the future.

15.2 BIOFERTILIZERS

The primary purpose of agriculture is to address the rising demand for food security in the expanding global human population, necessitating higher crop yields per unit area (Kumar et al., 2018). To achieve this, farmers adopt various practices, including soil fertilization, pest control, efficient irrigation, and improved agricultural processes (Savci, 2012). The use of intensive management practices has exponentially increased the consumption of synthetic fertilizers worldwide. This has resulted in environmental hazards in terms of degrading soil quality and adversely affecting the biogeochemical cycles (Sharma and Singhvi, 2017). Biofertilizers are considered a sustainable and eco-friendly alternative to chemical fertilizers and can improve soil health in the long term (Kaur and Kaur, 2018). Biofertilizers are natural or organic substances that contain living plant growth-promoting microorganisms (PGPM), such as bacteria, fungi, and algae, that, when applied to soils, seeds, or plant surfaces, help enhance soil fertility and promote plant growth by facilitating nutrient uptake. These microorganisms form a beneficial symbiotic relationship with plants and assist in nutrient cycling and availability. PGPM participates in biological processes, such as nitrogen, sulfur, or phosphorus fixation, and stimulates plant growth through the synthesis of growth-promoting substances in rhizosphere which contribute to the incorporation and enhancement of nutrients in the soil (Rojas-Sánchez et al., 2022). These nutrients are enriched by various organisms, including *Rhizobium* spp., *Sinorhizobium* spp., *Mesorhizobium* spp., *Azotobacter* spp., *Azospirillum* spp., *Pseudomonas* spp., *Bacillus* spp., *Aureobasidium pullulans, Epicoccum nigrum, Scolecobasidium constrictum, Myrothecium cinctum, and Acidianus* spp. (Ali et al., 2021; Bargaz et al., 2021; Barney, 2020; Fedele et al., 2020; Ke et al., 2019; Monachon et al., 2019; Pace et al., 2020). These microorganisms are natural dwellers in plant rhizospheres as well as internal compartments of plants, which offer several advantages. The microorganisms in biofertilizers can enhance soil structure, nutrient cycling, and microbial diversity, leading to healthier and more fertile soils and reducing environmental pollution and soil degradation. Thus, biofertilizers promote sustainable farming practices by reducing the reliance on non-renewable resources, minimizing environmental impacts, and reducing the cost of soil fertilization as well (Brahmaprakash and Sahu, 2012).

15.2.1 TYPES OF BIOFERTILIZERS AND THEIR MECHANISMS

Different plant growth-promoting microorganisms (PGPM) used as biofertilizers are characterized into various categories based on their functions and behaviors (Table 15.1).

TABLE 15.1 Different Types of Biofertilizers

Types of Biofertilizers	Crops	References
Nitrogen-fixing biofertilizers: *Azoarcus, Burkholderia, Herbaspirillum* *Azorhizobium,* *Azospirillum, Gluconacetobacter* *Azotobacter,* *Frankia,* *Rhizobia, Rhizobium, Bradyrhizobium* spp., *bacillus subtilis*	Rice, wheat, sugarcane, barley, oats, tea, coffee, maize, sunflower, beetroot, coconut, *Alnus*, legumes, soybean, sorghum	Abadi et al. (2020); Elbeltagy et al. (2001); Freitas et al. (2022); Govindarajan et al. (2007); Hungria and Mendes (2015); Kao et al. (2003); Muñoz-Rojas and Caballero-Mellado (2003); Reinhold-Hurek and Hurek (1998); Sabry et al. (1997); Sahoo et al. (2014); Wani et al. (2013); Yanni et al. (2001)
Phosphorus solubilizing biofertilizers: *Phyllobacterium,* *P. aeruginosa* *B. polymyxa, Enterobacter* sp., *Psuedomonas* sp., *B. thurinjensis,*	Strawberry, potato, tomato, wheat	Flores-Felix and Silva (2015); Ghadamgahi et al. (2022); Shintu and Jayaram (2015); Timmusk et al. (2014); Zuluaga et al. (2020)
Potassium solubilizing biofertilizers: *Bacillus,* *paenibacillus*	Pepper, cucumber, black pepper	Han and Lee (2005); Sangeeth et al. (2012)
Phytohormones synthesizing biofertilizers: *Azobacter, Acinetobacter* *Bacillus, P. aeruginosa,* *rhizobium* *Acinetobacter beijerinckii, P. putida* *Azospirillum brasilense,* *Enterobacter* sp., *Psuedomonas* sp., *Actinomyces*	Cucumber, potato, pepper, tomato, lettuce, carrot, soybean, wheat, maize, rice, palm tree, clover	Ahmed and Hasnain (2010); Aloni et al. (2006); Fukami et al. (2018); Garcia-Fraile et al. (2012); Ghadamgahi et al. (2022); Hussain et al. (2023); Joo et al. (2005); Kang et al. (2014); Zuluaga et al. (2020)

15.2.1.1 NITROGEN-FIXING BIOFERTILIZERS

These biofertilizers consist of primary nitrogen-fixing microorganisms that can convert atmospheric nitrogen (N_2) into ammonia (NH_3) or other inorganic forms of nitrogen that plants can readily absorb. They are particularly useful for crops like legumes (e.g., peas, beans, and clover) that have a symbiotic relationship with nitrogen-fixing bacteria and belong to different genera. The most common examples in this category are:

15.2.1.1.1 Rhizobium

The most common type of agricultural inoculants in the current scenario are rhizobial inoculants for leguminous crops, which are the first examples of commercially available microbial products in agriculture (Santos et al., 2019). These bacteria form symbiotic relationships with the roots of leguminous crops including soybeans, clover, beans, and peas. Rhizobium bacteria convert atmospheric nitrogen (N_2) into ammonia (NH_3) or ammonium ions (NH_4^+), which are forms of nitrogen that plants can readily absorb in exchange for sugars. Collectively, the group of bacteria comprising *Rhizobium, Bradyrhizobium, Sinorhizobium, Azorhizobium,* and *Mesorhizobium* is referred to as Rhizobia. According to estimates, various legume crops can fix nitrogen at a rate ranging from 40 to 250 kg of nitrogen per hectare per year. In due course, the presence of leguminous crops and their associated *Rhizobium* bacteria can enhance soil fertility by increasing nitrogen content and improving soil health.

15.2.1.1.2 Azatobacter

Azotobacter is another genus of nitrogen-fixing bacteria. Unlike *Rhizobium,* these are free-living nitrogen-fixing bacteria that are found in the soil and can associate with the roots of a wide range of non-leguminous plants. *Azotobacter* bacteria fix atmospheric nitrogen (N_2) into ammonium ions (NH_4^+) or other organic forms of nitrogen through the enzymatic activity of nitrogenase, which enables them to convert atmospheric nitrogen into a more accessible form. They contribute to the nitrogen pool in the soil by fixing approximately 20 kg of nitrogen per hectare annually without symbiosis with atmospheric nitrogen (Baweja et al., 2020). Farmers sometimes use *Azotobacter*-rich soil for crop rotation or intercropping, as it can enhance soil fertility and provide nitrogen benefits to subsequent crops. They are commonly used as biofertilizers for non-leguminous plants, particularly rice, cotton, and vegetables. Results showed a significant increase in yields over those obtained with chemical fertilizers in crops like potato, carrot, rice, maize, cauliflower, tomato, cotton, sugarcane, and sorghum, and so forth (Baweja et al., 2020).

15.2.1.1.3 Azospirillum

These N_2-fixing microorganisms excel at nitrogen enrichment through the production of growth-promoting substances. Apart from fixing nitrogen at rates of about 20–40 kg ha^{-1}, these microorganisms produce plant growth-regulating substances (Wahane et al., 2020) like indole acetic acid, gibberellins, and cytokinin, offer disease resistance, enhance plant nutrients (N, P, and K), and boost drought tolerance. Moreover, they significantly influence plants, improving shoot and root development and increasing water and mineral uptake by plant roots (Gonzalez et al., 2005). These benefits persist even in the presence of stress like drought. This suggests that the *Azospirillum* application could be expanded to arid soils as a strategy to enhance crop resilience against drought.

15.2.1.1.4 Cyanobacteria

Cyanobacteria (blue-green algae), a group of microscopic, photosynthetic bacteria such as *Anabaena, Nostoc, Cylindrospermum, Aulosira,* and *Tolypothrix,* are excellent nitrogen fixers, abundantly found in marine environments, and have great potential to be used for sustainable agriculture (Pisciotta et al., 2012). Both free-living as well as symbiotic cyanobacteria (blue-green algae) have been harnessed in rice cultivation in India (Kumar et al., 2017). They promote plant growth by producing auxin, indole acetic acid, and gibberellic acid and fix around 20–30 kg N/ha in submerged rice fields, as they are abundant in paddy fields (Mishra et al., 2013). In addition to nitrogen, cyanobacteria enrich soil with extracellular carbohydrates, hormones, and many secondary metabolites and improves soil health. It increases soil porosity and soil-water holding capacity.

15.2.1.2 PHOSPHORUS-MOBILIZING AND SOLUBILIZING BIOFERTILIZERS

15.2.1.2.1 Arbuscular mycorrhizal fungi (phosphate mobilizing biofertilizer)

An arbuscular mycorrhizal fungus (AMF), as a biofertilizer, has gained traction in agriculture due to its potential to enhance soil quality. AMF develops a symbiotic association with plants, which helps in the uptake of fixed N (by bacteria) from soil and also improves nodulation initiated by rhizobia (Surendran and Vani, 2013). The AMF can absorb N in both organic (as amino acids) (Whiteside et al., 2012) and inorganic (as nitrate and ammonium) forms (Govindarajulu et al., 2005). Several amino acids, such as glycine, cysteine, serine, arginine, aspartic acid, glutamine acids, and cysteine, and so forth, are absorbed by the fungi and then converted to ammonium for translocation at the symbiotic interface with the plant (Smith and Smith 2011).

15.2.1.2.2 Phosphate solubilizing biofertilizers

Phosphorus primarily exists in the soil in two insoluble forms: inorganic and organic, for example, $AlPO_4$, $FePO_4$, and $Ca_3(PO_4)_2$ (Zhu et al., 2018). Plants can take up phosphorus in two soluble forms: monobasic ($H_2PO_4^-$) and dibasic (HPO_4^{2-}) ions (Bargaz et al., 2021). Phosphate-solubilizing bacteria (PSB) are a group of beneficial microorganisms that play a crucial role in making phosphorus (P) more available to plants in the soil. Phosphorus is an essential nutrient for plant growth and development, and it is often present in insoluble forms in the soil. PSB has the ability to solubilize these insoluble phosphates, converting them into forms that plants can absorb and utilize. These contain phosphate-solubilizing microorganisms, such as mycorrhizal fungi and PSB (e.g., *Pseudomonas* and *Bacillus* species). These PSBs, also known as 'Phosphobacterium,' not only enhance P solubilization but have been reported to solubilize elements like zinc (Zn), potassium (K), iron (Fe), manganese (Mn), and more in the soil (Tariq et al., 2022). Additionally, they stimulate root proliferation through the release of growth-promoting hormones. PSBs like *Pseudomonas putida* and *Bacillus megaterium*, along with phosphate-solubilizing fungi (PSF) like *Aspergillus* and *Penicillium*, fulfil approximately 20–25% of a plant's phosphorus requirements (Kumar et al., 2018). Commercially, *Bacillus megaterium*, marketed as BioPhos, has been shown to reduce phosphate fertilizer (up to 75%) needs for plantation crops (Mehnaz, 2016). In India, *B. polymyxa*, *Pseudomonas striata*, and *B. megaterium* strains of P-solubilizing bacteria have been commercialized under different brands by different biofertilizer-producing companies (Mehnaz, 2016).

15.2.1.3 POTASSIUM-SOLUBILIZING BIOFERTILIZERS

Potassium-solubilizing biofertilizers are microbial products that contain beneficial microorganisms capable of solubilizing insoluble potassium

(K) in the soil and making it available for plant uptake. Potassium is one of the essential macronutrients required for plant growth and development (Zahoor et al., 2017). However, potassium may exist in many soil types that are difficult for plants to acquire. Potassium-solubilizing biofertilizers help improve potassium availability for plants by converting insoluble potassium compounds into soluble forms that plants can absorb. Saprophytic bacteria, actinomycetes, and fungi are known as K-solubilizing microbes that are found in the root rhizosphere (Archana et al., 2013; Saiyad et al., 2015). In soil bacterial groups, *Bacillus mucilaginosus, B. edaphicus,* and *B. circulans* are the most efficient K solubilizers. These biofertilizers contain microorganisms that improve potassium (K) availability in the soil. They do this by facilitating the release of potassium ions from mineral complexes in the soil, making it more accessible to plants (Sheng, 2005).

15.2.1.4 PHYTO-HORMONES PRODUCING BIOFERTILIZERS

Plant growth-promoting bacteria (PGPB) secrete valuable biostimulants (BS) and play a pivotal role in modulating plant stress responses. Plant growth-promoting rhizobacteria (PGPR) are known to produce various phytohormones for root uptake, including auxins, cytokinins, and gibberellins, which play essential roles in controlling the development and growth of plants. Additionally, they serve as chemical signals that react to outside conditions that restrict plant growth (Fahad et al., 2015). Species of the genera *Azospirillum, Pseudomonas, Xanthomonas, Rhizobium,* and *Bradyrhizobium* possess the capability to produce phytohormones (Mohamed and Gomaa, 2012). Furthermore, these microorganisms can induce the development of longer roots, increase root biomass, and reduce density and stomata size (Llorente et al., 2016). Additionally, they can activate genes associated with the auxin response, promoting enhanced plant growth (Ruzzi and Aroca, 2015). Under extreme conditions, such as drought stress, PGPR promotes more growth stimulation (Rubin et al., 2017). Numerous studies have demonstrated that PGPR inoculation, which generates ACC deaminase and lowers ethylene synthesis in plants, increases plants' ability to withstand stress (Glick, 2014; Vejan et al., 2016).

15.3 NANOFERTILIZERS: ADVANTAGES TO SIGNIFICANCE IN AGRICULTURE

Nanotechnology is recognized as a pivotal technology that contributes significantly to the sustainable advancement in the agriculture sector of crop growth, production, processing, storage, disease management, packaging, and transportation of agricultural products (Mittal et al., 2020). These nanomaterials consist of assemblies of atoms or molecules, characterized by having at least one dimension within the range of 1 to 100 nanometres. The high surface area-to-volume ratio of nanoparticles (NPs) gives them their unique surface properties, behaviors, and functionalities. Multiple patents in agricultural research are currently being submitted for nanomaterials, which are actively harnessed for various applications (Jeevanandam et al., 2018). These applications include the development of nanosensors, the formulation of nanopesticides, nanoherbicides, and nanofertilizers, the genetic modification of plants, and the production of nanoplant growth regulators (Singh and Gurjar, 2022). The fundamental components of these nanotools and nanodevices are composed of different types of nanomaterials, viz., carbon-based nanomaterials and metal and metal oxide NPs (Lowry et al., 2019). Nano-based devices have extensively improved existing plant breeding and genetic transformation technologies. It has also been reported that engineered nanomaterials (ENMs)-based agricultural practices can acclimatize to fluctuating environmental and climatic conditions (Mittal et al., 2020). Among these applications, the utilization of nanomaterials as nanofertilizers is becoming increasingly significant in agriculture. Nanofertilizers represent a promising innovation in agriculture, designed to enhance nutrient utilization efficiency while mitigating the environmental impact typically associated with conventional mineral fertilizers. These advanced materials, including silicon (Si), iron (Fe), titanium dioxide (TiO_2), silicon dioxide (SiO_2), zinc oxide (ZnO), iron oxide (Fe_2O_3), silver (Ag), carbon nanotubes (CNTs), and gold nanorods (Au nanorods), are being extensively utilized as nanofertilizers to optimize nutrient uptake by plants (Zulfiqar et al., 2019) (Figure 15.1).

Several studies have reported the profound effects of nanofertilizers on seed germination and early seedling growth. Nano-fertilizers possess the remarkable ability to penetrate seeds effectively, thereby enhancing nutrient accessibility during the crucial stages of seedling development (Babu et al., 2022; Singh et al., 2022). This translates into healthier seedlings characterized by augmented shoot and root growth. For instance, nanoZnO

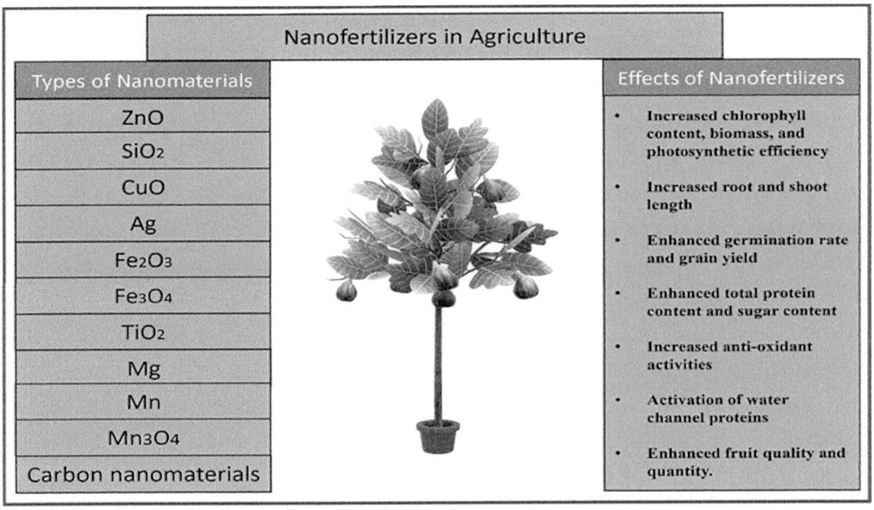

FIGURE 15.1 Different types of nanomaterials used as nanofertilizers and their effects in agriculture.

has demonstrated a superior capacity to promote peanut seed germination and stimulate root growth when compared to bulk zinc sulphate (Prasad et al., 2012). Similarly, the beneficial impact of nano-scale SiO_2 and TiO_2 on germination has been established in soybeans when compared to untreated controls (Qureshi et al., 2018). Recent research has also revealed that the application of nanofertilizers significantly elevates nutrient availability for plants. This, in turn, fosters the optimization of various growth parameters, including leaf area index, leaf area, plant height, the number of leaves per plant, dry matter production, chlorophyll synthesis, and the rate of photosynthesis (Dhoke et al., 2013). Consequently, these advancements culminate in an increase in the production and translocation of photosynthates to distinct parts of the plant. This augmented translocation of photosynthates from source tissues such as leaves to economically valuable sink tissues (e.g., grains, tubers, bulbs, stems, fibers, and leaves) translates into superior yield and improved quality parameters in plants treated with nanofertilizers (Sheoran et al., 2021).

Nanofertilizers have also shown immense potential for improving both biotic and abiotic stress tolerance in plants. Nanofertilizers, engineered with essential nutrients like nitrogen, phosphorus, and micronutrients, offer precise and controlled nutrient delivery to plants, thereby enhancing overall plant health and fortifying their resilience to stress. Furthermore, NPs can stimulate root growth and architecture, optimizing water and nutrient uptake, which in turn bolsters the plant's ability to endure water-related stress (Babu et al., 2022). Nanofertilizers have also been proven to be beneficial in tackling soil salinity issues. They facilitate ion exchange and diminish the harmful impact of excessive salts on plants, thereby improving salt stress tolerance and helping the plant maintain ion balance (Yadav et al., 2023). Application of hydrogel-based nanomaterials as a type of nanofertilizer has enhanced the soil water retention capacity in plants, alleviating water stress (Rizwan et al., 2021). CNTs and graphene-based nanofertilizers have the capacity to augment photosynthesis by enhancing the efficiency of light absorption and carbon dioxide assimilation, ultimately increasing plant growth and their resilience against abiotic stress factors (Safdar et al., 2022).

Moreover, nano-fertilizers are also being utilized as targeted and environmentally friendly solutions for pest, weed, and disease management. This approach minimizes the ecological footprint while effectively safeguarding plants from biotic stressors. Nevertheless, it is crucial to recognize that NPs exhibit a dualistic influence on plants, encompassing both advantageous and detrimental aspects depending upon the concentration used. It's vital to remember that the efficiency of uptake and utilization of NPs, as well as their impact on plant metabolism, can vary depending on the plant species, susceptibility, size, composition, and chemical characteristics of NMs, as well as the concentration utilized. In sum, nanofertilizers represent a cutting-edge approach with substantial potential for revolutionizing modern agriculture. Their ability to enhance nutrient utilization, promote seedling vigor, and boost overall plant growth and productivity underscores their significance in sustainable and efficient crop cultivation practices.

15.3.1 ADVANTAGES OF NANOFERTILIZERS OVER CONVENTIONAL FERTILIZERS

Nanofertilizers offer several advantages over conventional fertilizers, making their use an innovative and promising approach in agriculture (Figure 15.2).

1. **Enhanced Nutrient Delivery**: The substantial surface-to-volume ratio and smaller dimensions of nano-fertilizers enable improved

penetration, nutrient absorption, and enhanced accessibility for plants, thereby increasing the efficiency of nutrient utilization. Nano-fertilizers can be designed to release nutrients slowly and steadily over time. This controlled release improves nutrient utilization by plants, thereby reducing wastage and minimizing the risk of nutrient leaching into the environment (Gade et al., 2023).

2. **Increased Nutrient Availability:** Enhanced nanomaterial surface area offers increased points of contact for plant roots to absorb nutrients, resulting in more efficient nutrient uptake, leading to increased chlorophyll formation, higher photosynthesis rates, greater dry matter production, and overall improved plant growth (Ali et al., 2021).

3. **Reduced Environmental Impact:** The controlled release and improved nutrient uptake associated with nano-fertilizers can reduce the overuse of fertilizers, which is a common cause of nutrient runoff and water pollution. This contributes to more sustainable and environmentally friendly agriculture.

4. **Lower Application Rates:** Nanofertilizers are often more efficient, allowing for lower application rates compared to traditional fertilizers. This can lead to cost savings for farmers while achieving the same or better crop yields (Singh, 2017).

5. **Customizable Formulations:** Nano-fertilizers can be tailored to deliver specific nutrients and micronutrients needed for particular crops or soil conditions. This customization allows for precision agriculture, optimizing nutrient delivery for different plants and regions.

6. **Improved Soil Health:** Nanofertilizers can enhance soil structure and microbial activity, leading to improved soil health and the long-term sustainability of agricultural lands.

7. **Reduced Fertilizer Burns:** Traditional fertilizers can sometimes cause fertilizer burn, where the concentrated nutrients damage plant roots or leaves. Nano-fertilizers with controlled release mechanisms can reduce the risk of this issue.

8. **Increased Resistance to Environmental Stress:** Nanofertilizers can improve plant resilience to environmental stressors such as

drought, salinity, or disease, which can result in higher crop yields and better overall crop health (Ali et al., 2021).

9. **Potential for Targeted Delivery:** Nano-fertilizers can be engineered for specific delivery mechanisms, such as foliar spray, root application, or seed coating, allowing for precise nutrient delivery where it's needed most (Zulfiqar et al., 2019).

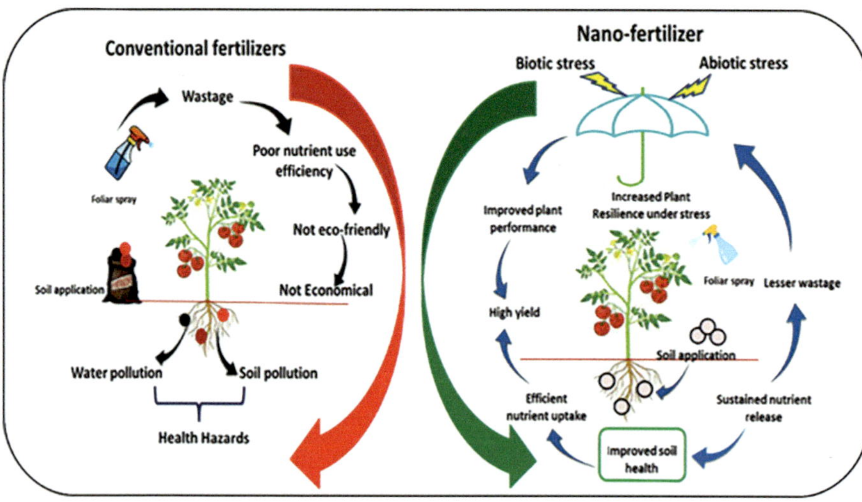

FIGURE 15.2 Advantages of nanofertilizers over conventional fertilizers.

15.3.2 UPTAKE OF NANOFERTILIZERS BY PLANTS

The uptake of nanofertilizers by plants involves a complex process influenced by various factors (Figure 15.3) (Gade et al., 2023). Nanofertilizers are typically applied to the soil or plant roots, whereas some nanofertilizers are designed for foliar application, to be sprayed directly on the leaves (Singh, 2017). For both applications, the size of NPs is a critical factor. Nanofertilizers have very small particle sizes, which facilitates their movement through soil and root cell walls. The entry of NPs through the cell wall depends on the pore diameter of the cell wall (5–20 nm) (Verma et al., 2022). Hence, NPs or NP aggregates with a diameter less than the pore size of the plant cell wall can easily enter through the cell wall and reach up to the plasma membrane. NPs enter plant cells by attaching to carrier proteins via aquaporins, endocytosis, or ion channels (Verma et al., 2022).

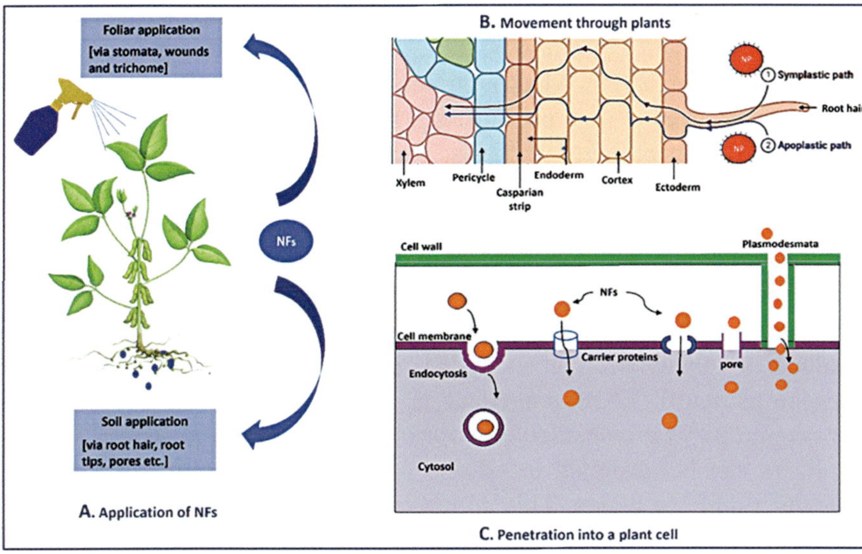

FIGURE 15.3 A schematic diagram shows the uptake of nanofertilizers (NFs) via various channels and their translocation paths across multiple plant sections. (A) Routes of nanofertilizer application in plants: (i) T.S. of maize leaf and (ii) T.S. of maize roots; (B) NFs may use apoplastic or simplistic pathways for moving up and down; and (C) internal distribution of NFs inside the cells through endocytosis and pore formation mediated by carrier proteins via plasmodesmata.

15.4 TYPES OF NANOFERTILIZERS

Nanofertilizers are a promising field in agricultural technology, offering enhanced nutrient delivery and efficiency. These types of nanofertilizers hold potential for improving agricultural productivity while minimizing environmental impacts. Figure 15.4 shows several types of nanofertilizers that are currently known and available.

15.4.1 *MACRONUTRIENT-BASED NANOFERTILIZERS*

Macronutrient-based nanofertilizers are innovative agricultural products designed to provide essential macronutrients to plants in a more efficient and targeted manner (Gade et al., 2023). These nanofertilizers contain NPs or nanomaterials that encapsulate or deliver macronutrients such as phosphorus (P), nitrogen (N), potassium (K), sulfur (S), magnesium (Mg), and calcium (Ca).

They have been combined with nanomaterials to efficiently deliver precise amounts of nutrients to crops while reducing bulk requirements (Preetha and Balakrishnan, 2017). Macronutrients, like nitrogen (N), play a pivotal role in plant development, serving as a fundamental component in energy metabolism and protein synthesis. Nitrogen-based nanofertilizers can encapsulate or deliver nitrogen in various forms, such as ammonium or nitrate ions, in nanoscale particles (Bernela et al., 2020). Phosphorus is another essential macronutrient for plant growth. Nanostructured materials, such as nanophosphates or nanocomposites, can be used to deliver phosphorus to plants. These nanofertilizers release phosphorus slowly, making it more available to plants over an extended period while reducing the risk of phosphorus runoff (Khatri and Bhateria, 2023). Potassium is vital for plant growth and plays a crucial role in various physiological processes. Nano-fertilizers can be designed to release potassium ions gradually, ensuring a steady supply to plants. This controlled release can improve potassium uptake efficiency and reduce the need for frequent potassium applications. Among other macronutrients like sulfur (S), calcium (Ca), and magnesium (Mg), calcium-based nano-fertilizers (NFs) also play a significant role in stabilizing cell walls, retaining minerals in the soil, facilitating mineral transport, neutralizing harmful substances, and aiding in seed formation (Anand et al., 2020; Thirunavukkarasu et al., 2018).

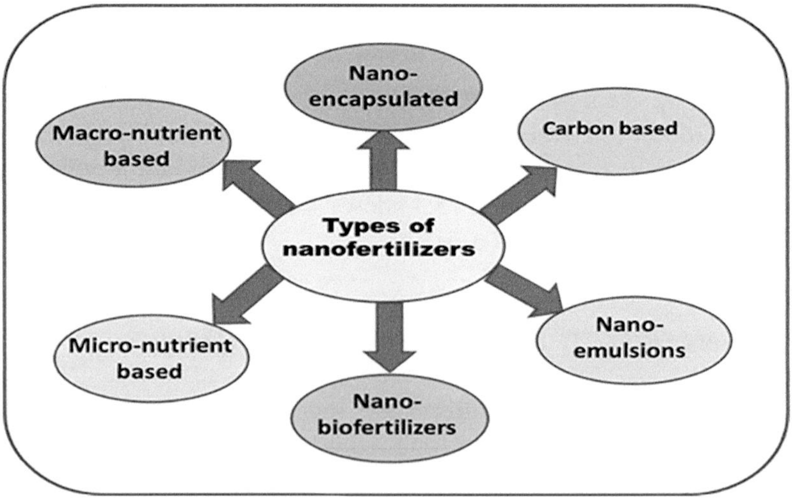

FIGURE 15.4 Various types of nanofertilizers (NFs) that are used in agriculture.

15.4.2 MICRONUTRIENT-BASED NANOFERTILIZERS

Micronutrient-based nanofertilizers are specialized agricultural products designed to deliver essential micronutrients to plants in a highly efficient and targeted manner. These nanofertilizers contain NPs or nanomaterials that encapsulate, carry, or release micronutrients such as iron (Fe), zinc (Zn), copper (Cu), manganese (Mn), boron (B), and molybdenum (Mo) (El-Saadony et al., 2021). Micronutrients are vital for various biochemical processes in plants, and nanofertilizers can ensure their availability while reducing waste and environmental impact. These are the most common types of nanofertilizers, mainly composed of nanomaterials such as metal oxides, carbon-based materials, and polymers. These NPs can be encapsulated with micronutrients, which favor the slow and targeted liberation of nutrients, minimizing nutrient loss due to volatilization and leaching (Solanki et al., 2015).

15.4.3 CARBON-BASED NANOFERTILIZERS

CNTs are artificial carbon nanomaterials with outstanding physicochemical characteristics and durability. CNTs' most important agricultural uses are seed germination, early plant development, and herbicides. CNTs can be used as carriers for nutrients. They offer a high surface area for nutrient adsorption and slow, controlled release of nutrients to plants (Tekam et al., 2023). CNT-infused nutrition media promotes longer shoots and increased seed germination in plants (Khodakovskaya et al., 2009). The penetration of the tomato seed coat by CNTs influences the seed germination and growth rates by increasing water intake within them. Khodakovskaya et al. (2012) reported that the use of CNTs is accelerating tobacco cell development. Comparable studies were conducted on date palm callus, embryo germination, embryogenesis, root elongation, and stages of rooting. No matter what stage of growth the cells were in, the data showed that the CNT treatment had an impact on their culture. The embryo germination rate and the quantity of roots increased (Taha et al., 2016).

15.4.4 NANO-ENCAPSULATED FERTILIZERS

In this type, nutrients are encapsulated within nanoscale carriers, such as polymer capsules. Polymers like albumin, chitosan, and alginate can all transport NPs. Chitosan is a cheap cationic biopolymer that is used to improve agricultural output, nutrient uptake, photosynthetic rate, seed germination, and plant growth (Kashyap et al., 2015; Kumaraswamy et al., 2018). When sprayed foliarly, Zn-chitosan NPs can stimulate wheat plant development (Deshpande et al., 2017). CuCNPs may be able to stimulate plant development in maize and tomato (Saharan et al., 2015, 2016). The encapsulation helps protect the nutrients from environmental factors like moisture, heat, and microbial activity, ensuring a controlled release to the plants. One such example of polymer-coated NPs is nanogels. These are three-dimensional networks of nanoscale polymer particles that can encapsulate and release nutrients gradually and also increase the water-holding capacity, which can help improve soil moisture retention.

15.4.5 NANO-EMULSIONS AND NANOSUSPENSIONS

Nano-emulsions and nanosuspensions are advanced delivery systems for nano-fertilizers, designed to enhance the dispersion and absorption of nutrients by plant roots. Nano-emulsions refer to systems where nutrients are dispersed as minuscule droplets or NPs within either an oil-in-water or water-in-oil medium (Mittal et al., 2020). On the other hand, nanosuspensions are comparable but involve solid NPs suspended within a water-based solution. These formulations are advantageous for enhancing nutrient absorption by plants because their tiny particle size and expanded surface area facilitate more efficient nutrient uptake. Nano-emulsions are stable, colloidal systems composed of tiny oil droplets dispersed in water or vice versa (Anjali et al., 2012). The droplets in nano-emulsions typically have a size range of 20 to 200 nanometres. Nano-emulsions are thermodynamically stable and can remain in a dispersed state for extended periods. In nano-fertilizers, nano-emulsions are used to encapsulate hydrophobic (water-repellent) nutrients, such as certain micronutrients or growth-promoting compounds. These nutrients are often insoluble or poorly soluble in water and may have low bioavailability when applied in their traditional forms. Nano-emulsions can improve the solubility and stability of hydrophobic nutrients, making them more accessible to plant roots. They can also protect nutrients from degradation due to environmental factors. Nanoemulsions can be applied to soil or plant roots, and they can be absorbed through the root system. They enable controlled nutrient release and can enhance nutrient uptake by plants (Mittal et al., 2020).

15.4.6 NANO-BIOFERTILIZERS

The progression of nano-biotechnology is marking a new era by amalgamating nanotechnology and biotechnology, giving rise to the exploration of next-generation hybrid fertilizers known as nano-biofertilizers (NBF), which hold significant promise to enhance the nutritional values of plants and soil systems. NBF is a new conjugated strategy, comprising a coupling of nanofertilizers and biofertilizers, and is considered one of the most advanced and eco-friendly approaches to subsiding the deleterious effects of chemical fertilizers on human health and agriculture productivity. These nanofertilizers are combined with beneficial microorganisms, or growth-promoting microorganisms (PGPM), to enhance their effectiveness in improving plant growth and health. Nanomaterials in NBF (Garg et al., 2023) can affect plant-microbe interactions by improving nutrient accessibility and stimulating microorganisms' growth. They offer benefits such as enhanced solubility, selectivity, permeability, and stability.

15.4.6.1 FORMULATIONS OF NANO-BIOFERTILIZERS AND THEIR ADVANTAGES OVER CONVENTIONAL FORMULATIONS

Commercial plant-microbe inoculation faces field survival challenges due to inadequate formulation protection against abiotic conditions (temperature, humidity, salinity, UV radiation, pH). Thus, the formulation step of fertilizer plays a key role in determining its success. Liquid bioinoculants of certain PGPM are commonly used as biofertilizers, but they are sensitive to contamination and prolonged storage (Chaudhary et al., 2020). Mixing them with carriers like peat can create solid formulations (Ijaz et al., 2019), favored for their stability, but peat can negatively affect some microorganisms and is susceptible to humidity, reducing cell survival (Chaudhary et al., 2020; Malusá et al., 2012). To meet the growing demand for PGPM-based bioinoculants in agriculture and overcome the limitations of traditional formulations (solid and liquid), research is now focused on creating new formulations that offer improved viability, enhanced ease of use, stability during transportation and storage, and superior field performance (Çakmakçı, 2019). While formulating NBF, it is crucial to ensure its successful application under harsh conditions like

UV radiation, saline or extreme pH soils, high temperatures, and drought. The encapsulation of live PGPM cells with NPs is the foundation of the NBF formulation. Encapsulation, an advanced and promising technology, offers significant advantages over other methods by creating micro- or macro-capsules that trap cells within non-toxic, biodegradable polymeric materials. These materials serve as permeable structures for nutrient and gas exchange, thus maintaining cell viability (Ding and Shah, 2009; John et al., 2011; Kim et al., 2012; Rathore et al., 2013; Schoebitz et al., 2013). Various polymers, such as alginate, starch, gelatin, chitosan, nano-zeolites, and nano-silica, among others, are utilized for this purpose. Alginate, derived from brown algae, is most commonly used for protecting and preserving the viability of bacterial cells (Malik et al., 2022).

These encapsulated formulations serve as a protective microenvironment for cells and their metabolites, shielding them from harsh external conditions before and during soil application. They offer several advantages, including controlled release through soil degradation, lower risk of contamination during storage and transport, extended shelf life, and preservation of biological activity for the required duration to stimulate plant growth (John et al., 2011). Furthermore, they ensure a uniform and gradual controlled release of cells near the intended site, facilitating effective colonization of the rhizosphere and roots (Bashan et al., 2016; Kim et al., 2012).

15.4.6.2 ENCAPSULATION AND OPTIMIZATION OF NANO-BIOFERTILIZERS

Encapsulation and optimization of nano-biofertilizers is a fascinating area of research at the intersection of nanotechnology and agriculture. Overall, the encapsulation and optimization of nano-biofertilizers hold great promise for addressing the challenges of modern agriculture, including improving crop yields, reducing nutrient runoff, and promoting sustainable farming practices. However, continued research and development are needed to overcome technical hurdles and ensure the responsible deployment of these innovative technologies (Rojas-Sánchez et al., 2022).

1. **Encapsulation:** This involves the process of enclosing or entrapping bioactive substances within a protective shell or matrix. In

the context of nano-biofertilizers, encapsulation serves several purposes:

2. **Optimization:** Optimization involves refining the formulation and delivery mechanisms of nano-biofertilizers to maximize their effectiveness and minimize any potential drawbacks.

Synthesizing NBF entails a stepwise procedure for their production and formulation (Figure 15.5). These steps include:

i. **Isolation and Cultivation:** The initial step involves the careful selection of specific beneficial microorganisms. These chosen microorganisms are then isolated and cultivated in a controlled environment to ensure their purity and viability.

ii. **Polymerization and Encapsulation:** In this step, a mixture of active microbial cell culture suspension and NPs is absorbed into a bio-polymeric matrix, which serves as the wall material, followed by drop-wise addition into a calcium chloride ($CaCl_2$) cross-linking solution while being constantly agitated. This results in the formation of solid particles resembling beads. Depending on their size, these particles can range from a few nanometres to a few millimetres, leading to macro-, micro-, or nano-encapsulation (Balla et al., 2022).

iii. **Optimization and Characterization:** The formulation is optimized to ensure uniform nutrient distribution and long-term stability (Panichikkal et al., 2021a). The resulting nano-biofertilizer is thoroughly characterized to assess its physical and chemical properties, ensuring that it meets the desired specifications.

iv. **Quality Control:** Quality control measures are implemented to maintain the effectiveness and safety of the NBF.

v. **Application and Optimization:** These NBF beads can then be applied to soil with nearby seeds or seedlings for agricultural purposes. Encapsulation techniques for NBF formulation are relatively rare in existing studies. Further research is needed to optimize various combinations of NPs and biofertilizers for NBF and conduct comprehensive physiological and molecular studies to understand their effects on plant growth under different conditions.

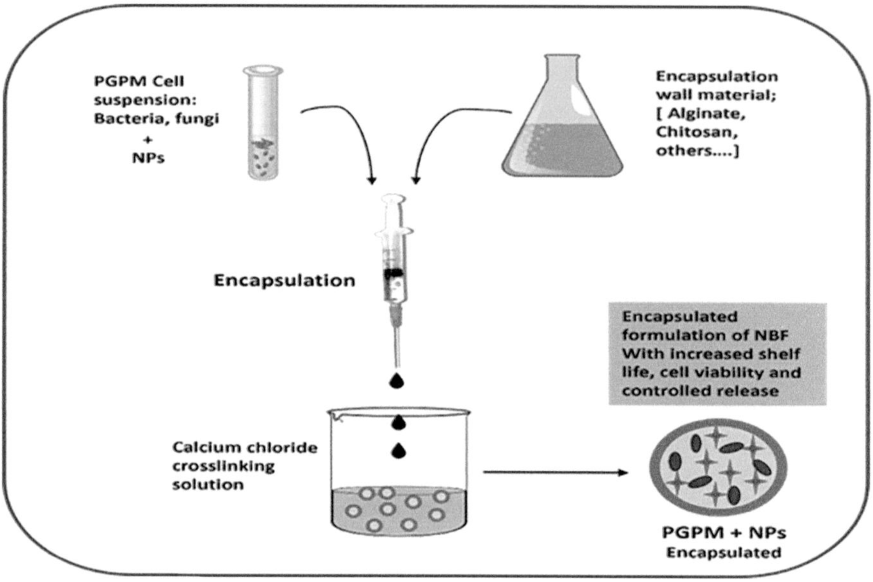

FIGURE 15.5 Formulation and optimization of nano-biofertilizers.
Source: Adapted from Rojas-Sánchez et al. (2022).

15.4.6.3 APPLICATION OF NANO-BIOFERTILIZERS TO ENHANCE AGRICULTURE PRODUCTIVITY

The application of NBF to enhance crop growth involves several mechanisms (Table 15.2). These mechanisms include optimizing photosynthesis, restoring soil microbiota, facilitating the dissolution and diffusion of insoluble nutrients in the soil, increasing the synthesis of plant growth-regulating hormones, and siderophore production. Additionally, NBF enhances the nutrient-use efficiency of plants. They achieve this by allowing for a slow and controlled release of nutrients, and their small size enables direct internalization and absorption by plant tissues (Garg et al., 2023). The genes that produce stress-related proteins, osmolytes, and antioxidants can all be upregulated by NBF. This up-regulation helps reduce the damaging effects of reactive oxygen species (ROS) on plants and helps maintain the structural and functional integrity of plant cells (Akhtar et al., 2022). NBF, encapsulated with growth-promoting microorganisms like *Bacillus subtilis*, *Pseudomonas fluorescens*, and *Paenibacillus elgii* within silver and gold NPs, greatly enhances crop growth and confers resistance to biotic and abiotic stresses (Sharma et al., 2023). This approach is more effective than directly applying NPs as fertilizer (Shafiei-Masouleh, 2021; Sharma et al., 2021).

The combined action of biofertilizers and NPs produces enhanced effects when applied to plants. This synergy activates various plant mechanisms, leading to improved growth and higher yields. In a pot experiment, the soil was treated with silicon dioxide nanoparticles (SiO_2), biofertilizer, and nanosilica before sowing. Nanosilica increased shoot length by 34.77% and root length by 16.88% under control conditions and by 30.58% and 21.56% under stress conditions, respectively. Additionally, it boosted RWC, photosynthetic pigments, membrane stability index, osmolyte content, and flavonoid and phenol content (Akhtar and Ilyas, 2022). *Pantoea agglomerans* and *Burkholderia caribensis* immobilized on nanofibers benefited seed germination and improved soybean root length and dry weight (De Gregorio et al., 2017). In another study, the addition of titanium dioxide NPs to *Bacillus subtilis* Vru1 strain microbeads composed of alginate, bentonite, and starch significantly increased its 90% inhibitory effect on *Rhizoctonia solani* and enhanced bean growth parameters (Saberi-Rise and Moradi-Pour, 2020). In an experiment, the impact of NPs on *S. lycopersicum* was assessed under the inoculation of *Azotobacter salinestris* in both control and treated conditions. *Azotobacter salinestris* produced exopolysaccharides (EPS), which effectively trapped NPs, retained their concentration in the rhizosphere, and mitigated the potentially harmful effects of high levels of certain NPs (Ahmed et al., 2021). The simultaneous use of SiNPs and PGPR on drought-stressed *Z. mays* resulted in several positive effects. These included enhanced photosynthesis rates, increased RWC, and elevated activity of antioxidant enzymes. Additionally, it led to improved yield attributes, increased nutrient uptake, and enhanced nutrient translocation in the plants (Hafez et al., 2021).

Moreover, NBF also elevates the levels of growth hormones, such as indole acetic acid and cytokinin, while decreasing the production of stress hormones like abscisic acid (ABA). These alterations enhance the plant's ability to withstand stress and increase the likelihood of successful crop establishment in adverse environmental conditions (Bibi et al., 2022). In the case of *Carthamus tinctorius* facing salinity, the consortium of biofertilizer and ZnO NPs significantly reduces the level of Na^+ by 60%

TABLE 15.2 Effects of Nano-Biofertilizers on Different Plants

Nano-Bio-Fertilizer	Plants	Plants Responses	References
ZnNPs + Organic fertilizer + Biofertilizer (*Rhizobium*)	*Phaseolus vulgaris L.*	• Elevated plant height, greater biomass, expanded leaf area, and an increased number of leaves per plant. • Improved nutrient absorption, higher pod yield, and increased levels of carbohydrates and proteins in the pods.	Morsy et al. (2017)
Burkholderia+ *Pantoea agglomerans* +Polyvinyl alcohol + *caribensis* + Nanofibers	*soybean*	• Promotion of overall Plant growth	De Gregorio et al. (2017)
Biofertilizer (*Pseudomonas, Azotobacter,* and *Azospirillum*) + Oxide NPs of Fe-Zn	*Triticum aestivum*	• Raised levels of soluble sugar, proline, and enzymatic activities. • Improved yield by 88% compared to the control (untreated drought-exposed plants).	Sharifi et al. (2020)
Biofertilizer (*Azotobacter chroococcum, A. caulinodans*, and *Azospirillum brasilense*) + TiNPs	*Triticum scale*	• Increased relative water content, 1000-grain weight, grain yield, and chlorophyll content. • Reduced leaf and seed cadmium (Cd) content.	Ghooshchi (2017)
Biofertilizer (Nitroxin) + AgNPs	*Solanum tuberosum L.*	• Enhanced tuber quantity, tuber diameter, and tuber weight per plant.	Davod et al. (2011)
Biofertilizer (*Pseudomonas fluorescens*) + Zero valent FeNPs	*Trifolium repens*	• Enhanced phytoremediation potential and plant development	Daryabeigi Zand et al. (2020)
Biofertilizer (*Bacillus spp.)* + Nanozeolite	*Z. mays*	• Increased protein content, chlorophyll, and plant length	Khati et al. (2018)
Biofertilizer (*Azetobacter*) + Nano-fertilizer	*Sorghum bicolor (L.) Moench.*	• Elevated levels of carbohydrates, carotene, and chlorophyll	Mir et al. (2015)
Bentonite + starch + titanium dioxide nanoparticles + *Bacillus subtilis*	*Beans*	• *Rhizoctonia solani* biocontrol and encouragement of plant growth	Saberi-Rise and Moradi-Pour (2020)

TABLE 15.2 *(Continued)*

Nano-Bio-Fertilizer	Plants	Plants Responses	References
Biofertilizer (*Panebacillus polymyxa*) + Acylated homoserine coated Fe-carbon nanofibers	*T. aestivum* and *Cicer aretianum* L.	• Development of resistance against pathogens • Improved biomass, length, and chlorophyll content	Najafi et al. (2021)
Biofertilizer+ Biochar and compost+ Zero-valent FeNPs	*Brassica juncea* L.	• Enhanced total carbon, total nitrogen, phosphorus, pH, and soil cation exchange capacity • Enhanced plant biomass and height	Baragaño et al. (2020)
Salicylic acid + zinc oxide nanoparticles + *Pseudomonas* sp.	*Oryza sativa*	• Plant growth enhancement and biological control of *Sclerotium rolfsii*	Panichikkal et al. (2021b)
Biofertilizer+ Nano-chelated Zn and B (boron)	*Zea mays*	• Enhanced yield attributes • Enhanced physiological and morphological characteristics, such as relative water content, chlorophyll content, leaf area, ear length, and diameter.	Akhtar and Ilyas (2022)
Biofertilizer + Composted biochar + Farmyard manure + ZnO NPs	*T. aestivum*	• Increased plant output, photosynthetic pigments, biomass, and antioxidants	Bashir et al. (2020)

(roots) and 74% (leaves). This treatment also leads to an increase in the number of pods per plant, the number of leaves, and the capitulum weight. These effects are attributed to the role of zinc (Zn), which is essential for auxin biosynthesis and chlorophyll production and affects soil fertility, pollen performance, and germination in an essential way. The combined use of ZnO NPs and biofertilizers in plants yields significant results, demonstrating their collaborative impact (Yasmin et al., 2021).

15.4.6.4 FUTURE PERSPECTIVES AND CHALLENGES OF NANO-BIOFERTILIZERS

The agricultural community is increasingly drawn to NPs for their rapid and beneficial effects. Incorporating nanotechnology enables precise farming practices and improved nutrient availability. Recent publications have touched upon NBF formulation and application, marking a significant stride towards an innovative and more efficient agricultural product. NBF offers a promising and sustainable approach to agricultural management, enhancing soil health and crop productivity (Figure 15.6). However, it's important to note that while nanofertilizers offer these advantages, their development and use also come with challenges and considerations related to environmental impact and human safety (Yadav et al., 2023; Zahra et al., 2022). Ongoing research and responsible use practices are essential to maximizing the benefits of NBF while minimizing potential risks. Biosafety regulations are crucial for the secure application of artificial and environmentally friendly NPs in the agriculture and food industries (Babu et al., 2021). The production of nano-fertilizers, which can be expensive compared to conventional fertilizers, is a concern for farmers, especially in low-income regions. Scaling up nanofertilizer production and developing cost-effective manufacturing methods is crucial to meeting agricultural demands. Further, microbe-based NBF faces limited adoption due to an incomplete understanding of their interactions with biofertilizers, NPs, and plants. Farmers and agricultural practitioners may need training and education on the proper use and benefits of nanofertilizers. Understanding how NPs interact with soil components and soil microbial communities is important for assessing their long-term impact on soil health and fertility (Mittal et al., 2020). One day, even a small amount of these fertilizers will deliver the required agricultural production. This will be made possible by the ease with which economically feasible nano-fertilizers can be produced utilizing the necessary techniques after these challenges are handled.

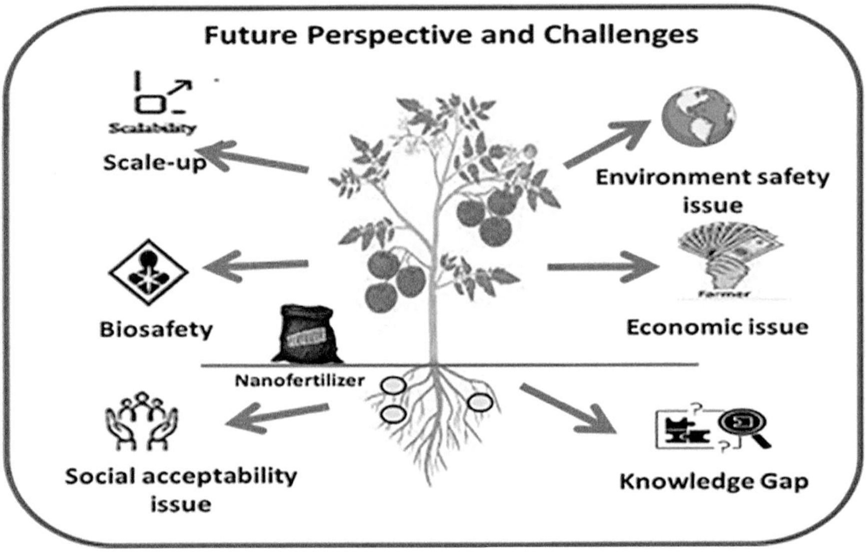

FIGURE 15.6 Future perspectives and challenges of nano-biofertilizers.

15.5 CONCLUSION

In recent years, the agriculture industry has witnessed the emergence of nanotechnology as a significant and promising tool. Nano-biofertilizers, in particular, play a crucial role in agriculture by enhancing crop productivity and improving plant resilience to abiotic stresses. These innovations promise more efficient resource utilization, such as water, nutrients, and land, while reducing the dependence on harmful chemicals that can disrupt ecosystems over time. While nanofertilizers offer exciting prospects for sustainable agriculture, addressing safety, regulatory, cost, and adoption challenges will be essential to realizing their full potential. Research and collaboration among scientists, policymakers, and farmers are crucial for harnessing the benefits of nanofertilizers while minimizing potential risks. More research needs to be focused on generating knowledge on the effect of these NPs on soil and microflora to recommend a specific nano-biofertilizer for a specific crop and soil type. Biosynthesized NPs-based fertilizers and nano-biofertilizers need further research to establish them as a promising, outstanding technology to enhance and optimize productivity while achieving sustainable agriculture.

KEYWORDS

- **arbuscular mycorrhizal fungus**
- **biostimulants**
- **nanobiofertilizers**
- **sustainable agriculture**
- **nanomaterials**
- **nanoparticles**

REFERENCES

Abadi, V. A. J. M., Sepehri, M., Rahmani, H. A., Zarei, M., Ronaghi, A., Taghavi, S. M., & Shamshiripour, M. (2020). Role of dominant phyllo sphere bacteria with plant growth-promoting characteristics on growth and nutrition of maize (*Zea mays* L.). *Journal of Soil Science and Plant Nutrition, 20*, 2348–2363.

Ahmed, A., & Hasnain, S. (2010). Auxin producing *Bacillus* sp.: Auxin quantification and effect on the growth *Solanum tuberosum*. *Pure and Applied Chemistry, 82*, 313–319.

Ahmed, B., Syed, A., Rizvi, A., Shahid, M., Bahkali, A. H., Khan, M. S., & Musarrat, J. (2021). Impact of metal-oxide nanoparticles on growth, physiology and yield of tomato (*Solanum lycopersicum* L.) modulated by *Azotobacter salinestris* strain ASM. *Environmental Pollution, 269*, 116218. https://doi.org/10.1016/j.envpol.2020.116218 (accessed on 6th November 2025).

Akhtar, N., & Ilyas, N. (2022). Role of nano silica to boost the activities of metabolites in *Triticum aestivum* facing drought stress. *Plant and Soil, 477*, 99–115.

Akhtar, N., Ilyas, N., Meraj, T. A., Pour-Aboughadareh, A., Sayyed, R. Z., Mashwani, Z. U. R., & Poczai, P. (2022). Improvement of plant responses by nano-biofertilizer: A step towards sustainable agriculture. *Nanomaterials, 12*(6), 965. https://doi.org/10.3390/nano12060965 (accessed on 6th November 2025).

Ali, M., Ahmad, Z., Ashraf, M. F., & Dong, W. (2021a). Maize endophytic microbial-communities revealed by removing PCR and 16S rRNA sequencing and their synthetic applications to suppress maize banded leaf and sheath blight. *Microbiological Research, 242*, 126639. https://doi.org/10.1016/j.micres.2020.126639 (accessed on 6th November 2025).

Ali, S. S., Darwesh, O. M., Kornaros, M., Al-Tohamy, R., Manni, A., El-Shanshoury, A. E., Metwally, M. A., Elsamahy, T., & Sun, J. (2021b). Nano-biofertilizers: Synthesis, advantages, and applications. In Rakshit, A., Vijay, S. M., Parihar, M., Singh, H. B., & Singh, A. K. (Eds.), *Biofertilizers: Advances in Bio-Inoculants* (pp. 359–370). Woodhead Publishing: United States.

Aloni, R., Aloni, E., & Langhans, M. (2006). Role of cytokinin and auxin in shaping root architecture: Regulating vascular differentiation, lateral root initiation, root apical dominance and root gravitropism. *Annals of Botany, 97*, 883–893.

Anand, K. V., Anugraga, A. R., Kannan, M., Singaravelu, G., & Govindaraju, K. (2020). Bio-engineered magnesium oxide nanoparticles as nano-priming agent for enhancing green gram seed germination and seedling vigor (*Vigna radiata* L.). *Materials Letters, 271*, 127792. https://doi.org/10.1016/j.matlet.2020.127792 (accessed on 6th November 2025).

Anjali, C. H., Sharma, Y., Mukherjee, A., & Chandrasekaran, N. (2012). Neem oil (*Azadirachta indica*) nanoemulsion – A potent larvicidal agent against *Culex quinquefasciatus*. *Pest Management Science, 68*, 158–163.

Archana, D. S., Nandish, M. S., Savalagi, V. P., & Alagawadi, A. R. (2013). Characterization of potassium solubilizing bacteria (KSB) from rhizosphere soil. *Bioinfolet–A Quarterly Journal of Life Sciences, 10*, 248–257.

Babu, S., Singh, R., Yadav, D., Rathore, S. S., Raj, R., Avasthe, R., Yadav, S. K., Das, A., Yadav, V., Yadav, B., Shekhawat, K., Upadhyay, P. K., Yadav, D. K., & Singh, V. K. (2022). Nano-fertilizers for agricultural and environmental sustainability. *Chemosphere, 292*, 133451. https://doi.org/10.1016/j.chemosphere.2021.133451 (accessed on 6th November 2025).

Balla, A., Silini, A., Cherif-Silini, H., Chenari Bouket, A., Alenezi, F. N., & Belbahri, L. (2022). Recent advances in encapsulation techniques of plant growth-promoting microorganisms and their prospects in the sustainable agriculture. *Applied Sciences, 12*(18), 9020. https://doi.org/10.3390/app12189020 (accessed on 6th November 2025).

Baragano, D., Forjan, R., Fernandez, B., Ayala, J., Afif, E., & Gallego, J. L. R. (2020). Application of biochar, compost and ZVI nanoparticles for the remediation of As, Cu, Pb and Zn polluted soil. *Environmental Science and Pollution Research, 27*, 33681–33691.

Bargaz, A., Elhaissoufi, W., Khourchi, S., Benmrid, B., Borden, K. A., & Rchiad, Z. (2021). Benefits of phosphate solubilizing bacteria on belowground crop performance for improved crop acquisition of phosphorus. *Microbiological Research, 252*, 126842. https://doi.org/10.1016/j.micres.2021.126842 (accessed on 6th November 2025).

Barney, B. M. (2020). Aerobic nitrogen-fixing bacteria for hydrogen and ammonium production: Current state and perspectives. *Applied Microbiology and Biotechnology, 104*, 1383–1399.

Bashan, Y., de-Bashan, L. E., & Prabhu, S. R. (2016). Superior polymeric formulations and emerging innovative products of bacterial inoculants for sustainable agriculture and the environment. In Singh, H., Sarma, B., & Keswani, C. (Eds.), *Agriculturally Important Microorganisms* (pp. 15–46). Springer: Singapore. https://doi.org/10.1007/978-981-10-2576-1_2 (accessed on 6th November 2025).

Bashir, A., Rizwan, M., Ali, S., Adrees, M., & Qayyum, M. F. (2020). Effect of composted organic amendments and zinc oxide nanoparticles on growth and cadmium accumulation by wheat; A life cycle study. *Environmental Science and Pollution Research, 27*, 23926–23936.

Baweja, P., Kumar, S., & Kumar, G. (2020). Fertilizers and pesticides: Their impact on soil health and environment. In *Soil health* (pp. 265–285).

Bernela, M., Rani, R., Malik, P., & Mukherjee, T. (2020). Nano fertilizers: Applications and future prospects. In Sindhu, R. K., Chitkara, M., & Sandhu, I. S. (Eds.), *Nanotechnology: Principles and Application* (pp. 287–330). Jenny Stanford Publishing: Dubai, United Arab Emirates.

Bibi, F., Ilyas, N., Arshad, A., Khalid, A., Saeed, M., Ansar, S., & Batley, J. (2022). Formulation and efficacy testing of bio-organic fertilizer produced through solid-state fermentation of agro-waste by *Burkholderia cenocepacia*. *Chemosphere, 291*, 132762. https://doi.org/10.1016/j.chemosphere.2021.132762 (accessed on 6th November 2025).

Brahmaprakash, G. P., & Sahu, P. K. (2012). Biofertilizers for sustainability. *Journal of the Indian Institute of Science, 92*, 37–62.

Cakmakci, R. (2019). A review of biological fertilizers: Current use, new approaches, and future perspectives. *International Journal of Innovative Studies in Science and Engineering Technology, 5*, 83–92.

Chaudhary, T., Dixit, M., Gera, R., Shukla, A. K., Prakash, A., Gupta, G., & Shukla, P. (2020). Techniques for improving formulations of bioinoculants. *3 Biotech, 10*, 199. https://doi.org/10.1007/s13205-020-02182-9 (accessed on 6th November 2025).

Daryabeigi Zand, A., Tabrizi, A. M., & Heir, A. V. (2020). The influence of association of plant growth-promoting rhizobacteria and zero-valent iron nanoparticles on removal of antimony from soil by *Trifolium repens*. *Environmental Science and Pollution Research, 27*, 42815–42829.

Davod, T., Reza, Z., Ali, V. A., & Mehrdad, C. (2011). Effects of nano silver and nitroxin biofertilizer on yield and yield components of potato Mini tubers. *International Journal of Agriculture and Biology, 13*, 986–990.

De Gregorio, P. R., Michavila, G., Ricciardi Muller, L., de Souza Borges, C., Pomares, M. F., Saccol de Sá, E. L., Pereira, C., & Vincent, P. A. (2017). Beneficial rhizobacteria immobilized in nanofibers for potential application as soybean seed bioinoculants. *PLoS ONE, 12*, e0176930. https://doi.org/10.1371/journal.pone.0176930 (accessed on 6th November 2025).

Deshpande, P., Dapkekar, A., Oak, M. D., Paknikar, K. M., & Rajwade, J. M. (2017). Zinc complexed chitosan/TPP nanoparticles: A promising micronutrient nanocarrier suited for foliar application. *Carbohydrate Polymers, 165*, 394–401.

Dhoke, S. K., Mahajan, P., Kamble, R., & Khanna, A. (2013). Effect of nanoparticles suspension on the growth of mung (*Vigna radiata*) seedlings by foliar spray method. *Nanotechnology Development, 3*(1), e1. https://doi.org/10.4081/nd.2013.e1 (accessed on 6th November 2025).

Ding, W. K., & Shah, N. P. (2009). An improved method of microencapsulation of probiotic bacteria for their stability in acidic and bile conditions during storage. *Journal of Food Science, 74*, 53–61.

Elbeltagy, A., Nishioka, K., & Sato, T. (2001). Endophytic colonization and in-plant nitrogen fixation by a *Herbaspirillum* sp. isolated from wild rice species. *Applied and Environmental Microbiology, 67*, 5285–5293.

El-Saadony, M. T., Al-Moshadak, A. S., Shafi, M. E., Al-Bqami, N. M., Saad, A. M., El-Tahan, A. M., El-Sayed, M. D., El-Nahal, A. S. M., Al-Makas, A., El-Mageed, T. A., Taha, A. E., Elrys, A. S., & Helmy, A. M. (2021). Vital roles of sustainable nano-fertilizers in improving plant quality and quantity–An updated review. *Saudi Journal of Biological Sciences, 28*(12), 7349–7359. https://doi.org/10.1016/j.sjbs.2021.08.032 (accessed on 6th November 2025).

Fahad, S., Hussain, S., Bano, A., Saud, S., Hassan, S., Shan, D., Khan, F. A., Khan, F., Chen, Y., Wu, C., Tabassum, M. A., Chun, M. X., Afzal, M., Jan, A., Jan, M. T., & Huang, J. (2015). Potential role of phytohormones and plant growth-promoting rhizobacteria in abiotic stresses: Consequences for changing environment. *Environmental Science and Pollution Research, 22*(7), 4907–4921. https://doi.org/10.1007/s11356-014-3754-2 (accessed on 6th November 2025).

Fedele, G., Brischetto, C., & Rossi, V. (2020). Biocontrol of *Botrytis cinerea* on grape berries as influenced by temperature and humidity. *Frontiers in Plant Science, 11*, 1232. https://doi.org/10.3389/fpls.2020.01232 (accessed on 6th November 2025).

Flores-Félix, J. D., Silva, L. R., & Rivera, L. P. (2015). Plant probiotics as a tool to produce highly functional fruits: The case of *Phyllobacterium* and vitamin C in strawberries. *PLoS ONE, 10*(4), e0122281. https://doi.org/10.1371/journal.pone.0122281 (accessed on 6th November 2025).

Freitas, V. F., Cerezini, P., Hungria, M., & Nogueira, M. A. (2022). Strategies to deal with drought-stress in biological nitrogen fixation in soybean. *Applied Soil Ecology, 172*, 104352. https://doi.org/10.1016/j.apsoil.2021.104352 (accessed on 6th November 2025).

Fukami, J., Cerezini, P., & Hungria, M. (2018). *Azospirillum*: Benefits that go far beyond biological nitrogen fixation. *AMB Express, 8*, 1–12.

Gade, A., Ingle, P., Nimbalkar, U., Rai, M., Raut, R., Vedpathak, M., Jagtap, P., & Abd-Elsalam, K. A. (2023). Nano fertilizers: The next generation of agrochemicals for long-term impact on sustainability in farming systems. *Agrochemicals*, 257–278.

Garcia-Fraile, P., Carro, L., & Robledo, M. (2012). *Rhizobium* promotes non-legumes growth and quality in several production steps: Towards a biofertilization of edible raw vegetables healthy for humans. *PLoS ONE, 7*, e38122. https://doi.org/10.1371/journal.pone.0038122 (accessed on 6th November 2025).

Garg, D., Sridhar, K., Stephen Inbaraj, B., Chawla, P., Tripathi, M., & Sharma, M. (2023). Nano-biofertilizer formulations for agriculture: A systematic review on recent advances and prospective applications. *Bioengineering, 10*(9), 1010. https://doi.org/10.3390/bioengineering10091010 (accessed on 6th November 2025).

Ghadamgahi, F., Tarighi, S., Taheri, P., Saripella, G. V., Anzalone, A., Kalyandurg, P. B., Catara, V., Ortiz, R., & Vetukuri, R. R. (2022). Plant growth-promoting activity of *Pseudomonas aeruginosa* FG106 and its ability to act as a biocontrol agent against potato, tomato and taro pathogens. *Biology, 11*(1), 140. https://doi.org/10.3390/biology11010140 (accessed on 6th November 2025).

Ghooshchi, F. (2017). Influence of titanium and bio-fertilizers on some agronomic and physiological attributes of triticale exposed to cadmium stress. *Global NEST Journal, 19*, 458–463.

Glick, B. R. (2014). Bacteria with ACC deaminase can promote plant growth and help to feed the world. *Microbiological Research, 169*, 30–39.

Gonzalez, L. J., Rodelas, B., Pozo, C., Salmeron, V., Martínez, M. V., & Salmeron, V. (2005). Liberation of amino acids by heterotrophic nitrogen fixing bacteria. *Amino Acids, 28*, 363–367.

Govindarajan, M., Balandreau, J., & Kwon, S. W. (2007). Effects of the inoculation of *Burkholderia vietnamensis* and related endophytic diazotrophic bacteria on grain yield of rice. *Microbial Ecology, 55*, 21–37.

Govindarajulu, M., Pfeffer, P. E., Jin, H., Abubaker, J., Douds, D. D., Allen, J. W., Bücking, H., Lammers, P. J., & Shachar-Hill, Y. (2005). Nitrogen transfer in the arbuscular mycorrhizal symbiosis. *Nature, 435*(7043), 819–823.

Hafez, E., Osman, H., Gowayed, S., Okasha, S., Omara, A., Sami, R., El-Monem, A. A., & El-Razek, U. A. (2021). Minimizing the adversely impacts of water deficit and soil salinity on maize growth and productivity in response to the application of plant growth promoting rhizobacteria and silica nanoparticles. *Agronomy, 11*, 676.

Han, H. S., & Lee, K. D. (2005). Phosphate and potassium solubilizing bacteria effect on mineral uptake, soil availability and growth of eggplant. *Research Journal of Agriculture and Biological Sciences, 1*, 176–180.

Hungria, M., & Mendes, I. C. (2015). Nitrogen fixation with soybean: The perfect symbiosis? In Bruijin, F. (Ed.), *Biological Nitrogen Fixation* (pp. 1009–1024). Wiley: Hoboken, NJ, USA.

Hussain, A., Shah, M., Hamayun, M., Iqbal, A., Qadir, M., Alataway, A., Dewidar, A. Z., Elansary, H. O., & Lee, I. (2023). Phytohormones producing rhizobacteria alleviate heavy metals stress in soybean through multilayered response. *Microbiological Research, 266*, 127237. https://doi.org/10.1016/j.micres.2022.127237 (accessed on 6th November 2025).

Ijaz, M., Ali, Q., Ashraf, S., Kamran, M., & Rehman, A. (2019). Development of future bioformulations for sustainable agriculture. In Kumar, V., Prasad, R., Kumar, M., & Choudhary, D. (Eds.), *Microbiome in Plant Health and Disease* (pp. 421–446). Springer, Singapore.

Jeevanandam, J., Barhoum, A., Chan, Y. S., Dufresne, A., & Danquah, M. K. (2018). Review on nanoparticles and nanostructured materials: History, sources, toxicity and regulations. *Beilstein Journal of Nanotechnology, 9*, 1050–1074.

John, R. P., Tyagi, R. D., Brar, S. K., Surampalli, R. Y., & Prevost, D. (2011). Bio-encapsulation of microbial cells for targeted agricultural delivery. *Critical Reviews in Biotechnology, 31*, 211–226.

Joo, G. J., Kim, Y. M., & Kim, J. T. (2005). Gibberellins-producing rhizobacteria increase endogenous gibberellins content and promote growth of red peppers. *Journal of Microbiology, 43*, 510–515.

Kang, S. M., Radhakrishnan, R., Khan, A. L., Kim, M. J., Park, J. M., Kim, B. R., Shin, D. H., & Lee, I. J. (2014). Gibberellin secreting rhizobacterium, *Pseudomonas putida* H-2-3 modulates the hormonal and stress physiology of soybean to improve the plant growth under saline and drought conditions. *Plant Physiology and Biochemistry, 84*, 115–124.

Kao, C. M., Chen, S. C., & Chen, Y. S. (2003). Detection of *Burkholderia pseudomallei* in rice fields with PCR-based technique. *Folia Microbiological, 48*, 521–524.

Kashyap, P. L., Xiang, X., & Heiden, P. (2015). Chitosan nanoparticle-based delivery systems for sustainable agriculture. *International Journal of Biological Macromolecules, 77*, 36–51.

Kaur, R., & Kaur, S. (2018). Biological alternates to synthetic fertilizers: Efficiency and future scopes. *Indian Journal of Agricultural Research, 52*, 587–585.

Ke, X., Feng, S., Wang, J., Lu, W., Zhang, W., Chen, M., & Lin, M. (2019). Effect of inoculation with nitrogen-fixing bacterium *Pseudomonas stutzeri* A1501 on maize plant growth and the microbiome indigenous to the rhizosphere. *Systematic and Applied Microbiology, 42*, 248–260.

Khati, P., Parul, Bhatt, P., Nisha, Kumar, R., & Sharma, A. (2018). Effect of nano zeolite and plant growth promoting rhizobacteria on maize. *3 Biotech, 8*, 1–12.

Khatri, A., & Bhateria, R. (2023). Efficacy of nano fertilizers over chemical fertilizers in boosting agronomic production. *Nature Environment and Pollution Technology, 22*, 767–776.

Khodakovskaya, M., Dervishi, E., Mahmood, M., Xu, Y., Li, Z., Watanabe, F., & Biris, A. S. (2009). Carbon nanotubes are able to penetrate plant seed coat and dramatically affect seed germination and plant growth. *ACS Nano, 3*, 3221–3227.

Khodakovskaya, M., Silva, K. D., Biris, A. S., Dervishi, E., & Villagarcia, H. (2012). Carbon nanotubes induce growth enhancement in tobacco cells. *ACS Nano, 6*, 2128–2135.

Kim, I. Y., Pusey, P. L., Zhao, Y., Korban, S. S., Choi, H., & Kim, K. K. (2012). Controlled release of *Pantoea agglomerans* E325 for biocontrol of fire blight disease of apple. *Journal of Controlled Release, 161*, 109–115.

Kumar, M. S., Reddy, G. C., Phogat, M., & Korav, S. (2018). Role of bio-fertilizers towards sustainable agricultural development: A review. *Journal of Pharmacognosy and Phytochemistry, 7*, 1915–1921.

Kumar, R., Kumawat, N., & Sahu, Y. K. (2017). Role of biofertilizers in agriculture. *Popular Kheti, 5*, 63–66.

Kumaraswamy, R. V., Kumari, S., Choudhary, R. C., Pal, A., Raliya, R., Biswas, P., & Saharan, V. (2018). Engineered chitosan-based nanomaterials: Bioactivities, mechanisms and perspectives in plant protection and growth. *International Journal of Biological Macromolecules, 113*, 494–506.

Llorente, B. E., Alasia, M. A., & Larraburu, E. E. (2016). Biofertilization with *Azospirillum brasilense* improves *in vitro* culture of *Handroanthus ochraceus*, a forestry, ornamental and medicinal plant. *New Biotechnology, 33*, 32–40.

Lowry, G. V., Avellan, A., & Gilbertson, L. M. (2019). Opportunities and challenges for nanotechnology in the agri-tech revolution. *Nature Nanotechnology, 14*, 517–522.

Malik, R., Saxena, R., & Warkar, S. G. (2022). Biopolymer-based biomatrices–Organic strategies to combat micronutrient deficit for dynamic agronomy. *ChemistrySelect, 7*, e202200006. https://doi.org/10.1002/slct.202200006 (accessed on 6th November 2025).

Malusa, E., Sas-Paszt, L., & Ciesielska, J. (2012). Technologies for beneficial microorganisms inocula used as biofertilizers. *The Scientific World Journal, 2012*, 491206. https://doi.org/10.1100/2012/491206 (accessed on 6th November 2025).

Mehnaz, S. (2016). An overview of globally available bioformulations. In Arora, N., Mehnaz, S., & Balestrini, R. (Eds.), *Bioformulations for Sustainable Agriculture* (pp. 267–281). Springer: New Delhi.

Mir, S., Sirousmehr, A., & Shirmohammadi, E. (2015). Effect of nano and biological fertilizers on carbohydrate and chlorophyll content of forage sorghum (Speedfeed hybrid). *International Journal of Biosciences, 6*, 157–164.

Mishra, D. J., Singh, R., Mishra, U. K., & Kumar, S. S. (2013). Role of bio-fertilizer in organic agriculture: A review. *Research Journal of Recent Sciences, 2*, 39–41.

Mittal, D., Kaur, G., Singh, P., Yadav, K., & Ali, S. A. (2020). Nanoparticle-based sustainable agriculture and food science: Recent advances and future outlook. *Frontiers in Nanotechnology, 2*, 579954. https://doi.org/10.3389/fnano.2020.579954 (accessed on 6th November 2025).

Mohamed, H. I., & Gomaa, E. Z. (2012). Effect of plant growth promoting *Bacillus subtilis* and *Pseudomonas fluorescens* on growth and pigment composition of radish plants (*Raphanus sativus*) under NaCl stress. *Photosynthetic, 50*(2), 263–272.

Monachon, M., Albelda-Berenguer, M., & Joseph, E. (2019). Biological oxidation of iron sulfides. *Advances in Applied Microbiology, 107*, 1–27.

Morsy, N. M., Shams, A. S., & Abdel-Salam, M. A. (2017). Zinc foliar spray on snap beans using nano-Zn with N-soil application using mineral, organic and biofertilizer. *Middle East Journal of Agriculture Research, 6*, 1301–1312.

Munoz-Rojas, J., & Caballero-Mellado, J. (2003). Population dynamics of *Gluconacetobacter diazotrophicus* in sugarcane cultivars and its effect on plant growth. *Microbial Ecology, 46*, 454–464.

Najafi, S., Nazari Nasi, H., Tuncturk, R., Tuncturk, M., Sayyed, R. Z., & Amirnia, R. (2021). Biofertilizer application enhances drought stress tolerance and alters the antioxidant enzymes in medicinal pumpkin (*Cucurbita pepo* convar. *pepo* var. *styriaca*). *Horticulture, 7*(12), 588. https://doi.org/10.3390/horticulturae7120588 (accessed on 6th November 2025).

Pace, L., Pellegrini, M., Palmieri, S., Rocchi, R., Lippa, L., & Del Gallo, M. (2020). Plant growth-promoting rhizobacteria for *in vitro* and *ex vitro* performance enhancement of Apennines' Genepi (*Artemisia umbelliformis* subsp. *eriantha*), an endangered phototherapeutic plant. *In Vitro Cellular & Developmental Biology–Plant, 56*, 134–142.

Panichikkal, J., Prathap, G., Nair, R. A., & Krishnankutty, R. E. (2021b). Evaluation of plant probiotic performance of *Pseudomonas* sp. encapsulated in alginate supplemented with salicylic acid and zinc oxide nano-particles. *International Journal of Biological Macromolecules, 166*, 138–143.

Panichikkal, J., Puthiyattil, N., Raveendran, A., Nair, R. A., & Krishnankutty, R. E. (2021a). Application of encapsulated *Bacillus licheniformis* supplemented with chitosan nanoparticles and rice starch for the control of *Sclerotium rolfsii* in *Capsicum annuum* (L.) seedlings. *Current Microbiology, 78*, 911–919.

Pisciotta, J. M., Zaybak, Z., Call, D. F., Nam, J. Y., & Logan, B. E. (2012). Enrichment of microbial electrolysis cell biocathodes from sediment microbial fuel cell bioanodes. *Applied and Environmental Microbiology, 78*, 5212–5219.

Prasad, T. N. V. K. V., Sudhakar, P., Sreenivasulu, Y., Latha, P., Munaswamy, V., Reddy, K. R., Sreeprasad, T. S., Sajanlal, P. R., & Pradeep, T. (2012). Effect of nanoscale zinc oxide particles on the germination, growth and yield of peanut. *Journal of Plant Nutrition, 35*, 905–927.

Preetha, P. S., & Balakrishnan, N. (2017). A review of nano fertilizers and their use and functions in soil. *International Journal of Current Microbiology and Applied Sciences, 6*, 3117–3133.

Qureshi, A., Singh, D. K., & Dwivedi, S. (2018). Nano-fertilizers: A novel way for enhancing nutrient use efficiency and crop productivity. *International Journal of Current Microbiology and Applied Sciences, 7*, 3325–3335.

Raliya, R., Franke, C., Chavalamane, S., Nair, R., Reed, N., & Biswas, P. (2016). Quantitative understanding of nanoparticle uptake in watermelon plants. *Frontiers in Plant Science, 7*, 1288. https://doi.org/10.3389/fpls.2016.01288 (accessed on 6th November 2025).

Rathore, S., Desai, P. M., Liew, C. V., Chan, L. W., & Heng, P. W. S. (2013). Microencapsulation of microbial cells. *Journal of Food Engineering, 116*, 369–381.

Reinhold-Hurek, B., & Hurek, T. (1998). Interactions of gramineous plants with *Azoarcus* spp. and other diazotrophs: Identification, localization, and perspectives to study their function. *Critical Reviews in Plant Sciences, 17*, 29–54.

Rizwan, M., Gilani, S. R., Durani, A. I., & Naseem, S. (2021). Materials diversity of hydrogel: Synthesis, polymerization process and soil conditioning properties in agricultural field. *Journal of Advanced Research, 33*, 15–40.

Rojas-Sánchez, B., Guzman-Guzman, P., Morales-Cedeno, L. R., Orozco-Mosqueda, M. D. C., Saucedo-Martinez, B. C., Sanchez-Yanez, J. M., Fadiji, A. E., Babalola, O. O., Glick, B. R., & Santoyo, G. (2022). Bioencapsulation of microbial inoculants: Mechanisms, formulation types and application techniques. *Applied Biosciences, 1*(2), 198–220. https://doi.org/10.3390/applbiosci1020013 (accessed on 6th November 2025).

Rubin, R. L., Van Groenigen, K. J., & Hungate, B. A. (2017). Plant growth promoting rhizobacteria are more effective under drought: A meta-analysis. *Plant and Soil, 416*, 309–323.

Ruzzi, M., & Aroca, R. (2015). Plant growth-promoting rhizobacteria act as bio stimulants in horticulture. *Scientia Horticulture, 196*, 124–134.

Saberi-Rise, R., & Moradi-Pour, M. (2020). The effect of *Bacillus subtilis* Vru1 encapsulated in alginate-bentonite coating enriched with titanium nanoparticles against *Rhizoctonia solani* on bean. *International Journal of Biological Macromolecules, 152*, 1089–1097.

Sabry, S. R. S., Saleh, S. A., & Batchelor, C. A. (1997). Endophytic establishment of *Azorhizobium caulinodans* in wheat. *Proceedings of the Royal Society B: Biological Sciences, 264*, 341–346.

Safdar, M., Kim, W., Park, S., Gwon, Y., Kim, Y. O., & Kim, J. (2022). Engineering plants with carbon nanotubes: A sustainable agriculture approach. *Journal of Nanobiotechnology, 20*, 1–30.

Saharan, V., Kumaraswamy, R. V., Choudhary, R. C., Kumari, S., Pal, A., Raliya, R., & Biswas, P. (2016). Cu-chitosan nanoparticle mediated sustainable approach to enhance seedling growth in maize by mobilizing reserved food. *Journal of Agricultural and Food Chemistry, 64*, 6148–6155.

Saharan, V., Sharma, G., Yadav, M., Choudhary, M. K., Sharma, S. S., Pal, A., Raliya, R., & Biswas, P. (2015). Synthesis and *in vitro* antifungal efficacy of Cu-chitosan nanoparticles against pathogenic fungi of tomato. *International Journal of Biological Macromolecules, 75*, 346–353.

Sahoo, R. K., Ansari, M. W., & Pradhan, M. (2014). Phenotypic and molecular characterization of native *Azospirillum* strains from rice fields to improve crop productivity. *Protoplasma, 251*, 943–953.

Saiyad, S. A., Jhala, Y. K., & Vyas, R. V. (2015). Comparative efficiency of five potash and phosphate solubilizing bacteria and their key enzymes useful for enhancing and improvement of soil fertility. *International Journal of Science and Research, 5*, 1–6.

Sangeeth, K. P., Bhai, R. S., & Srinivasan, V. (2012). *Paenibacillus glucanolyticus*, a promising potassium solubilizing bacterium isolated from black pepper (*Piper nigrum* L.) rhizosphere. *Journal of Spices and Aromatic Crops, 21*, 118–124.

Santos, M. S., Nogueira, M. A., & Hungria, M. (2019). Microbial inoculants: Reviewing the past, discussing the present and previewing an outstanding future for the use of beneficial bacteria in agriculture. *AMB Express, 9*, 1–22.

Savci, S. (2012). An agricultural pollutant: Chemical fertilizer. *International Journal of Environmental Science and Development, 3*(1), 77–80.

Schoebitz, M., Lopez, M. D., & Roldan, A. (2013). Bioencapsulation of microbial inoculants for better soil-plant fertilization: A review. *Agricultural Sciences and Development, 33*, 751–765.

Shafiei-Masouleh, S. S. (2021). Use of magnetic nano-chitosan as bio-fertilizer to reduce production period in three cyclamen cultivars. *Journal of Soil Science and Plant Nutrition, 22*, 281–293.

Sharifi, R. S., Khalilzadeh, R., Pirzad, A., & Anwar, S. (2020). Effects of biofertilizers and nano zinc-iron oxide on yield and physicochemical properties of wheat under water deficit conditions. *Communications in Soil Science and Plant Analysis, 51*, 2511–2524.

Sharma, A., Patel, S., & Menghani, E. (2021). Synthesis, application and prospects of nano-biofertilizers: A reappraisal. *Journal of Phytological Research, 34*, 79–85.

Sharma, B., Tiwari, S., Kumawat, K. C., & Cardinale, M. (2023). Nano-biofertilizers as bio-emerging strategies for sustainable agriculture development: Potentiality and their limitations. *Science of the Total Environment, 860*, 160476. https://doi.org/10.1016/j.scitotenv.2022.160476 (accessed on 6th November 2025).

Sharma, K., Sharma, S., & Prasad, S. R. (2019). PGPR: Renewable tool for sustainable agriculture. *International Journal of Current Microbiology and Applied Sciences, 8*, 525–530.

Sharma, N., & Singhvi, R. (2017). Effects of chemical fertilizers and pesticides on human health and environment: A review. *International Journal of Agriculture, Environment and Biotechnology, 10*, 675–680.

Sheng, X. F. (2005). Growth promotion and increased potassium uptake of cotton and rape by a potassium releasing strain of *Bacillus edaphicus*. *Soil Biology and Biochemistry, 37*, 1918–1922.

Sheoran, P., Grewal, S., Kumari, S., & Goel, S. (2021). Enhancement of growth and yield, leaching reduction in *Triticum aestivum* using biogenic synthesized zinc oxide nano fertilizer. *Biocatalysis and Agricultural Biotechnology, 32*, 101938.

Shintu, P. V., & Jayaram, K. M. (2015). Phosphate solubilizing bacteria (*Bacillus polymyxa*)– An effective approach to mitigate drought in tomato (*Lycopersicon esculentum* Mill.). *Tropical Plant Research, 2*, 17–22.

Singh, D., & Gurjar, B. R. (2022). Nanotechnology for agricultural applications: Facts, issues, knowledge gaps, and challenges in environmental risk assessment. *Journal of Environmental Management, 322*, 116033. https://doi.org/10.1016/j.jenvman.2022.116033 (accessed on 6th November 2025).

Singh, M. D. (2017). Nano-fertilizers is a new way to increase nutrients use efficiency in crop production. *International Journal of Agricultural Sciences, 9*, 3831–3833.

Smith, S. E., & Smith, F. A. (2011). Roles of arbuscular mycorrhizas in plant nutrition and growth: New paradigms from cellular to ecosystem scales. *Annual Review of Plant Biology, 62*, 227–250.

Solanki, P., Bhargava, A., Chhipa, H., Jain, N., & Panwar, J. (2015). Nano-fertilizers and their smart delivery system. In Rai, M., Ribeiro, C., Mattoso, L., & Duran, N. (Eds.), *Nanotechnologies in Food and Agriculture* (pp. 81–101). Springer: Cham, Switzerland.

Surendran, U., & Vani, D. (2013). Influence of arbuscular mycorrhizal fungi in sugarcane productivity under semiarid tropical agro ecosystem in India. *International Journal of Plant Production, 7*, 269–278.

Taha, R. A., Hassan, M. M., Ibrahim, E. A., Baker, N. H. A., & Shaaban, E. A. (2016). Carbon nanotubes impact on date palm *in vitro* cultures. *Plant Cell, Tissue and Organ Culture, 127*, 525–534.

Tariq, M. R., Shaheen, F., Mustafa, S., Ali, S., Fatima, A., Shafiq, M., Safdar, W., Sheas, M. N., Hameed, A., & Nasir, M. A. (2022). Phosphate solubilizing microorganisms isolated from medicinal plants improve growth of mint. *PeerJ, 10*, e13782. https://doi.org/10.7717/peerj.13782 (accessed on 6th November 2025).

Tekam, Y., Porwal, M., Verma, B., & Singh, D. (2023). Nano-fertilizers–A novel technology in modern agriculture. *Krishi Science*, 2583–4150.

Thirunavukkarasu, M., Subramanian, K. S., Kannan, P., & Balaji, T. (2018). Response of nano-Sulfur to the groundnut. *International Journal of Conservation Science, 6*, 2067–2072.

Timmusk, S., El-Daim, I. A. A., Copolovici, L., Tanilas, T., Kannaste, A., Behers, L., Nevo, E., Seisenbaeva, G., Stenstrom, E., & Niinemets, Ü. (2014). Drought-tolerance of wheat improved by rhizosphere bacteria from harsh environments: Enhanced biomass production and reduced emissions of stress volatiles. *PLoS ONE, 9*, e96086. https://doi.org/10.1371/journal.pone.0096086 (accessed on 6th November 2025).

Vejan, P., Abdullah, R., Khadiran, T., Ismail, S., & Nasrulhaq, B. A. (2016). Role of plant growth promoting rhizobacteria in agricultural sustainability–A review. *Molecules, 21*(5), 573. https://doi.org/10.3390/molecules21050573 (accessed on 6th November 2025).

Verma, K. K., Song, X. P., Joshi, A., Rajput, V. D., Singh, M., Sharma, A., Singh, R. K., Li, D. M., Arora, J., Minkina, T., & Li, Y. R. (2022). Nano fertilizer possibilities for healthy soil, water, food in future: An overview. *Frontiers in Plant Science, 13*, 865048. https://doi.org/10.3389/fpls.2022.865048 (accessed on 6th November 2025).

Wahane, M. R., Meshram, N. A., More, S. S., & Khobragade, N. H. (2020). Biofertilizer and their role in sustainable agriculture–A review. *Journal of Pharmacognosy and Phytochemistry, 9*, 127–130.

Wang, P., Lombi, E., Zhao, F. J., & Kopittke, P. M. (2016). Nanotechnology: A new opportunity in plant sciences. *Trends in Plant Science, 21*, 699–712.

Wani, S. A., Chand, S., & Ali, T. (2013). Potential use of *Azotobacter chroococcum* in crop production: An overview. *Current Agriculture Research Journal, 1*, 35–38.

Whiteside, M. D., Digman, M. A., Gratton, E., & Treseder, K. K. (2012). Organic nitrogen uptake by arbuscular mycorrhizal fungi in a boreal forest. *Soil Biology and Biochemistry, 55*, 7–13.

Yadav, A., Yadav, K., & Abd-Elsalam, K. A. (2023). Nano fertilizers: Types, delivery and advantages in agricultural sustainability. *Agrochemicals, 2*, 296–336.

Yanni, Y., Rizk, R., & Abd-El Fattah, F. (2001). The beneficial plant growth-promoting association of *Rhizobium leguminosarum* bv. *trifolii* with rice roots. *Australian Journal of Plant Physiology, 28*, 845–870.

Yasmin, H., Mazher, J., Azmat, A., Nosheen, A., Naz, R., Hassan, M. N., Noureldeen, A., & Ahmad, P. (2021). Combined application of zinc oxide nanoparticles and biofertilizer to induce salt resistance in safflower by regulating ion homeostasis and antioxidant defense responses. *Ecotoxicology and Environmental Safety, 218*, 112262. https://doi.org/10.1016/j.ecoenv.2021.112262 (accessed on 6th November 2025).

Zahoor, R., Zhao, W., Dong, H., Snider, J. L., Abid, M., Iqbal, B., & Zhou, Z. (2017). Potassium improves photosynthetic tolerance to and recovery from episodic drought stress in functional leaves of cotton (*Gossypium hirsutum* L.). *Plant Physiology and Biochemistry, 119*, 21–32.

Zahra, Z., Habib, Z., Hyun, H., & Shahzad, H. M. A. (2022). Overview on recent developments in the design, application, and impacts of nano-fertilizers in agriculture. *Sustainability, 14*(15), 9397. https://doi.org/10.3390/su14159397 (accessed on 6th November 2025).

Zhu, J., Li, M., & Whelan, M. (2018). Phosphorus activators contribute to legacy phosphorus availability in agricultural soils: A review. *Science of the Total Environment, 612*, 522–537.

Zulfiqar, F., Navarro, M., Ashraf, M., Akram, N. A., & Munne-Bosch, S. (2019). Nano fertilizer use for sustainable agriculture: Advantages and limitations. *Plant Science, 289*, 110270. https://doi.org/10.1016/j.plantsci.2019.110270 (accessed on 6th November 2025).

Zuluaga, M. Y. A., Milani, K. M. L., Miras-Moreno, B., Lucini, L., Valentinuzzi, F., Mimmo, T., Pii, Y., Cesco, S., Rodrigues, E. P., & de Oliveira, A. L. M. (2020). Inoculation with plant growth-promoting bacteria alters the rhizosphere functioning of tomato plants. *Applied Soil Ecology, 158*, 103784. https://doi.org/10.1016/j.apsoil.2020.103784 (accessed on 6th November 2025).

Index

C

D

H

I

J

K

L

M

N

Q

T

X

Y

Z

For Product Safety Concerns and Information please contact our EU representative GPSR@taylorandfrancis.com Taylor & Francis Verlag GmbH, Kaufingerstraße 24, 80331 München, Germany